Texts and
Monographs
in Physics

W0051150

W. Beiglböck
M. Goldhaber
E. H. Lieb
W. Thirring

Series Editors

Robert D. Richtmyer

Principles of Advanced Mathematical Physics

Volume I

Springer-Verlag
New York   Heidelberg   Berlin

Robert D. Richtmyer
Department of Physics and Astrophysics
University of Colorado
Boulder, Colorado 80309
USA

Editors:

Wolf Beiglböck
Institut für Angewandte Mathematik
Universität Heidelberg
Im Neuenheimer Feld 5
D-6900 Heidelberg 1
Federal Republic of Germany

Maurice Goldhaber
Department of Physics
Brookhaven National Laboratory
Associated Universities, Inc.
Upton, NY 11973
USA

Elliott H. Lieb
Department of Physics
Joseph Henry Laboratories
Princeton University
P O. Box 708
Princeton, NJ 08540
USA

Walter Thirring
Institut für Theoretische Physik
der Universität Wien
Boltzmanngasse 5
A-1090 Wien
Austria

With 45 Figures

ISBN-13: 978-3-642-46380-8   e-ISBN-13: 978-3-642-46378-5
DOI: 10.1007/978-3-642-46378-5

Library of Congress Cataloging in Publication Data

Richtmyer, Robert D.
  Principles of advanced mathematical physics.
  (Texts and monographs in physics)
  CONTENTS: v. 1. Hilbert and Banach spaces, distri-
butions, operators, probability, applications to quantum
mechanics, equations of evolution in physics.
  Includes index.
  1. Mathematical physics.   I. Title.
QC20.R56   530.1'5   78-16494

Contents

10   Ordinary Differential Operators   190

11   Some Partial Differential Operators of Quantum Mechanics   222

12   Compact, Hilbert–Schmidt, and Trace-Class Operators   241

Preface: On the Nature of Mathematical Physics

Reasoning in mathematics and reasoning in physics have very different textures. Mathematics is held together by short-range forces that bind each step in a deduction directly to the preceding steps, whereas physics is held together by the much longer-range forces of analogy and intuition and all sorts of indirect supporting evidence. The comparison of physical science with cryptanalysis ("deciphering the secrets of nature," etc.), though overworked, is apt. When one has solved a cipher and the message rings out loud and clear, one does not think of calling in a mathematician to provide a uniqueness proof, even though conceivably there might be a different solution, i.e., a different message. In physics, existence and uniqueness proofs are many decades behind current research (because of the inherent complexity of nature) and are often somewhat irrelevant, because they can be no more convincing than the hypotheses on which they are based, which in turn are matters of physics, while a large body of indirect evidence is often fully convincing. In mathematics, on the other hand, intuition and analogy are notoriously untrustworthy; although they often lead to useful conjectures, the conjectures never become part of the structure until proved. When one is proving a theorem in mathematics, one is not permitted to use any hypotheses except those present in the statement of the theorem.

A first consequence of this difference in texture concerns the attitude we must take toward some (or perhaps most) investigations in "applied mathematics," at least when the mathematics is applied to physics. Namely, those investigations have to be regarded as pure mathematics and evaluated as such. For example, some of my mathematical colleagues have worked in recent years on the Hartree–Fock approximate method for determining the structures of many-electron atoms and ions. When the method was introduced, nearly fifty years ago, physicists did the best they could to justify it, using variational principles, intuition, and other techniques within the texture of physical reasoning. By now the method has long since become part of the established structure of physics. The mathematical theorems that can be proved now (mostly for two- and three-electron systems, hence of limited interest for physics), have to be regarded as mathematics. If they are good mathematics (and I believe they are), that is justification enough. If they are not, there is no basis for saying that the work is being done to help the physicists. In that sense, applied mathematics plays no role in today's physics. In today's division of labor, the task of the mathematician is to create mathematics, in whatever area, without being much concerned about how the mathematics is used; that should be decided in the future and by physics.

Specialization has, of course, gone too far, but even with less of it, it would be out of the question for the methods of contemporary mathematics to be transplanted all the way over into the area of contemporary physics and produce significant results. The differences are just too great. Today's physicists know how to use mathematics; they know how to formulate problems, devise methods of solution, and perform long derivations and calculations, but they cannot create the mathematics. Experience has shown that the discovery and purification of abstract concepts and principles is peculiarly in the realm of mathematics. The division of labor is important and ought to be taken seriously.

There is no objection to a mathematician's working in the areas that have come to be designated as applied mathematics, and if he can derive inspiration for his mathematics from the physical world, that is very much to the good, but the value to physics of the fruits of his labor will be determined by their quality as mathematics.

There is also no objection to a mathematician's doing physics, provided he is qualified. The prime example was von Neumann—when he did physics, he talked, thought, and calculated like a physicist (but faster). He understood all branches of physics (including elementary particles as they were known then), and chemistry and astronomy, and he had a talent for introducing those and only those mathematical ideas that were relevant to the physics at hand. Anyone, regardless of professional affiliation, who can do physics one tenth that well should be encouraged to do it, but the objectives and methods are quite different from those of applied mathematics, whose purpose is to create mathematics.

Here are some quotations from Hardy's "A Mathematician's Apology":

1. "I said that a mathematician was a maker of patterns of ideas, and that beauty and seriousness were the criteria by which his patterns should be judged." (page 98)
2. "It is not possible to justify the life of any genuine professional mathematician on the ground of the 'utility' of his work." (page 119)
3. "One rather curious conclusion emerges, that pure mathematics is on the whole distinctly more useful than applied." (page 134)
4. "I hope I need not say that I am not trying to decry mathematical physics, a splended subject with tremendous problems where the finest imaginations have run riot." (page 135)

Another consequence of the difference in texture concerns the word "rigor," which is badly misused by both mathematicians and physicists and possibly ought to be banished from our language. Physicists think mathematicians spend an inordinate amount of time making sure that all i's are dotted and all t's crossed, and the mathematicians shake their heads and wonder how those sloppy physicists ever get anything right. Both attitudes result from failure to recognize the methodological difference between the two disciplines. The situation becomes a little clearer when one teaches mathematics to physicists, for it turns out that although the physicists are not to be deterred from their accepted and successful ways of investigating the physical world, they demand rigor in mathematics. They want to know exactly what is true and what is false and exactly why (although they are eager to be told lots of additional things without proof), and they want to see lots of examples and counterexamples, in order to delineate the areas of relevance of the theorems.

In one branch of physics, quantum field theory, the difference in texture has almost disappeared, owing to the failure of the traditional methods. In 1900 Max Planck said "let's quantize the electromagnetic field," and he showed what wonderful things would happen if we could do it. Einstein showed more. In a certain measure, all modern physics is based on that suggestion, but the task has proved to be enormously more difficult than was supposed. In many attempts during the first half of this century, based on the intuitive methods that had been so successful in other parts of quantum mechanics, emission and absorption rates and line breadths were successfully calculated, but only by arbitrary suppression of infinities and inconsistencies, and for the most part in cases where the required result was already known from experiment and cruder theories. In the 1950s, various physicists began to take seriously the suppression of the infinities by the introduction of precise new axioms ("renormalization"), and a flood of exciting new results came out (Lamb shift, precise magnetic moments, etc.). Still, we do not yet have a water-tight theory, and each new attempt to overcome the difficulties of previous attempts has involved the introduction of more precise and more powerful mathematical tools. It now seems that intuitive methods are just as untrustworthy in quantum field theory as in pure mathematics, and contemporary work in field theory has very much the same texture as pure mathematics; there is the feeling of "definition, lemma, proof, theorem, proof, etc." if not the actual words. Presumably, when success finally comes, it will be through interplay between physical intuition and the newly found mathematical rigor.

The consequence for mathematical physics is an increased relevance of the careful study of operators, distributions, Banach algebras, functions of several complex variables, representations of noncompact groups, and so on.

People in other areas are usually unaware of the wide range of mathematics now used in physics. They assume that physicists are interested only in analysis and specifically the part of analysis appropriate for nineteenth-century physics, as set forth in Courant and Hilbert. Most books, including recent ones, on "mathematical methods for physicists," and the like, contain no group theory, which has played an important role in physics since about 1925, and the authors give no indication they have ever heard of the mathematical principles and concepts basic to modern quantum mechanics, relativity, cosmology, scattering theory, quantum field theory, statistical mechanics, topological dynamical systems, and so on, to say nothing of the concepts and principles that have not yet found their way into physics, but are likely to do so in the near future and are likely to come from areas such such as algebra, logic, set theory, and topology. No part of mathematics is devoid of potential interest for physics.

For our purpose, mathematical concepts and principles are more important than methods, and the main goal of courses in mathematical physics, in my opinion, is to explain the concepts and principles in such a way that one can see their relevance for physics. Here is an example:

Manifolds in relativity: In 1916 Karl Schwarzschild derived the static spherical solution of the Einstein field equations in the form now known by his name. His formula appeared to indicate some sort of singularity at a radius now called the Schwarzschild radius. There followed forty-four years of confusion about the "Schwarzschild singularity." As time went by, it became gradually clear that Schwarzschild's formula described only a part of the relevant physical space-time, and in 1960 Martin Kruskal gave a description of the geodesically complete manifold of which Schwarzschild's formula determined a part. It was then seen that although certain interesting phenomena are associated with the Schwarzschild radius, there is no singularity there. Relativists now take the attitude that by a solution of the Einstein equations one has to understand not just a formula for a line element $ds^2 = \cdots$, but rather a complete manifold, and that the global topology of the manifold may be of cosmological significance. The introduction of the geometric notion of *manifold* into relativity is a prime example of mathematical physics. The theory of manifolds is set forth in Volume II.

An earlier example was the introduction of abstract Hilbert space theory into quantum mechanics, mainly by von Neumann, which made it possible to construct a solid theory on the basis of the powerful intuitive ideas of

Dirac and other physicists. No less important was the introduction of groups and group representations, mainly by Wigner and Weyl.

A recent example is the introduction by Ruelle and Takens of ideas from the topological theory of differentiable dynamical systems into the study of the onset of turbulence. These ideas are likely to play a role in other parts of physics, where nonlinear differential equations appear.

The basic mathematics of physics belongs in physics courses. The proper formulation of boundary-value problems, asymptotic expansions, consequences of symmetries, and so on are all matters of physics. Although the ideas are further clarified and analyzed in mathematical physics courses, their first introduction should appear as part of the physics. A physics instructor ought never to say to his students, "just how those things work will be explained in your mathematics courses." Physics and mathematics cannot be separated in that way, and it is not the purpose of courses in mathematical physics to relieve physics teachers of the responsibility of explaining their subject. In practice, however, the best physics courses cannot be adequate on all topics. For example, most books on quantum mechanics are hopelessly unclear about Hilbert spaces and operators, and students need to learn about those things after they have first studied quantum mechanics as an intuitive subject. It is not just a question of "rigor." Whether a given symmetric operator has self-adjoint extensions and, if so, how many different ones, is a matter of physics, because the self-adjoint operators are the observables. The probabilistic interpretation of the spectral family of a self-adjoint operator gives the physical interpretation of the observable even for states that are not in the domain of the operator, and so on. I interpret mathematical physics so as to include the explanation of these things.

Most good ideas turn out to be simple ones, and I believe it is important that they be so presented, without unnecessary ramification of other ideas. In my view, for example, distribution theory should be based (rigorously, of course) on the Riemann integral and advanced calculus, and L^2 spaces and the theory of differential operators should be based on distribution theory. The students can learn about measure theory and topological vector spaces later. It has been my intention in these two volumes to present the fundamental ideas in the most down-to-earth fashion possible. At the same time, I have not hesitated to introduce further ideas that have independent interest, for example transfinite cardinals in the chapter on Hilbert spaces.

Boulder, December 1978   Robert D. Richtmyer

CHAPTER 1

Hilbert Spaces

Connection with finite-dimensional spaces; Hilbert space axioms; the
Schwarz and triangle inequalities; the parallelogram law and the
connection with general Banach spaces; completeness of l^2;
transfinite cardinal numbers; equivalence of separable Hilbert spaces;
Hilbert spaces of larger dimensions; separability of Fock spaces;
completeness criteria for orthonormal sequences; linear functionals;
the Riesz–Fischer and Riesz–Fréchet theorems; strong and weak
convergence; polarization of quadratic functionals.

Prerequisite: Linear algebra

This chapter deals mainly with the geometry of (primarily abstract) Hilbert
spaces. In Chapter 5, Hilbert space theory is combined with distribution
theory to establish the theory of L^2 spaces, on which much of modern
functional analysis is based.

1.1 Review of Pertinent Facts About Matrices and Finite Dimensional Spaces

The reader is assumed to be familiar with the following material. An $n \times n$
real or complex matrix A determines a linear transformation $\mathbf{x} \to \mathbf{x}' = A\mathbf{x}$
in an n-dimensional real or complex space V^n; in terms of components,
$x'_j = \sum_{k=1}^{n} A_{jk} x_k$. If $\mathbf{x}$ is regarded as a column vector, i.e., an $n \times 1$ matrix
(in this case the column index is omitted), then $A\mathbf{x}$ is simply a matrix product.
The transpose and the Hermitian conjugate of a matrix M (square or not)
are denoted by M^T and M^*; that is, $(M^T)_{jk} = M_{kj}$ and $(M^*)_{jk} = \overline{M_{kj}}$;
in particular, $\mathbf{x}^T$ and $\mathbf{x}^*$ are row vectors. If $\mathbf{x}$ and $\mathbf{y}$ are any vectors in V^n,
then their Hermitian inner product is the 1×1 matrix or number $\mathbf{x}^*\mathbf{y} =$
$\sum_{j=1}^{n} \bar{x}_j y_j$, also denoted by $(\mathbf{x}, \mathbf{y})$. In the real case this is simply $\mathbf{x}^T\mathbf{y} = \sum x_j y_j$
and is often denoted by $\mathbf{x} \cdot \mathbf{y}$. [Note that $\mathbf{x}\mathbf{y}^*$ is not a scalar, but an $n \times n$
matrix (of rank 1); this provides some slight motivation for choosing $(\mathbf{x}, \mathbf{y})$

1

to be linear in the second factor and antilinear in the first, as usually done in physics, rather than conversely, as usually done in mathematics; that is, $(\mathbf{x}, a\mathbf{y}) = a(\mathbf{x}, \mathbf{y})$, whereas $(a\mathbf{x}, \mathbf{y}) = \bar{a}(\mathbf{x}, \mathbf{y})$.] The vectors $\mathbf{x}$ and $\mathbf{y}$ are said to be *orthogonal* if $(\mathbf{x}, \mathbf{y}) = 0$. The *length* of $\mathbf{x}$ is $\|\mathbf{x}\| = (\mathbf{x}, \mathbf{x})^{1/2}$, sometimes denoted simply by $|\mathbf{x}|$.

The main geometrical concepts in V^n are first those having to do with linear dependence and second those having to do with orthogonality. Vectors $\mathbf{x}^1, \ldots, \mathbf{x}^k$ are linearly *dependent* if there are constants $a_1, \ldots, a_k$, not all zero, such that $\sum_{j=1}^k a_j \mathbf{x}^j$ is the zero vector (they are always dependent if $k > n$). The set of all linear combinations $\sum a_j \mathbf{x}^j$ of k given vectors is called the (linear) *subspace* of V^n *spanned* by $\mathbf{x}^1, \ldots, \mathbf{x}^k$. If these vectors are independent, the subspace is *k-dimensional*; it is proved in linear algebra that then, if $\mathbf{y}^1, \ldots, \mathbf{y}^k$ are any other k linearly independent vectors lying in the subspace, then these other vectors also span the same subspace. If V^n is a real space, the constants $a_1, a_2, \ldots$ referred to are arbitrary real numbers; if V^n is complex, they are arbitrary complex numbers; the real or complex number system, respectively, (denoted by $\mathbb{R}$ or $\mathbb{C}$) is called the *field of scalars*. Fields other than $\mathbb{R}$ or $\mathbb{C}$ will not be used. A complex space V^n is sometimes said to have *2n real dimensions*.

A set $\{\mathbf{v}^j\}_1^k$ of vectors in V^n is *orthonormal* if $(\mathbf{v}^i, \mathbf{v}^j) = \delta_{ij} = 1$ when $i = j$ and $= 0$ when $i \neq j$. Clearly, then, $k \leq n$. If $k = n$, an arbitrary vector $\mathbf{x}$ can be written as $\sum_{j=1}^n a_j \mathbf{v}^j$, where $a_j = (\mathbf{v}^j, \mathbf{x})$.

If S is an arbitrary subspace of V^n, then $S^\perp$ denotes the *orthogonal complement* of S, defined as

$$S^\perp = \{\mathbf{x}: (\mathbf{x}, \mathbf{y}) = 0 \text{ for all } \mathbf{y} \text{ in } S\}; \qquad (1.1\text{-}1)$$

in words, $S^\perp$ consists of all $\mathbf{x}$ that are orthogonal to every $\mathbf{y}$ in S; furthermore, $(S^\perp)^\perp = S$, and $\dim S + \dim S^\perp = n$. The *projection theorem*, proved in linear algebra, says that if $\mathbf{x}$ is any vector in V^n, then $\mathbf{x}$ has a unique decomposition $\mathbf{x} = \mathbf{y} + \mathbf{z}$, where $\mathbf{y}$ and $\mathbf{z}$ are in S and $S^\perp$, respectively.

The ideas of the preceding three paragraphs are generalized to certain infinite-dimensional spaces, called *Hilbert spaces*, in this chapter. The infinity of dimensions leads to a few new concepts. For example, whereas, in a finite dimensional space, an inner product $\mathbf{x}^*\mathbf{y}$ is always available, there are infinite-dimensional spaces (See Chapter 15) in which a length $\|\mathbf{x}\|$ is defined, for all $\mathbf{x}$, but not an inner product, and in which, in fact, no inner product can be defined in a reasonable way. We shall start with Hilbert spaces, in which an inner product is defined, and which therefore constitute the natural generalization of the familiar Euclidean spaces.

The generalizations of matrices, or more precisely of the linear transformations to which they correspond, are the linear transformations or operators in Hilbert spaces, discussed in Chapters 7 through 12. A few of the definitions and facts to be generalized are these: If A is an $n \times n$ matrix, $\mathbf{v}$ a vector $\neq 0$, and λ a number, and if $A\mathbf{v} = \lambda\mathbf{v}$, then λ is an *eigenvalue* of A, and $\mathbf{v}$ is the corresponding *eigenvector*. A transformation $\mathbf{x} \to \mathbf{x}' = A\mathbf{x}$ has an inverse $\mathbf{x}' \to \mathbf{x} = A^{-1}\mathbf{x}'$ if and only if zero is not an eigenvalue of A. [Since the eigenvalues λ are the zeros of $\det(\lambda I - A) = 0$, A has an inverse if and only if $\det A \neq 0$. The inverse is obtained by means of cofactors, for purely theoretical purposes, and by Gauss elimination for practical purposes; these methods do not generalize.] If A is Hermitian, that is, if $A^* = A$, then its eigenvalues are all real, and it possesses a complete orthonormal set of eigenvectors $\mathbf{v}^{(1)}, \ldots, \mathbf{v}^{(n)}$. If U is the $n \times n$ matrix having these vectors as its columns (it is *unitary*: $UU^* = U^*U = I$), then U^*AU is a diagonal matrix D having the eigenvalues of A on the main diagonal, and $A = UDU^*$. A necessary and sufficient condition for a matrix A to be thus diagonalizable is that A be *normal*, i.e., that $AA^* = A^*A$. Commuting normal matrices A and B can be simultaneously diagonalized by one and the same unitary matrix U. A is called *positive definite* (semidefinite) if $\mathbf{x}^*A\mathbf{x}$ is > 0 (≥ 0) for all $\mathbf{x} \neq 0$. In this case, A is necessarily Hermitian. [The moment of inertia tensor is a positive definite real (hence symmetric) matrix.] These ideas all have analogues (though not always immediate and obvious ones) in Hilbert space.

1.2 Linear Space; Normed Linear Spaces

Although a vector was defined above as an *n*-tuple of numbers, vector spaces (also called *linear spaces*) can be defined abstractly by a set of axioms, as usually done in courses in linear algebra. The abstract method is preferred in the infinite-dimensional case, because there are many seemingly different concrete realizations. The axioms of a linear space V over a field $\mathbb{F}$ ($= \mathbb{R}$ or $\mathbb{C}$) are these:

1. V contains $a\mathbf{u} + b\mathbf{v}$ if it contains $\mathbf{u}$ and $\mathbf{v}$, for any a and b in $\mathbb{F}$
2. $\mathbf{u} + \mathbf{v} = \mathbf{v} + \mathbf{u}$, $\mathbf{u} + (\mathbf{v} + \mathbf{w}) = (\mathbf{u} + \mathbf{v}) + \mathbf{w}$
3. $a(b\mathbf{u}) = (ab)\mathbf{u}$
4. $a(\mathbf{u} + \mathbf{v}) = a\mathbf{u} + a\mathbf{v}$, $(a + b)\mathbf{u} = a\mathbf{u} + b\mathbf{u}$
5. There is a unique zero vector $\mathbf{0}$ such that $\mathbf{u} + \mathbf{0} = \mathbf{u}$, for all $\mathbf{u}$.
6. $1\mathbf{u} = \mathbf{u}$, $0\mathbf{u} = \mathbf{0}$

Generally, $-\mathbf{u}$ is written for $(-1)\mathbf{u}$, and $\mathbf{u} - \mathbf{v}$ for $\mathbf{u} + (-1)\mathbf{v}$. Also, one often writes simply 0 for $\mathbf{0}$, as in the equation $\mathbf{u} - \mathbf{u} = 0$. In the finite-dimensional case, where it is further assumed that n linearly independent vectors can be found, but not $n + 1$, the resulting space V is exactly equivalent to the V^n discussed earlier, as soon as a basis has been chosen in the space V.

EXERCISES

1. Show that $\mathbf{u} - \mathbf{u} = 0$, for all $\mathbf{u}$.

2. Show that the uniqueness of the zero vector $\mathbf{0}$ follows from the other axioms; i.e., if $\mathbf{u} + \mathbf{0}_1 = \mathbf{u}$, for all $\mathbf{u}$, and $\mathbf{u} + \mathbf{0}_2 = \mathbf{u}$, for all $\mathbf{u}$, then $\mathbf{0}_1 = \mathbf{0}_2$.

3. Show that, for given $\mathbf{u}$ and $\mathbf{w}$, the equation $\mathbf{u} + \mathbf{v} = \mathbf{w}$ has the unique solution $\mathbf{v} = \mathbf{w} + (-1)\mathbf{u}$.

A linear space V is a *normed linear space* if there is assigned to every **u** in V a real number $\|\mathbf{u}\|$, called the *norm* of **u**, such that

7. $\|\mathbf{u}\| > 0$, for all $\mathbf{u} \neq \mathbf{0}$
8. $\|\mathbf{0}\| = 0$
9. $\|a\mathbf{u}\| = |a| \|\mathbf{u}\|$
10. $\|\mathbf{u} + \mathbf{v}\| \leq \|\mathbf{u}\| + \|\mathbf{v}\|$

[In V^n, $\|\mathbf{u}\|$ is usually taken as the length of the vector **u**, that is, $\|\mathbf{u}\|^2 = \sum_{j=1}^{n} |u_j|^2$.]

EXERCISE

4. Show that the axiom pair (6) of a linear space can be derived from the other axioms and the axioms of the norm. *Hint*: Show in succession, for an arbitrary **u**, that $0 \cdot \mathbf{u} = \mathbf{0}, \mathbf{u} - \mathbf{u} = \mathbf{0}, 1 \cdot \mathbf{u} - \mathbf{u} = \mathbf{0}, 1 \cdot \mathbf{u} = \mathbf{u}$. On the other hand, if the axiom $0 \cdot \mathbf{u} = \mathbf{0}$ is retained, then Axiom 8, $\|\mathbf{0}\| = 0$, can be omitted.

1.3 Hilbert Space: Axioms and Elementary Consequences

A (real or complex) Hilbert space $\mathfrak{H}$ is a complete (real or complex) inner product space. That is, first, $\mathfrak{H}$ is a linear space (or vector space), as defined in the preceding section. Second, an inner product (u, v) is defined for all u and v in $\mathfrak{H}$; (u, v) is a function, with values in the field $\mathbb{F} = \mathbb{R}$ or $\mathbb{C}$ of scalars, which is linear in v and Hermitian symmetric $((v, u) = \overline{(u, v)})$, hence antilinear in u, and is such that the corresponding quadratic form is positive definite: $(u, u) \geq 0$ for all u, and $(u, u) = 0$ only for $u = \mathbf{0}$. Third, $\mathfrak{H}$ is complete with respect to the norm given by $\|u\| = (u, u)^{1/2}$; i.e., every Cauchy sequence in $\mathfrak{H}$ has a limit in $\mathfrak{H}$. The importance of completeness for quantum mechanics is discussed in Chapter 14.

The definition does not require that $\mathfrak{H}$ be infinite-dimensional, although this is usually assumed to be the case unless the contrary is stated. Dimensionality is discussed in Section 1.5.

A *real* Hilbert space is simply the infinite-dimensional analogue of ordinary n-dimensional Euclidean space; the inner product (u, v) (which is now symmetric—$(u, v) = (v, u)$—and linear in each factor) is the analogue of the scalar produce $\mathbf{u} \cdot \mathbf{v}$. In the complex case, (u, v) is the analogue of the Hermitian scalar product $\sum \bar{u}_j v_j$.

The positive-definiteness of the norm and the equation $\|au\| = |a| \|u\|$ follow immediately from the properties of the inner product and from the definition, $\|u\| = (u, u)^{1/2}$. The triangle inequality will now be derived.

If u and v are any elements of $\mathfrak{H}$, and a is any number, then

$$0 \leq (u + av, u + av) = (u, u) + (u, av) + (av, u) + (av, av);$$

therefore, since (av, u) is the complex conjugate of (u, av),

$$-2 \operatorname{Re}(u, av) \leq \|u\|^2 + |a|^2 \|v\|^2. \tag{1.3-1}$$

Choose

$$a = -\frac{\overline{(u, v)}\|u\|}{|(u, v)|\,\|v\|};$$

then, since $(u, av) = a(u, v)$, (1.3-1) gives

$$2|(u, v)|\frac{\|u\|}{\|v\|} \le 2\|u\|^2,$$

or

$$|(u, v)| \le \|u\|\,\|v\|, \tag{1.3-2}$$

which is the *Schwarz inequality*. [It is called the *Bunyakovskii inequality* in the Russian literature.] It implies the *continuity* of the inner product in each of its factors. For instance, if $v_n \to v$ (i.e., if $\|v_n - v\| \to 0$), then $|(u, v_n - v)| \le \|u\|\,\|v_n - v\| \to 0$; hence, $(u, v_n) \to (u, v)$. Lastly,

$$\begin{aligned} (u + v, u + v) &= \|u\|^2 + 2\,\mathrm{Re}(u, v) + \|v\|^2 \\ &\le \|u\|^2 + 2|(u, v)| + \|v\|^2 \\ &\le \|u\|^2 + 2\|u\|\,\|v\| + \|v\|^2 = (\|u\| + \|v\|)^2. \end{aligned}$$

That is,

$$\|u + v\| \le \|u\| + \|v\|,$$

which is the *triangle inequality*.

EXERCISE

1. Show that the inner product is jointly continuous in both factors, i.e., that if $u_n \to u$ and $v_n \to v$, then $(u_n, v_n) \to (u, v)$. *Hint*: Show first that $\|u_n\| \to \|u\|$ and $\|v_n\| \to \|v\|$.

Recall. A function $f(x, y)$ can be separately continuous (i.e., continuous as a function of x for each fixed y and of y for each x) without being jointly continuous. For example, the function

$$f(x, y) = \begin{cases} xy/(x^2 + y^2) & \text{for } x^2 + y^2 \ne 0 \\ 0 & \text{for } x = y = 0 \end{cases} \tag{1.3-3}$$

is separately but not jointly continuous at the origin, as one can see by letting x and y approach 0 along a 45° line in the x, y plane.

Two other forms of the triangle inequality can be obtained: first, replace v by $w - u$ in (1.3-3); second, interchange w and u. These forms are all summarized (after renaming the variables) in the formula

$$\big|\,\|u\| - \|v\|\,\big| \le \begin{Bmatrix} \|u + v\| \\ \text{or} \\ \|u - v\| \end{Bmatrix} \le \|u\| + \|v\|. \tag{1.3-4}$$

From the definition of the norm in terms of the inner product, it follows that

$$\|u + v\|^2 + \|u - v\|^2 = 2\|u\|^2 + 2\|v\|^2, \tag{1.3-5}$$

which is called the *parallelogram law*. Since u, v, $u + v$ and $u - v$ all lie in a two-dimensional subspace, the triangle inequalities and the parallelogram law merely express elementary theorems of plane Euclidean geometry.

Note. A Banach space $\mathfrak{B}$ (see Chapter 1.5) is a complete normed linear space, but the norm is not assumed to be derived from an inner product. It was proved by Jordan and von Neumann (1935) that if the parallelogram law (1.3-5) also holds, for all u and v in $\mathfrak{B}$, then an inner product can be defined so as to make $\mathfrak{B}$ into a Hilbert space. The inner product is then given in terms of the norm by the so-called *polarization* procedure,

$$(u, v) = \tfrac{1}{4} \sum_{(\alpha = 1, i, -1, -i)} \alpha \| \alpha u + v \|^2, \qquad (1.3\text{-}6)$$

which is easily seen to hold, once an inner product is known to exist; however, it is not completely easy to show in advance that the right member of this equation has all the properties of an inner product. (Equation (1.3-6) is generalized in Section 1.11.) Finally, there are many Banach spaces in which the parallelogram law does not hold. For example, consider the space $L^1(\mathbb{R})$ consisting of all functions (more precisely, distributions) $f(x)$ such that $\int_{-\infty}^{\infty} |f(x)| dx = \| f \| < \infty$; if $f(x)$ and $g(x)$ are continuous functions with disjoint supports (i.e., if, for every x, either $f(x) = 0$ or $g(x) = 0$), then, clearly

$$\| f \pm g \| = \int_{-\infty}^{\infty} |f(x)| dx + \int_{-\infty}^{\infty} |g(x)| dx$$

$$= \| f \| + \| g \|$$

so that

$$\| f + g \|^2 + \| f - g \|^2 = 2(\| f \| + \| g \|)^2,$$

which disagrees with (1.3-5). More generally, the so-called L^p norm of a function or distribution f on $\mathbb{R}$ is defined as $\{ \int |f(x)|^p \, dx \}^{1/p}$, for $p \geq 1$; only for $p = 2$ can an inner product be defined in such a way that $\| f \| = \sqrt{(f, f)}$.

EXERCISE

2. Verify the identity (1.3-6), assuming the existence of an inner product.

1.4 Examples of Hilbert Spaces

The spaces $L^2(a, b)$, $L^2(\mathbb{R}^n)$, etc., of quadratically integrable functions (rather, distributions), which will be discussed in Chapter 5, are Hilbert spaces. In them, the inner product is given by a formula of the type

$$(f, g) = \int_a^b \overline{f(x)} g(x) dx$$

or

$$(f, g) = \int_{-\infty}^{\infty} \cdots \int_{-\infty}^{\infty} \overline{f(\mathbf{x})} g(\mathbf{x}) d^n \mathbf{x}.$$

As a further example, consider the space l^2 consisting of all sequences $\{x_n\}$, $n = 1, 2, 3, \ldots$, of complex numbers such that

$$\sum_{n=1}^{\infty} |x_n|^2 < \infty. \tag{1.4-1}$$

The numbers x_n may be regarded as the coordinates of a point ξ; if $\xi = \{x_n\}$ and $\eta = \{y_n\}$ are two such sequences, then $\alpha\xi + \beta\eta$ is defined as the sequence

$$\alpha\xi + \beta\eta = \{\alpha x_n + \beta y_n\}, \tag{1.4-2}$$

and the inner product is defined by

$$(\xi, \eta) = \sum_{n=1}^{\infty} \bar{x}_n y_n. \tag{1.4-3}$$

EXERCISE

1. Show that if $\{x_n\}$ and $\{y_n\}$ are in l^2, then $\{\alpha x_n + \beta y_n\}$ is in l^2, and that the summation in (1.4-3) is convergent—in fact absolutely convergent. *Hint*: Use the Cauchy inequality

$$\left| \sum_{n=1}^{N} u_n v_n \right| \leq \sqrt{\sum_{n=1}^{N} |u_n|^2} \sqrt{\sum_{n=1}^{N} |v_n|^2}, \tag{1.4-4}$$

which is merely a discrete version of the Schwarz inequality and can be proved in the same way.

Assertion. The space l^2 is complete. To show this, let $\{\xi^j\}$ be a Cauchy sequence of elements of l^2, i.e., a sequence such that $\|\xi^j - \xi^k\| \to 0$, as $j, k \to \infty$, where, for each j, ξ^j is a sequence $\{x_n^j\}$, $n = 1, 2, \ldots$, of complex numbers. Then, for any $\varepsilon > 0$, there is an integer K such that

$$\|\xi^j - \xi^k\|^2 = \sum_{n=1}^{\infty} |x_n^j - x_n^k|^2 < \varepsilon, \quad \text{if } j \geq K \text{ and } k \geq K;$$

hence $|x_n^j - x_n^k|^2 < \varepsilon$, for any n. Therefore, for fixed n, $\{x_n^j\}$, $j = 1, 2, \ldots$, is a Cauchy sequence of numbers, and the quantity $y_n = \lim_{j \to \infty} x_n^j$ exists.

EXERCISE

2. Complete the proof by showing that $\{y_n\}$ is in l^2 and that, if $\eta = \{y_n\}$, then $\|\eta - \xi^j\| \to 0$, as $j \to \infty$, that is, that $\xi^j \to \eta$ in the space l^2. *Hint*: Show first that $\{\|\xi^j\|\}$ is a convergent, hence bounded, sequence of numbers.

Note. It might be supposed, on intuitive grounds, that these Hilbert spaces are of different sizes, in some sense, namely that l^2 is somehow of smaller infinite dimension than $L^2(a, b)$, and $L^2(a, b)$ of smaller dimension than

$L^2(\mathbb{R}^n)$. It will be seen that this is not so; these are merely three different representations of one and the same abstract Hilbert space and are isometrically isomorphic. That is, there are one-to-one correspondences among the three spaces that include all the elements of each, and under which all properties (inner product, norm, etc.) are preserved.

1.5 Cardinal Numbers; Separability; Dimension

The cardinal number (finite or transfinite) of a set (finite or infinite) tells how many elements the set contains. If there is a one-to-one correspondence $A \leftrightarrow B$ between two sets A and B, then A and B are said to *have the same cardinal number*; in symbols, $\bar{\bar{A}} = \bar{\bar{B}}$. If A is a finite set containing n elements, $\bar{\bar{A}} = n$. Transfinite cardinals are defined by attaching names (symbols) to particular examples of infinite sets and thence to other sets having the same cardinal number. For instance, if there is a one-to-one correspondence $A \leftrightarrow \{1, 2, 3, \ldots\}$ between the elements of A and the set of all positive integers, then A is called *countably* or *denumerably* infinite, and $\bar{\bar{A}} \overset{\text{def}}{=} \aleph_0$ ("aleph zero"). The elements of such a set can be arranged in a sequence.

EXAMPLES

The set $\{0, 1, -1, 2, -2, \ldots\}$ of all integers, the set $\{2, 4, 6, \ldots\}$ of even integers, the set $\{2, 3, 5, \ldots\}$ of prime numbers.

Assertion.   The union of a countable collection of countable sets is countable. [*Note*: "countable" means finite or countably infinite, but the statement is trivial unless we are concerned with infinitely many infinite sets.] To prove the assertion, arrange each set in a sequence horizontally and then arrange these sequences in a vertically descending sequence; then the one-to-one correspondence with $\{1, 2, 3, \ldots\}$ can be established by numbering the elements as follows:

$$
\begin{array}{llll}
1 & 2 & 4 & 7 \cdots \\
3 & 5 & 8 & \cdots \\
6 & 9 & & \cdots \\
10 & \text{etc.} & & \cdots \\
\vdots & \vdots & \vdots & \vdots
\end{array}
\tag{1.5-1}
$$

In this way it is proved that the rational numbers are countable:

$$
\begin{array}{llll}
\frac{1}{1} & \frac{1}{2} & \frac{1}{3} & \frac{1}{4} \cdots \\
\frac{2}{1} & \frac{2}{2} & \frac{2}{3} & \frac{2}{4} \cdots \\
\frac{3}{1} & \frac{3}{2} & \frac{3}{3} & \frac{3}{4} \cdots \\
\vdots & \vdots & \vdots & \vdots
\end{array}
$$

EXERCISE

1. Show that a subset of a countable set is countable. This principle has been used tacitly, in the foregoing example, because the given enumeration of the rational numbers contains many duplications, such as $\frac{2}{4}$ for $\frac{1}{2}$, and $\frac{2}{2}$ and $\frac{3}{3}$ for $\frac{1}{1}$, which must be eliminated before one can construct a one-to-one correspondence between the rational numbers and the positive integers.

An important countable set for the theory of function spaces is the set of all polynomials in n indeterminates $x_1, \ldots, x_n$ with rational coefficients: for each N, the set of all polynomials $p(x_1, \ldots, x_n)$ with degree $\leq N$ and with coefficients of the form r/s with $|r| \leq N$ and $1 \leq s \leq N$ is finite; therefore the union of all such sets can be arranged in a sequence.

As is well known, the set of all real numbers in $[0, 1]$ is not countable. The cardinal number of this set is denoted by c and is called the *power of the continuum.* (The cardinal number of a set is sometimes called its *power.*) The mapping $x \rightarrow \frac{1}{2}(1 + \tanh x)$ then shows that the cardinal number $\bar{\mathbb{R}}$ of the whole real line is also c. By means of a Peano curve, which maps $[0, 1]$ onto the entire unit square $(0 \leq x \leq 1, 0 \leq y \leq 1)$, and by similar curves in space, it is then seen that $\mathbb{R}^2$, $\mathbb{R}^3$, etc. all have the same cardinal number c.

An unending sequence of cardinals can be obtained by the following process: Let A and B be sets and let C be the set of all mappings of B into A; $\bar{A}$ and $\bar{B}$ are the cardinal numbers of A and B, we denote by $\bar{A}^{\bar{B}}$ the cardinal number of C; i.e. $\bar{C} = \bar{A}^{\bar{B}}$. [The reader should verify that this notation is correct if A and B are finite sets and also that this definition gives $\bar{A}^{\bar{B}}$ uniquely, i.e., that if there are one-to-one correspondences $A \leftrightarrow A_1$ and $B \leftrightarrow B_1$, then there is a one-to-one correspondence between the two sets of mappings: that of B into A and that of B_1 into A_1.] As an example, if we expand each real number x in $[0, 1)$ as a decimal $x = .a_1 a_2 a_3 \ldots$, then the correspondence $j \rightarrow a_j$, obtained from a given number x, is a mapping of the set B of all positive integers $j = 1, 2, \ldots$ into the set $A = (0, 1, \ldots, 9)$ of the ten digits; hence, each x corresponds to an element in the set of all such mappings. This last correspondence is one-to-one, except that $0.1999 \ldots$ and $0.2000 \ldots$ represent the same x, etc.; ignoring this slight complication, we see that c, the power of the continuum, is equal to $10^{\aleph_0}$. A binary (or n-ary) expansion of x shows similarly that c is also $= 2^{\aleph_0} = n^{\aleph_0}$. A continued fraction expansion shows that $c = \aleph_0^{\aleph_0}$.

In order to compare transfinite cardinals, we say that $\bar{A} < \bar{B}$ if there is a one-to-one mapping of A into B, but no such mapping of B into A. The uncountability of the continuum is then expressed by saying that $c > \aleph_0$. It can be shown (Zorn's lemma is used, in the form of the well-ordering principle) that if A and B are any two sets, then either A can be mapped one-to-one into B, or B one-to-one into A, or both. If both, then, according to the so-called *Bernstein* or *Schröder–Bernstein Theorem,* there exists a one-to-one mapping of each set onto the other. It follows that in any case either $\bar{A} < \bar{B}$ or $\bar{A} = \bar{B}$ or $\bar{A} > \bar{B}$. It can also be shown that if A has at least two elements ($\bar{\bar{A}} \geq 2$), and B is nonempty ($\bar{\bar{B}} \geq 1$), then $\bar{\bar{A}}^{\bar{\bar{B}}} > \bar{\bar{B}}$. Hence there are infinitely many transfinite cardinals.

A real-valued function of a real variable x is a mapping of $\mathbb{R}$ into $\mathbb{R}$ (not assumed continuous or measurable or anything), hence the set of all such functions has the cardinal number c^c, which is $> c$; that is, there are more functions than there are points on a line. On the other hand, the set of all *continuous* functions of a real variable has only the cardinal number $c^{\aleph_0}$, which is equal to c itself. To see this, first arrange the rational numbers in a sequence $x_1, x_2, x_3, \ldots$; corresponding to any continuous function $f(x)$ consider the sequence $f(x_1), f(x_2), \ldots$ of real numbers. Two different continuous functions cannot lead to the same sequence, so that the number of continuous functions cannot exceed the cardinal number of the set of all such sequences, which is clearly $c^{\aleph_0}$. To show that $c^{\aleph_0} = c$, let $\{\alpha, \beta, \gamma, \ldots\}$ be a sequence of reals and write, in decimal notation,

$$
\begin{aligned}
\alpha &= .a_1 a_2 a_3 \cdots \\
\beta &= .b_1 b_2 b_3 \cdots \\
\gamma &= .c_1 c_2 c_3 \cdots \\
&\quad \text{etc.}
\end{aligned}
\tag{1.5-2}
$$

[It is assumed that $0 \leq f(x) < 1$ for all x; for the general case the mapping $f(x) \leftrightarrow g(x) = \frac{1}{2}\{1 + \tanh f(x)\}$ is used.] Such a sequence is then put in correspondence with the single number ω given by

$$
\omega = .a_1 a_2 b_1 a_3 b_2 c_1 a_4 b_3 c_2 d_1 \ldots,
\tag{1.5-3}
$$

following the pattern (1.5-1); this correspondence of sequences $\{\alpha, \beta, \gamma, \ldots\}$ to numbers ω is one-to-one, except that ω's of the form $0.1999\ldots$ and $0.2000\ldots$ etc. are equal; this difficulty is overcome by supposing that (1.5-3) is a, say, duodecimal representation (without changing (1.5-2)), whereupon $0.1999\ldots$ and $0.2000\ldots$ become different numbers, and the set of all sequences $\{\alpha, \beta, \gamma, \ldots\}$ is mapped one-to-one onto a subset of $[0, 1)$. Therefore, there are no more continuous functions than there are real numbers. Also, there are at least as many (hence there are precisely as many), because the cardinal number of the set of constant functions is already $= c$.

A Hilbert space is called *separable* (an unfortunately chosen word) if it contains a countable dense subset. [Then, as will be shown in the next section, it contains a complete orthonormal sequence, in terms of which any element of $\mathfrak{H}$ can be expressed in the form of a generalized Fourier series.] The space l^2 is separable because it contains as a dense subset the set of all elements $\xi = \{x_1, x_2, \ldots\}$ such that (a) the x_i are rational and (b) only finitely many of the x_i are $\neq 0$. The space $L^2(a, b)$ is separable because any continuous function can be arbitrarily closely approximated by a polynomial in $[a, b]$, hence by a polynomial with rational coefficients, and any element of L^2 can be arbitrarily closely approximated in the mean by a continuous function (see Chapter 5). Thus, the set of all such polynomials, which is countable, is dense in $L^2(a, b)$. Similarly, $L^2(\mathbb{R})$ and $L^2(\mathbb{R}^n)$ are separable; in this case, the polynomials must be modified at large distances to make them quadratically integrable. Nonseparable Hilbert spaces occur in the theory of almost periodic functions but not, to my knowledge, in the usual physical applications. It will be proved in the next section that all infinite-dimensional separable real Hilbert spaces are identical, in the sense of isomorphism, and so are all the complex ones.

EXAMPLE

Let $\mathfrak{H}$ consist of all functions $f(x)$, defined on $\mathbb{R}$, each of which is equal to zero for all except countably many values of x and is such that $\sum |f(x)|^2 < \infty$; if f and g are any two such functions, define

$$(f, g) = \sum \overline{f(x)}g(x),$$

the summation being over those values of x for which the summand is $\neq 0$. This Hilbert space $\mathfrak{H}$ is nonseparable. If $\mathbb{R}$ is replaced by a set whose cardinal number $\aleph$ is $> c$, then a Hilbert space of still larger dimension results, and so on. The dimension of a Hilbert space is defined at the end of the next section.

EXERCISES

1. Prove that the space $\mathfrak{H}$ of this example is a Hilbert space and is nonseparable.

2. For each $n = 0, 1, 2, \ldots$, let $\varphi_n(\cdot, \cdots, \cdot)$ denote an element (distribution) in $L^2(\mathbb{R}^n)$—for $n = 0$, φ_0 is interpreted as a single complex number—and let Φ denote an infinite column vector

$$\Phi = \begin{bmatrix} \varphi_0 \\ \varphi_1(\cdot) \\ \varphi_2(\cdot, \cdot) \\ \vdots \\ \varphi_n(\cdot, \cdots, \cdot) \\ \vdots \end{bmatrix}$$

Let $\mathscr{F}$ denote the set of all such column vectors for which

$$|\varphi_0|^2 + \sum_{n=1}^{\infty} \|\varphi_n\|^2 < \infty$$

[in the nth term of the sum, $\|\cdot\|$ denotes the norm in $L^2(\mathbb{R}^n)$], and define the inner product in $\mathscr{F}$ by

$$(\Phi, \Psi) = \bar{\varphi}_0 \psi_0 + \sum_{n=1}^{\infty} (\varphi_n, \psi_n).$$

Show that $\mathscr{F}$ is a Hilbert space and is separable. (In quantum mechanics spaces like this one are called *Fock spaces*.)

1.6 Orthonormal Sequences

Two elements f and g of a Hilbert space $\mathfrak{H}$ are called *orthogonal* (in symbols, $f \perp g$) if $(f, g) = 0$. A sequence $\{\varphi_i\}$ of elements is *orthonormal* if $(\varphi_i, \varphi_j) = \delta_{ij}$

$(= 1$ for $i = j$ and $= 0$ for $i \neq j$). For example, orthogonal function systems provide such sequences in L^2 spaces. If $\{\varphi_i\}$ is an orthonormal sequence, and $c_1, c_2, \ldots$ are constants such that $\sum_{i=1}^{\infty} |c_i|^2 < \infty$, then the partial sums of the series $\sum_{i=1}^{\infty} c_i \varphi_i$ form a Cauchy sequence, and the series converges to some element of $\mathfrak{H}$. If $f \in \mathfrak{H}$, the numbers (φ_i, f) are called the (generalized) Fourier coefficients of f with respect to the orthonormal set $\{\varphi_i\}$. Since

$$0 \leq \left(f - \sum_{i=1}^{n} (\varphi_i, f)\varphi_i, f - \sum_{i=1}^{n} (\varphi_i, f)\varphi_i\right)$$

$$= \|f\|^2 - 2\sum_{i=1}^{n} |(\varphi_i, f)|^2 + \sum_{1}^{n} {}_{(i,j)} \overline{(\varphi_i, f)}(\varphi_j, f)(\varphi_i, \varphi_j)$$

$$= \|f\|^2 - \sum_{i=1}^{n} |(\varphi_i, f)|^2,$$

it follows that $\sum_{i=1}^{n} |(\varphi_i, f)|^2 \leq \|f\|^2$, for all n, hence

$$\sum_{i=1}^{\infty} |(\varphi_i, f)|^2 \leq \|f\|^2 \quad \text{(Bessel's inequality)}. \tag{1.6-1}$$

It follows that $\sum_{n=1}^{\infty} (\varphi_n, f)\varphi_n$ always converges (although not necessarily to f).

An orthonormal sequence $\{\varphi_i\}$ is called *complete* if no nonzero element ψ of $\mathfrak{H}$ is orthogonal to every φ_i.

Theorem 1.   *If $\{\varphi_i\}$ is an orthonormal sequence in $\mathfrak{H}$, then the following requirements are equivalent:*

(1) $\{\varphi_i\}$ *is complete*
(2) $f = \sum_{i=1}^{\infty} (\varphi_i, f)\varphi_i$ *for every f in $\mathfrak{H}$*
(3) $(f, g) = \sum_{i=1}^{\infty} (f, \varphi_i)(\varphi_i, g)$ *for every f, g in $\mathfrak{H}$*
(4) $\|f\|^2 = \sum_{i=1}^{\infty} |(f, \varphi_i)|^2$ *for every f in $\mathfrak{H}$.*

[The last two equations are called *Parseval relations*.]

PROOF.   The implications $(1) \Rightarrow (2) \Rightarrow (3) \Rightarrow (4) \Rightarrow (1)$ will be proved.
$(1) \Rightarrow (2)$ For any f in $\mathfrak{H}$, $f - \sum (\varphi_i, f)\varphi_i$ is orthogonal to every φ_j, hence is zero by (1).
$(2) \Rightarrow (3)$ Substitution of $\sum (\varphi_i, f)\varphi_i$ for f in (f, g) and use of the continuity of $(\cdot, \cdot)$ (the Schwarz inequality), to justify going to the limit, gives (3).
$(3) \Rightarrow (4)$ Set $g = f$ in (3).
$(4) \Rightarrow (1)$ If any f is orthogonal to all φ_j, then $\|f\|$ is zero by (4); hence, f is zero.

Theorem 2.   *A Hilbert space $\mathfrak{H}$ is separable if and only if it contains a complete orthonormal sequence.*

PROOF.   First, if $\mathfrak{H}$ is separable, then it contains a countable set $\{\psi_i\}$ $(i = 1, 2, \ldots)$ dense in $\mathfrak{H}$. From this set a complete orthonormal sequence is constructed by the *Gram–Schmidt* procedure:

1. Let ψ be the first nonzero element of $\{\psi_i\}$; call

$$\varphi_1 = \frac{1}{\|\psi\|} \psi \quad \left(\text{write as } \frac{\psi}{\|\psi\|}\right)$$

2. Let ψ' be the first element of $\{\psi_i\}$ that is not a multiple of φ_1 (hence, not a multiple of ψ). Call

$$\varphi_2 = \frac{\psi' - (\varphi_1, \psi')\varphi_1}{\|\psi' - (\varphi_1, \psi')\varphi_1\|}$$

$\vdots$

$n + 1$. Let $\psi^{(n)}$ be the first element of $\{\psi_i\}$ that is not a linear combination of $\varphi_1, \varphi_2, \ldots, \varphi_n$; Call

$$\varphi_{n+1} = \frac{\psi^{(n)} - \sum_{i=1}^{n} (\varphi_i, \psi^{(n)})\varphi_i}{\|\text{same}\|}$$

It is evident that $\{\varphi_i\}$ is an orthonormal set (it may be a finite set, for it may happen that from a certain point on all ψ_i are linear combinations of $\varphi_i, \ldots, \varphi_k$—in this case $\mathfrak{H}$ is finite-dimensional). To show the completeness of this set $\{\varphi_i\}$, suppose that ζ is an element of $\mathfrak{H}$ such that $\zeta \perp \varphi_i$ for all i; it will be shown that then $\zeta = 0$. Given $\varepsilon > 0$, one can choose an element ψ from $\{\psi_i\}$ such that $\|\psi - \zeta\| < \varepsilon$. From the Gram–Schmidt construction it is clear that ψ is a finite sum $\sum c_i \varphi_i$. Since $\zeta \perp$ all φ_i, it follows that $\zeta \perp \psi$: therefore

$$(\psi - \zeta, \psi - \zeta) = \|\psi\|^2 + \|\zeta\|^2 < \varepsilon^2$$

$$\therefore \quad \|\zeta\|^2 < \varepsilon^2 \quad \text{for every } \varepsilon > 0$$

$$\therefore \quad \|\zeta\| = 0,$$

$$\therefore \quad \zeta = 0.$$

Conversely, if $\mathfrak{H}$ contains a complete orthonormal sequence $\{\varphi_i\}$, then the set of all finite linear combinations of the φ_i with rational coefficients is a countable dense set, hence $\mathfrak{H}$ is separable.

Corollary. *A separable Hilbert space (real or complex) is either isomorphic to a finite-dimensional Euclidean space V^n (real or complex) or isomorphic to the Hilbert space l^2 (real or complex).*

Isomorphism is here intended to include isometry (preservation of the norm), so that isomorphic spaces have completely identical properties. [Preservation of the norm implies that the mapping and its inverse are continuous in the topologies of $\mathfrak{H}$ and l^2, so that the mapping is (among other things) a homeomorphism, in the language of topology.]

PROOF (for the infinite-dimensional case). Let $\{\varphi_i\}$ be a complete orthonormal sequence in $\mathfrak{H}$. For any f in $\mathfrak{H}$, the sequence of numbers $\{(\varphi_i, f)\}$ (Fourier coefficients) is, according to the Bessel inequality, an element of l^2. Conversely, if $\{x_i\}$ is in l^2, then $\sum_{i=1}^{\infty} x_i \varphi_i$ is in $\mathfrak{H}$. According to Theorem 1, the mapping $\mathfrak{H} \to l^2$ given by $f \to \{(\varphi_i, f)\}$ and the inverse mapping $l^2 \to \mathfrak{H}$ given by $\{x_i\} \to \sum_{i=1}^{n} x_i \varphi_i$ are isometric isomorphic onto mappings.

The statement that if $\{x_i\}$ is any sequence of real or complex numbers such that $\sum_{i=1}^{\infty} |x_i|^2$ converges, i.e., if $\{x_i\}$ is in l^2, then the series $\sum x_i \varphi_i$ converges to an element of $\mathfrak{H}$, is one form of the *Riesz–Fischer theorem*.

Theorem 2 and its corollary can be generalized to nonseparable Hilbert spaces, as follows (see Halmos, 1951): First, if a set S of vectors $\{\varphi_a : a \in A\}$ in a Hilbert space $\mathfrak{H}$ (A is a so-called *index set*, not necessarily countable, whose elements serve as subscripts to distinguish one vector of S from another) is such that

$$(\varphi_a, \varphi_b) = \begin{cases} 1 & \text{if } a = b, \\ 0 & \text{if } a \neq b, \end{cases}$$

then S is an *orthonormal set*. If the statement

$$(\varphi_a, \psi) = 0 \quad \text{for all } a \text{ in } A$$

implies that $\psi = 0$, then S is a *complete orthonormal set* or a *basis* in $\mathfrak{H}$. It can be proved that if $\{\varphi_a : a \in A\}$ and $\{\psi_b : b \in B\}$ are any two bases in $\mathfrak{H}$, then the index sets A and B have the same cardinal number $\bar{A} = \bar{B}$, which is called the *dimension* of $\mathfrak{H}$. If $\mathfrak{H}$ is separable, its dimension is $\aleph_0$ or finite. The dimension of the Hilbert space in the Example in the preceding section is equal to c, the power of the continuum. The generalization of the above corollary is that two Hilbert spaces are isomorphic if and only if they have the same dimension.

1.7 Subspaces; The Projection Theorem

A *closed linear manifold* or *subspace* $\mathfrak{M}$ of $\mathfrak{H}$ is a closed linear set of elements in $\mathfrak{H}$; $\mathfrak{M}$ is itself a Hilbert space. [If S is any set in $\mathfrak{H}$ (or, for that matter, in any metric space), and if $\{u_i\}_1^\infty$ is any convergent sequence of points of S, then $\lim u_i$ (which may not be in S) is called a *limit point* of S; if S contains all its limit points, it is *closed*.]

EXERCISES

Consider each of the following linear manifolds in l^2 and show whether it is closed:

1. The set of all points $\xi = \{x_n\}_1^\infty$ such that $x_n = 0$ for $n > 10$.
2. The set of all ξ such that $x_n = 0$ for $n > $ some n_0, which may depend on ξ.
3. The set of all ξ such that $x_n = 0$ for n even.
4. The set of all ξ such that $\sum_{n=1}^\infty n|x_n|^2$ is finite.
5. The set of all ξ such that $\sum_{n=1}^\infty (1/n)x_n = 0$.
6. The set of all ξ such that $\sum_{n=1}^\infty x_n = 0$.

If $\mathfrak{M}$ is a subspace of $\mathfrak{H}$, its *orthogonal complement* $\mathfrak{M}^\perp$ is defined as

$$\mathfrak{M}^\perp = \{\varphi \in \mathfrak{H} : (\psi, \varphi) = 0 \; \forall \psi \in \mathfrak{M}\}; \tag{1.7-1}$$

It is a closed linear set, hence it also is a subspace of $\mathfrak{H}$. The linearity of $\mathfrak{M}^\perp$ follows from the linearity of $(\cdot, \cdot)$ in the second factor: if $(\psi, \varphi_1) = 0$ and

$(\psi, \varphi_2) = 0$ for all ψ in $\mathfrak{M}$, then $(\psi, \alpha\varphi_1 + \beta\varphi_2) = 0$ for all ψ in $\mathfrak{M}$. The closure of $\mathfrak{M}$ follows from the continuity of $(\cdot, \cdot)$: If $\varphi_i \in \mathfrak{M}^{\perp}$ and $\varphi_i \to \tilde{\varphi}$ in $\mathfrak{H}$, then $(\psi, \tilde{\varphi}) = \lim_{i \to \infty} (\psi, \varphi_i) = 0$ for any ψ in $\mathfrak{M}$; therefore $\tilde{\varphi} \in \mathfrak{M}^{\perp}$.

Projection theorem. *If $\mathfrak{M}$ is a subspace of $\mathfrak{H}$ and ζ is any element of $\mathfrak{H}$, then ζ can be decomposed uniquely as $\zeta = \varphi + \psi$, where φ is in $\mathfrak{M}$ and ψ is in $\mathfrak{M}^{\perp}$.*

Remarks. (1) It follows that $(\mathfrak{M}^{\perp})^{\perp} = \mathfrak{M}$. (2) If $\mathfrak{M}$ is replaced by an arbitrary set S in the definitions (1.7-1), then $S^{\perp}$ is again a closed linear manifold, and $(S^{\perp})^{\perp}$ is the smallest closed linear manifold that contains S; it is called the *closed linear span* of S. An important application of the projection theorem is contained in the next section.

PROOF.   As in the finite-dimensional case, φ is that element of $\mathfrak{M}$ that minimizes the distance $\|\zeta - \varphi\|$. First we have to prove that the minimum can actually be attained. Call

$$d = \inf\{\|\zeta - \varphi\| : \varphi \in \mathfrak{M}\}.$$

Let $\{\varphi_i\}$ be a sequence of elements of $\mathfrak{M}$ such that $\|\zeta - \varphi_i\| \to d$. Given $\varepsilon > 0$; suppose that i and j are both large enough so that

$$\|\zeta - \varphi_i\|^2 \leq d^2 + \varepsilon$$
$$\|\zeta - \varphi_j\|^2 \leq d^2 + \varepsilon.$$

By the parallelogram law,

$$2\|\zeta - \varphi_i\|^2 + 2\|\zeta - \varphi_j\|^2 = \left\|2\left(\zeta - \frac{\varphi_i + \varphi_j}{2}\right)\right\|^2 + \|\varphi_i - \varphi_j\|^2;$$

the left member is $\leq 4d^2 + 4\varepsilon$; the first term on the right is $\geq 4d^2$, because $\frac{1}{2}(\varphi_i + \varphi_j)$ is in $\mathfrak{M}$; therefore $\|\varphi_i - \varphi_j\|^2 \leq 4\varepsilon$, so that $\{\varphi_i\}$ is a Cauchy sequence. Call $\varphi = \lim \varphi_j$ and $\psi = \zeta - \varphi$. Clearly, φ is in $\mathfrak{M}$, because $\mathfrak{M}$ is closed, and it will now be shown that ψ is in $\mathfrak{M}^{\perp}$. According to the minimum property that has just been established, if φ' is any element of $\mathfrak{M}$ and a is any positive number, then

$$(\psi, \psi) \leq (\psi \pm a\varphi', \psi \pm a\varphi'),$$

or

$$0 \leq \pm 2a \,\mathrm{Re}(\psi, \varphi') + a^2(\varphi', \varphi');$$

therefore,

$$|\mathrm{Re}(\psi, \varphi')| \leq \frac{a}{2}(\varphi', \varphi').$$

Letting $a \to 0$ shows that $\mathrm{Re}(\psi, \varphi') = 0$ for all φ' in $\mathfrak{M}$. Replacing φ' by $i\varphi'$ shows similarly that $\mathrm{Im}(\psi, \varphi') = 0$ for all φ' in $\mathfrak{M}$; hence, $\psi \in \mathfrak{M}^{\perp}$, as was to be shown. Uniqueness now follows easily: if $\varphi_1 + \psi_1$ and $\varphi_2 + \psi_2$ are two such decompositions of ζ, then the element $\varphi_1 - \varphi_2 = \psi_2 - \psi_1$ is in both $\mathfrak{M}$ and $\mathfrak{M}^{\perp}$, hence is orthogonal to itself, hence is equal to 0.

1.8 Linear Functionals; The Riesz–Fréchet Representation Theorem

A *linear functional* in $\mathfrak{H}$ is a scalar-valued function $l(\varphi)$ defined for all φ in $\mathfrak{H}$ and such that $l(\alpha\varphi + \beta\psi) = \alpha l(\varphi) + \beta l(\psi)$ for all φ, ψ in $\mathfrak{H}$ and all scalars α, β. A linear functional is *bounded* if there is a constant K such that $|l(\varphi)| \leq K\|\varphi\|$ for all φ in $\mathfrak{H}$. For fixed ψ_0, (ψ_0, φ) is a bounded linear functional, and it turns out that all bounded linear functionals are of this form:

Riesz–Fréchet representation theorem. *If $l(\varphi)$ is any bounded linear functional in $\mathfrak{H}$, then there is a unique element ψ_0 in $\mathfrak{H}$ such that $l(\varphi) = (\psi_0, \varphi)$ for all φ.*

PROOF.   The subset

$$\mathfrak{M} = \{\xi \in \mathfrak{H} : l(\xi) = 0\}$$

is clearly linear, and it is closed by virtue of the boundedness of $l(\cdot)$, hence is a subspace. [If $\xi_i \in \mathfrak{M}$ and $\xi_i \to \omega$ in $\mathfrak{H}$, then $|l(\omega) - l(\xi_i)| = |l(\omega - \xi_i)| \leq K\|\omega - \xi_i\| \to 0$, hence $l(\omega) = 0$, and $\omega \in \mathfrak{M}$.] Assuming $\mathfrak{M} \neq \mathfrak{H}$ (if $l(\varphi) \equiv 0$, take $\psi_0 = 0$), let ψ be any nonzero element of $\mathfrak{M}^\perp$: then if ψ_1 is any other element of $\mathfrak{M}^\perp$, $l(\psi - a\psi_1) = 0$ for $a = l(\psi)/l(\psi_1)$; hence $\psi - a\psi_1$ is in both $\mathfrak{M}$ and $\mathfrak{M}^\perp$, i.e., is zero, so that $\mathfrak{M}^\perp$ is one-dimensional. It is now seen that if

$$\psi_0 \stackrel{\text{def}}{=} \frac{\overline{l(\psi_1)}}{\|\psi_1\|^2}\,\psi_1,$$

then $l(\eta) = (\psi_0, \eta)$ for all η, as required; this is seen by decomposing η into its components in $\mathfrak{M}$ and $\mathfrak{M}^\perp$ and applying $l(\cdot)$ to each component separately; clearly, ψ_0 is unique.

1.9 Strong and Weak Convergence

In a finite-dimensional vector space, according to the Bolzano–Weierstrass theorem, any bounded sequence of vectors, say $\{u_k\}$, where $\|u_k\| \leq$ const. for all k, has a convergent subsequence. That is, the closed ball $\|u\| \leq$ const. is compact. In fact, any closed bounded set is compact, and the space is said to be *locally compact*. An infinite-dimensional Hilbert space is not locally compact: for instance, an infinite orthonormal sequence $\{u_k\}$ is bounded ($\|u_k\| = 1$ for all k), but it has no convergent subsequence.

Until now, convergence of a sequence $\{u_k\}$ to a limit u has been understood to mean that $\|u_k - u\| \to 0$, as $k \to \infty$. This is called *strong convergence*. As a generalization, the sequence $\{u_k\}$ is said to converge *weakly* to u if $(v, u_k) \to (v, u)$ for every v in $\mathfrak{H}$. [If the word "convergence" is unqualified, strong convergence is normally understood.]

Two theorems are here stated without proof (the proofs depend on the principle of uniform boundedness).

Theorem 1.   *A weakly convergent sequence is bounded.*

Theorem 2. $\mathfrak{H}$ *is weakly complete, in the sense that if* $\{u_n\}$ *is such that* $(u_k - u_l, v) \to 0$, *as* $k, l \to \infty$ *for all* v, *then there is a u such that* $(u_n - u, v) \to 0$, *as* $n \to \infty$, *for all* v.

EXERCISES

1.   Strong convergence implies weak convergence.

2.   In a finite-dimensional space, strong and weak convergence are equivalent.

3.   Any infinite orthonormal sequence is weakly convergent.

4.   Any bounded sequence in a Hilbert space has a weakly convergent subsequence. [i.e., the space is locally weakly compact.] *Hint*: If $\{u_n\}$ is the sequence, there is a subsequence $\{u_{1\,n}\}$ such that $(u_1, u_{1\,n})$ converges, then a subsequence of $\{u_{1\,n}\}$ such that $(u_2, u_{2\,n})$ converges, and so on. Consider the diagonal subsequence $\{u_{n\,n}\}$. Let $\mathfrak{M}$ be the linear manifold spanned by $\{u_n\}$, that is, the set of all finite linear combinations of the $\{u_n\}$. Show that $\{(u_{n\,n}, v)\}$ is a convergent sequence of numbers first for any v in $\mathfrak{M}$, then for any v in the closure $\bar{\mathfrak{M}}$, and finally for any v in $\mathfrak{M}^{\perp}$. Then apply Theorem 2 above.

5.   If $u_n \to u$ weakly and $\|u_n\| \to \|u\|$, then $u_n \to u$ strongly.

1.10 Hilbert Spaces of Analytic Functions

Even though separable Hilbert spaces are all structurally identical, the concrete realizations take many different forms. Spaces whose elements are functions analytic in a region Ω of the complex plane appear in some applications. If Ω is the unit disk, a Hilbert space is defined by the equations

$$\mathfrak{H} = \left\{ f(z) \text{ analytic}, |z| < 1 : \sup_{a<1} \int_{|z|=a} |f(z)|^2 \, ds < \infty \right\}$$

$$(f, g) = \lim_{a \uparrow 1} \int_{|z|=a} \overline{f(z)} g(z) ds.$$

It is easily seen that if c_n ($n = 0, 1, \ldots$) are the coefficients of the power-series expansion of $f(z)$, then the norm is simply $\|f\| = \text{const.} \sqrt{\sum_{n=0}^{\infty} |c_n|^2}$, so that the isomorphism of this space with l^2 is evident.

For applications of similar spaces to function theory, when Ω is a half plane, see Hille (1962), Chapter 19.

1.11 Polarization

A generalization of equation (1.3-6) is useful in many computations (see Chapter 9). Let $f(u, v)$ be a *sesquilinear* (sometimes less accurately called "bilinear") form, i.e., a numerical-valued function, defined for all u and v

in $\mathfrak{H}$; that is, $f(u, v)$ is linear in v and antilinear in u. [It is a consequence of the Riesz–Fréchet theorem that any such form can be written as (Au, v), also as (u, Bu), where A and B are linear operators; no use is made of this fact here.] Let $q(u)$ be the corresponding quadratic form $q(u) = f(u, u)$. Then, f can be computed from q by the *polarization formula*

$$f(u, v) = \tfrac{1}{4} \sum_{(\alpha = 1, i, -1, -i)} \alpha q(\alpha u + v). \qquad (1.11\text{-}1)$$

To verify this formula, note that

$$\begin{aligned}
\alpha q(\alpha u + v) &= \alpha f(\alpha u + v, \alpha u + v) \\
&= \alpha |\alpha|^2 f(u, u) + |\alpha|^2 f(u, v) + \alpha^2 f(v, u) + \alpha f(v, v); \quad (1.11\text{-}2)
\end{aligned}$$

the resulting sixteen terms on the right of (1.11-1) cancel in pairs, except for $f(u, v)$, which appears four times with the coefficient $|\alpha|^2 = 1$.

Often, a bound is available for q, namely $|q(u)| \le M\|u\|^2$, for all u. Then, polarization can be combined with the following device, to give an inequality of the Schwarz type for f. First, for $|\alpha| = 1$,

$$|\alpha q(\alpha u + v)| \le M\|\alpha u + v\|^2 \le M(\|u\| + \|v\|)^2.$$

Then, u and v in $f(u, v)$ are replaced by βu and $\beta^{-1} v$, where $\beta = \sqrt{\|v\|/\|u\|}$; this does not change the value of $f(u, v)$, but replaces $\|u\| + \|v\|$ by $2\sqrt{\|u\|\,\|v\|}$; hence,

$$|f(u, v)| \le 4M\|u\|\,\|v\|. \qquad (1.11\text{-}3)$$

CHAPTER 2

Distributions; General Properties

Linear functionals; test functions on $\mathbb{R}$ and $\mathbb{R}^n$; the bilinear form; convergence of test functions; continuous functionals; real and complex distributions; differentiation and integration; changes of independent variables; convergence of distributions; mollifiers; regularization of singularities.

Prerequisite: Advanced calculus

In functional analysis and its applications to physics, it is desirable to generalize the classical notion of function in the manner suggested by Dirac and carried out by Laurent Schwartz in his theory of "distributions" (called "generalized functions" by the Russian authors). In the presentation given here, in this chapter and the next few, I take the attitude that distribution theory is basically elementary and branches off from the classical development at the level of advanced calculus. The entire presentation is based on the Riemann integral, and the close relation between distributions and ordinary functions is stressed. From this point of view, L^2 spaces and their application to differential operators are also elementary, when based on distribution theory, and are so treated in Chapters 5, 6, 7, 10, and 11.

2.1 Origin of the Distribution Concept

In the early days of quantum mechanics, P. A. M. Dirac introduced the "function" $\delta(x)$, which was supposed to be $= 0$ for $x \neq 0$ and $= +\infty$ for $x = 0$ in such a way that its integral is $= 1$; hence, it was supposed,

$$\text{``} \int_{-\infty}^{\infty} \delta(x)\varphi(x)dx \text{''} = \varphi(0), \tag{2.1-1}$$

and more generally

$$\text{``} \int_{-\infty}^{\infty} \delta(x - a)\varphi(x)dx \text{''} = \varphi(a); \tag{2.1-2}$$

19

furthermore, the function $\delta'(x)$ was supposed to be such as to permit integration by parts:

$$ \text{``} \int_{-\infty}^{\infty} \delta'(x - a)\varphi(x)dx \text{''} = -\varphi'(a) \tag{2.1-3}$$

for any differentiable function $\varphi(x)$. Higher derivatives of $\delta(x)$ were similarly defined. Furthermore, if $f(x)$ is any, say piecewise continuous, function (not assumed differentiable in the ordinary sense), its derivative in a generalized sense was supposed to be a "function" $f'(x)$ such that

$$ \text{``} \int_{-\infty}^{\infty} f'(x)\varphi(x)dx \text{''} = -\int_{-\infty}^{\infty} f(x)\varphi'(x)dx \tag{2.1-4}$$

at least for any function $\varphi(x)$ that is continuously differentiable for all x and vanishes identically outside some finite interval. Dirac called $\delta(x)$, $\delta'(x)$, and $f'(x)$ "improper functions," and he said that although an improper function may not itself have well-defined values, nevertheless, when it occurs as a factor in an integral, the value of the integral can often be well defined—Dirac (1947), p. 59. That observation provided the basis for the theory of distributions of L. Schwartz (1950). The objects represented by the above symbols $\delta(x)$, $\delta'(x)$, $f'(x)$ are defined not by assigning values to them for specified values of x, but by assigning values to the integrals in quotation marks, for all functions $\varphi(x)$ in a certain class. Since each of the above expressions depends linearly on $\varphi(x)$, a generalized function or distribution is a linear functional on a suitable class of "test functions" $\varphi(x)$.

Operationally, a distribution is similar to a function. If $f(x)$ is an ordinary function, one puts in a value of the argument x and gets out a number; if $f(x)$ is a distribution, one puts in a test function $\varphi(x)$ and gets out a number. One cannot put in a precise value x_0 of x, but one can put in a test function that is very strongly peaked at x_0 and is $= 0$ outside a narrow interval containing x_0. That corresponds, in many physical situations, to the possibility of knowing x only to a finite accuracy.

An ordinary continuous function $f(x)$ is regarded as a special case of a distribution; the linear functional is then the integral

$$ \int_{-\infty}^{\infty} f(x)\varphi(x)dx; \tag{2.1-5}$$

The two points of view are equivalent, according to the following theorem, whose proof is left to the reader:

Theorem. *A continuous function $f(x)$, defined for $-\infty < x < \infty$, is completely determined if the value of the integral (2.1-5) is known for every continuous function $\varphi(x)$ that vanishes outside a finite interval (different such intervals being permitted for different functions $\varphi(x)$).*

Later, certain other ordinary functions (i.e., ones that need not be continuous) will be identified in a similar way with certain other distributions.

Many operations that can be applied to ordinary functions (e.g. differentiation, integration, and Fourier transformation) can be applied to distributions under quite general circumstances.

If f and g are two distributions, although one cannot say that $f(x_0) = g(x_0)$ or that $f(x_0) \geq g(x_0)$ for a particular x_0, a meaning is attached to the statement that $f = g$ in (a, b) or that $f \geq g$ in (a, b), where (a, b) is any open interval. (See Chapter 3). More generally, such statements can be made for any open set, but not for a set of Lebesgue measure zero. Sets of measure zero are usually nonphysical, because, in order to decide whether a point x belongs to such a set, one must know x to infinitely many decimal places. The essence of distribution theory is that by relinquishing the knowledge of functions on sets of Lebesgue measure zero, one opens the door to defining a large class of generalized functions, including various δ functions and their derivatives.

Owing to the operational similarity mentioned above, a distribution is in a sense just as concrete an object as a function. For instance, the elements of L^2 spaces (see Chapter 5) are distributions. In the earlier mathematical literature, an element of such a space is taken as an infinite equivalence class of functions, any two of which are permitted to differ arbitrarily on an arbitrary set of measure zero. For concreteness, one often imagines selecting a single function from such a class to represent it, at least for certain purposes, as an element of the L^2 space. We take the attitude that any such function is necessarily overdetermined and arbitrary, while the distribution is just the right thing. These particular distributions are sufficiently like ordinary functions that they can serve, for example, as wave functions in quantum mechanics.

In many physical applications, especially ones involving differential operators, the use of distributions avoids a lot of analysis that is necessary in the older approach, presumably because the avoidance of sets of measure zero makes distributions ideally suited to those applications.

The above discussion obviously applies, mutatis mutandis, to the cases of more than one independent variable.

2.2 Classes of Test Functions; Functions of Type C_0^∞

Different classes of test functions lead to different classes of distributions. In order to cover all the examples in the preceding section ($\delta(x)$ and all its derivatives and all the generalized derivatives of an arbitrary continuous $f(x)$), the test functions $\varphi(x)$ must be infinitely differentiable, and each of them must vanish identically outside some finite interval. Such functions are said to be *of class* C_0^∞ or $C_0^\infty(\mathbb{R})$; corresponding functions $\varphi(\mathbf{x}) = \varphi(x_1, \ldots, x_n)$ of n independent variables are said to be of class $C_0^\infty(\mathbb{R}^n)$; whenever it is desirable to distinguish between real-valued and complex-valued test functions, one writes $^{\text{real}}C_0^\infty$ and $^{\text{cpx}}C_0^\infty$. (The independent variables $x_1, \ldots, x_n$ are always real.) Generally, C^k denotes the class of continuous functions with continuous derivatives of order $\leq k$ (including mixed partial derivatives, for dimension $n > 1$). The *support* of a function $\varphi(\mathbf{x})$ is the closure of the set of points $\mathbf{x}$ for which $\varphi(\mathbf{x}) \neq 0$; the subscript 0 on C_0^∞ or C_0^k indicates that each function in the class has bounded support. Vector-valued test functions are often useful.

EXAMPLE

The function defined by

$$\varphi(x) = \begin{cases} \exp\{-1/(1 - x^2)\} & -1 < x < 1 \\ 0 & |x| \geq 1 \end{cases} \qquad (2.2\text{-}1)$$

is in $C_0^\infty(\mathbb{R})$.

EXERCISES

1.   Show that all derivatives of the function (2.2-1) are continuous for all x.

2.   Construct a C^∞ function that is $=1$ for $|x| \leq 1$ and is $=0$ for $|x| \geq 2$. *Hint*: Consider the properties of $\int^x \varphi(y)dy$, where φ is given by (2.2-1).

3.   Construct a spherically symmetric test function $\varphi(x_1, \ldots, x_n)$ of class $C_0^\infty(\mathbb{R}^n)$ that is >0 at $\mathbf{x} = 0$.

Any linear functional that is defined for all φ in $C_0^\infty(\mathbb{R}^n)$ and satisfies a certain so-called continuity requirement described in Section 2.4 below is a distribution *on* (or *in*) $\mathbb{R}^n$.

In Chapter 4, a wider class $\mathscr{S}$ or $\mathscr{S}(\mathbb{R}^n)$ of test functions (the so-called Schwartz class) is considered; a function $\varphi(\mathbf{x})$ in $\mathscr{S}(\mathbb{R}^n)$ is not required to be identically zero outside a bounded region or "support" but is required to $\rightarrow 0$ very rapidly as $\mathbf{x} \rightarrow \infty$; the corresponding functionals are called tempered distributions. Clearly any test function of type C_0^∞ is automatically of type $\mathscr{S}$ (i.e., $C_0^\infty \subset \mathscr{S}$), hence any tempered distribution is a distribution.

A subclass Z of $\mathscr{S}$, consisting of certain entire analytic functions in $\mathscr{S}$, is considered by Gel'fand et al. (1946 and subsequent volumes, in English). Z neither contains nor is contained in C_0^∞, because an analytic function cannot have compact support unless it is $\equiv 0$. The functionals on Z are generalized functions that include the tempered distributions, but will not be considered further in these Notes, except for a brief comment in Section 4.5.

2.3 Notations for Distributions; The Bilinear Form

It is recalled that a *linear functional* F on a function space, say $C_0^\infty = C_0^\infty(\mathbb{R}^n)$, is a correspondence that assigns to each $\varphi(\mathbf{x})$ in C_0^∞ a number, denoted by $F[\varphi]$, in such a way that if $\varphi_1(\mathbf{x})$ and $\varphi_2(\mathbf{x})$ are any functions in C_0^∞, and a_1 and a_2 are any constants, then

$$F[a_1\varphi_1 + a_2\varphi_2] = a_1 F[\varphi_1] + a_2 F[\varphi_2]. \qquad (2.3\text{-}1)$$

A distribution on $\mathbb{R}^n$ is such a functional. However, the expressions of the form "$\int f(x)\varphi(x)dx$" in equations (2.1-1 to 4) are clearly intended to be linear also in the generalized functions f being defined, hence we use the notation of bilinear forms: The linear functional that defines a particular distribution f is denoted by $\langle f, \cdot \rangle$, while $\langle f, \varphi \rangle$ denotes the value of the functional when a particular test function φ is put into it. Here, f may be regarded as an arbitrary symbol used to denote the distribution (functional) in question, but we adopt the convention that if f and g are two such symbols, and a and b are numbers, then the symbol $af + bg$ denotes the distribution given by

$$\langle af + bg, \varphi \rangle = a\langle f, \varphi \rangle + b\langle g, \varphi \rangle. \tag{2.3-2}$$

For example, if δ and δ' denote the Dirac "function" $\delta(x)$ and its "first derivative," then

$$\langle \delta, \varphi \rangle = \varphi(0), \tag{2.3-3}$$

$$\langle \delta', \varphi \rangle = -\varphi'(0), \tag{2.3-4}$$

for all φ, and $a\delta + b\delta'$ denotes the distribution (functional) given by

$$\langle a\delta + b\delta', \varphi \rangle = a\varphi(0) - b\varphi'(0).$$

In the complex case, where $C_0^\infty = {}^{cp x}C_0^\infty$, the complex conjugate $\bar{f}$ of a distribution f is the linear functional given by

$$\langle \bar{f}, \varphi \rangle \stackrel{\text{def}}{=} \overline{\langle f, \bar{\varphi} \rangle} \quad \text{for any} \quad \varphi \in C_0^\infty. \tag{2.3-5}$$

The interpretation of this definition is the following: If $\varphi(x)$ is a test function, then $\overline{\varphi(x)}$ is also a test function, hence $\langle f, \bar{\varphi} \rangle$ is a well-defined complex number when f is known, and $\overline{\langle f, \bar{\varphi} \rangle}$ is its complex conjugate, which obviously depends linearly on φ; thus is $\langle \bar{f}, \varphi \rangle$ defined for every φ. A distribution f is *real* if $f = \bar{f}$ (i.e., if f and $\bar{f}$ are the same distribution), or equivalently if $\langle f, \varphi \rangle$ is real for all real φ.

It is often convenient to have the independent variable x appear explicitly; then, distributions are denoted by symbols like those used for ordinary functions, for example $f(x)$, $g(x)$, $\delta(x)$. Anyone who knows what the distribution $\delta(x)$ is will know immediately, from Section 2.8 below, what distributions are denoted by $\delta(x - a)$ and $\delta(x^3 + x + 1)$ and will not need to introduce separate symbols for them.

In some mathematical discussions, it is customary to denote a function by a symbol like $f(\cdot)$, while the symbol $f(x)$ is reserved for the value of the function at a specified value x of the argument. In practical work that convention quickly becomes untenable: to denote the function x^x by $\cdot$ or the function $\sin(x^2 + y^3)$ of two variables by $\sin(\cdot^2 + \cdot^3)$ would be confusing, hence symbols like $f(x)$, x^x, $\sin(x^2 + y^3)$ are permitted to serve either purpose. But if one speaks of *a* (or *the*) *function* $f(x)$, it is clear that the entire functional relationship is being referred to. Thus, $J_\nu(z)$ can represent a number, a function of z, or a function of two variables ν and z, depending on the context. Hence, we take the attitude that symbols like $f(x)$, $\delta(x)$ can equally well represent generalized functions or distributions and that the presence of the letter x does not imply anything about the value of a generalized function (not even the existence of that value) for a specified value of x.

Once a distribution $f(x)$ has been defined, its value $\langle f, \varphi \rangle$ for a particular test function φ can be denoted symbolically as

$$\int_{-\infty}^{\infty} f(x)\varphi(x)dx, \tag{2.3-6}$$

the quotation marks that were used in Section 2.1 being omitted when the context makes clear just what sort of object each symbol stands for. That facilitates equations like

$$\int_{-\infty}^{\infty} \delta(3x - 6)\varphi(x)dx = \tfrac{1}{3}\varphi(2)$$

obtained by a formal change of variables $y = 3x - 6$ according to Section 2.8. But for much of this chapter and the next two, the bilinear notation is retained, in order to emphasize that a distribution is fundamentally a linear functional: you put in a test function and get out a number.

2.4 The Formal Definition; The Continuity of the Functionals

As noted earlier, a linear functional on a class of test functions is defined to be a *distribution* only if it satisfies a certain continuity requirement, which will now be discussed. In much of the elementary theory, that continuity requirement plays no role, but at some points it is essential. In the following several sections, and at the end of this one, we shall try to make clear just when the requirement is needed and when it is being carried along only for future use.

First, one defines a mode of convergence $\varphi_j \to \psi$ of test functions, and then one requires that the functionals $\langle f, \cdot \rangle$ be continuous with respect to that mode of convergence; i.e., one requires that

$$\langle f, \varphi_j \rangle \to \langle f, \psi \rangle \quad \text{whenever } \varphi_j \to \psi. \tag{2.4-1}$$

The stronger the mode of convergence of φ_j to ψ, the less is the restriction on the class of functionals. The strongest mode of convergence, denoted by $\overset{\mathscr{G}}{\to}$, is the following:

Definition. If ψ and φ_j are functions in $C_0^\infty(\mathbb{R}^n)$, then the statement $\varphi_j \overset{\mathscr{D}}{\to} \psi$ means that (1) there is a bounded set in $\mathbb{R}^n$ that contains the supports of all the φ_j, (2) $\varphi_j(\mathbf{x})$ converges to $\psi(\mathbf{x})$ uniformly with respect to $\mathbf{x}$ in $\mathbb{R}^n$, as

$j \rightarrow \infty$, and (3) similarly every partial derivative of $\varphi_j(x)$ (of any order, mixed or not) converges uniformly to the corresponding derivative of $\psi(x)$ in $\mathbb{R}^n$.

Then (2.4-1) is the requirement that makes a linear functional $\langle f, \cdot \rangle$ a distribution. Experience indicates that the requirement is always satisfied in the cases that arise in practice, but, just to show that it is not an empty requirement, an example of a discontinuous linear functional on the space $C_0^\infty(\mathbb{R})$ is given in the Appendix to this chapter. We note that it was found necessary to use the axiom of choice to construct that example.

The symbol $\mathscr{D}$ was introduced by Laurent Schwartz to denote the space $C_0^\infty(\mathbb{R}^n)$ when made into a topological space by means of the above mode of convergence. The space $\mathscr{D}'$ dual to $\mathscr{D}$, hence the set of all distributions on $\mathbb{R}^n$, can also be made into a topological space in various ways. For a discussion of these topological questions, the reader is referred to Schwartz or to Gel'fand et al.

Weaker modes of convergence in C_0^∞ lead to more restricted or "milder" classes of distributions. A slightly weaker mode, discussed in Chapter 4, leads to the so-called tempered distributions. At the other extreme, root-mean-square convergence in C_0^∞ leads to elements of the L^2 spaces discussed in Chapter 5; they are so mild that they are often called simply "functions."

EXAMPLE 1

It was mentioned in Section 2.1 that a continuous function $f(x)$ has derivatives of all orders in the distribution sense. A partial converse was proved by Schwartz: given any distribution g on $\mathbb{R}$ and any finite interval (a, b), there is a continuous function $f(x)$ and an integer $k \geq 0$ such that g is equal to $f^{(k)}(x)$ in (a, b), that is, $\langle g, \varphi \rangle = \langle f^{(k)}, \varphi \rangle$ for any test function φ with support in (a, b). The continuity of the functional $\langle g, \cdot \rangle$ is necessary for this result.

The following two exercises contain the main result on integration and differentiation with respect to a parameter. The continuity of the functional $\langle f, \cdot \rangle$ is needed in both of them.

EXERCISE

1. Assume that for each y in a finite interval $[a, b]$ the function $\varphi(x, y)$, as a function of x, is a test function (i.e., is in $C_0^\infty(\mathbb{R})$), and hence for each y all derivatives $\partial_x^k \varphi$ ($k = 0, 1, 2, \ldots$) are continuous and vanish outside some interval $|x| < c = c(y)$. Assume furthermore, however, that the dependence on the parameter y is "reasonable" in the sense that c can be taken independent of y and that the derivatives $\partial_y \partial_x^k \varphi$ are also continuous for all x in $\mathbb{R}$ and all y in $[a, b]$ and also vanish outside the interval $|x| < c$. Show that if f is any distribution on $\mathbb{R}$, then

$$\int_a^b \langle f, \varphi \rangle dy = \left\langle f, \int_a^b \varphi \, dy \right\rangle. \tag{2.4-2}$$

Hint: Show that $\psi(x) \overset{\text{def}}{=} \int_a^b \varphi(x, y) dy$ is a test function and that the Riemann sum for $\int_a^b \varphi \, dy$ converges to $\psi(x)$ in the sense of convergence in $\mathscr{D}$.

Comments   1. The result is valid under somewhat weaker assumptions, which the reader may wish to investigate. 2. If $\langle f, \varphi \rangle$ is written symbolically as an integral, as in (2.3-6), then (2.4-2) becomes

$$\int_a^b \int_{-\infty}^{\infty} f(x)\varphi(x, y)dx \, dy = \int_{-\infty}^{\infty} f(x) \int_a^b \varphi(x, y)dy \, dx \qquad (2.4\text{-}3)$$

EXERCISE

2.   Assume additionally that the derivatives $\partial_y^2 \partial_x^k \varphi$ ($k = 0, 1, 2, \ldots$) are continuous and vanish outside the interval $|x| < c$. Show that

$$\partial_y \langle f, \varphi \rangle = \langle f, \partial_y \varphi \rangle.$$

This is also valid under weaker assumptions.

These ideas are pursued further in Section 2.6 in connection with mollifiers.

2.5 Examples of Distribution

EXAMPLE 1

The example of equation (2.1-5) can be extended to n dimensions. If $f(\mathbf{x})$ is any continuous function on $\mathbb{R}^n$, we identify it with the distribution defined by

$$\langle f, \varphi \rangle = \int_{\mathbb{R}^n} f(\mathbf{x})\varphi(\mathbf{x})d^n\mathbf{x} \quad \text{for every } \varphi \text{ in } C_0^{\infty}(\mathbb{R}^n). \qquad (2.5\text{-}1)$$

The continuity of the functional $\langle f, \cdot \rangle$ is easily proved. Let $\{\varphi_j\}$ be a sequence of test functions such that $\varphi_j \overset{\mathscr{G}}{\to} \psi$ according to the definition in the preceding section. Let $\mathscr{K}$ be a cube large enough to contain the supports of ψ and all the φ_j. Then,

$$\langle f, \varphi_j \rangle - \langle f, \psi \rangle = \int_{\mathscr{K}} f(\mathbf{x})[\varphi_j(\mathbf{x}) - \psi(\mathbf{x})]d^n\mathbf{x}.$$

The uniform convergence of $\varphi_j(\mathbf{x})$ to $\psi(\mathbf{x})$ means that

$$M_j \overset{\text{def}}{=} \max_{\_(\mathbf{x})} |\varphi_j(\mathbf{x}) - \psi(\mathbf{x})| \to 0, \quad \text{as } j \to \infty.$$

Therefore, since

$$|\langle f, \varphi_j \rangle - \langle f, \psi \rangle| \le \int_{\mathscr{K}} |f(\mathbf{x})| d^n\mathbf{x} M_j,$$

we have

$$\langle f, \varphi_j \rangle \to \langle f, \psi \rangle,$$

as required.

The continuity of the *function* $f(\mathbf{x})$ is not really necessary. The foregoing discussion remains valid if $f(\mathbf{x})$ is any integrable function such that the integral of $|f(\mathbf{x})|$ over every cube $\mathscr{K}$ is finite. (Such functions are said to belong to the class $L_{\text{loc}}^1(\mathbb{R}^n)$—but see *Note* below.) $f(\mathbf{x})$ can have simple jumps and it can also become infinite at certain singularities, provided those singularities are integrable. For example, in dimension $n \ge 2$, the equation

$$\langle f, \varphi \rangle = \int_{\mathbb{R}^n} \frac{1}{\|\mathbf{x}\|} \varphi(\mathbf{x})d^n\mathbf{x} \qquad (2.5\text{-}2)$$

defines a distribution f, which can be identified with the function $f(\mathbf{x}) = 1/\|\mathbf{x}\|$.

Unless $f(\mathbf{x})$ is continuous, the identification of functions with distributions is not quite unique, because the value of $f(\mathbf{x})$ at a discontinuity can be altered without altering the corresponding distribution. If $f(\mathbf{x})$ has *nonintegrable* singularities, the identification becomes nonunique in the other direction, too; in that case the above argument breaks down, and a given function, like $f(\mathbf{x}) = 1/x$ in one dimension, may correspond to many different distributions, as explained in Section 2.10 below on regularization.

Note.   As explained in Section 2.9 below, we prefer to restrict the definition (2.5-1) to functions that are integrable in the Riemann sense; if the Lebesgue integral is used, certain inconsistencies can arise.

The next three examples concern potential theory, where $\rho(\mathbf{x})$ usually denotes the spatial density of charge. For a so-called volume distribution of charge, $\rho(\mathbf{x})$ is an ordinary function, and is usually piecewise continuous, but for a surface, line, or point distribution of charge the spatial density is infinite, and $\rho(\mathbf{x})$ has to be regarded as a generalized function, i.e., as a distribution in the sense of the present discussion.

EXAMPLE 2: Single layer of charge
Let $\mathbb{S}$ be a smooth closed surface in $\mathbb{R}^3$ and $\sigma(\mathbf{x})$ a continuous function on $\mathbb{S}$. Then a distribution $\rho(\mathbf{x})$ is defined on $\mathbb{R}^3$ by the equation

$$\langle \rho, \varphi \rangle = \int_{\mathbb{S}} \sigma(\mathbf{x})\varphi(\mathbf{x})d\mathscr{A} \qquad \forall \varphi \text{ in } C_0^\infty(\mathbb{R}^3), \tag{2.5-3}$$

where $d\mathscr{A}$ is the element of area on $\mathbb{S}$. This distribution resembles the Dirac δ-function at each point of $\mathbb{S}$.

EXAMPLE 3: Double layer of charge
With $\mathbb{S}$ and $\sigma(\mathbf{x})$ as above, let $\mathbf{n}(\mathbf{x})$, defined for $\mathbf{x}$ in $\mathbb{S}$, be the outward unit vector normal to the surface $\mathbb{S}$ at $\mathbf{x}$. Then, a distribution $\rho(\mathbf{x})$ is defined by the equation

$$\langle \rho, \varphi \rangle = \int_{\mathbb{S}} \sigma(\mathbf{x})\nabla\varphi(\mathbf{x}) \cdot \mathbf{n}(\mathbf{x})d\mathscr{A} \qquad \forall \varphi \text{ in } C_0^\infty(\mathbb{R}^3). \tag{2.5-4}$$

This distribution resembles the derivative of the δ-function at each point of $\mathbb{S}$.

EXAMPLE 4: Monopole and multiple point charges
The distribution $\rho(\mathbf{x})$ on $\mathbb{R}^3$ given by $\langle \rho, \varphi \rangle = \varphi(0)$ for any $\varphi(\mathbf{x})$ in $C_0^\infty(\mathbb{R}^3)$ represents the density of a point charge at the origin. In accordance with principles to be encountered later, it can be written as $\rho(\mathbf{x}) = \delta(x)\delta(y)\delta(z)$, although this expression fails to exhibit explicitly the spherical symmetry of $\rho(\mathbf{x})$—see Section 2.8. Similarly, the distribution

$$\langle \rho, \varphi \rangle = \partial_x^p \partial_y^q \partial_z^r \varphi(\mathbf{x}) \Big|_{\mathbf{x}=0} \qquad \forall \varphi \tag{2.5-5}$$

represents the density of a multipole of order 2^{p+q+r} at the origin. It can be written schematically as

$$(-1)^{p+q+r}\delta^{(p)}(x)\delta^{(q)}(y)\delta^{(r)}(z). \tag{2.5-6}$$

EXAMPLE 5

One of the Lorentz-invariant δ-functions appearing in quantum electrodynamics and written by Dirac (1947) either as

$$\delta(t^2 - \|\mathbf{x}\|^2) \tag{2.5-7}$$

or as

$$\frac{1}{2\|\mathbf{x}\|}[\delta(t - \|\mathbf{x}\|) + \delta(t + \|\mathbf{x}\|)] \tag{2.5-8}$$

(here, $\mathbf{x}$ is a 3-vector, and the speed of light is taken as unity), is a distribution on $\mathbb{R}^4$, called D or $D(t, \mathbf{x})$, given by

$$\langle D, \varphi \rangle = \int_{\mathbb{R}^3} \frac{1}{2\|\mathbf{x}\|}[\varphi(\|\mathbf{x}\|, \mathbf{x}) + \varphi(-\|\mathbf{x}\|, \mathbf{x})]d^3\mathbf{x} \quad \text{for all } \varphi(t, \mathbf{x}) \text{ in } C_0^\infty(\mathbb{R}^4).$$

$$\tag{2.5-9}$$

This distribution is concentrated on the light cone

$$|t| = \|\mathbf{x}\|;$$

that is, $\langle D, \varphi \rangle$ is $= 0$ for every $\varphi(t, \mathbf{x})$ whose support does not intersect that cone. The Lorentz invariance of $D(t, \mathbf{x})$ is discussed in Section 2.8.

EXERCISE

1. Show that (2.5-9) can be obtained formally from either (2.5-7) or (2.5-8) by writing $\langle D, \varphi \rangle$ symbolically as $\int_{\mathbb{R}^4} D\varphi \, dt \, d^3\mathbf{x}$ and integrating with respect to t.

EXAMPLE 6: Measures

If the function $\sigma(x)$ of a real variable x has bounded variation in every finite interval, then the functional defined by a Stieltjes integral as

$$\langle f, \varphi \rangle = \int_{-\infty}^{\infty} \varphi(x)d\sigma(x) \tag{2.5-10}$$

is a distribution; distributions of this kind are called *measures*—see Chapter 13.

EXAMPLE 7: A product

The product of two distributions is not generally defined (but see Section 6.5, where the so-called direct product is defined for distributions $f(x)$ and $g(y)$ having different independent variables). However, if f is any distribution on $\mathbb{R}^n$ and $\alpha(\mathbf{x})$ is any C^∞ function on $\mathbb{R}^n$ (not necessarily of bounded support), then a distribution called $f\alpha$ or αf is defined by

$$\langle f\alpha, \varphi \rangle \overset{\text{def}}{=} \langle f, \alpha\varphi \rangle. \tag{2.5-11}$$

Note that if $\varphi(\mathbf{x})$ is any test function, then $\alpha(\mathbf{x})\varphi(\mathbf{x})$ is also a test function, hence this definition makes sense.

2.6 Distributions as Limits of Sequences of Functions; Convergence of Distributions

The distribution $\delta(x)$ can be regarded as the limit of the sequence $\{f_n(x)\}$ of functions, where

$$f_n(x) = \frac{n}{\sqrt{\pi}} e^{-n^2 x^2}, \qquad (2.6\text{-}1)$$

because if $\varphi(x)$ is any test function (or even any bounded continuous function), then

$$\lim_{n \to \infty} \int_{-\infty}^{\infty} f_n(x)\varphi(x)dx = \varphi(0). \qquad (2.6\text{-}2)$$

One writes

$$\frac{n}{\sqrt{\pi}} e^{-n^2 x^2} \to \delta(x). \qquad (2.6\text{-}3)$$

More generally, if $\{f_n(x)\}$ is any sequence of functions such that the limit of $\int_{-\infty}^{\infty} f_n(x)\varphi(x)dx$ exists for every test function $\varphi(x)$, the sequence is said to converge to the distribution f defined by

$$\langle f, \varphi \rangle = \lim_{n \to \infty} \int_{-\infty}^{\infty} f_n(x)\varphi(x)dx, \qquad \forall \varphi \qquad (2.6\text{-}4)$$

(see *Note* below).

Still more generally, if $\{f_n\}$ is any sequence of distributions, say on $\mathbb{R}^n$, and f is another distribution such that $\langle f_n, \varphi \rangle \to \langle f, \varphi \rangle$ for every test function φ, then the distributions f_n are said to *converge* to f. In symbols, $f_n \to f$, or $f_n(\cdot) \to f(\cdot)$, or even $f_n(x) \to f(x)$, but if the last form is used one must bear in mind that it is not pointwise convergence that is meant.

As a rather trivial special case, if continuous functions $f_n(x)$ converge uniformly to $f(x)$, then as distributions they also converge to $f(x)$. Mere pointwise convergence is not sufficient.

Note. Schwartz proved that if the f_n are distributions (i.e. *continuous* linear functionals) such that the limit of $\langle f_n, \varphi \rangle$ exists for every test function φ, then the limit, which is obviously a linear functional on the space of test functions, is automatically continuous in the sense of Section 2.4, hence is a distribution.

Exercises

1. Verify (2.6-3), i.e., (2.6-2), when the f_n are given by (2.6-1).
2. Verify similarly

$$\frac{1 - \cos nx}{n\pi x^2} \to \delta(x), \quad \text{as } n \to \infty \qquad (2.6\text{-}5)$$

3.   (compare Exercise 2 in Section 2.4)   Let f be a distribution on $\mathbb{R}^n$ and $\varphi(\mathbf{x}, \mathbf{y})$ a function in $C_0^\infty(\mathbb{R}^{2n})$. We think of the components of $\mathbf{y}$ as parameters and write $\varphi(\mathbf{x}, \mathbf{y}) = \varphi_\mathbf{y}(\mathbf{x})$. Show that $\langle f, \varphi_\mathbf{y} \rangle$ is a C_0^∞ function of $y_1, \dots, y_n$. Show that if $\varphi_\mathbf{y}(\mathbf{x}) = \psi(\mathbf{x} - \mathbf{y})$, where $\psi \in C_0^\infty(\mathbb{R}^n)$, then $\langle f, \varphi_\mathbf{y} \rangle$ is a C^∞ function of $y_1, \dots, y_n$.

This result can be applied to the problem of *smoothing*. Let $\rho(\mathbf{x})$ be a spherically symmetric nonnegative function in $C_0^\infty(\mathbb{R}^n)$ with support in the unit sphere and so normalized that $\int_{\mathbb{R}^n} \rho(\mathbf{x})d^n\mathbf{x} = 1$. For any $\delta > 0$ call

$$\rho_{\mathbf{y}\,\delta}(\mathbf{x}) = \rho\left(\frac{1}{\delta}(\mathbf{x} - \mathbf{y})\right)\left(\frac{1}{\delta}\right)^n.$$

Then the function

$$f_\delta(\mathbf{y}) = \langle f, \rho_{\mathbf{y}\,\delta} \rangle$$

is said to be the result of *smoothing* the distribution f by averaging over a distance δ. We write $f_\delta = J_\delta\, f$; the operator J_δ is called a *mollifier*.

EXERCISES

4.   Show that as $\delta \to 0$ the function f_δ converges to f in the sense of convergence of distributions, i.e.,

$$J_\delta f \to f. \tag{2.6-6}$$

Furthermore, if D represents any partial differentiation $\partial/\partial x_j$, then

$$DJ_\delta f = J_\delta Df. \tag{2.6-7}$$

Hence, the derivatives of f_δ converge to the derivatives of f. *Hint*: First show that

$$\langle J_\delta f, \varphi \rangle = \langle f, J_\delta \varphi \rangle, \tag{2.6-8}$$

using the result of Exercise 1 in Section 2.4 on integration with respect to a parameter. Then show that if φ is any test function in C_0^∞, then $J_\delta \varphi \to \varphi$ in the sense of convergence of test functions. For that purpose, it is convenient to prove that $J_\delta D\varphi = DJ_\delta \varphi$, from which (2.6-7) follows.
   It is also noted that for $\delta_1 > 0$, $\delta_2 > 0$, $J_{\delta_1} J_{\delta_2} = J_{\delta_2} J_{\delta_1}$.
   5.   Show that if $f_j \to f$ in the sense of convergence of distributions, then $J_\delta f_j \to J_\delta f$ in that same sense.

These exercises show that any distribution can be approximated by a C^∞ function. Examples of that are provided by the approximations (2.6-3 and 5) to the δ-function.
   The result of smoothing can be regarded as a convolution of the distribution with the test function $\rho(\mathbf{x}/\delta)(1/\delta)^n$, since it can be written symbolically as

$$f_\delta(\mathbf{x}) = \int_{\mathbb{R}^n} f(\mathbf{y})\rho\left(\frac{\mathbf{x} - \mathbf{y}}{\delta}\right)\left(\frac{1}{\delta}\right)^n d^n\mathbf{y}.$$

The convolution of two distributions is defined and studied in Chapter 6, below.

2.7 Differentiation and Integration

If f is any distribution on $\mathbb{R}$, then another distribution, denoted by f' and called the *derivative* of f, is defined by

$$\langle f', \varphi \rangle \stackrel{\text{def}}{=} -\langle f, \varphi' \rangle, \quad \text{for all } \varphi \text{ in } C_0^\infty. \qquad (2.7\text{-}1)$$

If $\varphi = \varphi(x)$ is any test function, then $\varphi' = \varphi'(x)$ is also a test function, hence (2.7-1) makes sense as a definition of the functional $\langle f', \cdot \rangle$. If f and f' are ordinary functions, (2.7-1) is just integration by parts ($\varphi(x)$ vanishes identically for both positive and negative x outside the support of φ).

Higher derivatives are similarly defined, by further formal integration by parts, and also partial derivatives; for example, if $f = f(x, y)$ is any distribution on $\mathbb{R}^2$, a distribution denoted by $\partial_x \partial_y f$ is defined by

$$\langle \partial_x \partial_y f, \varphi \rangle = \langle f, \partial_x \partial_y \varphi \rangle, \quad \text{for all } \varphi \in C_0^\infty. \qquad (2.7\text{-}2)$$

Remarks on continuity of the functionals (1) First, if $f(x)$ is an ordinary differentiable function on $\mathbb{R}$, then

$$\frac{1}{h}[f(x + h) - f(x)] \to f'(x), \quad \text{as } h \to 0. \qquad (2.7\text{-}3)$$

The same is true for any distribution f on $\mathbb{R}$ (for a distribution, f' is defined by (2.7-1), not (2.7-3)), but to prove this, the continuity of the functional $\langle f, \cdot \rangle$ in the sense of Section 2.4 is needed. Namely, $f(x + h)$ is the distribution defined by

$$\langle f(x + h), \varphi(x) \rangle = \langle f(x), \varphi(x - h) \rangle, \qquad (2.7\text{-}4)$$

hence (2.7-3) holds if we can show that

$$\lim_{h \to 0} \left\langle f, \frac{\varphi(x - h) - \varphi(x)}{h} \right\rangle = \left\langle f, \lim_{h \to 0} \frac{\varphi(x - h) - \varphi(x)}{h} \right\rangle. \qquad (2.7\text{-}5)$$

This equation follows from the continuity of the functional $\langle f, \cdot \rangle$, if we can show that

$$\frac{\varphi(x - h) - \varphi(x)}{h} \xrightarrow{\ \ } -\varphi'(x), \quad \text{as } h \to 0. \qquad (2.7\text{-}6)$$

See Exercise 1 below.

(2) All this reminds us that we are obliged to show that if f is any distribution on $\mathbb{R}$, then the linear functional f' defined by (2.7-1) is continuous in the sense of Section 2.4, hence is a distribution. See Exercise 2 below.

(3) If the distribution f is translation-invariant, i.e. $f(x + h)$ is the same distribution as $f(x)$ for all h, it follows from the above that f' is the function $f'(x) \equiv 0$, hence f is the function $f(x) = $ constant. In this argument the continuity of the functional $\langle f, \cdot \rangle$ was assumed in advance. But G. H.

Meisters showed in 1971 by a rather complicated method that a translation-invariant linear functional on $\mathbb{R}$ is *necessarily* continuous in the sense of Section 2.4, hence is a distribution.

(4) It follows similarly from the continuity of the functional $\langle f, \cdot \rangle$, where f is any distribution on $\mathbb{R}^2$, that if $\partial_x f = 0$, then f is independent of x in the sense that $f(x + h, y) = f(x, y)$ for any h, because $\varphi(x - h, y) - \varphi(x, y)$ can always be written as $\partial_x \psi$ for some test function ψ, hence $f(x + h, y)$ and $f(x, y)$ are the same distribution on $\mathbb{R}^2$.

Integration

It will now be shown that if g is any distribution on $\mathbb{R}$, then there is a distribution f, which is called the *primitive* of g or the *indefinite integral* of g, such that $f' = g$. Furthermore, f is unique up to an additive constant. To show this, the linear functional $\langle f, \cdot \rangle$ will be defined in such a way that if ψ is $= -\varphi'$ for some test function φ, then $\langle f, \psi \rangle = \langle f, -\varphi' \rangle = \langle g, \varphi \rangle$. For any ψ in C_0^∞, call

$$\varphi(x) = -\int_{-\infty}^{x} \psi(x')dx'.$$

Then φ is always in C^∞, and $\varphi' = -\psi$. Clearly, φ is also in C_0^∞ if and only if $\int_{-\infty}^{\infty} \psi(x')dx' = 0$, and in this case we define $\langle f, \psi \rangle = \langle g, \varphi \rangle$, as required. Then, to define $\langle f, \psi \rangle$ for an arbitrary ψ in C_0^∞, a fixed test function ψ_1 is now chosen such that $\int_{-\infty}^{\infty} \psi_1(x)dx = 1$; an arbitrary constant c is also chosen, and $\langle f, \psi_1 \rangle$ is defined to be $=c$. That fixes the functional $\langle f, \cdot \rangle$ completely, because any ψ can be written as $\psi = \psi_0 + a\psi_1$, where

$$\int_{-\infty}^{\infty} \psi_0(x)dx = 0, \qquad a = \int_{-\infty}^{\infty} \psi(x)dx;$$

then

$$\begin{aligned}
\langle f, \psi \rangle &= \langle f, \psi_0 \rangle + \langle f, a\psi_1 \rangle \\
&= \langle f, \psi_0 \rangle + a\langle f, \psi_1 \rangle \\
&= \langle f, \psi_0 \rangle + \int_{-\infty}^{\infty} \psi(x)dx\, c \\
&= \langle f, \psi_0 \rangle + \langle c, \psi \rangle,
\end{aligned}$$

of which the last term represents the additive constant function c.

EXERCISES

1. Prove (2.7-6) by verifying the conditions for convergence in $\mathscr{D}$ as defined in Section 2.4.

2. Show that $\langle f' \cdot \rangle$ as defined by (2.7-1) is a continuous functional by verifying that if $\varphi_n \xrightarrow{\mathscr{D}} \varphi$, then $\varphi'_n \xrightarrow{\mathscr{D}} \varphi'$.

If f is the indefinite integral of g, one writes symbolically

$$f = \int^x g \quad \text{or} \quad \int^x g \, dx,$$

or even

$$f(x) = \int g(x) dx,$$

but it must of course be remembered that f and g are distributions.

EXERCISES

3.   Let $f(x)$ be the function on $\mathbb{R}$ defined by $f(x) = |x|$, regarded as a distribution. Find the distributions f' and f'', using the definition (2.7-1).

4.   If f is any distribution on $\mathbb{R}$, and α is a C^∞ function, and if the product αf is defined as in (2.5-11), show that $(\alpha f)' = \alpha' f + \alpha f'$.

5.   Let $f = f(x)$ be a nondecreasing function (not necessarily differentiable or even continuous). Then it is Riemann-integrable over any finite interval, and Example 1 of Section 2.5 applies. Show that $f' \geq 0$ on all $\mathbb{R}$, which means, according to the next chapter, that $\langle f', \varphi \rangle \geq 0$ for every test function $\varphi(x)$ that is ≥ 0 for all x. *Hint*: Show that $\int f \varphi' \, dx$ can be approximated by a Riemann sum of the form $\sum_{j=1}^{N} f(\xi_j) [\varphi(x_j) - \varphi(x_{j-1})]$.

6.   Show that if $g = g(x, y)$ is any distribution on $\mathbb{R}^2$, there is a distribution $f = f(x, y)$ such that $\partial_x f = g$. Show that f is unique up to an additive distribution h which is independent of x in the sense of Remark 4 above.

7.   If a distribution $f(x, y)$ on $\mathbb{R}^2$ is independent of x, show that there is a distribution $g(y)$ on $\mathbb{R}$ such that for any test function of the form $\varphi(x, y) = \psi(x)\chi(y)$,

$$\langle f, \varphi \rangle = a \langle g, \chi \rangle, \tag{2.7-7}$$

where

$$a = \int_{-\infty}^{\infty} \psi(x) dx. \tag{2.7-8}$$

Note: This is formally equivalent to writing $f(x, y) = g(y)$ and integrating first with respect to x in the equation

$$\langle f, \varphi \rangle = \iint f(x, y)\varphi(x, y) dx \, dy. \tag{2.7-9}$$

8.   Show that if $g = g(x, y)$ is any distribution on $\mathbb{R}^2$, then there is a distribution $f = f(x, y)$ such that $\partial_x \partial_y f = g$; show that f is unique except for an additive distribution of the form $h_1 + h_2$, where h_1 is independent of x, and h_2 is independent of y.

2.8 Changes of Independent Variable; Symmetries

Expressions such as $f(\alpha(x))$ can be defined, under limited circumstances. If α is a real-valued C^∞ function on $\mathbb{R}$ such that the equation $y = \alpha(x)$ has a unique solution $x = \beta(y)$, also of class C^∞, then if $f(x)$ is any continuous function,

$$\int_{-\infty}^{\infty} f(\alpha(x))\varphi(x)dx = \int_{-\infty}^{\infty} f(y)\varphi(\beta(y))|\beta'(y)|dy. \qquad (2.8\text{-}1)$$

Note. $\beta'(y)$ cannot change sign, so that $|\beta'(y)|$ is either $= \beta'(y)$ for all y or $= -\beta'(y)$ for all y.

The same equation holds for any distribution $f(x)$ on $\mathbb{R}$ if we define a distribution $g(x) = f(\alpha(x))$ on $\mathbb{R}$ by writing

$$\langle g, \varphi \rangle = \langle f, \psi \rangle, \qquad \forall \varphi, \qquad (2.8\text{-}2)$$

where

$$\psi(y) = \varphi(\beta(y))|\beta'(y)|, \qquad (2.8\text{-}3)$$

which is a test function if $\varphi(x)$ is.

A simple example is $\delta(ax) = \delta(x)/|a|$, where $\alpha(x) = ax$, $\beta(y) = y/a$ $(a \neq 0)$.

Note. If $x \to \alpha(x)$ is a one-to-one C^∞ mapping of $\mathbb{R}$ onto itself, the inverse mapping $y \to \beta(y)$ is not automatically of class C^∞, as shown by the example $\alpha(x) = x^3$.

Invertible C^∞-transformations of variables can also be made in distributions on $\mathbb{R}^n$. In place of $|\beta'|$ in (2.8-3) there appears $|J|$, where J is the Jacobian of the inverse transformation. If the transformation is linear, i.e., if $\alpha(\mathbf{x}) = A\mathbf{x} + \mathbf{x}_0$, A being a non-singular matrix, then

$$\langle f(\alpha(\mathbf{x})), \varphi(\mathbf{x}) \rangle = |\det A|^{-1}\langle f(\mathbf{y}), \varphi(A^{-1}(\mathbf{y} - \mathbf{x}_0)) \rangle. \qquad (2.8\text{-}4)$$

Two important cases are:

(1) $n = 2$ or 3, $\mathbf{x}_0 = 0$, while A is a rotation matrix (hence det $A = 1$). Then

$$\langle f(A\mathbf{x}), \varphi(\mathbf{x}) \rangle = \langle f(\mathbf{y}), \varphi(A^{-1}\mathbf{x}) \rangle. \qquad (2.8\text{-}5)$$

$f(A\mathbf{x})$ is the distribution that results from $f(\mathbf{x})$ by a rotation in space. If $f(\mathbf{x})$ is the density of a point charge at the origin, as in Example 4 of Section 2.5, namely

$$\rho(x, y, z) = \delta(x)\delta(y)\delta(z),$$

given more precisely by

$$\langle \rho, \varphi \rangle = \varphi(0, 0, 0), \qquad (2.8\text{-}6)$$

it follows from this last equation that $\rho(A\mathbf{x}) = \rho(\mathbf{x})$, i.e., this distribution has spherical symmetry about the origin. But one should not conclude that any distribution concentrated at the origin has spherical symmetry; the multipole distribution (2.5-5) or (2.5-6) does not.

(2) $n = 4$, $x_0 = 0$, A a Lorentz matrix (again, $\det A = 1$). In this case $f(Ax)$ is obtained from $f(x)$ by a (homogeneous) Lorentz transformation.

EXERCISE

1. Show that the distribution $D(t, x)$ given by (2.5-9) is invariant under a Lorentz transformation.

If the substitution $x \to \alpha(x)$ is not one-to-one, or is not defined in all $\mathbb{R}$ or is not onto all of $\mathbb{R}$, special definitions can often be made in such a way as to preserve formal rules of operation. For example, Dirac defines $\delta(x^2 - a^2)$ for $a \neq 0$ by

$$\delta(x^2 - a^2) \stackrel{\text{def}}{=} \frac{1}{2|a|} [\delta(x + a) + \delta(x - a)]. \qquad (2.8\text{-}7)$$

Then, a formal substitution $y = x^2 - a^2$ in $\int \delta(x^2 - a^2)\varphi(x)dx$ gives, if we note that, as x goes from $-\infty$ to ∞, y goes from ∞ to $-a^2$ and then back to ∞,

$$\int_{-\infty}^{\infty} \delta(x^2 - a^2)\varphi(x)dx = \int_{-a^2}^{\infty} \delta(y) \left[\varphi(\sqrt{y + a^2}) + \varphi(-\sqrt{y + a^2}) \right]$$
$$\times \frac{1}{2} \frac{dy}{\sqrt{y + a^2}} \qquad (2.8\text{-}8)$$

which agrees with (2.8-7).

2.9 Restrictions, Limitations, and Warnings

Distributions are in many ways like ordinary functions and are subject to most of the ordinary rules of operation. In fact, many of the limitations of those rules disappear in the context of distributions. For example, every distribution can be differentiated and integrated, and integration is always the inverse of differentiation, in either order, for distributions (fundamental theorem of calculus). For distributions on $\mathbb{R}^2$ or $\mathbb{R}^n$, ∂_x always commutes with ∂_y (see example below). As will be seen a little later, every tempered distribution has a Fourier transform, and any distributions f and g one of which has compact support always have a convolution $f * g$; the Fourier transforms and convolutions satisfy all the usual rules. Poisson's equation $\nabla^2 \varphi = -4\pi\rho$ in $\mathbb{R}^3$ or a bounded region Ω can always be solved by a Green's function for any distribution ρ with compact support; and so on.

More important, distribution theory provides a very natural setting for the differential operators of interest in physics, as will be seen in Chapters 10 and 11.

Nevertheless, distribution theory has certain restrictions, some of which are discussed in this section. Blind manipulation of formulas is no safer here than in any other branch of mathematics or physics.

Limitations on $\langle f, g \rangle$

If f is a distribution, $\langle f, g \rangle$ is not generally defined unless g is a test function, but for distributions in certain classes, the definition can be extended to include functions or distributions g in a wider class than C_0^∞. At the extreme, if f and g are both in one of the L^2 spaces discussed in Chapter 5, $\langle f, g \rangle$ is defined, and $\langle \bar{f}, g \rangle$ is the inner product that makes L^2 a Hilbert space.

Limitations on $f(x)g(x)$

If f is a distribution and g is a function of class C^∞, fg $(=gf)$ was defined in Section 2.5, Example 7. For distributions f in certain classes, the product makes sense for functions or distributions g in a wider class than C^∞. At the extreme, if f and g are in an L^2 space, the product fg is well defined as a distribution (it is generally in L^1, not L^2). If f, g, f', and g' are all in $L^2(\mathbb{R})$, the formula for integration by parts holds. See Chapter 5. $\delta(x)^2$ is meaningless.

Limitations on $f(g(x))$

$f(g(x))$ was defined in Section 2.8 when $g(x)$ is a C^∞ function with a C^∞ inverse. It is almost never defined otherwise unless g and f are ordinary functions, or f is linear. Expressions like $\delta(x^2)$, $e^{\delta(x)}$ are meaningless.

Nonlinear Problems

Because of the foregoing limitations, distributions are useful mainly in linear analysis, e.g. in the study of linear differential equations or the linear aspects of quantum mechanics; in nonlinear problems their application can be ambiguous. The differential equation

$$\partial_t u + u\partial_x u = 0 \tag{2.9-1}$$

for a function $u(t, x)$ is often studied as a simplified prototype of the equations of fluid dynamics. Its solutions can develop shock-like discontinuities in a finite time. To study such solutions, (2.9-1) is first written in so-called conservation law form

$$\partial_t u + \partial_x(\tfrac{1}{2}u^2) = 0, \tag{2.9-2}$$

and then a function $u(t, x)$ is said to be a *weak solution* of this equation if

$$\iint (u\partial_t \varphi + \tfrac{1}{2}u^2\partial_x \varphi)dt\, dx = 0 \tag{2.9-3}$$

for all test functions $\varphi(t, x)$. [Then $u(t, x)$ is a function such that the distribution derivative of u with respect to t and the distribution derivative of u^2

with respect to x are related by (2.9-2).] The weak solutions of the corresponding conservation laws of fluid dynamics are known to correspond to physical reality; in particular, their jump discontinuities satisfy the so-called Rankine–Hugoniot jump condition for hydrodynamic shock waves. However, the correct form of the conservation law has to be determined by physical considerations; (2.9-1) can also be written as

$$\partial_t(\tfrac{1}{2}u^2) + \partial_x(\tfrac{1}{3}u^3) = 0, \tag{2.9-4}$$

but the weak solutions of this equation are not the same as those of (2.9-2). No amount of mathematical reasoning can tell us which set of weak solutions has a right to be called generalized solutions of the original equation (2.9-1). A weak solution of (2.9-2) may agree with a weak solution of (2.9-4) until shocks appear (i.e. as long as they are strong solutions) and differ from it thereafter.

Mixed Partial Derivatives

The following example illustrates the principle that one should not generally try to assign a value to a distribution at a particular point. Consider the continuous function on $\mathbb{R}^2$ defined by

$$f(x, y) = \frac{x^3 y - xy^3}{x^2 + y^2} \quad \text{for } x^2 + y^2 \neq 0 \tag{2.9-5}$$

$$f(0, 0) = 0$$

The first and second partial derivatives exist in the classical sense for all x and all y. In particular,

$$\partial_x f = -y \quad \text{for } x = 0, \text{ all } y,$$
$$\partial_y f = +x \quad \text{for } y = 0, \text{ all } x,$$

hence

$$\partial_y \partial_x f = -1 \quad \text{for } x = y = 0,$$
$$\partial_x \partial_y f = +1 \quad \text{for } x = y = 0,$$

whereas the *distributions* $\partial_y \partial_x f$ and $\partial_x \partial_y f$ are equal (they are the same distribution, simply because $\partial_y \partial_x \varphi = \partial_x \partial_y \varphi$ in the classical sense for any test function. The distribution $\partial_y \partial_x f$ has no value assigned to it (neither -1 nor $+1$) at $x = y = 0$, and that is reasonable in most applications because the function $\partial_y \partial_x f$ has a wildly singular behavior if (x, y) approaches $(0, 0)$ from any direction other than along the x or y axis.

Identification of Functions With Distributions

In Section 2.5, a function $f(x)$ was identified with the distribution f given by

$$\langle f, \varphi \rangle = \int_{-\infty}^{\infty} f(x)\varphi(x)dx \qquad \forall \varphi \in C_0^{\infty}$$

only if the integral exists as a Riemann integral. Now let $f(x)$ be the Cantor function to be described in Chapter 13; $f(x)$ is continuous and rises from $f(0) = 0$ to $f(1) = 1$ in such a way that its derivative $f'(x)$ exists for almost all x (i.e., except on a set of measure zero) and is $=0$ where it exists. Set $f(x) = 0$ for $x < 0$ and $=1$ for $x > 1$. Then $f'(x)$ is integrable in the Lebesgue sense, and

$$\int_{-\infty}^{\infty} f'(x)\varphi(x)dx = 0 \qquad \forall \varphi \in C_0^{\infty}.$$

Hence, if we admitted Lebesgue integrals, the distribution identified with $f'(x)$ would not be the derivative of the distribution identified with $f(x)$.

Avoidance of Cabala

(1) In Dirac's book there is the formula

$$\frac{d}{dx} \log x = \frac{1}{x} - i\pi\delta(x),$$

which Dirac uses in the theory of collision processes. The formula must surely be subject to some special restrictions, because all mathematics is invariant under the substitution $-i \to i$. (2) One of the books on distribution theory says that the convolution $f * g$ is defined as a distribution for any distributions f and g (say on $\mathbb{R}$). That is false already for functions; if $f(x) = g(x) = 1$ for all x, then $f * g$ would have to have the value $+\infty$ for all x, hence would be neither a function nor a distribution nor any kind of generalized function known so far. (3) Gel'fand and Shilov give a formula for $\delta(x)$ for complex values of x. They base their formula on definitions that go beyond the realm of distribution theory in the sense of Schwartz, and the formula would be meaningless in any context that does not include those definitions.

Operator-Valued Distributions

In quantum field theory one often defines an operator-valued distribution $T(\mathbf{x})$ by saying that for every test function $\varphi(\mathbf{x})$, $\langle T, \varphi \rangle$ is a bounded linear operator on some Hilbert space. The linearity of the functional $\langle T, \cdot \rangle$ is exactly as above, but its continuity requires that a decision be made. Namely, if a sequence $\{\varphi_n\}$ converges, $\varphi_n \overset{\mathcal{G}}{\to} \varphi$, as in Section 2.4, then we require $\langle T, \varphi_n \rangle \to \langle T, \varphi \rangle$, and the question arises whether this last should mean weak, strong, or uniform convergence of the operators. (See Chapter 7.)

Convergence of Functions

As pointed out in Section 2.6, the pointwise convergence of functions $f_n(x)$ to $f(x)$ is neither necessary nor sufficient for the convergence $f_n \to f$ as distributions. That is not surprising; already in the elementary theory of Fourier series and orthogonal systems of functions it was found that convergence in the mean (i.e. in L^2), which is a special case of convergence in the distribution sense but for which pointwise convergence is neither necessary nor sufficient, is usually physically more relevant than pointwise convergence.

If $f_n \to f$ in the distribution sense, then $f'_n \to f'$ etc.

2.10 Regularization

So far no method has been given for associating distributions with functions like $f(x) = 1/x$ or $1/x^2$ having nonintegrable singularities. Generally, if $f(x)$ has a singularity of some kind at $x = x_0$, but is continuous otherwise, we seek a distribution f, called the *regularization* of $f(x)$, such that

$$\langle f, \varphi \rangle = \int_{-\infty}^{\infty} f(x)\varphi(x)dx \qquad (2.10\text{-}1)$$

for every test function φ whose support does not include the point x_0. Since $\langle f, \varphi \rangle$ must be specified for *all* test functions (though not necessarily by the above integral), the regularization is not unique.

Let us take $x_0 = 0$, for convenience. If $x^m f(x)$ is bounded, for some integer m, in some neighborhood of the origin (then $x = 0$ is called an *algebraic singularity*) one regularization is given by

$$\langle f, \varphi \rangle = \int_{|x|>a} f(x)\varphi(x)dx$$

$$+ \int_{|x|<a} f(x)\left[\varphi(x) - \varphi(0) - x\varphi'(0) - \cdots - \frac{1}{m!} x^m \varphi^{(m)}(0) \right] dx,$$

$$\forall \varphi \in C_0^{\infty}, \qquad (2.10\text{-}2)$$

for some fixed given $a > 0$, because (1) this is clearly a linear functional on C_0^{∞}, and (2) if the support of φ does not include the origin, then $\varphi(0) = \varphi'(0) = \cdots = \varphi^{(n)}(0) = 0$, so that (2.10-2) agrees with (2.10-1).

EXERCISE

1. Show that the functional $\langle f, \cdot \rangle$ given by (2.10-2) is continuous in the sense of Section 2.4.

Nonuniqueness

If f denotes the above regularization, other regularizations can be obtained by adding to f any distribution g that is concentrated at the origin, i.e., that vanishes for any test function whose support does not include the origin, for example any distribution of the form

$$g(x) = \sum_{j=0}^{k} a_j \delta^{(j)}(x),$$

which is equivalent to adding to (2.10-2) the linear functional

$$\sum_{j=0}^{k} a_j (-1)^j \varphi^{(j)}(0).$$

Any function $f(x)$ having at most finitely many algebraic singularities in any finite interval can be similarly regularized. The nonuniqueness raises a question discussed at length for this class of functions by Gel'fand and Shilov. They show that a *canonical* regularization can be chosen for each function in the class in such a way that (1) the regularization of $f(x) + g(x)$ is the sum of the regularizations of $f(x)$ and $g(x)$, (2) the regularization of $f'(x)$, if it also belongs to the class, is the derivative of the regularization of $f(x)$, and (3) if $\alpha(x)$ is any C^∞ function on $\mathbb{R}$, the regularization of $\alpha(x)f(x)$ is $\alpha(x)$ times the regularization of $f(x)$.

EXERCISE

2. Let f be the function

$$f(x) = \begin{array}{ll} \sqrt{x} & \text{if } x \geq 0, \\ 0 & \text{if } x \leq 0, \end{array}$$

regarded as a distribution. Find f' and f'' and compare them with the regularizations of $f'(x)$ and $f''(x)$ given by (2.10-2).

Appendix to Chapter 2—A Discontinuous Linear Functional

A linear functional on the space $C_0^\infty = C_0^\infty(\mathbb{R})$ of test functions $\varphi(x)$ on $\mathbb{R}$ will be defined, which is not continuous in the sense of Section 2.4, hence is not a distribution in the sense of Schwartz.

Let $\rho(x)$ be a given function in C_0^∞; ρ will be fixed from now on, and we suppose that the support of ρ is contained in an interval (a, b). Call

$$X = \{\psi(x): \exists \varphi \in C_0^\infty, \psi(x) = \rho(x)\varphi(x)\}; \tag{2.A-1}$$

X is a linear space. We shall define first a discontinuous linear functional $\langle f, \cdot \rangle$ on X. In X, as in any linear space, one can choose a Hamel basis (see Dunford and Schwartz 1958), i.e., a set $B = \{\psi_a: a \in A\}$ of elements of X, where A is an (uncountable) index set, such that any ψ in X has a unique representation

$$\psi = c\psi_a + c'\psi_{a'} + \cdots + c^{(m)}\psi_{a^{(m)}} \qquad (2.A\text{-}2)$$

as a finite linear combination of elements of B (m depends on ψ). We can define $\langle f, \psi_a \rangle$ arbitrarily for each ψ_a in the Hamel basis B, and then, when ψ is given by (2.A-2), $\langle f, \psi \rangle$ is given by

$$\langle f, \psi \rangle = c\langle f, \psi_a \rangle + c'\langle f, \psi_{a'} \rangle + \cdots + c^{(m)}\langle f, \psi_{a^{(m)}} \rangle. \qquad (2.A\text{-}3)$$

We choose arbitrarily a denumerable sequence $\{\psi_{a_k}\}_{k=1}^{\infty}$ of distinct elements of B. For each k, we choose a C_0^{∞} function φ_k such that (1) $\psi_{a_k}(x) = \rho(x)\varphi_k(x)$, and (2) supp $\varphi_k \subset (a, b)$. Clearly, that is always possible because the support of ρ is a closed set inside (a, b), and requirement (1) fixes φ_k only on supp ρ. Call

$$N_k = \sup\{|\psi_{a_k}^{(j)}(x)|, |\varphi_k^{(j)}(x)| : x \in \mathbb{R}, j = 1, \ldots, k\}.$$

The functional $\langle f, \cdot \rangle$ is now defined on the elements of B by the equations

$$\langle f, \psi_{a_k} \rangle = kN_k \qquad (k = 1, 2, \ldots)$$
$$\langle f, \psi_a \rangle = 0 \quad \text{for } \psi_a \notin \{\psi_{a_k}\}.$$

We define

$$\tilde{\psi}_k = \frac{1}{kN_k} \psi_{a_k}, \qquad \tilde{\varphi}_k = \frac{1}{kN_k} \varphi_k,$$

and we note that

$$\tilde{\psi}_k \to 0 \quad \text{and} \quad \tilde{\varphi}_k \to 0, \quad \text{as } k \to \infty,$$

in the sense of convergence of test functions (convergence in $\mathscr{D}$). Furthermore,

$$\langle f, \tilde{\psi}_k \rangle = 1 \quad \text{for all } k = 1, 2, \ldots,$$

hence $\langle f, \cdot \rangle$ is not continuous. Lastly, we define a functional g on C_0^{∞} by writing

$$\langle g, \varphi \rangle = \langle f, \rho\varphi \rangle \qquad \forall \varphi \in C_0^{\infty}.$$

Since $\langle g, \tilde{\varphi}_k \rangle = \langle f, \tilde{\psi}_k \rangle = 1$, g is not continuous either.

Comment. The notion of a Hamel basis in an infinite-dimensional space is regarded by many as lying on an undesirably high level of abstraction, because it depends on the uncountable axiom of choice, and there is no known way of constructing such a basis. However, that axiom, in its un-countable form, is part of the classical Zermelo–Fraenkel axiom system, on which modern mathematics is based, and has proved to be very fruitful in many branches of mathematics. So long as we work in that system, the continuity of the various linear functionals that we introduce as distributions cannot be taken for granted, but must be proved if that continuity is to be made use of (and it is often indispensable). Recently another system was introduced as a possible alternative logical foundation of mathematics and has been discussed in detail by Solovay (1970) and others; it is based on a model of set theory which contains among other things the Zermelo–Fraenkel system modified by including only the countable axiom of choice, i.e., the axiom as applied to a countable collection of sets. In that model, many so-called pathological features of the present foundation are absent; every set in $\mathbb{R}^n$ is Lebesgue measurable, every linear operator defined everywhere in a Hilbert space is bounded, and so on. It follows from a theorem of J. D. M. Wright (1973) that in that model every linear functional on a space $C_0^\infty(\mathbb{R}^n)$ or $\mathscr{S}(\mathbb{R}^n)$ of test functions (and other spaces, too) is automatically continuous, in the sense of Section 2.4.

The Solovay model uses difficult techniques of modern set theory and cannot be described here. It is apparently not yet known to what extent the uncountable axiom of choice can be abandoned or replaced by the countable axiom in various applications; hence, it is probably best to take a somewhat tentative attitude toward the Solovay model at the present time.

Local Properties of Distributions

Topology of open and closed sets in Euclidean space; coverings;
Bolzano–Weierstrass and Heine–Borel theorems; shrinkage principle;
partitions of unity; comparison of distributions in an arbitrary open
set; piecing-together principle; support of a distribution; derivative as
a local property.

Prerequisite: Chapter 2

Although a distribution does not have a definite value at a specified value x
of its argument, one can discuss its properties in any arbitrarily small
neighborhood of x. Such *local* properties are discussed in this chapter.

3.1 Quick Review of Open and Closed Sets in $\mathbb{R}^n$

A point $\mathbf{x}$ in a set $\mathbb{S}$ is called an *interior* point of $\mathbb{S}$ if, for some sufficiently
small $\varepsilon > 0$, the ball $B = \{\mathbf{y}: \|\mathbf{y} - \mathbf{x}\| < \varepsilon\}$ with center at $\mathbf{x}$ and radius ε
lies in $\mathbb{S}$ (i.e., if every point $\mathbf{y}$ of the ball B is a point of $\mathbb{S}$). A set $\mathbb{S}$ is called
open if every point of $\mathbb{S}$ is an interior point. For example, the ball B itself
is an open set, because if $\mathbf{y}$ is any point of B, and $\varepsilon' = \varepsilon - \|\mathbf{y} - \mathbf{x}\|$, then the
ball $B' = \{\mathbf{z}: \|\mathbf{z} - \mathbf{y}\| < \varepsilon'\}$ lies in B. See Figure 3-1.

A set is called *closed* if it contains all its limit points. The closed ball
$B = \{\mathbf{y}: \|\mathbf{y} - \mathbf{x}\| \leq \varepsilon\}$ is a closed set (the points on the surface have now been
included). A set $\mathbb{S}$ is closed if its complement $\mathbb{R}^n - \mathbb{S}$ is open, and conversely.
A set $\mathbb{S}$ together with all its limit points is a closed set, called the *closure*
of $\mathbb{S}$, and is denoted by $\bar{\mathbb{S}}$. If $\mathbb{S}$ itself is closed, then $\bar{\mathbb{S}} = \mathbb{S}$.

EXERCISES

1. Show that if $f(\mathbf{x})$ is a continuous real function on $\mathbb{R}^n$, then the sets

$$\{\mathbf{x}: f(\mathbf{x}) > 0\}, \qquad \{\mathbf{x}: f(\mathbf{x}) \neq 0\}, \qquad \{\mathbf{x}: a < f(\mathbf{x}) < b\}$$

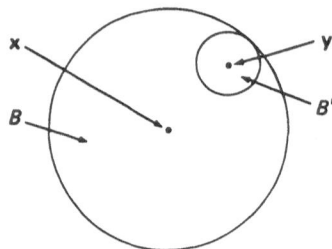

Figure 3.1 Open sets (see text).

are open sets, while the sets

$$\{x : f(x) \geq 0\}, \qquad \{x : f(x) = 0\}, \qquad \{x : a \leq f(x) \leq b\}$$

are closed.

The union of an arbitrary collection of open sets is open, and so is the intersection of a finite collection of open sets. In the corresponding statements about closed sets, the words "arbitrary" and "finite" have to be interchanged.

2. Prove the foregoing statements and discuss the unions and intersections of the following collections of intervals on $\mathbb{R}$ (in each case, $k = 1, 2, \ldots$): (1) $|x| < 1 - (1/k)$, (2) $|x| \leq 1 - (1/k)$, (3) $|x| < 1 + (1/k)$, (4) $|x| \leq 1 + (1/k)$.

For any function f, the closure of the set $\{x : f(x) \neq 0\}$ is called the *support* of f.

According to the *Bolzano–Weierstrass* theorem, any sequence $\{x_j\}_1^\infty$ in a bounded set $\mathbb{S}$ in $\mathbb{R}^n$ has a convergent subsequence; if $\mathbb{S}$ is also closed, the limit of the subsequence lies in $\mathbb{S}$.

For any set $\mathbb{S}$, if there is given a collection of open sets $\{\Omega, \Omega', \Omega'', \ldots\}$—possibly an infinite or even uncountable collection—such that every point x of $\mathbb{S}$ lies in at least one of them, then the collection is called an *open covering* of $\mathbb{S}$. According to the *Heine–Borel* theorem, if furthermore $\mathbb{S}$ is a closed bounded set in $\mathbb{R}^n$, then there is a finite subcollection, which will be called $\{\Omega_i : i = 1, \ldots, N\}$, of the above collection that also covers $\mathbb{S}$; that is, every point of $\mathbb{S}$ lies in at least one of the sets $\Omega_i (i = 1, \ldots, N)$. (The number N generally depends, for a given $\mathbb{S}$, on the particular open covering in question.) See Natanson (1955), where, however, the theorem is called the *Borel covering theorem*.

[In any topological space, a set K is called *compact* if it has the above property that every open covering of K contains a finite covering of K, and it is called *sequentially compact* if every sequence lying in K has a subsequence that converges to a limit in K. In any metric space, the two concepts are equivalent. In $\mathbb{R}^n$, a set is compact if and only if it is closed and bounded.]

Lemma 1. *If K is a closed bounded set in $\mathbb{R}^n$ contained in an open set Ω, then the distance d from K to the complement of Ω is positive. I.e., there is a margin around K in Ω, whose width is nowhere less than d.*

PROOF. The distance d is given by

$$d = \inf\{\|x - y\| : x \in K, y \notin \Omega\}. \tag{3.1-1}$$

Cleary $d > 0$, for if it were not, there would be a sequence $\{x_j\}$ in K such that distance $\{x$, complement of $\Omega\}$ would $\to 0$. According to the Bolzano–Weierstrass theorem, $\{x_j\}$ would have a convergent subsequence whose limit would be in K, hence in Ω, but would not be an interior point of Ω, and that would be a contradiction, because Ω is open.

Lemma 2.   *If a bounded closed set K lies in an open set Ω, then an inter-mediate open set Ω' can always be found that also contains K and is such that the closure $\tilde{\Omega}'$ of Ω' lies in Ω.*

PROOF.   The set

$$\Omega' = \{x : \text{distance } (x, K) < \tfrac{1}{2}d\},$$

where d is given by (3.1-1), has the required property.

3.2 Local Properties Defined

If f and g are ordinary functions on $\mathbb{R}^n$, and S is an arbitrary set in $\mathbb{R}^n$, the statement "$f = g$ in S" clearly means that $f(x) = g(x)$ for each x in S. If f and g are distributions, one cannot make such a statement for an arbitrary set S (in particular, one cannot, when S consists of a single point), but for an open set the statement can be given a well-defined meaning.

Definition 1.   If f and g are distributions on $\mathbb{R}^n$, and Ω is any open set in $\mathbb{R}^n$, then the statement "$f = g$ in Ω" means that $\langle f, \varphi \rangle = \langle g, \varphi \rangle$ for every test function φ whose support lies in Ω.

Definition 2.   If f and g are real distributions, the statement "$f \geq g$ in Ω" means that $\langle f, \varphi \rangle \geq \langle g, \varphi \rangle$ for every nonnegative test function φ whose support lies in Ω.

Note that since the support of any given such φ is a closed set, and Ω is an open set, then, according to Lemma 1 in the preceding Section, there is always a margin in Ω separating the support of φ from the boundary of Ω. The above statements say nothing about f and g in the margin; however, every point in the margin lies in the support of some other test function φ with support in Ω.

That the definitions given above are appropriate ones will be apparent from the theorems in this chapter; in particular, that the definitions are consistent with customary usage for ordinary functions is shown by the following theorems:

Theorem 1.   *If f and g are continuous functions $f(x)$ and $g(x)$, regarded as distributions, then $f = g$ in Ω, according to the above definition, if and only if $f(x) = g(x)$ for all x in Ω.*

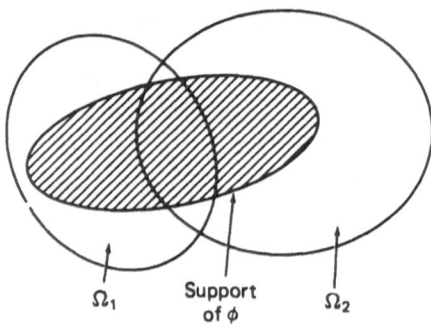

Ω_1 Support Ω_2
of ϕ

Figure 3.2   Covering of a compact set by two open sets.

PROOF.   The "if" part is obvious. To prove "only if," suppose the contrary: suppose that $f(\mathbf{x}_0)$ were $\neq g(\mathbf{x}_0)$ for some $\mathbf{x}_0$ in Ω; then either $\operatorname{Re} f(\mathbf{x}_0) \neq \operatorname{Re} g(\mathbf{x}_0)$ or $\operatorname{Im} f(\mathbf{x}_0) \neq \operatorname{Im} g(\mathbf{x}_0)$; assume the former. Then, in some neighborhood Ω_0 of $\mathbf{x}_0$, $\operatorname{Re} f(\mathbf{x}) - \operatorname{Re} g(\mathbf{x})$ would be of one sign; let $\varphi(\mathbf{x})$ be a test function which is ≥ 0 for all $\mathbf{x}$ and > 0 at $\mathbf{x}_0$ and whose support lies in Ω_0; then the quantity

$$\operatorname{Re}(\varphi, f - g) = \int_{-\infty}^{\infty} \cdots \int_{-\infty}^{\infty} \varphi(\mathbf{x}) \operatorname{Re}[f(\mathbf{x}) - g(\mathbf{x})]dx_1 \ldots dx_n$$

would be $\neq 0$, thus contradicting the hypothesis that $f = g$ in Ω.

Theorem 2 (The Proof is Now Evident).   *If f and g are continuous real functions, then $f \geq g$ in Ω if and only if $f(\mathbf{x}) \geq g(\mathbf{x})$ for all $\mathbf{x}$ in Ω.*

EXAMPLES
The distribution $\delta(x - x_0)$ on $\mathbb{R}$ is $= 0$ in any open interval not containing the point x_0, and so are $\delta'(x - x_0)$, $\delta''(x - x_0)$, etc. Furthermore, $\delta(x - x_0) \geq 0$ in any interval, while $\delta'(x - x_0)$, $\delta''(x - x_0)$, etc. are *not* ≥ 0 in an interval, if the interval contains x_0.

To show that these definitions have a truly local character, it must be proved that if Ω is the union of two or more open sets $\Omega_1, \Omega_2, \ldots$, then $f = g$ in Ω (or $f \geq g$ in Ω) if and only if $f = g$ (or $f \geq g$) in each Ω_i separately. That this is not quite a trivial matter is seen by considering two overlapping sets Ω_1 and Ω_2. If φ is a test function whose support lies in $\Omega_1 \cup \Omega_2$, but not in either set Ω_i alone (see Figure 3-2), then, to show that the equation $\langle f, \varphi \rangle = \langle g, \varphi \rangle$ follows from the hypothesis that $f = g$ in each Ω_i, it is necessary to decompose φ as $\varphi = \varphi_1 + \varphi_2$, where φ_1 and φ_2 are test functions and have their supports in Ω_1 and Ω_2, respectively. Then, the equation $\langle f, \varphi \rangle = \langle g, \varphi \rangle$ will follow from $\langle f, \varphi_i \rangle = \langle g, \varphi_i \rangle$ $(i = 1, 2)$ by the linearity of the functionals $\langle f, \cdot \rangle$ and $\langle g, \cdot \rangle$. Such decompositions of test functions and related subjects are discussed in the next two sections.

3.3 A Theorem on Open Coverings

Evidently, if two open sets overlap, they must overlap by a margin, and the margin can always be narrowed a little without altering the union of the sets. See Figure 3-3. This fact, suitably generalized, is now proved.

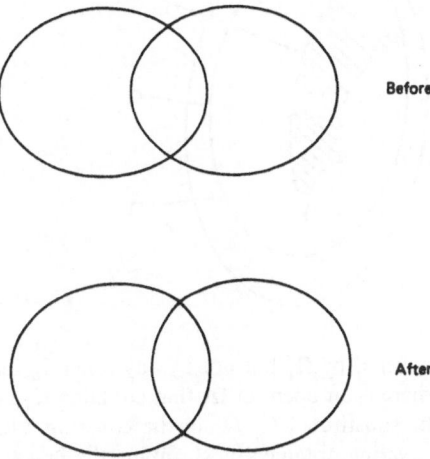

Figure 3.3   Reduction of margin of overlap.

Theorem 1 (Shrinkage Principle).   *Let Z be a closed set in $\mathbb{R}^n$ (possibly all of $\mathbb{R}^n$). Let $\{\Omega_i\}$ be a countable (i.e. finite or countably infinite) collection of bounded open sets that covers Z, i.e., such that $Z \subset \bigcup_{(i)} \Omega_i$. See Figure 3-4. Then it is possible to shrink all the sets a little, without destroying the covering property of the collection. Specifically, there is a collection $\{\Omega_i'\}$ of open sets such that the closure $\tilde{\Omega}_i'$ of Ω_i' is contained in the corresponding Ω_i, for each i, while $Z \subset \bigcup_{(i)} \Omega_i'$.*

PROOF.   Note first that the intersection of Z with the complement of $\bigcup_{i=2}^{\infty} \Omega_i$ (here, Ω_1 has been omitted) is a bounded closed set K_1 lying in Ω_1. [K_1 is the part of Z that was

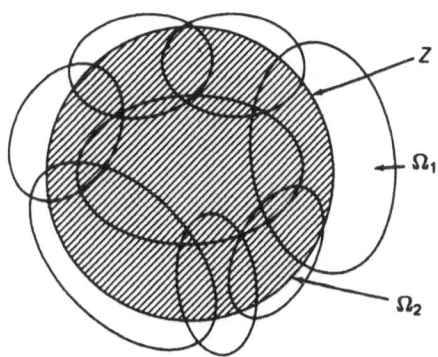

Figure 3.4   An open covering of a compact set.

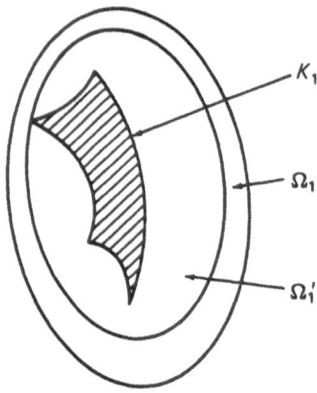

Figure 3.5　Shrinkage of Ω_1.

covered by Ω_1 but not by any other Ω_i—see Figure 3-5.] By Lemma 2 of Section 3.1 there is an open set Ω_1' that contains K_1 and whose closure lies in Ω_1. Clearly, Ω_1' can be substituted for Ω_1 in the covering. Now suppose that $\{\Omega_1', \ldots, \Omega_{k-1}', \Omega_k \ldots\}$ is a covering obtained by shrinking the first $k-1$ of the open sets. Then, the intersection of Z with the complement of $\Omega_1' \cup \cdots \cup \Omega_{k-1}' \cup \Omega_{k+1} \cup \cdots$ (now Ω_k has been omitted) is a bounded closed set K_k contained in Ω_k; Ω_k' can be constructed as Ω_1' was, an induction on k proves the theorem.

3.4 Theorems on Test Functions; Partitions of Unity

[The theorems in this section are useful in many parts of analysis.]

> **Theorem 1.**　*If K is a bounded closed set in $\mathbb{R}^n$, contained in an open set Ω, then there is a test function φ (a function in C_0^∞) such that (1) $\varphi(\mathbf{x}) = 1$ for $\mathbf{x}$ in K, (2) the support of φ lies in Ω, and (3) $0 \le \varphi(\mathbf{x}) \le 1$ for all $\mathbf{x}$. [This is a C^∞ version of a special case of Uryson's lemma—see Kelley (1955) or Thron (1966).] See Figure 3-6.*

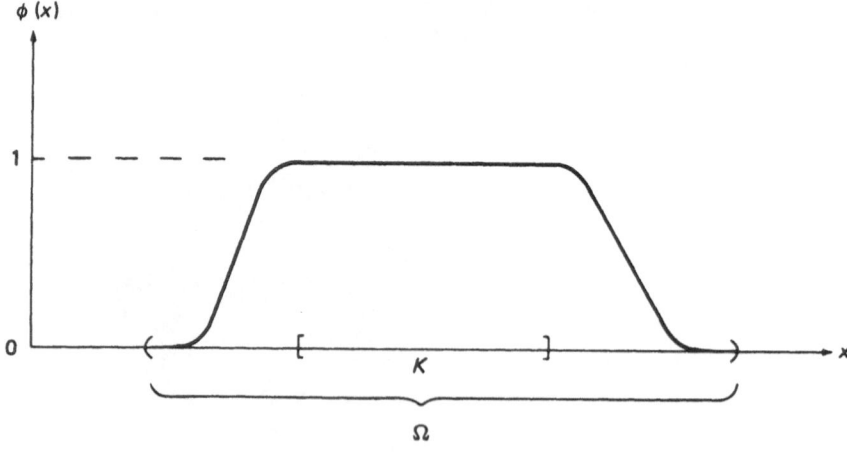

Figure 3.6　The function $\varphi(x)$.

PROOF.   Call

$$\delta = \tfrac{1}{3} \text{ distance } (K, \text{complement of } \Omega)$$
$$= \tfrac{1}{3} \inf\{\|\mathbf{x} - \mathbf{y}\| : \mathbf{x} \in K, \mathbf{y} \notin \Omega\}.$$

By Lemma 1 of Section 3.1, $\delta > 0$. For any x, call $d(\mathbf{x}) = $ distance $(\mathbf{x}, \text{complement of } \Omega)$, and define a function f by

$$f(\mathbf{x}) = \begin{cases} 0, & \text{for } d(\mathbf{x}) \le \delta \\ [d(\mathbf{x}) - \delta]/\delta, & \text{for } \delta \le d(\mathbf{x}) \le 2\delta \\ 1, & \text{for } 2\delta \le d(\mathbf{x}) \end{cases};$$

See Figure 3.7.

It is intuitively evident that $f(\mathbf{x})$ is continuous, and the proof is a good exercise. Clearly, $f(\mathbf{x})$ has all the properties required of the function $\varphi(\mathbf{x})$, except that it needs smoothing, to make it a C^∞ function. The factor $\tfrac{1}{3}$ in the definition of δ provides margins within the region between K and the complement of Ω, to allow for this smoothing. The smoothing is done by means of a C^∞ mollifier $\alpha(\mathbf{x})$ so chosen that

1.   the support of $\alpha(\mathbf{x})$ lies in the ball $\|\mathbf{x}\| < \delta$,
2. $\int \cdots \int \alpha(\mathbf{x}) dx_1 \ldots dx_n = 1$,
3. $\alpha(\mathbf{x}) \ge 0$, for all $\mathbf{x}$;

Then, φ, defined by

$$\varphi(\mathbf{x}) = \int \cdots \int f(\mathbf{y})\alpha(\mathbf{x} - \mathbf{y}) dy_1 \ldots dy_n,$$

has the properties stated in the theorem. (See Section 2.6.)

Theorem 2.   *Let $\{\Omega_i\}$ be a countable covering of $\mathbb{R}^n$ by bounded open sets. Assume that any bounded region in $\mathbb{R}^n$ intersects only a finite number of the sets $\{\Omega_i\}$. [This assumption could be weakened, but is adequate for present purposes.] Then there is a collection $\{\alpha_i(\mathbf{x})\}$ of C^∞ functions such that*

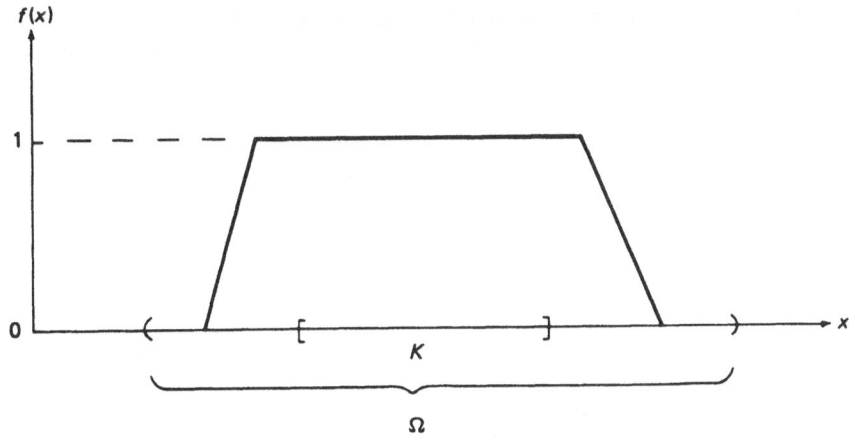

Figure 3.7   The function $f(x)$.

(1) *for each i, the support of α_i lies in Ω_i, and* (2) $\sum_{i=1}^{\infty} \alpha_i(\mathbf{x}) = 1$ *for all* $\mathbf{x}$. *Such a collection of functions is called a* partition of unity.

PROOF. Let $\{\Omega_i'\}$ be a covering obtained from $\{\Omega_i\}$ by shrinkage according to the theorem of the preceding section. For each i, let $\varphi_i(\mathbf{x})$ be a C^∞ function $=1$ inside Ω_i', with support in Ω_i, and ≥ 0 generally, as in Theorem 1, above. Define

$$\alpha_i(\mathbf{x}) = \frac{\varphi_i(\mathbf{x})}{\sum_{(i)} \varphi_i(\mathbf{x})}.$$

The denominator cannot vanish, because $\{\Omega_i'\}$ is a covering of $\mathbb{R}^n$; it contains only finitely many nonzero terms in any bounded region of $\mathbb{R}^n$, hence $\alpha_i(\mathbf{x})$ is of class C^∞. Lastly, $\sum \alpha_i(\mathbf{x}) = 1$ for all $\mathbf{x}$, by construction.

Theorem 2'.   *Let* $\{\Omega_1, \ldots, \Omega_N\}$ *be a finite covering of a compact (i.e., closed bounded) set* K *in* $\mathbb{R}^n$ *by open sets. Then there is a corresponding collection* $\{\alpha_1(\mathbf{x}), \ldots, \alpha_N(\mathbf{x})\}$ *of* C^∞ *functions such that* (1) *for each i, the support of* $\alpha_i(\mathbf{x})$ *lies in* Ω_i, *and* (2)$\sum_{i=1}^{N} \alpha_i(\mathbf{x}) = 1$ *for all* $\mathbf{x}$ *in* K.

This is a simple corollary of Theorem 2, obtained by first assuming the sets Ω_i to be bounded (there is no loss of generality here, because K is bounded) and by then extending the collection $\{\Omega_i\}$ in an obvious way, by adding further bounded open sets, to cover all $\mathbb{R}^n$, so that Theorem 2 applies.

3.5 The Main Theorems on Local Properties

Theorem 1.   *If f and g are distributions on* $\mathbb{R}^n$, *and* Ω *is an open set in* $\mathbb{R}^n$, *then* $f = g$ *in* Ω *(according to the definition in Section 3.2) if and only if* $f = g$ *in some neighborhood (open set) of each point of* Ω.

PROOF. The "only if" part is obvious. What has to be proved is therefore that if $f = g$ in a neighborhood of every point, and if ψ is any test function whose support K lies in the union of all these neighborhoods, then $\langle f, \psi \rangle = \langle g, \psi \rangle$. By the Heine–Borel theorem, K is covered by a finite subset $\{\Omega_1, \ldots, \Omega_N\}$ of the neighborhoods. Let $\{\Omega_1', \ldots, \Omega_N'\}$ be a covering of K by shrunken neighborhoods, as in Theorem 2' of Section 3.3, and let $\{\alpha_1(\mathbf{x}), \ldots, \alpha_N(\mathbf{x})\}$ be a corresponding partition of unity. Then, for each i, $\alpha_i(\mathbf{x})\psi(\mathbf{x})$ is a test function whose support lies in Ω_i. Therefore

$$\langle f - g, \alpha_i \psi \rangle = 0, \quad \text{for each } i;$$

hence,

$$\sum_{(i)} \langle f - g, \alpha_i \psi \rangle = \langle f - g, \sum \alpha_i \psi \rangle = \langle f - g, \psi \rangle = 0,$$

as was to be proved.

Theorem 2.   *Under the conditions of Theorem 1,* $f \geq g$ *in* Ω *if and only if* $f \geq g$ *in some neighborhood of each point of* Ω.

The proof is similar to that of Theorem 1. The nonnegativity of the $\alpha_i(\mathbf{x})$ is used.

EXERCISES

1. By a similar use of a partition of unity, prove the *piecing-together principle*: Let $\{\Omega_i\}$ be a family of open sets whose union is Ω, and let $\{f_i\}$ be a corresponding family of distributions with the property that whenever Ω_j and Ω_k have a nonempty intersection, then $f_j = f_k$ in $\Omega_j \cap \Omega_k$. Then there is a unique distribution f such that, for each i, $f = f_i$ in Ω_i. [By "unique" is meant that if f and g are any two such distributions, then $f = g$ in Ω. Ω may of course be the whole space $\mathbb{R}^n$.]

2. If $f = f(\mathbf{x})$ is any distribution on $\mathbb{R}^n$, and $\alpha = \alpha(\mathbf{x})$ is any C^∞ function in $\mathbb{R}^n$ the distribution αf was defined in Section 2.5. Show that, if $\alpha f = 0$ (i.e., is the zero distribution), and if Ω is any open set of $\mathbb{R}^n$ in which $\alpha(\ )$ is nowhere $= 0$, then $f = 0$ in Ω.

This result is used in the next chapter in connection with the Fourier transform of a periodic distribution.

EXERCISE

3. Show that if f and g are distributions on $\mathbb{R}$, if Ω is an open set, and if $f = g$ in Ω, then $f' = g'$ in Ω. (This shows that differentiation is a local operation.) What is the converse of this statement?

3.6 The Support of a Distribution

The support of a distribution f on $\mathbb{R}^n$, like that of a continuous function, is the complement of the largest open set on which $f = 0$. That is, we consider the union of all open sets $\{\Omega_a\}$ (if any) such that $f = 0$ in each Ω_a. That union is an open set Ω such that $f = 0$ in Ω, and $\mathbb{R}^n - \Omega$ is the *support* of f.

EXERCISE

1. Suppose that the support of a distribution f on the real line $\mathbb{R}$ consists of a single point, which we can take as $x = 0$. According to Schwartz's theorem quoted without proof in Example 1 of Section 2.4, f is equal in $-1 < x < 1$ to the kth derivative, for some k, of a continuous function g. Since $g^{(k)} = 0$ in $(-1, 0)$ and in $(0, 1)$, we can write

$$g(x) = \begin{cases} p_1(x) & \text{if } x > 0, \\ p_2(x) & \text{if } x < 0, \end{cases} \tag{3.6-1}$$

where p_1 and p_2 are polynomials of degree $\leq k - 1$, with the same constant term, which is $g(0)$. (It is easy to see that f is the kth derivative of (3.6-1) not merely in $(-1, 1)$ but but also in all $\mathbb{R}$.) Show from this that f is equal to a finite linear combination of $\delta(x)$ and some of its derivatives.

CHAPTER 4

Tempered Distributions and Fourier Transforms

The class $\mathscr{S}$ of test functions that decrease rapidly at ∞; tempered distributions; growth at ∞; a tempered distribution as some derivative of a continuous function of slow growth; Fourier transforms in $\mathscr{S}$; inversion by Fejér's method; Fourier transforms of tempered distributions; power spectrum of a perpetually oscillating function.

Prerequisites: Chapters 2 and 3; concept of a Fourier integral.

Tempered distributions, as functionals, are continuous with respect to a slightly weaker mode of convergence than that described in Section 2.4, hence are slightly milder, as a class, than the class of all Schwartzian distributions. The mildness is not a local matter (the δ function and all its derivatives are tempered distributions), but has to do with the behavior as $\|\mathbf{x}\| \to \infty$. The Fourier transform of a tempered distribution is easily defined; it is also a tempered distribution.

4.1 The Space $\mathscr{S}$

When attention is restricted to a class of distributions having a certain degree of mildness (continuity with respect to a specified mode of convergence in the space of test functions), the domain of definition of the functionals $\langle f, \cdot \rangle$, etc., can be extended to a wider class of test functions, without loss of continuity of the functionals. The milder the distributions, the wider is the class of admissible test functions. For tempered distributions on $\mathbb{R}$, the appropriate class of test functions is the space $\mathscr{S} = \mathscr{S}(\mathbb{R})$ defined as follows: The functions in $\mathscr{S}$ are of class C^∞ but do not necessarily have bounded support. Instead, they are required to go to zero, as $|x| \to \infty$, faster than any inverse power of x, and so are all their derivatives. That is, a C^∞ function

φ is in $\mathscr{S}$ if there are constants K_{pk} such that

$$|x^p\varphi^{(k)}| < K_{pk}, \quad \text{for } p, k = 0, 1, \ldots \text{ and for all } x. \qquad (4.1\text{-}1)$$

In other words, for each p and each k,

$$\sup_{(x)} |x^p\varphi^{(k)}(x)| < \infty. \qquad (4.1\text{-}2)$$

For tempered distributions on $\mathbb{R}^n$, the space $\mathscr{S}(\mathbb{R}^n)$ of test functions is similarly defined, the above inequality being replaced by

$$\sup_{(x)} \|\mathbf{x}\|^p \left| \frac{\partial^{k_1 + \cdots + k_n}}{\partial x_1^{k_1} \ldots \partial x_n^{k_n}} \varphi(\mathbf{x}) \right| < \infty, \qquad (4.1\text{-}3)$$

for all p, all $\mathbf{k}$. [In both cases, the supremum is actually the maximum; the boundedness of $x^{p+1}\varphi^{(k)}$ shows that $x^p\varphi^{(k)} \to 0$, as $x \to \pm\infty$.]

4.2 Tempered Distributions

Tempered distributions on $\mathbb{R}$ will now be defined—the extension to $\mathbb{R}^n$ will be obvious.

Definition 1 (Convergence in $\mathscr{S}$).   If ψ and φ_j $(j = 1, 2, \ldots)$ are test functions (in C_0^∞ or in $\mathscr{S}$), then $\varphi_j \overset{\mathscr{S}}{\to} \psi$ if, for all p and all k,

$$\sup_{(x)} |x^p\{\varphi_j^{(k)}(x) - \psi^{(k)}(x)\}| \to 0, \quad \text{as } j \to \infty. \qquad (4.2\text{-}1)$$

It is easily seen that this is equivalent to the two requirements (1) that there exist constants K_{pk}^0 such that

$$|x^p\varphi_j^{(k)}(x)| < K_{pk}^0 \quad \text{for } p, k = 0, 1, \ldots \text{ and for all } x, \qquad (4.2\text{-}2)$$

where the constants K_{pk}^0 do not depend on j, and (2) that, as $j \to \infty$, $\varphi_j^{(k)}(x)$ converges to $\psi^{(k)}(x)$ uniformly with respect to x on $\mathbb{R}$, for each k.

EXERCISE

1. Show that these requirements are equivalent to (4.2-1).

Note   It is easily verified that, for any p and k, the function

$$\|\varphi\|_{pk} \overset{\text{def}}{=} \sup_{(x)} |x^p\varphi^{(k)}(x)| \qquad (4.2\text{-}3)$$

has all the properties of a norm. These norms determine a topology in the space $\mathscr{S}$, and (4.2-1) shows that the convergence mode $\overset{\mathscr{S}}{\to}$ is convergence with respect to this topology, i.e., (4.2-1) is equivalent to

$$\|\varphi_j - \psi\|_{pk} \to 0, \quad \text{as } j \to \infty, \text{ for all } p, k. \qquad (4.2\text{-}4)$$

The space C_0^∞ is dense in $\mathscr{S}$; i.e., for ψ in $\mathscr{S}$, there is a sequence $\{\varphi_j\}$ in C_0^∞ such that $\varphi_j \overset{\mathscr{S}}{\to} \psi$.

Definition 2.   A tempered distribution f on $\mathbb{R}$ is a linear functional on $\mathscr{S} = \mathscr{S}(\mathbb{R})$ which is continuous with respect to the convergence mode just defined; that is, $\langle f, \varphi_j \rangle \to \langle f, \psi \rangle$ whenever $\varphi_j \overset{\mathscr{S}}{\to} \psi$.

For functions in C_0^∞, $\varphi_j \overset{\mathscr{S}}{\to} \psi$ implies $\varphi_j \overset{\mathscr{D}}{\to} \psi$; hence, if f is a tempered distribution, the restriction of $\langle f, \cdot \rangle$ to C_0^∞ is a distribution in the sense of Chapter 2. The Exercise below shows that if two tempered distributions agree on C_0^∞, they agree on all of $\mathscr{S}$. Therefore, for f tempered, no distinction will be made between $\langle f, \cdot \rangle$ and its restriction to C_0^∞, and a distribution f in the sense of Chapter 2 will be called "tempered" if $\langle f, \cdot \rangle$ has an extension to $\mathscr{S}$ which is continuous with respect to convergence in $\mathscr{S}$.

EXERCISE

2.   Let φ be any function in $\mathscr{S}$ and let ψ be a function in C_0^∞ such that $\psi(0) = 1$; call $\varphi_\varepsilon(x) = \varphi(x)\psi(\varepsilon x)$. Then $\varphi_\varepsilon \in C_0^\infty$. Show that $\varphi_\varepsilon \overset{\mathscr{S}}{\to} \varphi$ as $\varepsilon \to 0$ and conclude that if f is tempered,

$$\langle f, \varphi \rangle = \lim_{\varepsilon \to 0} \langle f, \varphi_\varepsilon \rangle.$$

4.3 Growth at Infinity

A result of the theory of tempered distributions is that they are of slow growth at infinity. A real *function* $f(x)$ is said to be of slow growth if there are positive constants X and p such that

$$-|x|^p \le f(x) \le |x|^p, \quad \text{for } x < -X \text{ and for } x > X.$$

This statement (which says merely $|f(x)| \le |x|^p$ for $|x| > X$) has been so phrased that its translation into the language of distributions is obvious; however, that is *not* what was meant in the above statement about tempered distributions. What *was* meant is this: Let f_δ be the result of smoothing the distribution f over a distance δ with a mollifier, as explained in Section 2.6; then Schwartz's result is that f is a tempered distribution if and only if the function $f_\delta(x)$ is of slow growth at $\pm \infty$ for every positive δ.

A further result (Schwartz, p. 95) is that f is a tempered distribution if and only if there are integers p and k such that f is the pth derivative of a continuous function $g(x)$ which is $O(|x|^k)$, as $x \to \pm \infty$.

EXAMPLE 1

The function e^x is not a tempered distribution, because it grows too rapidly at $+\infty$, and smoothing does not alter this. However, $e^x \cos(e^x)$ is a tempered distribution. Although it is not a function of small growth, any amount of smoothing, no matter how little, reduces the rate of growth by mutual cancellation of the positive and negative excursions at large positive x; note that this function is the derivative of the bounded function $\sin(e^x)$.

4.4 Fourier Transforms in $\mathscr{S}$

Let φ be a function in the set $\mathscr{S} = \mathscr{S}(\mathbb{R})$. [Here, too, the generalization to test functions on $\mathbb{R}^n$ is obvious.] Since $\varphi(x) \to 0$, as $x \to \pm\infty$, more rapidly than any inverse power of x, the Fourier transform

$$\hat{\varphi}(y) = \frac{1}{\sqrt{2\pi}} \int_{-\infty}^{\infty} \varphi(x) e^{-iyx}\, dx \qquad (4.4\text{-}1)$$

clearly exists, for all real y. Similarly, for any integer $q \geq 0$, the integral

$$\frac{1}{\sqrt{2\pi}} \int_{-\infty}^{\infty} (-ix)^q \varphi(x) e^{-iyx}\, dx = \left(\frac{d}{dy}\right)^q \hat{\varphi}(y) \qquad (4.4\text{-}2)$$

exists, for all real y. [The indicated differentiation of (4.4-1) can be performed under the integral sign, because the resulting integrand, which appears in (4.4-2), is continuous and tends to zero rapidly at infinity.] Therefore, $\hat{\varphi}(y)$ is a C^∞ function of y. Also, for any integer $p \geq 0$,

$$|(iy)^p \hat{\varphi}(y)| = \frac{1}{\sqrt{2\pi}} \left| \int_{-\infty}^{\infty} \varphi^{(p)}(x) e^{-iyx}\, dx \right|$$

$$\leq \frac{1}{\sqrt{2\pi}} \int_{-\infty}^{\infty} |\varphi^{(p)}(x)|\, dx < \infty;$$

that is, $\hat{\varphi}(y) \to 0$ as $\to \pm\infty$, faster than any inverse power of y. Finally,

$$|y^p \hat{\varphi}^{(q)}(y)| \leq \frac{1}{\sqrt{2\pi}} \int_{-\infty}^{\infty} |x^q \varphi^{(p)}(x)|\, dx < \infty;$$

hence, the Fourier transform of any function in $\mathscr{S}$ is a function in $\mathscr{S}$.

The familiar inverse of (4.4-1), namely,

$$\varphi(x) = \frac{1}{\sqrt{2\pi}} \int_{-\infty}^{\infty} \hat{\varphi}(y) e^{iyx}\, dy, \qquad (4.4\text{-}3)$$

will now be established by a modification of Fejér's method (1904) for Fourier series: The right member of the above equation is equal to

$$\lim_{n \to \infty} \frac{1}{\sqrt{2\pi}} \int_{-n}^{n} \hat{\varphi}(y) e^{iyx} \left(1 - \frac{|y|}{n}\right) dy$$

$$= \lim_{n \to \infty} \frac{1}{2\pi} \int_{-n}^{n} \left[\int_{-\infty}^{\infty} \varphi(x') e^{-iyx'}\, dx' \right] e^{iyx} \left(1 - \frac{|y|}{n}\right) dy.$$

Here, the order of the integrations can be reversed, because the absolute value of the integrand can be integrated over the same intervals, with a finite result. Now,

$$\frac{1}{2\pi} \int_{-n}^{n} e^{iy(x-x')} \left(1 - \frac{|y|}{n}\right) dy = \frac{1 - \cos n(x - x')}{\pi n(x - x')^2}; \tag{4.4-4}$$

According to (2.6-5), this function converges to $\delta(x - x')$, as $n \to \infty$; equation (4.4-3) follows.

It is thus seen that the operation of taking the Fourier transforms of the functions φ maps the space $\mathscr{S}$ of test functions onto itself. The mapping is continuous:

Theorem. If $\varphi_n \overset{\mathscr{S}}{\to} \psi$, then $\hat{\varphi}_n \overset{\mathscr{S}}{\to} \hat{\psi}$.

PROOF. Without loss of generality, we can take $\psi = \hat{\psi} = 0$. Then:

Assertion 1. For any sequence $\{\chi_n\}$, if $\chi_n \overset{\mathscr{S}}{\to} 0$, then $\sup_{(y)} |\hat{\chi}_n(y)| \to 0$. To prove this, recall that by definition of convergence in $\mathscr{S}$, $\sup_{(x)} |x^p \chi_n^{(k)}(x)| \to 0$. By taking $k = 0$, and $p = 0$ and 2, we see that

$$a_n \overset{\text{def}}{=} \sup_{(x)} |(1 + x^2)\chi_n(x)| \to 0;$$

hence,

$$|\hat{\chi}_n(y)| = \frac{1}{\sqrt{2\pi}} \left| \int_{-\infty}^{\infty} e^{-iyx} \chi_n(x) dx \right| \le \frac{1}{\sqrt{2\pi}} \int_{-\infty}^{\infty} \frac{a_n}{1 + x^2} dx \to 0.$$

The third member here is independent of y, and the assertion follows.

Assertion 2. The hypothesis $\varphi_n \overset{\mathscr{S}}{\to} 0$ implies that $x^p \varphi^{(k)}(x)$ also $\overset{\mathscr{S}}{\to} 0$, for any p, k. To show this, differentiate $x^p \varphi^{(k)}(x)$ l times and multiply by x^r, for arbitrary l and r; the result is a finite linear combination of terms that $\to 0$ uniformly, and this proves the assertion. Now $y^k \hat{\varphi}_n^{(p)}(y)$ is (except for a factor i to some power) the Fourier transform of $\chi_n(x) \overset{\text{def}}{=} x^p \varphi^{(k)}(x)$, which $\overset{\mathscr{S}}{\to} 0$ by Assertion 2. By Assertion 1, then, $\sup_{(y)} |y^k \varphi_n^{(p)}(y)| \to 0$; that is, $\hat{\varphi}_n \overset{\mathscr{S}}{\to} 0$, as was to be proved.

4.5 Fourier Transforms of Tempered Distributions

First, let $f(x)$ be a continuous function whose Fourier transform $\hat{f}(y)$ exists in the ordinary sense and is continuous (for example, f might be in $\mathscr{S}$). Regarded as distributions, f and $\hat{f}$ are the functionals $\langle f, \varphi \rangle$ and $\langle \hat{f}, \varphi \rangle$, respectively; the relation between them is given by Perseval's identity, one form of which is

$$\langle \hat{f}, \varphi \rangle = \int_{-\infty}^{\infty} \hat{f}(x)\varphi(x)dx = \int_{-\infty}^{\infty} \hat{f}(x) \frac{1}{\sqrt{2\pi}} \int_{-\infty}^{\infty} \varphi(y)e^{iyx}\, dy\, dx$$

$$= \int_{-\infty}^{\infty} \frac{1}{\sqrt{2\pi}} \int_{-\infty}^{\infty} \hat{f}(x)e^{iyx}\, dx\, \hat{\varphi}(y)dy = \int_{-\infty}^{\infty} f(y)\hat{\varphi}(y)dy$$

$$= \langle f, \hat{\varphi} \rangle. \tag{4.5-1}$$

The Fourier transform of a tempered distribution is defined by analogy:

Definition. If f is any tempered distribution, its Fourier transform $\hat{f}$ is defined as the distribution (functional) given by $\langle \hat{f}, \varphi \rangle = \langle f, \hat{\varphi} \rangle$, for all φ in $\mathscr{S}$.

The theorem in the preceding section says that if φ_n ($n = 1, 2, \ldots$) and ψ are test functions such that $\varphi_n \overset{\mathscr{S}}{\to} \psi$, then $\hat{\varphi}_n \overset{\mathscr{S}}{\to} \hat{\psi}$, and conversely. Therefore, since then $\langle f, \hat{\varphi}_n \rangle \to \langle f, \hat{\psi} \rangle$, because f is a tempered distribution, and hence $\langle \hat{f}, \varphi_n \rangle \to \langle \hat{f}, \psi \rangle$, it follows that *the Fourier transform of a tempered distribution is a tempered distribution.*

If Fourier transformation is applied twice to a tempered distribution f, the result is the distribution $\hat{\hat{f}}$ given by $\hat{\hat{f}}(x) = f(-x)$, because evidently $\hat{\hat{\varphi}}(x) = \varphi(-x)$, and therefore

$$\langle \hat{\hat{f}}, \varphi \rangle = \langle \hat{f}, \hat{\varphi} \rangle = \langle f, \hat{\hat{\varphi}} \rangle = \langle f(x), \varphi(-x) \rangle,$$

which is equal to $\langle f(-x), \varphi(x) \rangle$ by the rules for changes of independent variable in a distribution—see Section 2.8.

It follows from the above definition that if $f_n \to f$ in the sense of convergence of distributions, then $\hat{f}_n \to \hat{f}$ in the same sense. That is, Fourier transformation is a continuous mapping in $\mathscr{S}'$.

The entire development for the n-dimensional case proceeds by obvious generalization, starting with the space $\mathscr{S}(\mathbb{R}^n)$ of test functions described at the end of Section 4.1. A function $f(\mathbf{x})$ is of slow growth if it is bounded by const $+ |\mathbf{x}|^p$ for some p. Schwartz's theorems of Section 4.3 take the following forms:

1. A distribution f on $\mathbb{R}^n$ is tempered if and only if $J_\delta f$ is a function of slow growth for each $\delta > 0$.

2. f is tempered if and only if it is some (pure or mixed) partial derivative of some continuous function of slow growth.

Corollary. *If $D^q f$ denotes any derivative of f, $D^q f$ is tempered if and only if f is.*

In place of (4.4-1) for the Fourier transform of a test function, we have

$$\hat{\varphi}(\mathbf{y}) = \frac{1}{(2\pi)^{n/2}} \int_{-\infty}^{\infty} \cdots \int_{-\infty}^{\infty} \varphi(\mathbf{x}) e^{-i\mathbf{y}\cdot\mathbf{x}}\, dx_1 \ldots dx_n.$$

The transform variable $\mathbf{y}$ is often denoted by $\mathbf{k}$.

It is recalled that for ordinary Fourier integrals, the rate at which $\hat{f} \to 0$ at infinity depends on the smoothness of f, and conversely. The same is true in a sense for distributions, as shown for example by the following theorem.

Theorem. *If f is a tempered distribution with bounded support in $\mathbb{R}^n$, then $\hat{f}$ is an entire analytic function $f(\mathbf{k})$, i.e., is analytic in each component k_j of $\mathbf{k}$ in the entire complex k_j plane.*

PROOF. Since f has bounded support, the functional $\langle f, \varphi \rangle$ is well defined for any φ in C^∞, regardless of *its* support. We define

$$e_{\mathbf{k}}(\mathbf{x}) = e^{-i\mathbf{k}\cdot\mathbf{x}},$$

$$F(\mathbf{k}) = \frac{1}{(2\pi)^{n/2}} \langle f, e_{\mathbf{k}} \rangle,$$

and we claim that $\hat{f}$ is the function $F(\mathbf{k})$, which is clearly analytic in each component of $\mathbf{k}$ by Exercise 2 in Section 2.4 on differentiation with respect to a parameter. Namely, for any φ in C_0^∞,

$$\int F(\mathbf{k})\varphi(\mathbf{k})d^n\mathbf{k} = \frac{1}{(2\pi)^{n/2}} \int \langle f, e_{\mathbf{k}} \rangle \varphi(\mathbf{k})d^n\mathbf{k}.$$

By Exercise 1 of the same Section 2.4 on integration with respect to a parameter, this is $\langle f, \hat{\varphi} \rangle$; i.e., $\langle F, \varphi \rangle = \langle f, \hat{\varphi} \rangle$, hence $F = \hat{f}$, as required.

The Fourier transform of a not necessarily tempered distribution is defined in Gel'fand and Shilov, Vol. 1, Chapter 2. It is in general not a distribution in the sense of Laurent Schwartz but rather a continuous linear functional on the space Z of test functions referred to at the end of Section 2.2. We shall sketch the situation for distributions on $\mathbb{R}$; for distributions on $\mathbb{R}^n$ and other details, see Gel'fand and Shilov. It is recalled that the class $\mathcal{S} = \mathcal{S}(\mathbb{R})$ of test functions is mapped onto itself by Fourier transformation and that $C_0^\infty = C_0^\infty(\mathbb{R})$ is a subset of $\mathcal{S}$. Hence C_0^∞ is mapped onto some other subset of $\mathcal{S}$, and in fact onto the space $Z = Z(\mathbb{R})$ defined as follows: a function $\varphi(y)$ is in Z if

1. $\varphi(y)$ is an entire function of y
2. There are positive constants $a, C_0, C_1, \ldots$, such that

$$|y^q\varphi(y)| \leq C_q e^{a|\mathrm{Im}\, y|} \qquad q = 0, 1, 2, \ldots, y \in \mathbb{C}.$$

EXERCISE

1. Show that if $\varphi(x)$ is in C_0^∞ and the support of $\varphi(x)$ is contained in $[-a, a]$, then the Fourier transform $\hat{\varphi}(y)$ satisfies (1) and (2) above, for suitable constants C_q. This shows that C_0^∞ is mapped into Z by Fourier transformation; that conversely Z is mapped into C_0^∞ is proved by Gel'fand and Shilov, loc. cit.

A mode of convergence $\overset{Z}{\to}$ of test functions in Z is defined in such a way that $\hat{\varphi}_j \overset{Z}{\to} \hat{\varphi}$ if and only if $\varphi_j \overset{Z}{\to} \varphi$, and then a class of generalized functions is defined as the class of continuous linear functionals on Z, hence of elements of the dual space Z'. Therefore, by means of the generalized Parseval relation $\langle \hat{f}, \varphi \rangle = \langle f, \hat{\varphi} \rangle$, which was used for defining the Fourier transform of a distribution, it is seen that the elements of Z' are the Fourier transforms of the elements of $\mathcal{D}'$, i.e. of distributions in the sense of Schwartz.

Whereas the elements of $\mathcal{D}'$ are to be thought of as generalized functions defined only on $\mathbb{R}$, the elements of Z' are defined on the entire complex plane. Gel'fand and Shilov show, for example, that the Fourier transform of the function $f(x) = e^x$ (which, as observed above, is not a *tempered* distribution) is $= 2\pi\delta(y - i)$, i.e., is the functional on Z given, for any $\varphi(y)$, by

$$\langle \hat{f}, \varphi \rangle = 2\pi\varphi(i)$$

(which is well defined because φ is an entire function). For further details, see Gel'fand and Shilov.

Examples of Fourier Transforms of Distributions

EXAMPLE 1
$f(x) = \delta(x - x_0); \hat{f}(y) = (2\pi)^{-1/2} e^{-iyx_0}$

PROOF. For any test function φ,

$$\langle \hat{f}, \varphi \rangle = \langle f, \hat{\varphi} \rangle = \hat{\varphi}(x_0) = \frac{1}{\sqrt{2\pi}} \int_{-\infty}^{\infty} \varphi(y) e^{-iyx_0} \, dy$$

$$= \left\langle \frac{1}{\sqrt{2\pi}} e^{-iyx_0}, \varphi \right\rangle.$$

EXAMPLE 2
$f(x) = \delta'(x - x_0); \hat{f}(y) = (2\pi)^{-1/2} iy e^{-iyx_0}.$

EXAMPLE 3
$f(x) = a_0 + a_1 x + \cdots + a_n x^n; \hat{f}(y) = \sqrt{2\pi}[a_0 \delta(y) + ia_1 \delta'(y) + \cdots + i^n a_n \delta^{(n)}(y)].$

EXAMPLE 4
$f(x) =$ periodic function given by a convergent Fourier series $\sum_{(p)} c_p e^{ipx}$; $\hat{f}(y) = \sqrt{2\pi} \sum_{(p)} c_p \delta(y - p).$

EXAMPLE 5

$f(x)$ is a distribution with period 2π; i.e., $f(x)$ and $f(x + 2\pi)$ are the same distribution. [In this case, $f(x)$ is automatically tempered, according to Schwarz's theorem on the rate of growth—see Section 4.3]. Then, for any test function φ,

$$\langle \hat{f}(y), \varphi(y) \rangle = \langle f(x), \hat{\varphi}(x) \rangle$$
$$= \langle f(x + 2\pi), \hat{\varphi}(x) \rangle$$
$$= \langle f(x), \hat{\varphi}(x - 2\pi) \rangle;$$

but $\hat{\varphi}(x - 2\pi)$ is the Fourier transform of $e^{2\pi i y}\varphi(y)$; therefore,

$$\langle \hat{f}(y), \varphi(y) \rangle = \langle \hat{f}(y), e^{2\pi i y}\varphi(y) \rangle$$

for all φ in $\mathscr{S}$; i.e., $(1 - e^{2\pi i y})\hat{f}(y)$ is the zero distribution. By Exercise 2 of Section 3.5, then, $\hat{f} = 0$ in Ω, where Ω is the real axis with the points $y = 0, \pm 1, \pm 2, \ldots$ deleted; that is, $\hat{f}$ is concentrated at the integers. Hence, by the Exercise in Section 3.6, the part of $\hat{f}$ at integer k is a linear combination of $\delta(y - k)$ and some of its derivatives. Exercise 2 below shows however that the derivatives do not in fact appear, hence $\hat{f}$ becomes

$$\hat{f}(y) = \sum_{-\infty}^{\infty} {}_{(k)} c_k \delta(y - k), \qquad (4.5\text{-}2)$$

and this is the general form of the Fourier transform of a periodic distribution.

EXERCISES

2.   Using the Exercise in Section 3.6, write $\hat{f}(y)$ first as

$$\hat{f}(y) = \sum_{-\infty}^{\infty} {}_{(k)} \sum_{j=0}^{J_k} c_{kj} \delta^{(j)}(y - k),$$

then show that since $\hat{f}(y)$ is a tempered distribution, there is some integer p such that $J_k \leq p$ for all k. Then show from the periodicity of f that p is $= 0$.

3.   Show that since $\hat{f}(y)$ is a tempered distribution, the coefficients c_k in (4.5-2) cannot grow faster than some power of $|k|$ as $k \to \pm\infty$.

4.   Find the distribution whose Fourier transform is

$$\hat{f}(y) = \sum_{k=0}^{\infty} \frac{1}{k!} \delta'(y - k)$$

and generalize the result.

4.6 The Power Spectrum

Let $f(t)$ be a bounded continuous function that oscillates more or less irregularly for all t in $(-\infty, \infty)$; it may be thought of as one of the components of the electric field at some point of space in the radiation from a light source or a velocity component at some point in a turbulence field. (Then, $f(t)$ would be real-valued, but that restriction is not essential.) Clearly $f(t)$ cannot be written as a classical Fourier series, because it is not periodic, nor as a classical Fourier integral, because it is not quadratically integrable, but it is a tempered distribution, hence has a Fourier transform $\hat{f} = \hat{f}(\omega)$, which is easily seen to be the second derivative $F''(\omega)$, in the distribution sense, of the continuous function

$$F(\omega) = \frac{1}{\sqrt{2\pi}} \int_{-\infty}^{\infty} f(t) \frac{1}{t^2} \left(-e^{-i\omega t} + \frac{1 - i\omega t}{1 + t^2} \right) dt. \qquad (4.6\text{-}1)$$

Namely, an elementary calculation first shows that $F(\omega)$ is even Lipschitz continuous, and then, for any test function φ,

$$\langle F'', \varphi \rangle = \langle F, \varphi'' \rangle = \int_{-\infty}^{\infty} F(\omega)\varphi''(\omega)d\omega;$$

substitution of (4.6-1) for F and two integrations by parts with respect to ω gives $\langle f, \hat{\varphi} \rangle$, which shows that $F'' = \hat{f}$.

We wish to determine how the intensity or power associated with the function $f(t)$ is distributed with respect to the frequency ω, i.e., the *spectrum* of $f(t)$. In order for that notion to have a meaning, we must assume that the long-term statistical properties of $f(t)$ are well defined. Specifically, we assume that the *autocovariance function*

$$R(\tau) \overset{\text{def}}{=} \lim_{T \to \infty} \frac{1}{2T} \int_{-T}^{T} \overline{f(t + \tau)} f(t)dt \qquad (4.6\text{-}2)$$

exists for all τ. ($R(\tau)/R(0)$ is then the *autocorrelation* of $f(t)$ for time difference τ.)

We assume furthermore that the convergence indicated in (4.6-2) is uniform with respect to τ in any finite interval, hence that $R(\tau)$ is a continuous function. Some of the consequences of these assumptions are explored in the examples below. It is easily shown that $R(\tau)$ is also given by

$$R(\tau) = \lim \frac{1}{2T} \int_{-T}^{T} \overline{f(t + s + \tau)} f(t + s)dt \qquad (4.6\text{-}3)$$

for any real s; in other words, if the long-term statistical properties are well defined at all, they are unaltered by shifting the function $f(t)$ forward or backward in time. We note that $R(-\tau) = \overline{R(\tau)}$. From Schwarz's inequality it follows that $|R(\tau)| \leq R(0)$.

If we suppose that $f(t)$ represents the voltage applied to a unit resistive load, then $R(0)$ is the time-averaged power input to the load. We seek a nondecreasing function $S(\omega)$, called the *power spectrum* of $f(t)$, such that, for $\omega_2 > \omega_1$, $S(\omega_2) - S(\omega_1)$ is the power associated with frequencies in the interval (ω_1, ω_2). If such a function can be found, a jump of $S(\omega)$ will correspond to a spectral line, and $S'(\omega)$, whenever it exists in the classical sense, will be the intensity of the continuous spectrum at frequency ω.

Operationally, $S(\omega)$ is obtained by inserting an electrical filter between the source and the load, as indicated schematically in Figure 4-1. (The source is supposed to have infinite impedance, so that the signal $f(t)$ is not influenced by the presence of the filter.) Let the transmission characteristic of the filter be given by a (generally complex-valued) function $\psi(\omega)$: if a unit sinusoidal voltage at frequency ω is applied to the filter, then $\psi(\omega)$ gives the amplitude and phase of the output. If the filter were an ideal band-pass filter, i.e., if $|\psi(\omega)|$ were $=1$ for ω in an interval (ω_1, ω_2) and $=0$ otherwise, then the power input to the load would be $S(\omega_2) - S(\omega_1)$. Otherwise, the power input is the Stieltjes integral

$$\int_{-\infty}^{\infty} |\psi(\omega)|^2 \, dS(\omega), \qquad (4.6\text{-}4)$$

because $|\psi(\omega)|^2$ is the power attenuation factor for the filter at frequency ω. Hence, $S(\omega)$ must be so chosen that for any test function $\psi(\omega)$ the above integral is equal to

$$\lim_{T \to \infty} \frac{1}{2T} \int_{-T}^{T} |g(t)|^2 \, dt, \qquad (4.6\text{-}5)$$

where $g(t)$ is the voltage output of the filter (see Figure 4-1).

We assert that $S(\omega)$ is given by the formula

$$S(\omega) = \int_{-\infty}^{\infty} R(\tau) \frac{e^{i\omega\tau} - 1}{2\pi i \tau} \, d\tau, \qquad (4.6\text{-}6)$$

which is the fundamental formula for the power spectrum of $f(t)$. It has been used extensively in fluid dynamics, ergodic theory, statistical mechanics, and the theory of dynamical systems. In turbulence theory it is applied for example to the dependence of a velocity component on a cartesian coordinate, at a given time, which is then called $v(x)$ instead of $f(t)$; in this case, $R(\tau)$ represents spatial rather than temporal correlation.

The function $R(\tau)$, although bounded, does not in general $\to 0$ as $\tau \to \infty$, as the simple examples $f(t) = $ constant and $f(t) = \sin t$ show; hence the convergence of the above integral is not a priori obvious, but was established by Norbert Wiener in 1926 under quite general circumstances (Wiener 1930). Sometimes, as in Example 3 below (if one of the frequencies ω_n happens to be $=0$), the equation must be interpreted as

$$S(\omega) = \lim_{T \to \infty} \int_{-T}^{T} R(\tau) \frac{e^{i\omega\tau} - 1}{2\pi i \tau} \, d\tau. \qquad (4.6\text{-}7)$$

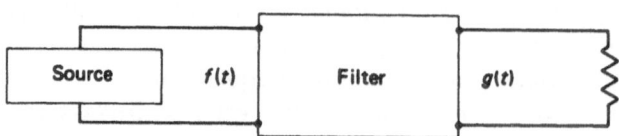

Figure 4.1 Schematic circuit.

In the remainder of this discussion, we shall assume that the right member of this last equation is meaningful as an ordinary function.

In order to establish the formula for $S(\omega)$ by equating (4.6-4) and (4.6-5), we need a formula for the output voltage $g(t)$ of the filter. Clearly, $g(t)$ is the function whose Fourier transform is $\hat{f}(\omega)\psi(\omega)$. That is,

$$\langle \hat{g}, \varphi \rangle = \langle \hat{f}\psi, \varphi \rangle = \langle \hat{f}, \psi\varphi \rangle = \langle f, \widehat{\psi\varphi} \rangle$$

for any test function φ. The Fourier transform of the product $\psi\varphi$ is given by the convolution of the Fourier transforms $\hat{\psi}$ and $\hat{\varphi}$, namely,

$$(\widehat{\psi\varphi})(s) = \frac{1}{\sqrt{2\pi}} \int_{-\infty}^{\infty} \hat{\psi}(s - t)\hat{\varphi}(t)dt. \tag{4.6-8}$$

Therefore,

$$\langle g, \hat{\varphi} \rangle = \langle \hat{g}, \varphi \rangle = \int_{-\infty}^{\infty} f(s) \frac{1}{\sqrt{2\pi}} \int_{-\infty}^{\infty} \hat{\psi}(s - t)\hat{\varphi}(t)dt \, ds,$$

which shows that

$$g(t) = \frac{1}{\sqrt{2\pi}} \int f(s)\hat{\psi}(s - t)ds$$

$$= \frac{1}{\sqrt{2\pi}} \int f(t + s)\hat{\psi}(s)ds. \tag{4.6-9}$$

This equation describes how the filter transforms the signal $f(t)$ into $g(t)$; the interpretation of the function $\hat{\psi}(t)$ that appears in it will be discussed below.

From the above equation for $g(t)$, we find that

$$\int_{-T}^{T} |g(t)|^2 \, dt = \frac{1}{2\pi} \int_{-T}^{T} dt \int_{-\infty}^{\infty} \overline{f(t + s')\hat{\psi}(s')}ds' \cdot \int_{-\infty}^{\infty} f(t + s)\hat{\psi}(s)ds.$$

Here we take $\tau = s' - s$ as a new variable in place of s', for fixed s; we divide by $2T$, and we let $T \to \infty$; the uniform convergence of the expression (4.6-3) for the autocovariance permits taking the limit under the integral sign; hence

$$\lim_{T \to \infty} \frac{1}{2T} \int_{-T}^{T} |g(t)|^2 \, dt = \frac{1}{2\pi} \int\int_{-\infty}^{\infty} R(\tau)\overline{\hat{\psi}(s + \tau)}\hat{\psi}(s)ds \, d\tau.$$

We use again the relation between Fourier transform and convolution in the form

$$\int_{-\infty}^{\infty} \overline{\hat{\psi}(s + \tau)}\hat{\psi}(s)ds = \int_{-\infty}^{\infty} |\psi(\omega)|^2 e^{i\omega\tau} \, d\omega,$$

and we have:

$$\lim_{T \to \infty} \frac{1}{2T} \int_{-T}^{T} |g(t)|^2 \, dt = \frac{1}{2\pi} \int_{-\infty}^{\infty} R(\tau) \int_{-\infty}^{\infty} |\psi(\omega)|^2 e^{i\omega\tau} \, d\omega \, d\tau$$

$$= \frac{1}{2\pi} \int_{-\infty}^{\infty} R(\tau) \int_{-\infty}^{\infty} |\psi(\omega)|^2 \, d_{\omega}\left(\frac{e^{i\omega\tau} - 1}{i\tau}\right) d\tau$$

$$= \int_{-\infty}^{\infty} |\psi(\omega)|^2 \, dS(\omega), \tag{4.6-10}$$

on the assumption that the convergence in (4.6-7) is uniform with respect to ω.

Lastly, the nonnegativity of the first member of (4.6-10) shows that $S(\omega)$ is a nondecreasing real function.

EXAMPLES

For some of these examples, it is convenient to require $f(t)$ to be only piecewise continuous; the theory is unchanged, and in fact Wiener considered a still wider class of functions. The reader can easily round off the corners of $f(t)$ if he wishes.

EXAMPLE 1

$f(t) = 1$ for $1 \leq |t| \leq 2$, $4 \leq |t| \leq 8$, $16 \leq |t| \leq 32$, etc., while $f(t) = 0$ otherwise. Then the quantity $(1/2T)\int_{-T}^{T} f(t + \tau)f(t)dt$ oscillates between $\approx \frac{1}{3}$ and $\approx \frac{2}{3}$, as $T \to \infty$; hence the limit in (4.6-2) doesn't exist for any τ. This function $f(t)$ does not have well-defined long-range statistical properties.

EXAMPLE 2

$f(t) = e^{it^2}$. Here $R(0) = 1$, while $R(\tau) = 0$ for $\tau \neq 0$. That is, $R(\tau)$ is not continuous as we have required, and $S(\omega) \equiv 0$. Owing to the rapid oscillation of e^{it^2} for large t, we may say roughly that here all the power is at infinite frequencies. $\sin t^2$ and $\cos t^2$ behave similarly.

EXAMPLE 3

Let $f(t)$ be almost periodic; then it can be expanded in the series

$$\sum_{n=0}^{\infty} c_n e^{i\omega_n t}, \tag{4.6-11}$$

where the ω_n are real constants and the c_n complex ones, and the series converges to $f(t)$ in the norm $\|f\|$ obtained from the inner product

$$(f, g) = \lim_{T \to \infty} \frac{1}{2T} \int_{-T}^{T} \overline{f(t)}g(t)dt; \tag{4.6-12}$$

the coefficients c_n satisfy the Parseval relation

$$\sum |c_n|^2 = \|f\|^2 = R(0)$$

(Riesz–Nagy 1953). Note that $\|f\|$ is not the ordinary L^2 norm on $\mathbb{R}$ because of the factor $1/2T$ in (4.6-12). For the sake of simplicity let us assume that $f(t)$ is such that the series (4.6-11) converges absolutely. Then,

$$R(\tau) = \sum |c_n|^2 e^{i\omega_n \tau}, \tag{4.6-13}$$

and it is easily seen from (4.6-7) that

$$S(\omega) = \text{const.} + \sum_{(\omega_n < \omega)} |c_n|^2. \tag{4.6-14}$$

(This assumes that ω is not equal to any of the ω_n; if $\omega = \omega_m$ for some m, then $\frac{1}{2}|c_m|^2$ has to be added to the above sum.) Hence, the function $f(t)$ has a pure line spectrum; the line at frequency ω_n has intensity $|c_n|^2$.

This example includes the case of a periodic function, where the ω_n are integer multiples of some fundamental frequency and (4.6-11) is an ordinary Fourier series.

EXAMPLE 4
Suppose that $f(t)$ is given by an ordinary Fourier integral

$$f(t) = \frac{1}{\sqrt{2\pi}} \int_{-\infty}^{\infty} \hat{f}(\omega) e^{i\omega t} \, d\omega,$$

where f and $\hat{f}$ are quadratically integrable ordinary functions. It is easily seen that in this case $R(0) = 0$, hence $R(\tau) \equiv 0$. The function $f(t)$ corresponds to a finite *energy*, but when that energy is averaged over all time, the average power is zero.

EXAMPLE 5
It might be supposed that a Fourier–Stieltjes integral

$$f(t) = \int e^{i\omega t} \, d\sigma(\omega), \tag{4.6-15}$$

where $\sigma(\omega)$ is a function of bounded variation, would provide a suitable generalization of Fourier series and Fourier integrals, but we get again only a line spectrum, with lines at frequencies ω, if any, where $\sigma(\omega)$ has jumps. The Fourier transform of (4.6-15) is the first derivative of the ordinary function $\sigma(\omega)$, and we have already seen that the transform $\hat{f}$ of a function of interest to us is the *second* distribution derivative of an ordinary function $F(\omega)$ (4.6-1).

EXAMPLE 6
We now describe a function $f(t)$, given by Wiener, that *does* have a continuous spectrum. It takes on only the values ± 1 and is constant in each interval between two successive integers. In any string of m such intervals, $f(t)$ can be described by a sequence of m + or − signs, which we call a *pattern* of length m. The function is so constructed that if N_1 is the number of times that a given such pattern occurs in an interval of length N (which we think of as $\gg m$), then

$$\frac{N_1}{N} \to 2^{-m} \quad \text{as } N \to \infty.$$

We can say that asymptotically any one of the 2^m patterns of length m has the same probability of occurring as any other. Now consider

$$\frac{1}{2T} \int_{-T}^{T} f(t + \tau) f(t) dt,$$

where τ is an integer $\neq 0$. The contribution of an interval $n < t < n + 1$ to the integral is $= \pm 1$ according as the pattern of length $\tau + 1$ of $f(t)$ in the interval $n < t < n + \tau + 1$ has the same or opposite signs at its two ends. Since those two eventualities have equal probabilities, we see, on letting $T \to \infty$, that $R(\tau)$ is $= 0$

when τ is an integer $\neq 0$. Clearly, $R(0) = 1$, and it is easily seen that $R(\tau)$ varies linearly as τ varies between successive integers, hence

$$R(\tau) = \begin{cases} 1 - |\tau| & \text{for } |\tau| < 1, \\ 0 & \text{otherwise,} \end{cases} \qquad (4.6\text{-}16)$$

from which we find via (4.6-6) that

$$S(\omega) = \int_0^1 (1 - \tau)\frac{\sin \omega\tau}{\pi\tau}\, d\tau. \qquad (4.6\text{-}17)$$

Therefore the spectrum is continuous in this case, and the density of the spectrum is

$$S'(\omega) = \frac{1 - \cos \omega}{\pi\omega^2} \qquad (4.6\text{-}18)$$

It remains to construct $f(t)$, which we do for $t > 0$, with the understanding that $f(-t) = f(t)$. The sequence of signs for $f(t)$, $t > 0$, is

$$\left.\begin{array}{l} +,\ - \\ + +,\ + -,\ - +,\ - -, \text{ repeated twice,} \\ + + +,\ + + -,\ + - +,\ + - -,\ - + +,\ - + -,\ - - +,\ - - -, \text{ repeated 4 times,} \\ + + + +,\ + + + -, \text{ etc., repeated 8 times,} \\ \cdots\cdots\cdots \end{array}\right\}$$

$$(4.6\text{-}19)$$

The commas serve no purpose except to separate groups of signs for the purpose of discussion. Now consider patterns of length m, and consider a row of the above scheme in which the distance between commas is $\gg m$; among patterns that do not straddle a comma, all patterns of length m occur equally often, and the probability of straddling a comma $\to 0$ as the distance between commas increases. Furthermore, owing to the repetitions, each row of the scheme dominates all preceding rows, and it is seen that $f(t)$ has the required property.

The importance of this example is independent of the somewhat artificial scheme (4.6-19). If the succession of $+$ and $-$ signs is determined by some physical process that is completely random, the same continuous spectrum (4.6-17) with density (4.6-18) is obtained. Presumably, the radiation from an incandescent body and white noise have this general character.

Lastly, we should mention another example quoted by Wiener (1930) and attributed by him to K. Mahler, leading to a so-called singular continuous spectrum (but we omit the description, owing to its length). We recall that a nondecreasing function $S(\omega)$ can be represented as the sum of three parts: a pure jump function with countably many jumps, an absolutely continuous function, and a continuous function having derivative $=0$ almost everywhere, like the Cantor function described in Chapter 13. The first part gives the line spectrum, the second the continuous spectrum in the usual physical sense, and the third the singular continuous spectrum, which is not normally found in physical problems, although Mahler's example shows that it can occur in principle.

The function $\hat{\psi}(t)$ appearing in (4.6-9) represents the response of the filter to a unit impulse $\delta(t)$ applied at time $t = 0$. Hence, whatever the detailed mechanism of the filter, $\hat{\psi}(t)$ must be $= 0$ for $t < 0$, owing to causality. However, that puts no essential restriction on our argument. If $\hat{\psi}_1(t)$ is any C_0^∞ function and is the Fourier transform of $\psi_1(\omega)$, then for some real a the Fourier transform of $\psi(\omega) \stackrel{\text{def}}{=} e^{i\omega a}\psi_1(\omega)$ has its support in $\{t \geq 0\}$, hence satisfies the causality condition. Hence, since $|\psi(\omega)|^2 = |\psi_1(\omega)|^2$, the function $|\psi(\omega)|^2$ in (4.6-4) can be regarded as an arbitrary nonnegative test function whose Fourier transform is in C_0^∞.

CHAPTER 5

L^2 Spaces

Mean convergence; quadratically integrable functions and
distributions and their properties. Spaces of type L^2, L^1, L^p, L^∞,
L^2_δ; Fourier transforms and mollifiers in L^2 spaces; Sobolev spaces;
boundary values in Sobolev spaces.

Prerequisite: Chapters 1–4.

The concepts of Hilbert space, distribution, and mean convergence are
combined to construct function spaces suitable for the analysis of differential
operators. From the quantum-mechanical point of view, the elements or
"points" in such a space are the wave functions that represent the states of a
physical system.

5.1 Mean Convergence; Completeness of Function Systems

We first describe several familiar cases of mean convergence. Let $f(x)$ be
a continuous periodic function with period 2π. Its Fourier coefficients are

$$c_k = \frac{1}{2\pi} \int_{-\pi}^{\pi} f(x)e^{-ikx}\, dx, \qquad k = 0, \pm 1, \pm 2, \ldots, \tag{5.1-1}$$

and we denote by $S_m(x)$ the mth partial sum of the Fourier series of $f(x)$:

$$S_m(x) = \sum_{k=-m}^{m} c_k e^{ikx}. \tag{5.1-2}$$

Then the functions $S_m(x)$ *converge in the mean* or *in L^2* to $f(x)$, by which is
meant that

$$\int_{-\pi}^{\pi} |S_m(x) - f(x)|^2\, dx \to 0, \quad \text{as } m \to \infty. \tag{5.1-3}$$

It will be seen later in this chapter that the same is true for periodic functions $f(x)$ in a much wider class, whereas even if $f(x)$ is continuous the point-wise convergence $S_m(x) \to f(x)$ for each x cannot be proved without some further restriction on $f(x)$, such as the requirement of bounded variation.

More generally, let $f(\mathbf{x})$ or $f(x_1, \ldots, x_n)$ be a continuous function periodic with period 2π in each of the variables $x_1, \ldots, x_n$. Let $\mathbf{k}$ denote a vector with integer components $k_1, \ldots, k_n$; then the Fourier coefficients of $f(\mathbf{x})$ are

$$c_{\mathbf{k}} = (2\pi)^{-n} \int f(\mathbf{x}) e^{-i\mathbf{k}\cdot\mathbf{x}} \, d^n\mathbf{x}, \qquad (5.1\text{-}4)$$

where the integration is over the cube

$$K = \{|x_i| \leq \pi, i = 1, \ldots, n\}. \qquad (5.1\text{-}5)$$

We denote the mth partial sum of the multiple Fourier series for $f(\mathbf{x})$ by

$$S_m(\mathbf{x}) = \sum_{\mathbf{k} \in L_m} c_{\mathbf{k}} e^{i\mathbf{k}\cdot\mathbf{x}}, \qquad (5.1\text{-}6)$$

where L_m is the set of lattice points in a cube of edge $2m$ in $\mathbf{k}$ space:

$$L_m = \{\mathbf{k}: (k_i = -m, \ldots, +m), i = 1, \ldots, n\}. \qquad (5.1\text{-}7)$$

Then,

$$\int_K |S_m(\mathbf{x}) - f(\mathbf{x})|^2 \, d^n\mathbf{x} \to 0, \quad \text{as } m \to \infty. \qquad (5.1\text{-}8)$$

More general expansions in orthogonal systems of functions (of which the Fourier series expansions are special cases) also converge in the mean. We consider systems of polynomials in one variable. Let $\rho(x)$ be a positive continuous weight function for a finite interval $[a, b]$, and suppose that $\{p_k(x)\}_{k=0}^{\infty}$ is a corresponding orthonormal system of polynomials; i.e.,

$$\int_a^b p_k(x)p_l(x)\rho(x)dx = \delta_{kl}, \qquad (5.1\text{-}9)$$

where it is understood that for each $k = 0, 1, 2, \ldots, p_k(x)$ is a polynomial of degree k.

[It is recalled that the $p_k(x)$ can be constructed by applying the Gram–Schmidt orthonormalization procedure (Section 1.6) to the functions $1, x, x^2, \ldots$, using the inner product (p_k, p_l) given by the left member of (5.1-9). The construction can be generalized to the case of an open or even infinite interval (a, b) if $\int_a^b \rho(x)dx$ is finite. In this way, the familiar families of polynomials of Legendre, Hermite, Laguerre, Chebyshev, and so on are obtained. The following discussion is restricted to the case in which $\rho(x)$ is continuous in a finite closed interval $[a, b]$.]

In this case, the generalized Fourier coefficients are

$$c_k = \int_a^b p_k(x)f(x)\rho(x)dx, \qquad k = 0, 1, 2, \ldots, \qquad (5.1\text{-}10)$$

the partial sum of the *generalized Fourier series* is

$$S_m(x) = \sum_{k=0}^{m} c_k p_k(x), \qquad (5.1\text{-}11)$$

and then

$$\int_a^b |S_m(x) - f(x)|^2 \rho(x) dx \to 0, \quad \text{as } m \to \infty. \qquad (5.1\text{-}12)$$

We sketch the proof of mean convergence of the generalized Fourier series of a continuous function in these cases. In each case we have a compact region $\mathbb{R}$ (interval or cube) of the independent variable x or $\mathbf{x}$, a weight function ρ, an inner product of the form

$$(f, g) = \int_{\mathbb{R}} \overline{f(\mathbf{x})} g(\mathbf{x}) \rho(\mathbf{x}) d^n \mathbf{x} \qquad (5.1\text{-}13)$$

and a corresponding norm $\|f\| = (f, f)^{1/2}$ in the function space. We have also an orthonormal system $\{\varphi_k\}$. In the polynomial case, $\varphi_k(x) = p_k(x)$; in the Fourier series case the $\varphi_k(\mathbf{x})$ are some convenient enumeration of the functions $e^{i\mathbf{k} \cdot \mathbf{x}}$ and the weight function is $\rho(\mathbf{x}) = (2\pi)^{-n}$. Convergence in the mean is convergence with respect to the norm $\|f\|$, hence what must be shown is that the orthonormal system is complete in the sense of Section 1.6.

It will be shown first that in each case an arbitrary continuous function $f(\mathbf{x})$ can be arbitrarily closely approximated, uniformly in the region $\mathfrak{R}$, by a finite linear combination of the functions $\{\varphi_k\}$. For this, the *Weierstrass approximation theorem* (see Courant–Hilbert, vol. 1) is used: If $F(\mathbf{X})$ is continuous on a compact (i.e. closed bounded) set S in an N-dimensional space, then for any $\varepsilon > 0$ there is a polynomial $P(\mathbf{X})$ such that

$$|F(\mathbf{X}) - P(\mathbf{X})| < \varepsilon \qquad \forall \mathbf{X} \text{ in } S.$$

Consider first the problem of expansion in the orthogonal polynomials $\{\varphi_k(x)\}$; then $X = x$. Here, the compact set S is the interval $[a, b]$ on the real line, and $F(X) = f(x)$. Now $P(x)$ is a linear combination of $1, x, x^2, \dots, x^q$, where $q = $ degree of $P(X)$; each x^k can be written as a linear combination of $\varphi_0(x), \dots, \varphi_k(x)$, hence $P(x)$ can be written as

$$P(x) = \sum_{j=0}^{q} a_j \varphi_j(x).$$

Consider next the one-variable Fourier series, and call $X = \cos x$, $Y = \sin x$. Then the equation $F(X, Y) = f(x)$ defines F on the unit circle $S: X^2 + Y^2 = 1$ in the X, Y plane. S is a compact set in the plane. The corresponding polynomial $P(X, Y)$ is a polynomial in $\cos x$ and $\sin x$, hence in e^{ix} and e^{-ix}, i.e. is a linear combination

$$P(X, Y) = \sum_{k=-q}^{q} a_k e^{ikx}.$$

Lastly, for the n-variable Fourier series, the unit circle S is replaced by an n-dimensional torus. Namely, we write

$$e^{ix_k} = X_{2k-1} + iX_{2k} \qquad (k = 1, \ldots, n)$$

and the equation $f(\mathbf{x}) = F(\mathbf{X})$, where $\mathbf{X}$ is a $2n$-vector, defines F on the torus

$$S: X_1^2 + X_2^2 = 1, \ldots, X_{2n-1}^2 + X_{2n}^2 = 1,$$

which is a compact set in $\mathbb{R}^{2n}$. The corresponding polynomial $P(\mathbf{X})$ or $P(X_1, \ldots, X_{2n})$ is a polynomial in $e^{\pm ix_1}$, $e^{\pm ix_2}$, etc., hence can be written as

$$\sum_{k \in L_q} a_k e^{ik \cdot \mathbf{x}}$$

for some integer q.

It has been shown in each case that if $f(\mathbf{x})$ is continuous in the region $\mathfrak{R}$ of $\mathbb{R}^n$ and ε is > 0, then for some q and some coefficients a_j,

$$\left| f(\mathbf{x}) - \sum_{j=0}^{q} a_j \varphi_j(\mathbf{x}) \right| < \varepsilon, \quad \text{for all } \mathbf{x} \text{ in } \mathfrak{R}. \tag{5.1-14}$$

Therefore,

$$\int_{\mathfrak{R}} \left| f(\mathbf{x}) - \sum_{j=0}^{q} a_j \varphi_j(\mathbf{x}) \right|^2 \rho(\mathbf{x}) d^n \mathbf{x} \leq \varepsilon^2 M,$$

where

$$M = \int_{\mathfrak{R}} \rho(\mathbf{x}) d^n \mathbf{x};$$

that is,

$$\| f - \sum a_j \varphi_j \| \leq \varepsilon \sqrt{M}, \tag{5.1-15}$$

where $\| \cdot \|$ denotes the norm obtained from the inner product (5.1-13), with respect to which the φ_j form an orthonormal system.

The a_j are not necessarily equal to the generalized Fourier coefficients

$$c_j = (\varphi_j, f), \tag{5.1-16}$$

but it can be shown that the *best* approximation in the mean to f by an expression of the form $\sum_{j=0}^{q} a_j \varphi_j$ for given q is obtained by setting $a_j = c_j (j = 0, \ldots, q)$. Namely, if we minimize the expression

$$\| f - \sum a_j \varphi_j \|^2 = \| f \|^2 - \sum (\bar{a}_j c_j + a_j \bar{c}_j) + \sum |a_j|^2$$

with respect to the quantities $\operatorname{Re} a_j$ and $\operatorname{Im} a_j$, we find $\operatorname{Re} a_j = \operatorname{Re} c_j$ and $\operatorname{Im} a_j = \operatorname{Im} c_j (j = 0, \ldots, q)$. Replacement of the a_j by the c_j improves the mean approximation (5.1-15) (possibly at the expense of making the uniform approximation (5.1-14) worse). Hence, for given $\varepsilon > 0$,

$$\left\| f - \sum_{j=0}^{q} c_j \varphi_j \right\| \leq \varepsilon \sqrt{M}$$

for large enough q, hence the generalized Fourier series $\sum_{j=0}^{\infty} c_j \varphi_j(\mathbf{x})$ converges in the mean, i.e. in the L^2 norm, to $f(\mathbf{x})$.

Comments on Pointwise Convergence.   (1) One cannot conclude from the above argument that $f(\mathbf{x})$ is the pointwise limit of a series of the form $\sum_{j=1}^{\infty} a_j \varphi_j(x)$, because each coefficient a_j in (5.1-14) depends generally on q and need not have a limit as $q \to \infty$.

(2) It is known that if a sequence of functions $S_m(\mathbf{x})$ converges in L^2 to a limit $f(\mathbf{x})$, then there is a subsequence of those functions that converges for almost all $\mathbf{x}$ to $f(\mathbf{x})$.

(3) In the special case of a Fourier series in one independent variable, it was proved recently by Carleson (1966) that the partial sums $S_m(x)$ given by (5.1-2) converge to $f(x)$ for almost all x; i.e., in this case it is not necessary to select a subsequence.

(4) For multiple Fourier series it was proved by Fefferman 1971 that $S_m(\mathbf{x})$ converges to $f(\mathbf{x})$ for almost all $\mathbf{x}$, provided that the series is summed in the order indicated in (5.1-6), but not necessarily if summed for example as

$$\sum_{-\infty}^{\infty} {}_{k_1} \left(\sum_{-\infty}^{\infty} {}_{k_2} (\cdots) \right) c_{\mathbf{k}} e^{i\mathbf{k} \cdot \mathbf{x}}.$$

In fact Fefferman gave an example of a function $f(x, y)$ in $L^2((-\pi, \pi)^2)$ for which the series does not converge for any point x, y when summed as above.

It should be noted that the n-torus was not chosen arbitrarily as the set S for the multiple Fourier series, but is the topologically appropriate structure for carrying the values of a multiply periodic continuous function. In the case $n = 2$, it would not do, for example, to try to represent $f(x, y)$ on the sphere by setting $x = 2\theta - \pi$, $y = \varphi$, where θ and φ are the polar angles, for then the values $f(\pm\pi, y)$ would all be bunched together at the north and south poles.

In general, different methods are needed for proving completeness of different kinds of orthonormal function systems. Systems derived from Sturm–Liouville problems are discussed in Chapter 10.

EXERCISES

1. Show that the functions

$$g_n(x) = n^2 x e^{-nx}$$

converge pointwise to the function $f(x) = 0$ for all $x \geq 0$ but do not converge in $L^2(0, 1)$.

2. Let $\{\xi_n\}_1^{\infty}$ be the numbers

$$\tfrac{1}{2}, \tfrac{1}{4}, \tfrac{3}{4}, \tfrac{1}{8}, \tfrac{3}{8}, \tfrac{5}{8}, \tfrac{7}{8}, \tfrac{1}{16}, \dots,$$

and let $g_n(x)$ be the functions

$$g_n(x) = \begin{cases} 1 - \dfrac{n}{4}|x - \xi_n|, & \text{for } |x - \xi_n| < \dfrac{4}{n}, \\ 0 & \text{otherwise.} \end{cases}$$

Show that the $g_n(x)$ converge to the function $f(x) = 0$ in $L^2(0, 1)$ but not pointwise for any x in $(0, 1)$. *Hint*: Sketch graphs of the functions $g_n(x)$.

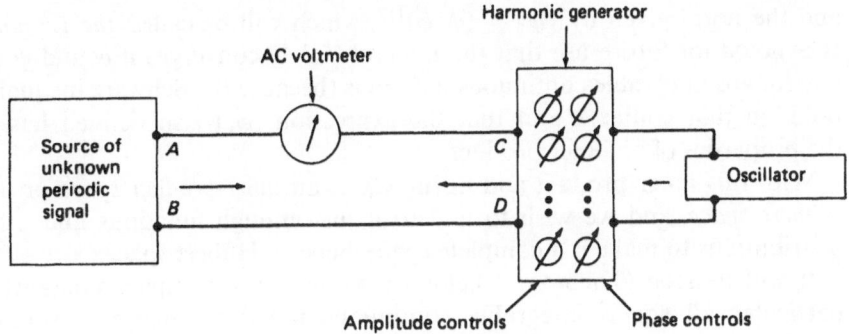

Figure 5.1 Schematic circuit for RMS Fourier analysis.

5.2   A Physical Example of Approximation in the Mean

Consider the idealized electrical circuit shown in Figure 5-1. Between terminals A and B is a periodic voltage $f(t)$ with an unknown wave form, whose period can be taken as 2π by suitable choice of the unit of time. It is to be matched as closely as possible by the generated wave form

$$g(t) = \sum_{k=0}^{n} a_k \cos k(t + \alpha_k) \tag{5.2-1}$$

having the same frequency, by adjusting the controls of the harmonics generator (to adjust the values of the a_k and the α_k) until the voltmeter reading is minimized. (In practice, there would need to be some coupling from the source to the oscillator to provide synchronization.) Since an idealized AC voltmeter reads the root-mean-square of the voltage applied to it (which is here equal to $f(t) - g(t)$), the adjustment is such as to minimize the integral

$$\int |f(t) - g(t)|^2 \, dt,$$

integrated over a period, that is, such as to approximate $f(t)$ in the mean by (5.2-1).

Approximation and convergence in the mean are often the appropriate concepts for physical applications because, as in this example, energy or power is a quadratic expression in certain functions representing the primary variables.

5.3 The Spaces $L^2(\mathbb{R}^n)$ and $L^2(\Omega)$

Let the space of all test functions, $C_0^\infty(\mathbb{R}^n)$, be endowed with the inner product given by

$$(\varphi, \psi) = \int \cdots \int \overline{\varphi(\mathbf{x})} \psi(\mathbf{x}) dx_1 \ldots dx_n \tag{5.3-1}$$

and the norm given by $\|\varphi\| = (\varphi, \varphi)^{1/2}$, which will be called *the L^2 norm*. It is noted for future use that the integral (5.3-1) converges if φ and ψ are any square-integrable continuous functions (because the Schwarz inequality holds in that context), and that the expression (φ, ψ) so defined has all the properties of an inner product.

With this inner product and norm, C_0^∞ is an *inner-product space* or *pre-Hilbert space*, and we wish to add to it just enough functions and other distributions to make it a complete space, hence a Hilbert space.

It will be seen (Theorem 2 below) that the resulting space contains in particular all square-integrable continuous functions, hence is suitable as a Hilbert space of wave functions for quantum mechanics.

Remark on the Completion of Metric Spaces.   The space C_0^∞ with the L^2 norm $\|\cdot\|$ defined in it is an example of a normed linear space, and any normed linear space is an example of a metric space. A metric space has defined in it a distance function or *metric* $d(x, y)$ such that for all points x, y, z, (1) $d(x, y) = d(y, x)$, (2) $d(x, z) \le d(x, y) + d(y, z)$, (3) $d(x, y) > 0$ for $x \ne y$, (4) $d(x, x) = 0$. In a normed linear space we take $d(x, y) = \|x - y\|$.

A sequence $\{x_j\}$ in a metric space is a *Cauchy sequence* if $d(x_j, y_k) \to 0$ as j and $k \to \infty$ independently; it is a *convergent* sequence if there is a point y such that $d(x_j, y) \to 0$ as $j \to \infty$, and y is the *limit* of the sequence. The space is *complete* if every Cauchy sequence converges, i.e., has a limit. A very general theorem says that any metric space S can be completed by adding to it certain so-called ideal elements or points, just as one adds irrational numbers to the rationals to form the real number system. Each ideal point is defined as an equivalence class of Cauchy sequences of points of S, and the result is a complete space S_1 in which S is densely embedded. Furthermore, the completion is unique in the sense that if S_2 is any other complete space in which S is densely embedded, then S_1 and S_2 are isomorphic.

Certain more general topological spaces, the so-called uniform spaces, can be similarly completed. See Thron 1966.

In the present context, it is not necessary to construct ideal elements to be added to C_0^∞ because the necessary objects are already present in the theory—they are distributions. Still, some of the techniques of the standard completion theorem are used in the proof of the theorem below.

[Here, and elsewhere in this chapter, all functions are complex-valued. The corresponding treatment of real-valued functions, which leads to L^2 spaces of real distributions, is evident.]

As in any normed space (whether complete or not), the triangle inequality, in the form $|\|\varphi_k\| - \|\varphi_l\|| \le \|\varphi_k - \varphi_l\|$, shows that if $\{\varphi_k\}$ is a Cauchy sequence, then $\{\|\varphi_k\|\}$ is a Cauchy sequence of numbers and hence has a limit, even though $\{\varphi_k\}$ may not have a limit, in the L^2 norm, in the given space, unless the space is complete. Furthermore, if $\{\varphi_k\}$ is a Cauchy sequence in C_0^∞, then $\lim_{k \to \infty} (\psi, \varphi_k)$ exists for all test functions ψ (regardless of whether the sequence $\{\varphi_n\}$ has a limit in C_0^∞), because

$$|(\psi, \varphi_k) - (\psi, \varphi_l)| = |(\psi, \varphi_k - \varphi_l)| \le \|\psi\| \|\varphi_k - \varphi_l\|,$$

which $\to 0$ as $k, l \to \infty$ for any given ψ; hence $\{(\psi, \varphi_k)\}$ is a Cauchy sequence of numbers. Clearly, the limit is antilinear in ψ, hence determines a distribution f by the equation

$$\langle f, \psi \rangle = \lim(\bar{\psi}, \varphi_k), \quad \text{for all } \psi \text{ in } C_0^\infty. \qquad (5.3\text{-}2)$$

Lemma. *Two Cauchy sequences $\{\varphi_k\}$ and $\{\tilde{\varphi}_k\}$ determine the same distribution if and only if they are equivalent. [The sequences are called equivalent if $\|\varphi_n - \tilde{\varphi}_n\| \to 0$ as $n \to \infty$.]*

PROOF. If the sequences are equivalent, then $|(\psi, \varphi_k) - (\psi, \tilde{\varphi}_k)| \le \|\psi\| \, \|\varphi_k - \tilde{\varphi}_k\| \to 0$ as $k \to \infty$, hence the sequences determine the same distribution. Conversely, if $(\psi, \chi_k) \to 0$ for every ψ, where $\chi_k = \varphi_k - \tilde{\varphi}_k$, then

$$\lim_{k \to \infty} \|\chi_k\|^2 = \lim(\chi_k, \chi_k) = \lim(\chi_k - \psi, \chi_k) \le \lim\|\chi_k - \psi\| \, \|\chi_k\|.$$

Now, the sequence $\{\|\chi_k\|\}$ has a limit, because $\{\chi_k\}$ is a Cauchy sequence; $\{\chi_k - \psi\}$ is also a Cauchy sequence, so $\{\|\chi_k - \psi\|\}$ also has a limit. Therefore

$$(\lim|\chi_k\|)^2 \le (\lim\|\chi_k - \psi\|)(\lim\|\chi_k\|);$$

hence

$$\lim\|\chi_k\| \le \lim\|\chi_k - \psi\|,$$

but this can be made arbitrarily small by taking $\psi = \chi_{k_0}$, where k_0 is large enough. Therefore $\chi_k \to 0$ in norm, and the two given sequences are equivalent.

Definition. The space $L^2(\mathbb{R}^n)$ is the set of all distributions determined by Cauchy sequences C_0^∞, with the inner product in $L^2(\mathbb{R}^n)$ defined by

$$(f, g) = \lim_{k \to \infty} (\varphi_k, \psi_k), \qquad (5.3\text{-}3)$$

where $\{\varphi_k\}$ and $\{\psi_k\}$ are Cauchy sequences that determine f and g, respectively. From the continuity of the inner product (Schwarz inequality) it is evident that the limit always exists and is unaltered if either of the Cauchy sequences is replaced by an equivalent one. From (5.3-3) it follows that

$$\|f\| = \lim_{k \to \infty} \|\varphi_k\|. \qquad (5.3\text{-}4)$$

It is evident that (f, g) as defined, is linear in g, Hermitian symmetric, and positive definite. That is, $L^2(\mathbb{R}^n)$ is an inner product space. Since the elements of this space are distributions, they have integrals, derivatives, and local properties, as described in Chapters 2 and 3.

Clearly, if f is in L^2 and φ is a test function, then $\langle f, \varphi \rangle = (\bar{f}, \varphi)$. The continuity of the inner product then shows that if $f_j \to f$ in the L^2 norm, as $j \to \infty$, then $f_j \to f$ in the sense of convergence of distributions.

As is intuitively evident, the elements of $L^2(\mathbb{R}^n)$ are tempered distributions, because their growth at infinity is restricted, in a sense, by quadratic integrability. To prove that, we note first that for f in L^2 the definition of $\langle f, \varphi \rangle$ can be extended to all φ in $\mathscr{S}$ by merely defining it to be $(\bar{f}, \varphi)$. Hence, we must show that if φ_k converges to ψ in the sense of convergence in $\mathscr{S}$, then $(f, \varphi_k) \to (f, \psi)$, but that is evident, because

$$\sup |\mathbf{x}|^l |\varphi_k(\mathbf{x}) - \psi(\mathbf{x})| \to 0$$

as $k \to \infty$, for any l, hence

$$\int |\varphi_k(\mathbf{x}) - \psi(\mathbf{x})|^2 \, d^n\mathbf{x} \to 0,$$

i.e., φ_k converges to ψ in L^2; consequently, $(f, \varphi_k) \to (f, \psi)$ because of the continuity of the inner product in L^2. That is, f is a tempered distribution.

Remarks. If f and g happen both to be elements of the original space C_0^∞ then the new inner product defined by (5.3-3) is the same as the old inner product in C_0^∞, because then the sequences $\{\varphi_k\}$ and $\{\psi_k\}$ can be taken as $\{f, f, f, \ldots\}$ and $\{g, g, g, \ldots\}$, whereupon the quantity (φ_k, ψ_k) in (5.3-3) is equal to the old inner product for all k. A similar remark applies to the new norm, given by (5.3-4).

Theorem 1.   *The space* $L^2 = L^2(\mathbb{R}^n)$ *is complete (i.e., is a Hilbert space).*

Remark. If f is determined by the Cauchy sequence $\{\varphi_k\}$, then, for any m, $f - \varphi_m$ is determined by the Cauchy sequence $\{\varphi_k - \varphi_m : k = 1, 2, \ldots\}$. By (5.3-4), then,

$$\|f - \varphi_m\| \to 0, \qquad \text{as } m \to \infty;$$

i.e., the Cauchy sequence $\{\varphi_k\}$ has a limit in L^2, namely f. What has to be proved is that *any* Cauchy sequence $\{f_k\}$ in L^2, not necessarily in C_0^∞, also has a limit in L^2.

PROOF. Suppose that $\{f_k\}$ is a Cauchy sequence of elements of L^2. An element g of L^2 will be found (in fact constructed) such that $\|f_k - g\| \to 0$, as $k \to \infty$. First let

$$\delta_j \overset{\text{def}}{=} \sup_{k > j} \|f_k - f_j\| \qquad (j = 1, 2, \ldots), \tag{5.3-5}$$

and note that $\delta_j \to 0$ as $j \to \infty$. Each f_j is a distribution determined by a Cauchy sequence $\{\varphi_{j,l} : l = 1, 2, \ldots\}$ of functions in C_0^∞. For each j, let $N(j)$ be such that

$$\|\varphi_{j\,l} - \varphi_{j\,m}\| < \delta_j, \quad \text{if } l, m \geq N(j), \tag{5.3-6}$$

and furthermore let the $N(j)$ be chosen so that $N(j + 1)$ is always $\geq N(j)$. From (5.3-6) it follows in the limit $m \to \infty$ that

$$\|\varphi_{j\,l} - f_j\| < \delta_j, \quad \text{if } l \geq N(j), \tag{5.3-7}$$

because $\varphi_{j\,m} \to f_j$ by the above Remark. It will now be shown that if ψ_j is defined to be $\varphi_{j\,N(j)}$, then $\{\psi_j\}$ is a Cauchy sequence (of elements of C_0^∞) that determines the desired element g of L^2: For $k > j$,

$$\|\psi_k - \psi_j\| \leq \|\varphi_{k\,N(k)} - \varphi_{j\,N(k)}\| + \|\varphi_{j\,N(k)} - \varphi_{j\,N(j)}\|; \tag{5.3-8}$$

the last term is $< \delta_j$ by (5.3-6), since $N(k) \geq N(j)$. The other term on the right of (5.3-8) is less than or equal to

$$\|\varphi_{k\,N(k)} - f_k\| + \|\varphi_{j\,N(k)} - f_j\| + \|f_j - f_k\|,$$

which is less than $\delta_k + \delta_j + \delta_j$, by (5.3-7), used twice, and (5.3-5), used once. Therefore,

$$\|\psi_k - \psi_j\| \leq 3\delta_j + \delta_k \to 0 \quad \text{as } j, k \to \infty,$$

so that $\{\psi_k\}$ is a Cauchy sequence in C_0^∞. If g is the element of L^2 determined by $\{\psi_k\}$, then

$$\|f_k - g\| \leq \|f_k - \varphi_{kN(k)}\| + \|\psi_k - g\|$$

(because $\varphi_{k\,N(k)} = \psi_k$); as $k \to \infty$, the first term on the right tends to zero by (5.3-7) and the second term tends to zero by the Remark above. Therefore, $f_k \to g$, as was to be proved; hence, $L^2(\mathbb{R}^n)$ is complete.

Theorem 2. *Any square-integrable continuous function over $\mathbb{R}^n$ is in $L^2(\mathbb{R}^n)$.*

PROOF. Suppose that $f(x)$ is continuous and that

$$\int_{\mathbb{R}^n} |f(x)|^2 \, d^n x < \infty. \tag{5.3-9}$$

We denote the left member of this inequality as usual by $\|f\|^2$. According to the Remark following Theorem 1 and the definition of L^2, a distribution f is in L^2 if it is determined by a Cauchy sequence $\{\varphi_k\}$ of C_0^∞ functions, and then $\varphi_k \to f$ in the L^2 norm. Hence, to show that $f(x)$ is in L^2, we must find a sequence such that $\|f - \varphi_k\| \to 0$ as $k \to \infty$ (see Remark 2, below), that is, we must show that for any $\varepsilon > 0$ there is a φ in C_0^∞ such that

$$\|f - \varphi\| < \varepsilon. \tag{5.3-10}$$

Let $f_R(x)$ be the function $= f(x)$ for $|x| < R$ and $= 0$ for $|x| > R$. If R is large enough, $\|f - f_R\| < \frac{1}{2}\varepsilon$, because $\|f - f_R\|^2$ is the contribution of the region $|x| > R$ to the convergent integral (5.3-9). Then let $f_{R,\delta}$ be the result of smoothing f_R with a mollifier J_δ of width δ, as described in Section 2.6. Since $f(x)$ is continuous in the compact region $|x| \leq R$, the quantity

$$M_\delta = \sup\{|f(x) - f(x + y)| : |x| \leq R - \delta, |y| \leq \delta\}$$

tends to zero as $\delta \to 0$. Therefore,

$$\begin{aligned} &\text{for } |x| < R - \delta, \quad &|f_{R\,\delta}(x) - f_R(x)| \leq M_\delta; \\ &\text{for } |x| > R + \delta, \quad &f_{R\,\delta}(x) = f_R(x) = 0. \end{aligned}$$

Since also both $f_{R\,\delta}$ and f_R are bounded in the thin spherical shell $R - \delta < |x| < R + \delta$, it is clear that $\|f_{R\,\delta} - f_R\|$ can be made $< \frac{1}{2}\varepsilon$ by taking δ small enough. Then (5.3-10) is satisfied by taking

$$\varphi(x) = f_{R\,\delta}(x),$$

and we see that f is in L^2, as stated.

Remark 1. If f and g are square-integrable functions, their inner product, as elements of the space L^2, has been defined in terms of approximating test functions by equation (5.3-3), but it is clearly also given by the familiar expression

$$(f, g) = \int_{\mathbb{R}^n} \overline{f(\mathbf{x})} g(\mathbf{x}) d^n \mathbf{x},$$

because it follows from (5.3-1) and the Schwarz inequality for functions that (φ_k, ψ_k) converges to the above integral as $k \to \infty$.

Remark 2. If $\{\varphi_k\}$ is a sequence of continuous functions that converge in L^2 to a continuous function f, then $\{\varphi_k\}$ is a Cauchy sequence, because the Schwarz and triangle inequalities hold for integrals of continuous functions over $\mathbb{R}^n$, hence $\|\varphi_k - \varphi_l\| \leq \|\varphi_k - f\| + \|\varphi_l - f\|$.

If $C^{(2)}$ denotes the space of continuous square integrable functions, what has been shown is that

$$C_0^\infty \subset C^{(2)} \subset L^2;$$

both inclusions are dense because each f in L^2 can be arbitrarily closely approximated (in L^2) by a test function.

Remark 3. It is easy to see that if the elements of a Cauchy sequence $\{\varphi_k\}$ that determine a distribution f in L^2 had only been assumed to be in $\mathscr{S}$, not necessarily in C_0^∞, the same space L^2 would have resulted. More generally, the φ_k could have been taken as any square-integrable continuous functions. Hence, L^2 can be regarded as the completion, with respect to the L^2 norm, of any of the spaces C_0^∞, $\mathscr{S}$, $C^{(2)}$, i.e., as the smallest complete space that contains C_0^∞, $\mathscr{S}$, or $C^{(2)}$.

Remark 4. If square-integrable functions $f_k(\mathbf{x})$ converge in L^2 as $k \to \infty$ and also converge pointwise uniformly, then the two limits are the same; i.e., the limit in L^2 is the distribution that we identify with the continuous function $\lim f_k(\mathbf{x})$.

[Other functions are also in L^2, such as the function $|x|^{-1/4} e^{-|x|}$, in one dimension. When the Lebesgue integral is used, as in many books, then all measurable and Lesbesgue-square-integrable functions are included, but the one-to-one relation between functions and elements of L^2 is lost, and each element of L^2 is an infinite equivalence class of such functions, any two of which, in a given class, differ arbitrarily on an arbitrary set of measure zero. Nevertheless, the elements of L^2 are usually called simply "functions," although, from our point of view, they are really distributions.]

Functions or distributions that are quadratically integrable over a region Ω of $\mathbb{R}^n$ appear in many applications. Let $C_0^\infty(\Omega)$ be the set of C^∞ functions $\varphi(\mathbf{x})$ with support contained in Ω. Then a *distribution on Ω* is defined as a linear functional on $C_0^\infty(\Omega)$ that is continuous in the sense of Section 2.4.

The theory of the space $L^2(\Omega)$ is the same as that of $L^2(\mathbb{R}^n)$ except for replacement of $\mathbb{R}^n$ by Ω throughout. In particular, the inner product (φ, ψ) of two functions in $C_0^\infty(\Omega)$ is given by (5.3-1) as before; the integral is formally over all $\mathbb{R}^n$, but of course the integrand is zero outside Ω.

An element u of $L^2(\Omega)$ can also be regarded as an element u_0 of $L^2(\mathbb{R}^n)$. Namely, if u is determined by a Cauchy sequence $\{\varphi_k\}$ of functions in $C_0^\infty(\Omega)$, then u_0 is given by

$$\langle u_0, \psi \rangle = \lim_{(k \to \infty)} \int \varphi_k \psi \, d^n\mathbf{x}, \tag{5.3-11}$$

for all ψ in $C_0^\infty(\mathbb{R}^n)$, not merely in $C_0^\infty(\Omega)$. The norm of u in $L^2(\Omega)$ is equal to the norm of u_0 in $L^2(\mathbb{R}^n)$. If supp ψ lies entirely outside of Ω, then (5.3-11) gives zero, hence supp $u_0 \subset \tilde{\Omega}$. It can be shown conversely, at least if Ω is a region with a reasonable boundary, that if u_0 is in $L^2(\mathbb{R}^n)$ and has its support in $\tilde{\Omega}$, then it can be regarded as an element u in $L^2(\Omega)$; that is, there is a sequence $\{\varphi_k\}$ of functions in $C_0^\infty(\Omega)$ such that (5.3-11) holds for all ψ.

More generally, any distribution u_0 on $\mathbb{R}^n$ has a unique *restriction* u on Ω obtained by considering only test functions in $C_0^\infty(\Omega)$. If furthermore u_0 is in $L^2(\mathbb{R}^n)$, then u is in $L^2(\Omega)$, and $\|u\| \leq \|u_0\|$. What was shown above is that the restriction mapping, although usually many-to-one, is one-to-one for elements of L^2, and in fact is an isomorphism when applied to the subspace of $L^2(\mathbb{R}^n)$ consisting of distributions with supports in $\tilde{\Omega}$.

It is necessary to distinguish between u and u_0 in connection with differentiation. If u is in $L^2(\Omega)$ and D denotes ∂/∂_k, then Du is the distribution on Ω given by

$$\langle Du, \varphi \rangle = -\langle u, D\varphi \rangle \qquad \forall \varphi \in C_0^\infty(\Omega).$$

It may then happen that Du is in $L^2(\Omega)$, while Du_0 is not in $L^2(\mathbb{R}^n)$ because of δ-function-like contributions concentrated on the boundary of Ω; that happens, for example if u_0 is a nonzero constant in Ω and zero outside. If Du is extended, as described above, to a distribution $(Du)_0$ in $L^2(\mathbb{R}^n)$ with support in $\tilde{\Omega}$, then Du_0 and $(Du)_0$ are equal in Ω and also in $\mathbb{R}^n - \tilde{\Omega}$ (where they are both zero), but not generally in an open set that intersects the boundary $\partial\Omega$. [It should be noticed in this connection that if $\{\varphi_k\}$ is a Cauchy sequence that represents u, then $\{D\varphi_k\}$ represents Du in the sense of (5.3-11) but is not in general a Cauchy sequence.]

The simplest space of this kind is $L^2(a, b)$, where $n = 1$, and where Ω is an open interval (a, b) on $\mathbb{R}$.

Any function $f(\mathbf{x})$ that is continuous in Ω and is such that

$$\int_\Omega \cdots \int |f(\mathbf{x})|^2 \, dx_1 \ldots dx_n < \infty \tag{5.3-12}$$

determines a distribution u in $L^2(\Omega)$ by means of the usual formula

$$\langle u, \varphi \rangle = \int_{\mathbb{R}^n} \cdots \int f(\mathbf{x})\varphi(\mathbf{x})dx_1 \ldots dx_n, \qquad \forall \varphi \in C_0^\infty(\Omega).$$

In most cases (e.g., if Ω is an interval, a cube, or a sphere), it is evident what is meant by the (Riemann) integral (5.3-12) over Ω; otherwise, it can be taken as the supremum

$$\sup \int_{\mathbb{R}^n} \cdots \int \psi(\mathbf{x})|f(\mathbf{x})|^2 \, dx_1 \ldots dx_n$$

for all continuous functions $\psi(\mathbf{x})$ with support in Ω and with values in $[0, 1]$. If Ω is bounded, (5.3-12) is satisfied by *any* function $f(\mathbf{x})$ continuous in $\bar{\Omega}$.

If S is any reasonable m-dimensional surface (e.g. a sphere, cylinder, or torus) in $\mathbb{R}^n$, a space $L^2(S)$ can be defined in an obvious way. For example, let S be the 2-sphere, i.e. the surface $x^2 + y^2 + z^2 = 1$ in $\mathbb{R}^3$. A function f on S is said to be *of class* $C^\infty(S)$, if, in a neighborhood of any point of S where $z \neq 0$ (i.e. any point not on the equator), f is a C^∞ function of x and y, and it is a C^∞ function of x and z near any point where $y \neq 0$ and of y and z near any point where $x \neq 0$. Equivalently, f is a C^∞ function of ϑ and φ, for $0 < \vartheta < \pi$, $-\pi < \varphi < \pi$, in both of two polar coordinate systems so arranged that the singularities of each are covered by the nonsingular parts of the other. A *distribution* on S is a linear functional on $C^\infty(S)$; for two functions in $C^\infty(S)$, the inner product is

$$(\chi, \psi) = \int_{-\pi}^{\pi} \int_0^\pi \overline{\chi(\vartheta, \varphi)}\psi(\vartheta, \varphi)\sin \vartheta \, d\vartheta \, d\varphi,$$

and from here on the discussion is the same as for the L^2 spaces described above.

L^2 spaces of this kind appear in the theory of group representations; S is the group manifold or a so-called homogeneous space (see Volume 2). It will be evident that $L^2(\mathfrak{M})$ can be similarly defined, if $\mathfrak{M}$ is any manifold, as there defined.

If Ω is a region with a sufficiently smooth boundary $\partial\Omega$, the boundary values of a distribution f in $L^2(\Omega)$ can sometimes be thought of as distributions in $L^2(\partial\Omega)$.

L^2 spaces of periodic distributions can be defined in an obvious way. If a function $f(\mathbf{x}) = f(x_1, \ldots, x_n)$ is such that $f(\mathbf{x} + \mathbf{p}) = f(\mathbf{x})$, for all $\mathbf{x}$, where $\mathbf{p} = (p_1, \ldots, p_2)$ is a fixed vector $\neq 0$, then $f(\mathbf{x})$ is *periodic*, and $\mathbf{p}$ is a *period* of $f(\mathbf{x})$. We consider functions and distributions having periods of the form $\mathbf{p} = (0, \ldots, 0, 2\pi, 0, \ldots, 0)$. That is, it is assumed that, for each j,

$$\begin{aligned} &f(x_1, \ldots, x_{j-1}, x_j + 2\pi, x_{j+1}, \ldots, x_n) \\ = &f(x_1, \ldots, x_{j-1}, x_j, x_{j+1}, \ldots, x_n) \qquad (j = 1, \ldots, n). \end{aligned} \qquad (5.3\text{-}13)$$

It follows that if $\mathbf{p}$ is any vector all of whose components are integer multiples of 2π, then $f(\mathbf{x} + \mathbf{p}) = f(\mathbf{x})$.

Let C_{per}^∞ be the space of C^∞ functions that are periodic as in (5.3-13), endowed with an inner produce defined by

$$(\varphi, \psi) = \int_0^{2\pi} \cdots \int_0^{2\pi} \overline{\varphi(\mathbf{x})}\psi(\mathbf{x})dx_1 \ldots dx_n.$$

The corresponding complete space, which will be denoted by $L^2(\mathbb{R}^n, 2\pi)$ is defined as follows. If $\{\varphi_k\}$ is a Cauchy sequence in C_{per}^∞, then a distribution $\langle f, \cdot \rangle$ is defined by

$$\langle f, \psi \rangle = \lim_{k \to \infty} \int_{\mathbb{R}^n} \cdots \int \varphi_k(\mathbf{x})\psi(\mathbf{x})dx_1 \ldots dx_n, \qquad \text{for any } \psi \text{ in } C_0^\infty.$$

The set of all such distributions is the space $L^2(\mathbb{R}^n, 2\pi)$, with the inner product given by (5.3-3).

5.4 Multiplication in L^2 Spaces

If f and g are elements of L^2 and are the limits of the Cauchy sequences $\{\varphi_k\}$ and $\{\psi_k\}$ in C_0^∞, the product fg is defined as a distribution (*not* generally in L^2), as follows: Given any test function $\chi(\mathbf{x})$, the sequence $\{\chi\varphi_k\}$ is also a Cauchy sequence, hence $(\bar{\chi}\bar{\varphi}_k, \psi_k)$ has a limit as $k \to \infty$, and the functional $\langle fg, \cdot \rangle$ is defined as

$$\langle fg, \chi \rangle \overset{\text{def}}{=} \lim_{k \to \infty} (\bar{\chi}\bar{\varphi}_k, \psi_k) = \lim_{k \to \infty} \int \varphi_k \psi_k \chi \, d^n\mathbf{x} \qquad (5.4\text{-}1)$$

This defines the product fg. It will be seen in the next section that the integral of $\bar{f}g$ over all space is equal to the inner product (f, g). Clearly, if f and g are ordinary functions, fg is their ordinary product.

On the other hand, if $h = h(\mathbf{x})$ is any bounded continuous function, and $\varphi_k \to f$ as above, then the sequence $\{h(\mathbf{x})\varphi_k(\mathbf{x})\}$ is a Cauchy sequence in L^2, whose limit is defined as the distribution hf, which *is* an element of L^2.

If $h(\mathbf{x})$ is a function in C^m ($m = $ integer ≥ 0), and f is in $L^2(\mathbb{R})$, then a distribution $hf^{(m)}$, not necessarily in L^2, is defined by the equation

$$\langle hf^{(m)}, \psi \rangle \overset{\text{def}}{=} (-1)^m ((\bar{\psi}h)^{(m)}, f), \qquad \forall \psi \text{ in } C_0^\infty; \qquad (5.4\text{-}2)$$

each term of the expansion of $(\bar{\psi}h)^{(m)}$ is clearly in L^2, and the inner product on the right is therefore a linear functional defined for all test functions ψ, i.e., is a distribution. This product is useful in connection with the domain of a differential operator

$$Af = \sum_{(m)} h_m f^{(m)},$$

where the coefficient functions $h_m(x)$ are not assumed to be C^∞, but only differentiable as many times as needed. Namely, attention is restricted to distributions f in L^2 such that Af, as a distribution, is also in L^2.

Lastly, let $h(\mathbf{x})$ be continuous, but not necessarily bounded. If $\psi(\mathbf{x})$ is a test function, then $\psi(\mathbf{x})h(\mathbf{x})$ is in L^2; hence, for f in L^2, $(\bar{\psi}h, f)$ exists, and a distribution hf is defined as the functional

$$\langle hf, \psi \rangle = (\bar{\psi}h, f), \quad \text{for all } \psi \text{ in } C_0^\infty(\mathbb{R}^n).$$

Summary. 1. The product of two distributions in $L^2(\mathbb{R}^n)$ is a distribution, not generally in L^2 (it is actually in L^1—see Section 5.7). 2. The product of a distribution in $L^2(\mathbb{R}^n)$ by an arbitrary continuous function h is a distribution; it is in L^2 if h is a bounded function. 3. In the case of one variable, the product of the mth derivative of any distribution in $L^2(\mathbb{R})$ and any function in C^m is a distribution. [The first two of these statements, when rephrased in classical terms, are as follows:

1. The product of two functions in L^2 (two measurable and quadratically integrable functions) is a measurable function whose absolute value is integrable.
2. The product of a function in L^2 by a bounded continuous function is in L^2. Alteration of the factors on sets of measure zero causes the product to be altered only on a set of measure zero, hence the product is in a uniquely determined equivalence class (element of L^1 or L^2). The third statement obviously has no classical analogue, because it involves the distribution derivative.]

EXERCISE

1. Show that any distribution f in $L^2(\mathbb{R}^n)$ can be arbitrarily closely approximated in the L^2 norm by a distribution ψf, $\psi \in C_0^\infty$, i.e. by a distribution having bounded support.

5.5 Integration in L^2 Spaces; Definite Integrals

The case of one independent variable is considered first. It will be shown that for any f and g in L^2, the indefinite integrals $\int^x f\, dx$ and $\int^x fg\, dx$ are continuous functions. The Schwarz inequality and the usual formula for integration by parts will follow.

Let $\{\varphi_k\}$ be a sequence in $C_0^\infty(\mathbb{R})$ converging (in the L^2 norm) to an element f in $L^2(\mathbb{R})$. Since, for functions, the Schwarz inequality holds in any interval,

$$\left| \int_0^x [\varphi_k(y) - \varphi_l(y)] \cdot 1 \, dy \right|^2 \le \int_0^x |\varphi_k(y) - \varphi_l(y)|^2 \, dy \int_0^x 1 \, dy$$

$$\le \|\varphi_k - \varphi_l\|^2 |x| \qquad (5.5\text{-}1)$$

(note that if $x < 0$ the sign has to be changed in both factors in the second member), and it follows that $\int_0^x \varphi_k(y) dy$ converges uniformly in any finite interval, hence coverages to a continuous function $F(x)$. The indefinite integral $\int f(x) dx$ of a distribution was defined in Section 2.7 as a distribution whose derivative is equal to f. Now,

$$-\langle f, \psi \rangle = -\lim \langle \varphi_k, \psi \rangle = \lim \left\langle \int \varphi_k(x) dx, \psi' \right\rangle$$

$$= \lim \int_{-\infty}^\infty \int_0^x \varphi_k(y) dy \, \psi'(x) dx \qquad (5.5\text{-}2)$$

$$= \int_{-\infty}^\infty F(x) \psi'(x) dx = \langle F, \psi' \rangle;$$

In the next to last step above, it is permissible to take the limit under the first integral sign, because of the uniform convergence of the inner integral to $F(x)$ on the support of ψ. Hence, by Section 2.7,

$$\int f(x)dx = F(x) + \text{const.} \tag{5.5-3}$$

We have thus proved the first part of the following:

Theorem. *If f is a distribution in $L^2(\mathbb{R})$, its indefinite integral $\int^x f\,dx$ is a continuous function. If $\{f_n\}$ is a sequence in L^2 that converges in L^2 to a distribution f in L^2, then*

$$\int^x f_n\,dx \rightarrow \int^x f\,dx \qquad (n \rightarrow \infty)$$

in the sense of pointwise convergence uniform in any finite interval of x.

The proof of the second part is left to the reader.

In more dimensions, certain higher-order partial derivatives of a distribution must also be in $L^2(\mathbb{R}^n)$ for it to be continuous. See Section 5.13.

Similarly, if f and g are the limits of $\{\varphi_n\}$ and $\{\psi_n\}$, then

$$\left| \int_{-\infty}^x [\varphi_k(y)\psi_k(y) - \varphi_l(y)\psi_l(y)]dy \right|$$

$$= \left| \int_{-\infty}^x [\varphi_k\psi_k - \varphi_k\psi_l + \varphi_k\psi_l - \varphi_l\psi_l]dy \right|$$

$$\leq \left\{ \int_{-\infty}^x |\varphi_k|^2\,dy \int_{-\infty}^x |\psi_k - \psi_l|^2\,dy \right\}^{1/2} \tag{5.5-4}$$

$$+ \left\{ \int_{-\infty}^x |\varphi_k - \varphi_l|^2\,dy \int_{-\infty}^x |\psi_l|^2\,dy \right\}^{1/2} \rightarrow 0 \quad \text{as } k, l \rightarrow \infty.$$

Hence, $\int_{-\infty}^x \varphi_k(y)\psi_k(y)dy$ converges uniformly on $\mathbb{R}$, hence converges to a continuous function $G(x)$. By the same reasoning as in (5.5-2), it is seen that

$$\int f(x)g(x)dx = G(x) + \text{const.} \tag{5.5-5}$$

For distributions in general only indefinite integrals or primitives are defined, but for distributions in L^2 spaces (also L^p) certain definite integrals can also be defined. For distributions in $L^2(\mathbb{R})$, since the indefinite integrals $\int^x f\,dx$ and $\int^x fg\,dx$ are continuous functions, say $F(x)$ and $H(x)$, we can define

$$\int_a^b f\,dx = F(b) - F(a)$$

$$\int_a^b fg\,dx = H(b) - H(a).$$

Some multidimensional cases are discussed below.

1 Show that the Schwarz inequality

$$\left| \int_a^b f(x)g(x)dx \right|^2 \le \int_a^b |f(x)|^2 \, dx \int_a^b |g(x)|^2 \, dx \tag{5.5-6}$$

holds for arbitrary elements f and g of L^2 (if $f = \lim \varphi_k$, then $\bar{f}$ is equal to $\lim \bar{\varphi}_k$, and $|f|^2 = f\bar{f}$), and also that

$$(f, g) = \int_{-\infty}^{\infty} \overline{f(x)}g(x)dx. \tag{5.5-7}$$

2. Show that if the distributions f, g, and their first derivatives are all in $L^2(\mathbb{R})$, then the formula for integration by parts holds:

$$\int_a^b (f'g + fg')dx = f(x)g(x)\Big|_a^b. \tag{5.5-8}$$

[Note first that all terms here are well-defined; in particular, the right member is, because f and g are ordinary continuous functions, since f' and g' are in L^2.]

Since $\int_a^b \varphi_k \psi_k \, dx$ converges to $\int_a^b fg \, dx$, and hence $\int_a^b |\varphi_k|^2 \, dx$ converges to $\int_a^b |f|^2 \, dx$, it is seen that any distribution $f(x)$ in $L^2(\mathbb{R})$ is a fortiori in $L^2(a, b)$ for any interval (a, b). That is, the restriction of $f(x)$ to the interval (a, b), namely the linear functional $(\bar{\psi}, f)$ defined for all test functions ψ with support in (a, b), is a distribution in $L^2(a, b)$.

Note that if f and g are in $L^2(a, b)$, where (a, b) is a finite interval, then the functions $\int f \, dx$ and $\int fg \, dx$ are continuous in the *closed* interval $[a, b]$.

For quantum mechanics, where the elements f, g of $L^2(\mathbb{R}^n)$ are wave functions, and $\mathbb{R}^n$ is the configuration space ($n = 3 \times$ no. of particles), it is of interest to consider integrals of the form $\int_\Omega |f(x)|^2 \, dx_1 \ldots dx_n$ or, more generally, $\int_\Omega g(x)f(x)dx_1 \ldots dx_n$, where Ω is any open set in $\mathbb{R}^n$. The first integral is the probability that the point x representing the configuration of the system lies in Ω (assuming the wave function to be normalized so that the integral is $=1$ when Ω is all of $\mathbb{R}^n$). If f and g are the limits of sequences $\{\varphi_k\}$ and $\{\psi_k\}$ in $C_0^\infty(\mathbb{R}^n)$, then (5.5-4) still holds, with the single integration over the interval $(-\infty, x)$ of $\mathbb{R}$ replaced by volume integration over the region Ω of $\mathbb{R}^n$. It follows that

$$\int_\Omega \psi_k(y)\varphi_k(y)dy_1 \ldots dy_n \tag{5.5-10}$$

has a limit, as $k \to \infty$, for any Ω; the limit is denoted by

$$\int_\Omega g(y)f(y)dy_1 \ldots dy_n. \tag{5.5-11}$$

If Ω is the rectangular cell given by $0 < y_i < x_i$ ($i = 1, \ldots, n$), then the convergence of (5.5-10) to (5.5-11) is uniform with respect to x in any bounded region of the quadrant $x_i > 0$ ($i = 1, \ldots, n$), hence the limit is a continuous function, say $G(\mathbf{x})$. For distributions, differentiation and integration are always inverse processes, hence

$$g(\mathbf{x})f(\mathbf{x}) = \frac{\partial^n}{\partial x_1 \dots \partial x_n} G(\mathbf{x}); \qquad (5.5\text{-}12)$$

this is an extension of the one-dimensional result (5.5-5), which can be written as $g(x)f(x) = dG(x)/dx$. The extension of (5.5-12) from the positive quadrant to the other quadrants of $\mathbb{R}^n$ is obtained by obvious redefinitions of the rectangular cell and corresponding changes in sign.

If $g(\mathbf{x}) = \overline{f(\mathbf{x})}$, it is easily seen that $G(\mathbf{x})$ is nondecreasing, in the sense of probability theory (see Section 13.3), hence $\int_\Omega |f(\mathbf{x})|^2 \, dx_1 \dots dx_n$ can be interpreted as a probability. Alternatively, the integral of $|f(\mathbf{x})|^2$ has the obvious properties

$$\int_{\Omega_1 \cup \Omega_2} = \int_{\Omega_1} + \int_{\Omega_2} \quad \text{if } \Omega_1 \text{ and } \Omega_2 \text{ are disjoint,} \qquad (5.5\text{-}13)$$

$$\int_{\Omega_1} \leq \int_{\Omega_2} \quad \text{if } \Omega_1 \subset \Omega_2, \qquad (5.5\text{-}14)$$

which are needed for the probability interpretation.

If the distributions f and g are continuous functions and Ω is bounded, (5.5-11) is an ordinary Riemann integral; if Ω is unbounded, it is an improper Riemann integral, but is convergent for f and g in L^2. In any case, by generalization of Exercise 1,

$$(f, g) = \int_{\mathbb{R}^n} \overline{f(\mathbf{x})} g(\mathbf{x}) dx_1 \dots dx_n \qquad (5.5\text{-}15)$$

for distributions in $L^2(\mathbb{R}^n)$, and

$$(f, g) = \int_{\Omega} \overline{f(\mathbf{x})} g(\mathbf{x}) dx_1 \dots dx_n \qquad (5.5\text{-}16)$$

for ones in $L^2(\Omega)$.

The following is a simple corollary of the nonnegativity of $\int_\Omega |f|^2 \, d^n\mathbf{x}$:

Lemma. *If f is in L^2, and if h_1 and h_2 are bounded continuous functions such that $|h_1(\mathbf{x})| \leq |h_2(\mathbf{x})|$ for all $\mathbf{x}$, then*

$$\|h_1 f\| \leq \|h_2 f\|,$$

where the norm is that of $L^2(\Omega)$ or $L^2(\mathbb{R}^n)$.

5.6 On Vanishing at Infinity I

One sometimes hears it said that a quadratically integrable function $f(x)$ must tend to zero as $x \to \pm\infty$. That that is not true is shown by the examples $f(x) = \exp\{-x^4 \sin^2 x\}$ and $f(x) = x^2 \exp\{-x^8 \sin^2 x\}$, the second of

which is even unbounded. However, *if f and its derivative f' are both in*
$L^2(\mathbb{R})$ (in which case, f is a continuous function), *then* $f(x) \to 0$ as $x \to \pm\infty$.

PROOF.   According to (5.5-6), the Schwarz inequality holds for distributions in L^2;
hence,

$$\left| \int_a^b f(x)f'(x)dx \right|^2 \leq \int_a^b |f(x)|^2 \, dx \int_a^b |f'(x)|^2 \, dx.$$

If *a* and *b* both tend independently to $+\infty$ (note sign), then the right member tends to
zero, because the corresponding integrals over all R converge. According to (5.5-8),
the left member is

$$\tfrac{1}{2}|f(b)^2 - f(a)^2|;$$

therefore $f(x)^2$ converges to a constant, as $x \to +\infty$, but the constant must be zero,
for otherwise f would not be in $L^2(\mathbb{R})$; the limit as $x \to -\infty$ is also zero, by analogous
reasoning.

Vanishing at ∞ for functions in $L^2(\mathbb{R}^n)$ is discussed in Section 5.13; higher derivatives
are also required to be in L^2.

EXERCISE

1.   Find a function $f(x, y, z)$ such that f, $\partial f/\partial x$, $\partial f/\partial y$, and $\partial f/\partial z$ are in $L^2(\mathbb{R}^3)$,
but $f \nrightarrow 0$ as $|\mathbf{x}| \to \infty$.

5.7 Spaces of Type L^1, L^p, L^∞

Although the case $p = 2$ is by far the most important in physics, it seems
desirable to summarize the properties of general L^p spaces. Let p be any
real number in the interval $1 \leq p < \infty$. The linear space consisting of
elements of C_0^∞ with the norm given by

$$\|\varphi(\cdot)\|_p = \left(\int_{-\infty}^\infty |\varphi(x)|^p \, dx \right)^{1/p} \tag{5.7-1}$$

is an incomplete normed space which can be made complete by extending it
to include certain rather mild distributions, which are again called simply
"functions" in most of the literature. Except for $p = 2$, this norm cannot be
obtained from an inner product, because the parallelogram law (1.2-5)
does not hold.

The main results on L^p spaces will be stated without proof. Most of them
can be found in Riesz and Sz. Nagy (1952), where they are expressed in terms
of Legesgue theory. These results and others have been obtained recently by
distribution theory methods by P. Werner (1969).

For p in the interval $1 < p < \infty$, a second real number q in the same
interval is defined by the equation

$$\frac{1}{p} + \frac{1}{q} = 1; \tag{5.7-2}$$

the analogue of the inner product involves one element of L^p and one element of L^q. First, the *Hölder inequality*

$$\left| \int_{-\infty}^{\infty} \varphi(x)\psi(x)dx \right| \leq \|\varphi\|_p \|\psi\|_q, \qquad (5.7\text{-}3)$$

where φ and ψ are in C_0^∞, generalizes the Schwarz inequality. From it can be derived the triangle inequality

$$\|\varphi + \psi\|_p \leq \|\varphi\|_p + \|\psi\|_p, \qquad (5.7\text{-}4)$$

which is sometimes called the *Minkowski* inequality when applied to functions.

A direct consequence of the Hölder inequality is that if $\{\varphi_i\}$ is a Cauchy sequence of functions with respect to the L^p norm in C_0^∞, then

$$\lim_{i \to \infty} \int_{-\infty}^{\infty} \varphi_i(x)\psi(x)dx \qquad \text{exists } \forall \psi \in C_0^\infty; \qquad (5.7\text{-}5)$$

the limit is a linear functional on C_0^∞, that is, a distribution, which will be denoted by f, so that the above limit is $\langle f, \psi \rangle$. In the terminology of distributions, $\varphi_i \to f$. As in the case $p = 2$, the following holds:

Lemma. *Two Cauchy sequences $\{\varphi_i\}$ and $\{\tilde{\varphi}_i\}$ with respect to the L^p norm in C_0^∞ determine the same distribution if and only if they are equivalent, i.e., if $\|\varphi_i - \tilde{\varphi}_i\|_p \to 0$, as $i \to \infty$.*

I know of no simple proof like the corresponding proof in Section 5.3 for $p = 2$. An elegant, but slightly lengthy proof was communicated privately to the author by Norman Rehner, and a similar proof appeared later in the extensive paper of Werner (1969) referred to above.

In any normed linear space (complete or not), if $\{\varphi_i\}$ is a Cauchy sequence, then $\|\varphi_i\|$ has a limit as $i \to \infty$. This follows from the triangle inequality in the form

$$|\,\|\varphi_i\| - \|\varphi_j\|\,| \leq \|\varphi_i - \varphi_j\|, \qquad (5.7\text{-}6)$$

which shows that $\{\|\varphi_i\|\}$ is a Cauchy sequence of real numbers. If φ_i is in C_0^∞, and $f = \lim \varphi_i$, then the norm of the distribution f is defined as

$$\|f\|_p = \lim_{i \to \infty} \|\varphi_i\|_p. \qquad (5.7\text{-}7)$$

The space $L^p = L^p(\mathbb{R})$ is the set of all such distributions; i.e.,

$$L^p = \left\{ f = \lim_{i \to \infty} \varphi_i : \{\varphi_i\} \text{ Cauchy in the } L^p \text{ norm} \right\} \qquad (5.7\text{-}8)$$

Theorem 1. *L^p, with the norm $\|\cdot\|_p$, is a complete space, hence a Banach space, and C_0^∞ is dense in L^p (proof omitted).*

If $\{\varphi_i\}$ is a Cauchy sequence with limit f in L^p, and if $\{\psi_i\}$ is a Cauchy sequence with limit g in L^q, where $(1/p) + (1/q) = 1$ (as will be assumed throughout this section), then the Hölder inequality shows that the limit

$$\lim_{i \to \infty} \int_{-\infty}^{\infty} \varphi_i(x)\psi_i(x)dx \overset{\text{def}}{=} \langle f, g \rangle \tag{5.7-9}$$

exists, and that

$$|\langle f, g \rangle| \le \|f\|_p \|g\|_q. \tag{5.7-10}$$

This is the general Hölder inequality. The domain of definition of the linear functional $\langle \cdot, g \rangle$ has thereby been extended from C_0^∞ to L^p.

For a given g in L^q, $\langle \cdot, g \rangle$ is a bounded linear functional on L^p. The following theorem of F. Riesz (1910) generalizes the Riesz–Fréchet theorem of 1907:

Theorem 2. *If $F(\cdot)$ is any bounded linear functional on L^p, then there is a unique element g in L^q such that $F(f) = \langle f, g \rangle$, for all f in L^p. Furthermore, the bound of the functional $F(\cdot)$, defined as*

$$\|F(\cdot)\| = \sup_{f \neq 0} \frac{|F(f)|}{\|f\|_p}, \tag{5.7-11}$$

is equal to $\|g\|_q$.

The set of all bounded linear functionals on a Banach space constitutes another Banach space, called the *dual* space. Riesz's theorem shows that the space dual to L^p is isomorphic to L^q.

In the limiting case $p = 1$, the norm in C_0^∞ is $\|\varphi\|_1 = \int_{-\infty}^{\infty} |\varphi(x)|dx$. The space L^1 is defined exactly as L^p, for $p > 1$, although different proofs are needed to establish the Lemma and Theorem 1, above—see Werner (1969).

The analogue of theorem 2 of Section 5.3 holds, for any $p \ge 1$: any continuous function $f(x)$ such that $\int_{-\infty}^{\infty} |f(x)|^p \, dx$ is finite is in L^p.

In the other limiting case, $p = \infty$, the appropriate method of defining L^∞ is suggested by the fact that L^p is the dual of L^q, and that $p \to \infty$, as $q \to 1$, by (5.7-2). In fact, an alternative characterization of L^p (the one used by Werner) is that a distribution f is in L^p if and only if there is a constant M such that

$$|\langle f, \varphi \rangle| \le M\|\varphi\|_q, \quad \text{for all } \varphi \text{ in } C_0^\infty. \tag{5.7-12}$$

Definition. L^∞ is the space of all distributions f such that, for some constant $M = M(f)$,

$$|\langle f, \varphi \rangle| \le M\|\varphi\|_1, \quad \text{for all } \varphi \text{ in } C_0^\infty; \tag{5.7-13}$$

the norm $\|f\|_\infty$ is the smallest possible value of $M(f)$, which obviously has all the properties of a norm.

With this norm, L^∞ is a complete space, hence a Banach space (Werner).

It will be shown that a distribution f is in L^∞ if and only if it is bounded. The real case is considered first. If φ is a nonnegative test function, then $\|\varphi\|_1$ is $= \int_{-\infty}^\infty \varphi(x)dx = \langle 1, \varphi \rangle$, where 1 is the function $\equiv 1$. Therefore, (5.7-13) says that $\langle M \pm f, \varphi \rangle \geq 0$; i.e., the distributions $M + f$ and $M - f$ are ≥ 0 on $\mathbb{R}$, and $\|f\|_\infty$ is the smallest M for which that is true. In the complex case, $\|f\|_\infty$ is the smallest M such that

$$M - \text{Re}(fe^{i\alpha}) \geq 0 \quad \text{on } \mathbb{R}, \text{ for all real } \alpha. \tag{5.7-14}$$

Any bounded continuous function $f(x)$ is in L^∞, and then $\|f\|_\infty = \sup\{|f(x)| : x \in \mathbb{R}\}$.

L^∞ is the space dual to L^1, which is almost obvious from its definition, but L^1 is not the dual of L^∞. If the dual of a space is denoted by a prime, as in Section 2.8, then a Banach space $\mathfrak{B}$ is called *reflexive* if $(\mathfrak{B}')' = \mathfrak{B}$. Therefore L^p is reflexive for $p > 1$, but not for $p = 1$. The dual of L^∞ is a space of measures; see Section 13.9.

Spaces $L^p(\mathbb{R}^n)$, $L^p(a, b)$, and $L^p(\Omega)$ can be treated in an exactly analogous way.

EXERCISE

1. Show that if f is any distribution in $L^p(\mathbb{R})$ $(1 < p < \infty)$, then the integral $\int f$ is a continuous function. (cf. Section 5.5)

5.8 Fourier Transforms in L^1, Riemann–Lebesgue Lemma, Luzin's Theorem

Elements of L^1 spaces are tempered distributions, hence they have Fourier transforms. The Riemann–Lebesgue Lemma, proved below, says that in this case the transforms are continuous functions and tend to zero (perhaps very slowly) at ∞. The original form of the Lemma said that if $f(x)$ is measurable in $(0, 2\pi)$ and $\int_0^{2\pi} |f(x)|dx$ is $< \infty$, then the Fourier coefficients

$$c_n = \frac{1}{2\pi} \int_0^{2\pi} f(x)e^{-inx} \, dx$$

tend to zero as $n \to \pm\infty$. (For practical purposes that is less important than the theorem that if $f(x)$ is of bounded variation in $(0, 2\pi)$ the coefficients $c_n \to 0$ at least like const./$|n|$, but the former has many theoretical applications.)

According to the ideas of Section 5.7, the space $L^1(\mathbb{R}^n)$ is defined as follows: If $\{\varphi_j(\mathbf{x})\}_{j=1}^\infty$ is a Cauchy sequence of test functions in the L^1 norm, i.e., if

$$\int_{\mathbb{R}^n} |\varphi_j(\mathbf{x}) - \varphi_k(\mathbf{x})| d^n\mathbf{x} \to 0, \quad \text{as } j, k \to \infty, \tag{5.8-1}$$

then the sequence determines a distribution f by the equation

$$\langle f, \psi \rangle = \lim_{j \to \infty} \int_{\mathbb{R}^n} \varphi_j(\mathbf{x}) \psi(\mathbf{x}) d^n \mathbf{x} \qquad \forall \psi \in C_0^\infty(\mathbb{R}^n),$$

and $L^1(\mathbb{R})$ is the set of all such distributions f.

Lemma (Riemann–Lebesgue).   *If f is a distribution in $L^1(\mathbb{R}^n)$, its Fourier transform $\hat{f}$ is a continuous function $\hat{f}(\mathbf{y})$ and $\to 0$ as $|\mathbf{y}| \to \infty$.*

PROOF.   We show first that $\hat{f}$ is a continuous function. Let $\{\varphi_j\}$ be as above. From the formula for the Fourier transform of a test function it follows that for any $\mathbf{y}$,

$$|\hat{\varphi}_j(\mathbf{y}) - \hat{\varphi}_k(\mathbf{y})| \leq (2\pi)^{-n/2} \int_{\mathbb{R}^n} |\varphi_j(\mathbf{x}) - \varphi_k(\mathbf{x})| d^n \mathbf{x},$$

because $|e^{i\mathbf{x} \cdot \mathbf{y}}| = 1$. Since the right member is independent of $\mathbf{y}$ and $\to 0$ as j and $k \to \infty$, we see that $\hat{\varphi}_j(\mathbf{y})$ converges uniformly as $j \to \infty$. Hence the limit,

$$\hat{f}(\mathbf{y}) = (2\pi)^{-n/2} \lim_{j \to \infty} \int_{\mathbb{R}^n} \varphi_j(\mathbf{x}) e^{-i\mathbf{y} \cdot \mathbf{x}} d^n \mathbf{x}$$

is a continuous function. Since $e^{i\pi} = -1$, $\hat{f}(y)$ is also given by

$$\hat{f}(\mathbf{y}) = -(2\pi)^{-n/2} \lim_{j \to \infty} \int_{\mathbb{R}^n} \varphi_j(\mathbf{x}) e^{-i\mathbf{y} \cdot (\mathbf{x} + (\pi \mathbf{y}/|\mathbf{y}|^2))} d^n \mathbf{x}.$$

The average of the two expressions is

$$\hat{f}(\mathbf{y}) = \tfrac{1}{2}(2\pi)^{-n/2} \lim_{j \to \infty} \int_{\mathbb{R}^n} \left[\varphi_j(\mathbf{x}) - \varphi_j\left(\mathbf{x} - \frac{\pi \mathbf{y}}{|\mathbf{y}|^2} \right) \right] e^{-i\mathbf{y} \cdot \mathbf{x}} d^n \mathbf{x}.$$

Therefore,

$$|\hat{f}(\mathbf{y})| \leq \tfrac{1}{2}(2\pi)^{-n/2} \lim_{j \to \infty} \int_{\mathbb{R}^n} \left| \varphi_j(\mathbf{x}) - \varphi_j\left(\mathbf{x} - \frac{\pi \mathbf{y}}{|\mathbf{y}|^2} \right) \right| d^n \mathbf{x},$$

and we shall show that the right member of this inequality goes to zero as $|\mathbf{y}| \to \infty$. Let ε be any positive number. First choose K such that

$$\int_{\mathbb{R}^n} |\varphi_j(\mathbf{x}) - \varphi_k(\mathbf{x})| d^n \mathbf{x} < \varepsilon \quad \text{for } j, k \geq K. \tag{5.8-2}$$

Then choose R so that

$$\int_{\mathbb{R}^n} \left| \varphi_K(\mathbf{x}) - \varphi_K\left(\mathbf{x} - \frac{\pi \mathbf{y}}{|\mathbf{y}|^2} \right) \right| d^n \mathbf{x} < \varepsilon \quad \text{for } |\mathbf{y}| > R.$$

If we combine this inequality with (5.8-2) for $k = K$ twice, once as written and once with $\mathbf{x}$ translated by $\pi \mathbf{y}/|\mathbf{y}|^2$, we see that

$$\int_{\mathbb{R}^n} \left| \varphi_j(\mathbf{x}) - \varphi_j\left(\mathbf{x} - \frac{\pi \mathbf{y}}{|\mathbf{y}|^2} \right) \right| d^n \mathbf{x} < 3\varepsilon \quad \text{for } j > K \text{ and } |\mathbf{y}| > R.$$

Therefore $|\hat{f}(\mathbf{y})| \to 0$ as $|\mathbf{y}| \to \infty$.

Note 1. A distribution in L^1 need not be in L^2, and one in L^2 need not be in L^1. Fourier transforms of elements of L^2 are discussed in Section 5.10, below; they are not in general ordinary functions.

The following slightly specialized version of *Luzin's theorem* shows how close distributions in L^1 (and L^2, see Note 2) come to being ordinary functions. In the original version of the theorem (see Natanson 1955), f is taken as a Lebesgue-measurable function.

EXERCISE (Luzin's Theorem)

Suppose that a distribution f is in $L^1(\Omega)$, $\Omega \subset \mathbb{R}^n$. Show that for any $\delta > 0$ there is an open set Ω_δ in Ω of volume $<\delta$ and a continuous function $g_\delta(\mathbf{x})$, defined in $\Omega - \Omega_\delta$, such that

$$ f = g_\delta \quad \text{inside } \Omega - \Omega_\delta. $$

Hint: Let $\{\varphi_k\}$ be a series of test functions converging in $L^1(\Omega)$ to f. Show that for any $\delta > 0$ there are integers $k_1, k_2, \ldots$ (all depending on δ) such that

$$ \text{volume} \left\{ \mathbf{x} : |\varphi_k(\mathbf{x}) - \varphi_l(\mathbf{x})| > \frac{1}{n^2} \right\} < \frac{6\delta}{\pi^2 n^2} \quad \text{for all } k, l \geq k_n, $$

then define

$$ \Omega_\delta = \bigcup_{n=1}^{\infty} \left\{ \mathbf{x} : |\varphi_{k_n}(\mathbf{x}) - \varphi_{k_{n-1}}(\mathbf{x})| > \frac{1}{n^2} \right\}, $$

and consider the sequence $\{\varphi_{k_n}\}_{n=1}^{\infty}$.

Note 2. If f is a distribution on $\mathbb{R}^n$ and is in $L^1(\Omega)$ for every *bounded* Ω in $\mathbb{R}^n$, f is said to be *locally in* L^1. A distribution locally in L^2 is also locally in L^1, because the Schwarz inequality holds for elements of L^2, hence

$$ \int_\Omega |f \cdot 1| d^n\mathbf{x}^2 \leq \int |f|^2 \, d^n\mathbf{x} \text{ volume } (\Omega); $$

hence, Luzin's theorem applies to f.

5.9 Spaces of Type L_σ^2

In the spectral representation of self-adjoint operators, the spaces described in this section appear. Let $\sigma(x)$ be a non-decreasing function of the real variable x. For any two functions φ and ψ in $C_0^\infty(\mathbb{R})$, define

$$ (\varphi, \psi)_\sigma = \int_{-\infty}^{\infty} \overline{\varphi(x)}\psi(x)d\sigma(x) \tag{5.9-1} $$

and

$$ \|\varphi\|_\sigma = \sqrt{(\varphi, \varphi)_\sigma}. $$

The functional $\|\cdot\|_\sigma$ is in general only positive *semidefinite* on the space C_0^∞, for if $\varphi(x)$ is zero except in an interval of constancy of $\sigma(x)$, then $\|\varphi\|_\sigma^2 = \int |\varphi(x)|^2 \, d\sigma(x) = 0$, even if $\varphi(x)$ is not identically zero; $\|\cdot\|$ is called a *seminorm*. The Schwarz and triangle inequalities still hold. [The proofs of these and other statements in this section are very nearly the same as the proofs of the corresponding statements in Section 5.3 and are omitted.]

If $\{\varphi_i\}$ is a Cauchy sequence with respect to this seminorm, then the limit of $(\psi, \varphi_i)_\sigma$, as $i \to \infty$, exists for all test functions ψ; a distribution f is defined as the functional

$$\langle f, \psi \rangle = \lim_{i \to \infty} (\bar{\varphi}_i, \psi)_\sigma, \quad \text{for all } \psi \text{ in } C_0^\infty.$$

Two Cauchy sequences $\{\varphi_i\}$ and $\{\tilde{\varphi}_i\}$ determine the same distribution if and only if they are equivalent, i.e., $\|\varphi_i - \tilde{\varphi}_i\|_\sigma \to 0$.

The space $L_\sigma^2 = L_\sigma^2(\mathbb{R})$ is defined as the set of all distributions determined in this way by Cauchy sequences. An inner product is defined in $L_\sigma^2(\mathbb{R})$ by

$$(f, g)_\sigma = \lim_{i \to \infty} (\varphi_i, \psi_i)_\sigma,$$

where $\{\varphi_i\}$ and $\{\psi_i\}$ are Cauchy sequences that determine f and g, respectively, and a norm is defined by

$$\|f\|_\sigma = \sqrt{(f, f)_\sigma} = \lim_{i \to \infty} \|\varphi_i\|_\sigma.$$

$L_\sigma^2(\mathbb{R})$ is a complete space—a Hilbert space.

The following remarks show some minor new features of this space.

1. If (a, b) is any interval in which $\sigma(x)$ is constant, and f is in L_σ^2, then $f = 0$ in (a, b). [For the meaning of "$f = 0$ in (a, b)", where f is a distribution, see Chapter 3.]
2. C_0^∞ is not necessarily contained in L_σ^2; in fact, a continuous function $f(x)$ is in L_σ^2 if and only if

    (a) $\int_{-\infty}^{\infty} |f(x)|^2 \, d\sigma(x) < \infty$, and
    (b) $f(x) = 0$ in each interval of constancy of $\sigma(\cdot)$.
3. Although $\|\cdot\|_\sigma$ is only semidefinite in C_0^∞, it is definite in L_σ^2. That is, if $\|\varphi_i\|_\sigma \to 0$, as $i \to \infty$, so that $\|f\|_\sigma = 0$, then the Cauchy sequence $\{\varphi_i\}$ determines the zero distribution: $f = 0$. Therefore $\|\cdot\|_\sigma$ is a norm in L_σ^2, not merely a seminorm.

Spaces $L_\sigma^2(\mathbb{R}^n)$ are similarly treated, where σ is a nondecreasing function of several real variables, in the sense of Section 13.3, and multiple Stieltjes integrals are used.

Exercise

1. Interpret the space $L_\sigma^2(\mathbb{R})$ in the following cases: (1) $\sigma(\cdot)$ is a step function; (2) $\sigma(\cdot)$ has a derivative $\sigma'(x) > 0$ for all x; (3) $\sigma(x) = x$.

5.10 Fourier Transforms and Mollifiers in L^2 Spaces

Since any f in $L^2(\mathbb{R}^n)$ is a tempered distribution, it has a Fourier transform $\hat{f}$, which is also tempered. It will be shown that $\hat{f}$ is also in L^2: Suppose $\varphi_k \to f$ in L^2, as $k \to \infty$; then $\{\varphi_k\}$ is a Cauchy sequence. By the Parseval relation, $\|\hat{\varphi}_k - \hat{\varphi}_j\|^2 = \|\varphi_k - \varphi_j\|^2$, hence $\{\hat{\varphi}_k\}$ is also a Cauchy sequence. For any ψ in $\mathscr{S}$,

$$\langle \hat{f}, \psi \rangle = \langle f, \hat{\psi} \rangle = \lim_{k \to \infty} \langle \varphi_k, \hat{\psi} \rangle = \lim_{k \to \infty} \langle \hat{\varphi}_k, \psi \rangle.$$

That is, $\hat{f}$ is equal to the distribution determined by the sequence $\{\hat{\varphi}_k\}$ in the sense of Section 5.3, hence is in $L^2(\mathbb{R}^n)$. By similarly letting ψ_k converge in L^2 to another element g, it is easily seen that $(\hat{f}, \hat{g}) = (f, g)$. Therefore the mapping $f \to \hat{f}$ is an isomorphism (sometimes called an isometric isomorphism, because it preserves norm and inner product as well as linearity) of the Hilbert space L^2 onto itself. That is, this mapping is a unitary transformation in the space.

In quantum-mechanical terms, if f represents a state of an N-particle system in the coordinate representation (where $n = 3N$, and where spins are ignored), then $\hat{f}$ represents the same state in the momentum representation, and the isometric isomorphism shows that the two representations are completely equivalent.

It was shown in Chapter 2 that if f is a distribution and J_δ a mollifier (Section 2.6), then $J_\delta f$ and f are close, as distributions, if δ is small; i.e., $J_\delta f \to f$ in the sense of convergence of distributions, as $\delta \to 0$. We show that if $f \in L^2$, then they are also close in the L^2 norm.

Theorem 1. *If f is a distribution in $L^2(\mathbb{R}^n)$ and has bounded support, then*

$$\|J_\delta f\| \le \|f\| \qquad (5.10\text{-}1)$$

for all δ.

Corollary. *The same is true even if the support of f is not bounded.*

PROOF OF THE THEOREM. Since f has bounded support, its Fourier transform $\hat{f}$ is a continuous (in fact analytic) function, by the theorem in Section 4.5; hence

$$\|f\|^2 = \|\hat{f}\|^2 = \int |\hat{f}(\mathbf{k})|^2 \, d^n\mathbf{k}.$$

Let g be the function $J_\delta f$; that is,

$$g(\mathbf{y}) = \langle f, \rho_{\mathbf{y}\,\delta} \rangle, \qquad (5.10\text{-}2)$$

where $\rho_{\mathbf{y}\,\delta}(\mathbf{x})$ is the test function introduced in Section 2.6 in the definition of the mollifier J_δ. To get $\hat{g}(\mathbf{k})$, we multiply the last equation by $(2\pi)^{-n/2} e^{-i\mathbf{k}\cdot\mathbf{y}}$ and integrate over $\mathbb{R}^n$ with respect to $\mathbf{y}$. Using Exercise 1 of Section 2.4 on integration with respect to a parameter, we easily find that

$$\hat{g}(\mathbf{k}) = \langle f, e^{-i\mathbf{k}\cdot\mathbf{x}} \rangle \hat{\rho}(\delta\mathbf{k})$$
$$= \hat{f}(\mathbf{k})\hat{\rho}(\delta\mathbf{k})$$

(which was to be expected, because (5.10-2) is a convolution). Now $\rho(\mathbf{x})$ is a nonnegative function that integrates to 1, hence

$$|\hat{\rho}(\mathbf{k})| \le |\hat{\rho}(0)| = 1,$$

hence $\|\hat{g}\|^2 \le \|\hat{f}\|^2$, and (5.10-1) follows.

PROOF OF THE COROLLARY.   Let $\{f_j\}$ be a sequence of distributions, each having bounded support, that converges in L^2 to f (see Exercise in Section 5.4). Then,

$$\|J_\delta f_j\| \le \|f_j\| \tag{5.10-3}$$

for each j. Clearly $\{J_\delta f_j\}$ is a Cauchy sequence, and its limit is $J_\delta f$ (because, according to Exercise 5 in Section 2.6, that is its limit in the sense of convergence of distributions, and the two limits are the same whenever the L^2 limit exists). Then (5.10-1) follows from (5.10-3) in the limit $j \to \infty$.

Theorem 2.   *If f is in $L^2(\mathbb{R}^n)$, then $J_\delta f \to f$ in L^2, i.e.*

$$\|J_\delta f - f\| \to 0 \quad \text{as } \delta \to 0. \tag{5.10-4}$$

PROOF.   Let $\varepsilon > 0$ and let ψ be a function in C_0^∞ such that

$$\|f - \psi\| < \varepsilon.$$

Then, by Theorem 1,

$$\|J_\delta f - J_\delta \psi\| < \varepsilon$$

for any δ. Since ψ is smooth, it is intuitively clear that $J_\delta \psi$ can differ little from ψ if δ is small enough, and it is left as an exercise for the reader to show that δ can be so chosen that

$$\|J_\delta \psi - \psi\| < \varepsilon.$$

These three inequalities together show that $\|J_\delta f - f\|$ is $< 3\varepsilon$, which is arbitrary, hence (5.10-4) follows.

5.11 The Sobolev Spaces; The Space W^1

If $\psi(\mathbf{x})$ or $\psi(\mathbf{x}, t)$ is a quantum-mechanical wave function, the physically relevant Hilbert space is often one in which $\psi(\mathbf{x})$ and its first partial derivatives are all in L^2, so that for example the expectation of the kinetic energy, which is

$$\frac{1}{2} \int |\nabla \psi|^2 \, d^3\mathbf{x}$$

in suitable units, is finite. Such spaces are called Sobolev spaces and are useful in the study of partial differential operators. Functions in them have well-defined boundary values on hypersurfaces, in a certain sense. It is recalled that if a distribution f of one variable and its derivative f' are both in L^2, then f is an ordinary continuous function $f(x)$, hence has a well-defined value at each x in $[a, b]$. (In particular, $f(x) \to 0$ at infinity if $[a, b]$ is an unbounded interval.) In more dimensions, if $f(\mathbf{x})$ is in the corresponding Sobolev space, then, although it is not in general an ordinary function, its "values" on a simple closed surface S constitute a distribution on S, as will be explained in the next section.

Let Ω be either all of $\mathbb{R}^n$ or a region in $\mathbb{R}^n$ bounded by a piecewise smooth simple closed surface $\partial\Omega$, and let L^2 be the Hilbert space $L^2(\Omega)$, whose inner product and norm will be denoted by $(u, v)_0$ and $\|u\|_0$. Let $W^1 = W^1(\Omega)$ be the linear manifold

$$W^1 = \{u \in L^2 : \partial_m u \in L^2 \quad (m = 1, \ldots, n)\}, \tag{5.11-1}$$

where

$$\partial_m = \frac{\partial}{\partial x_m} \quad (m = 1, \ldots, n). \tag{5.11-2}$$

W^1 is not a *closed* linear manifold in L^2, hence is not a subspace of L^2, but we shall show that it is a complete space (hence a Hilbert space) with respect to a new inner product and norm given by

$$(u, v)_1 = (u, v)_0 + (\nabla u, \nabla v)_0, \tag{5.11-3}$$

$$\|u\|_1^2 = \|u\|_0^2 + \|\nabla u\|_0^2, \tag{5.11-4}$$

where we have abbreviated:

$$(\nabla u, \nabla v)_0 = \sum_{m=1}^{n} (\partial_m u, \partial_m v)_0, \tag{5.11-5}$$

$$\|\nabla u\|_0^2 = \sum_{m=1}^{n} \|\partial_m u\|_0^2. \tag{5.11-6}$$

Theorem. *W^1 is complete in the norm $\|\ \ \|_1$.*

PROOF. Suppose $\{u_j\}_1^\infty$ is a Cauchy sequence in W^1, i.e.,

$$\|u_j - u_k\|_1 \to 0 \quad \text{as } j, k \to \infty; \tag{5.11-7}$$

we must show there is a v in W^1 such that

$$\|u_j - v\|_1 \to 0 \quad \text{as } j \to \infty. \tag{5.11-8}$$

Because of (5.11-7), equation (5.11-4), with u replaced by $u_j - u_k$, shows that $\{u_j\}$ and $\{\partial_m u_j\}$ $(m = 1, \ldots, n)$ are all Cauchy sequences in L^2. Call v and $w_m(m = 1, \ldots, n)$ their respective limits in L^2. For any test function φ in $C_0^\infty(\Omega)$,

$$(w_m, \varphi)_0 = \lim_{(j \to \infty)} (\partial_m u_j, \varphi)_0$$

$$= \lim(u_j, -\partial_m \varphi)_0$$

$$= (v, -\partial_m \varphi)_0 = \langle \partial_m \bar{v}, \varphi \rangle,$$

by definition of the distribution derivative $\partial_m \bar{v}$. Therefore w_m and $\partial_m v$ are the same distribution in Ω, and it follows that (1) $\partial_m v$ is in L^2 for each m, hence v is in W^1, and (2) $\partial_m v$ is the limit in L^2 of $\partial_m u_j$ for each m, hence, by (5.11-6),

$$\|u_j - v\|_1 \to 0, \quad \text{as } j \to \infty.$$

Therefore W_1 is complete.

Further Sobolev spaces are often defined. In $W^{l\,p}$, the L^2 norm is replaced by a norm of the L^p type involving the function u and its partial derivatives of order $\leq l$. See Friedman 1969. We shall discuss only the space $W^1 = W^{1\,2}$.

5.12 Boundary Values in W^1; The Subspace W_0^1

Let u be in $W^1(\Omega)$, where Ω is a bounded region with a piecewise smooth boundary, and let $\psi(\mathbf{x})$ be any function that is in $C^2(\tilde{\Omega})$. Call $u_\delta(\mathbf{x}) = J_\delta u$, where J_δ is the mollifier defined in Section 2.6; u_δ may be nonzero at points outside Ω, but it is a continuous function in $\tilde{\Omega}$ and as such is an element of $L^2(\Omega)$. Integrating by parts (Green's formula) gives

$$(\nabla\psi, \nabla u_\delta)_0 = \int_\Omega \nabla\bar{\psi} \cdot \nabla u_\delta \, d^n\mathbf{x}$$

$$= \int_{\partial\Omega} \nabla\bar{\psi} \cdot \mathbf{n} u_\delta \, d\mathscr{A} - \int_\Omega (\nabla^2\bar{\psi}) u_\delta \, d^n\mathbf{x}.$$

Now, J_δ commutes with ∇, so $\nabla(u_\delta) = (\nabla u)_\delta$; furthermore, for any f in $L^2(\Omega)$, $f_\delta \to f$ in $L^2(\mathbb{R}^n)$ (as $\delta \to 0$), hence a fortiori in $L^2(\Omega)$. Therefore,

$$\lim_{(\delta \to 0)} \int_{\partial\Omega} u_\delta \nabla\bar{\psi} \cdot \mathbf{n} \, d\mathscr{A} = (\nabla\psi, \nabla u)_0 + (\nabla^2\psi, u)_0 \qquad (5.12\text{-}1)$$

The limit depends on the values of $\nabla\bar{\psi}$ on $\partial\Omega$ but not otherwise on ψ. We define a function χ on $\partial\Omega$ by writing

$$\chi(\mathbf{x}) = \nabla\bar{\psi} \cdot \mathbf{n} \quad \text{for } \mathbf{x} \in \partial\Omega, \qquad (5.12\text{-}2)$$

and we think of χ as a test function. Then the equation

$$\langle \tilde{u}, \chi \rangle = \lim_{(\delta \to 0)} \int_{\partial\Omega} u_\delta \chi \, d\mathscr{A}$$

defines a linear functional $\langle \tilde{u}, \cdot \rangle$ on the space of the test functions χ, i.e., a distribution $\tilde{u}$ on $\partial\Omega$, which may be thought of as consisting of the boundary values of u in $W^1(\Omega)$.

It can be proved that $\tilde{u}$ is in the space $L^2(\partial\Omega)$ defined on the surface $\partial\Omega$ (see Section 5.3); that is a special case of a more general theorem of Sobolev (see Sobolev 1950, 1963). If u is an ordinary function in W_1, continuous in $\tilde{\Omega}$, then $\tilde{u}$ is the distribution on $\partial\Omega$ to be identified with the function $u(\mathbf{x})$ for $\mathbf{x} \in \partial\Omega$, and is given by

$$\langle \tilde{u}, \chi \rangle = \int_{\partial\Omega} \tilde{u}(\mathbf{x})\chi(\mathbf{x}) d\mathscr{A}.$$

We shall be concerned with the case in which the boundary values $\tilde{u}$ vanish. Then, by (5.12-1),

$$(\nabla\psi, \nabla u)_0 + (\nabla^2\psi, u)_0 = 0, \qquad (5.12\text{-}3)$$

which is the formula for integration by parts in Ω when the boundary values of u are zero. Hence, we define

$$W_0^1 = \{u \in W^1 : (5.12\text{-}3) \text{ holds } \forall \psi \in C^2(\tilde{\Omega})\}. \qquad (5.12\text{-}4)$$

Owing to the continuity of the inner product it follows from (5.12-3) that W_0^1 is a *closed* linear manifold, hence a *subspace* of W^1.

In the next chapter, W_0^1 will play the role of the Hilbert space that contains the eigenfunctions of the Laplacian in Ω subject to the boundary condition of vanishing on $\partial\Omega$.

The integration-by-parts formula (5.12-3) holds in a slightly more general context. Let v and its first partial derivatives and $\nabla^2 v$ all be in $L^2(\Omega)$. Then take ψ to be $v_\delta = J_\delta v$. Since J_δ commutes with differentiation, (5.12-3) gives

$$(J_\delta \nabla u, \nabla u)_0 + (J_\delta \nabla^2 v, u)_0 = 0.$$

Since for any w in L^2, $J_\delta w \to w$ in $L^2(\mathbb{R}^n)$ hence a fortiori in $L^2(\Omega)$, we have

$$(\nabla v, \nabla u)_0 + (\nabla^2 v, u)_0 = 0 \quad \text{for } u \in W_0^1, v \in W^1, \nabla^2 v \in L^2. \quad (5.12\text{-}5)$$

EXERCISES

1. Show that $C^\infty(\tilde{\Omega})$ is dense in $W^1(\Omega)$. *Hint*: Apply the mollifier J_δ to the elements of $W^1(\Omega)$.

2. Let Ω be the unit cube in $\mathbb{R}^n$. Show that there is a constant K such that $\|\nabla\varphi\| \geq K\|\varphi\|$ for all φ in $C_0^\infty(\Omega)$, and show that it follows that $C_0^\infty(\Omega)$ is not dense in $W^1(\Omega)$.

5.13 On Vanishing at Infinity II

The discussion started in Section 5.4 can now be completed for L^2 spaces.

Theorem. *If a distribution f and all its partial derivatives of order l are in $L^2(\mathbb{R}^n)$, where l is an integer $> n/2$, then f is a continuous function $f(\mathbf{x})$, which $\to 0$ as $|\mathbf{x}| \to \infty$.*

PROOF. First suppose that l is even. Then

$$f + \left(\sum_{j=1}^{n} \partial_{x_j}^2\right)^{l/2} f \in L^2.$$

Fourier transformation gives

$$\chi \hat{f} \overset{\text{def}}{=} \hat{g} \in L^2,$$

where

$$\chi(\mathbf{y}) = 1 + |\mathbf{y}|^l = 1 + (y_1^2 + \cdots + y_n^2)^{l/2}.$$

(It follows incidentally that each partial derivative of order $< l$ is also in L^2, because its Fourier transform is $q(y_1, \ldots, y_n)\hat{f}$, where q is a monomial of degree $< l$, hence the transform can be written as $(q/\chi)\hat{g}$, but q/χ is a bounded continuous function.) Now let $\hat{\psi}_j$ be test functions that $\to \hat{g}$ in L^2. Then the test functions

$$\hat{\phi}_j(y) \stackrel{\text{def}}{=} \chi(y)^{-1}\hat{\psi}_j(y)$$

converge to $\hat{f}$ in L^2. From the Schwarz inequality,

$$\left(\int_{\mathbb{R}^n} |\hat{\phi}_j(y) - \hat{\phi}_k(y)|\, d^n y\right)^2 \leq \int \chi(y)^{-2}\, d^n y \int |\hat{\psi}_j(y) - \hat{\psi}_k(y)|^2\, d^n y.$$

The first integral on the right is finite because its integrand goes to zero at least as $|y|^{-n-1}$ at infinity; hence,

$$\|\hat{\phi}_j - \hat{\phi}_k\|_{L^1} \leq \text{const.}\|\hat{\psi}_j - \hat{\psi}_k\|_{L^2}.$$

Therefore $\{\hat{\phi}_j\}$ converges in L^1, $\hat{f}$ is in L^1, hence, by the Riemann–Lebesgue Lemma in Section 5.8, f is a continuous function $f(\mathbf{x})$ which $\to 0$ as $|\mathbf{x}| \to \infty$. Lastly, if l is odd, $\chi\hat{f}$, as defined above, is still in L^2, because (1) if y_j is any component of $\mathbf{y}$ and if $y_j\hat{h}$ is in L^2 for any $\hat{h}$ in L^2, then $|y_j|\hat{h}$ is also in L^2, and (2),

$$(y_1^2 + \cdots + y_n^2)^{l/2} \leq (|y_1| + \cdots + |y_n|)^l;$$

hence the conclusion of the theorem holds also for l odd.

Remark.   For $n = 2$ and $n = 3$ it follows from the proof of the theorem, with $l = 2$, that if f and $\nabla^2 f$ are in L^2, then f is a continuous function and $\to 0$ as $|\mathbf{x}| \to \infty$.

CHAPTER 6

Some Problems Connected with the Laplacian

Vibration eigenfunctions in a bounded domain; variational methods;
the Dirichlet integral; the potential due to a given charge distribution;
Poisson's equation; convolutions; the direct product; Schwartz's
nuclear theorem; the Cauchy-Riemann equations; harmonic functions.

Prerequisite: Chapter 5

The Laplacian is in many respects of a more classical nature than many of the
differential operators to be discussed in Chapters 10 and 11. One of the basic
problems is to find the eigenfunctions $u(\mathbf{x})$ of the equation $\nabla^2 u + \lambda u = 0$
in a region Ω of n-dimensional space with the boundary condition $u(\mathbf{x}) = 0$
on the boundary $\partial\Omega$. For $n = 2$ that is the classical problem of a vibrating
membrane. More generally, for both $n = 2$ and $n = 3$ the eigenfunctions and
the variational methods that determine them are useful in problems of
vibration, heat flow, electromagnetic fields, and hydrodynamic stability.
That is the main subject of this chapter.

Some additional topics illustrate the universality of distribution-theory
methods. First, the validity of Poisson's equation for the potential $V(\mathbf{x})$ due
to a charge with density $\rho(\mathbf{x})$ is established, when $\rho(\mathbf{x})$ is an arbitrary distribution with bounded support in $\mathbb{R}^3$. Second, it is shown that the Cauchy–
Riemann equations for distributions u and v on $\mathbb{R}^2$ imply the analyticity of
$u + iv$ more generally than in classical theory, if the derivatives in the
Cauchy–Riemann equations are interpreted in the distribution theory sense.
In that connection it is shown that any harmonic distribution in $\mathbb{R}^n$ is a
harmonic function in $\mathbb{R}^n$.

Convolutions of distributions are discussed briefly, because they are
needed in the discussion of Poisson's equation, and then Schwartz's Nuclear
Theorem is discussed because it is needed for a full understanding of convolutions.

6.1 The Potential; Poisson's Equation

It is recalled that in electrostatics the potential $V(\mathbf{x})$ due to a distributed charge with density $\rho(\mathbf{x})$ is given by

$$V(\mathbf{x}) = \int_{\mathbb{R}^3} \frac{1}{|\mathbf{x} - \mathbf{y}|} \rho(\mathbf{y}) d^3\mathbf{y} \qquad (6.1\text{-}1)$$

and that this potential satisfies Poisson's equation

$$\nabla^2 V = -4\pi\rho. \qquad (6.1\text{-}2)$$

(Single vertical bars are used in this chapter to denote the length of a vector, to permit use of the double bars to denote the L^2 norm of a vector field when regarded as an element of a Banach or Hilbert space.)

These equations will be generalized, in the next two sections, to the case in which ρ is any distribution of bounded support on $\mathbb{R}^3$. We discuss briefly also the modified problem in which the charge is contained in a region Ω bounded by a simple closed surface $\partial\Omega$ on which $V(\mathbf{x}) = 0$. Then the first factor in the integrand of (6.1-1) is replaced by a Green's function $G(\mathbf{x}, \mathbf{y})$.

6.2 Convolutions

According to (6.1-1), $V(\mathbf{x})$ is the three-dimensional convolution of the functions $1/|\mathbf{x}|$ and $\rho(\mathbf{x})$, hence the first problem is to define the convolution $f * g$ of two distributions f and g on $\mathbb{R}^n$. If f and g are ordinary functions and g has bounded support, then the convolution is also a function $(f * g)(\mathbf{x})$; as a distribution, it is given by

$$\begin{aligned} \langle f * g, \varphi \rangle &= \iint f(\mathbf{x} - \mathbf{y})g(\mathbf{y})d^n\mathbf{y}\varphi(\mathbf{x})d^n\mathbf{x} \\ &= \iint f(\mathbf{w})g(\mathbf{y})\varphi(\mathbf{y} + \mathbf{w})d^n\mathbf{y} \, d^n\mathbf{w}, \end{aligned} \qquad (6.2\text{-}1)$$

and we shall proceed by direct imitation of this formula. If g is a distribution with bounded support, then the inner integration in the last member of (6.2-1) (the integration with respect to $\mathbf{y}$) is to be interpreted as

$$\langle g, \varphi_{\mathbf{w}} \rangle, \qquad (6.2\text{-}2)$$

where

$$\varphi_{\mathbf{w}}(\mathbf{y}) = \varphi(\mathbf{w} + \mathbf{y}). \qquad (6.2\text{-}3)$$

According to Exercise 3 in Section 2.6 and the accompanying discussion of mollifiers, $\langle g, \varphi_{\mathbf{w}} \rangle$ is a C^∞ function of $\mathbf{w}$. It is clear from (6.2-3) that if $|\mathbf{w}|$ is large enough, the supports of g and $\varphi_{\mathbf{w}}$ do not overlap, and $\langle g, \varphi_{\mathbf{w}} \rangle$ is zero; that is, the function $\langle g, \varphi_{\mathbf{w}} \rangle$ has bounded support, hence is a test function. The outer integration in (6.2-1) can be interpreted as the result of putting that test function into $\langle f, \cdot \rangle$, hence we define

$$\langle f * g, \varphi \rangle \stackrel{\text{def}}{=} \langle f, \langle g, \varphi_w \rangle \rangle. \tag{6.2-4}$$

From this last equation and Exercise 2 in Section 2.4 on differentiation with respect to a parameter, it is evident that if ∂_j denotes $\partial/\partial w_j$, where w_j is one of the components of $\mathbf{w}$ ($j = 1, \ldots, n$), then

$$\partial_j(f * g) = (\partial_j f) * g = f * (\partial_j g), \tag{6.2-5}$$

and similarly for higher derivatives, just as for ordinary functions.

Consider the case in which one of f, g is the n-dimensional δ-function $\delta_n(\mathbf{x})$ defined by the equation

$$\langle \delta_n, \varphi \rangle = \varphi(0) \qquad \forall \varphi(\mathbf{x}) \text{ in } C_0^\infty(\mathbb{R}^n). \tag{6.2-6}$$

If $f = \delta_n$, then the right member of (6.2-4) is

$$\langle \delta_n, \langle g, \varphi_w \rangle \rangle = \langle g, \varphi_0 \rangle = \langle g, \varphi \rangle,$$

hence

$$\delta_n * g = g. \tag{6.2-7}$$

Similarly

$$f * \delta_n = f. \tag{6.2-8}$$

These results generalize the symbolic relation

$$\int \delta_n(\mathbf{x} - \mathbf{y}) g(\mathbf{y}) d^n \mathbf{y} = g(\mathbf{y})$$

from functions g to distributions g.

Further properties of convolutions are given in Section 6.5.

EXERCISES

1. Show how to modify the discussion of this section for distributions f and g on $\mathbb{R}$ whose supports are both bounded from below (or both from above). What can you say about the support of $f * g$ in that case?

2. Show that for test functions φ and ψ on $\mathbb{R}^n$ the Fourier transform of $\varphi * \psi$ is $(2\pi)^{n/2}$ times the ordinary product $\hat{\varphi}\hat{\psi}$.

3. Let f and g be tempered distributions on $\mathbb{R}^n$, of which g has compact support. Show that the Fourier transform of $f * g$ is $(2\pi)^{n/2}$ times the ordinary product $\hat{f}\hat{g}$, after noting that the product is well defined because $\hat{g}$ is a C^∞ function (see Theorem in Section 4.5).

4. Let f and g be tempered distributions on $\mathbb{R}$ whose supports are both bounded from below (or both from above). For what (complex) values of the transform variable k are $\hat{f}$ and $\hat{g}$ ordinary functions? Show that for those values of k, $\hat{h} = \hat{f}\hat{g}$, where $h = f * g$.

This exercise is related to the Laplace transform, which is discussed in Section 9.5.

6.3 Proof of Poisson's Equation

Let $k = k(\mathbf{x})$ denote the distribution $k(\mathbf{x}) = |\mathbf{x}|^{-1}$ (first example in Section 2.5; see equation (2.5-2)). Then, the potential $V = V(\mathbf{x})$ is defined to be the distribution

$$V = k * \rho, \tag{6.3-1}$$

which, when written symbolically, is just (6.1-1).

We shall now show that the Laplacian of the distribution k is given by

$$\nabla_y^2 k(\mathbf{x} - \mathbf{y}) = \nabla_x^2 k(\mathbf{x} - \mathbf{y}) = -4\pi\delta_3(\mathbf{x} - \mathbf{y}), \tag{6.3-2}$$

and then it will follow from the rule (6.2-5) for differentiating a convolution that

$$\nabla^2 V = (\nabla^2 k) * \rho = -4\pi\delta_3 * \rho = -4\pi\rho,$$

as required.

To establish (6.3-2), let $\varphi(\mathbf{x})$ be any test function in $C_0^\infty(\mathbb{R}^3)$. Then

$$\langle \nabla_x^2 k(\mathbf{x} - \mathbf{y}), \varphi(\mathbf{y}) \rangle = \langle k(\mathbf{x} - \mathbf{y}), \nabla^2 \varphi(\mathbf{y}) \rangle$$

$$= \int \frac{1}{|\mathbf{x} - \mathbf{y}|} \nabla^2 \varphi(\mathbf{y}) d^3 \mathbf{y}$$

$$= \int \frac{1}{|\mathbf{w}|} \nabla_x^2 \varphi(\mathbf{x} - \mathbf{w}) d^3 \mathbf{w}$$

$$= \nabla_x^2 \int \frac{1}{|\mathbf{w}|} \varphi(\mathbf{x} - \mathbf{w}) d^3 \mathbf{w}$$

$$= \nabla_x^2 \int \frac{1}{|\mathbf{x} - \mathbf{y}|} \varphi(\mathbf{y}) d^3 \mathbf{y}$$

$$= -4\pi\varphi(\mathbf{x}),$$

where the last step follows from classical potential theory (6.1-1 and 2), with $\rho(\mathbf{x})$ replaced by $\varphi(\mathbf{x})$. This establishes (6.3-2).

The same method can be used in higher dimensions, where the classical equations (6.1-1 and 2) are replaced by

$$V(\mathbf{x}) = \int \frac{1}{|\mathbf{x} - \mathbf{y}|^{n-2}} \rho(\mathbf{y}) d^n \mathbf{y} \tag{6.3-3}$$

and

$$\nabla^2 V = -\frac{2(n - 2)\pi^{n/2}}{\Gamma(n/2)} \rho \tag{6.3-4}$$

$$\overset{\text{def}}{=} -c_n \rho.$$

The constant c_n is equal to $(n-2)$ times the area of the unit sphere S^{n-1} in $\mathbb{R}^n$;

$$\text{area } (S^{n-1}) = \frac{2\pi^{n/2}}{\Gamma(n/2)}. \tag{6.3-5}$$

EXERCISES

1.   Derive (6.3-5) by evaluating the integral of $\exp\{-x_1^2 - \cdots - x_n^2\}$ over $\mathbb{R}^n$ first in polar coordinates and then as an iterated integral in Cartesian coordinates.

2.   Find the Fourier transform of the function $f(\mathbf{x}) = |\mathbf{x}|^{-n+2}$ in $\mathbb{R}^n$ by transforming the equation $\nabla^2(f * \rho) = -c_n\rho$, where ρ is a test function.

6.4 The Classical Potential-Theory Problems of Poisson, Dirichlet, Green, and Neumann

Suppose that electric charge is contained in a region of Ω of $\mathbb{R}^n$ with a piecewise smooth simple closed boundary $\partial\Omega$ (grounded conducting hypersurface) on which $V(\mathbf{x}) = 0$. Then the classical *Poisson problem* is, given that $\rho(\mathbf{x})$ is a reasonable function, say continuous in $\tilde{\Omega} = \Omega \cup \partial\Omega$, to find a function $V(\mathbf{x})$ of class C^2 in Ω and continuous in $\tilde{\Omega}$ such that

$$\nabla^2 V = -c_n\rho \quad \text{(given) in } \Omega \tag{6.4-1}$$

$$V(\mathbf{x}) = 0 \quad \text{for } \mathbf{x} \text{ on } \partial\Omega. \tag{6.4-2}$$

In the classical *Dirichlet problem*, the inhomogeneity appears at the boundary:

$$\nabla^2 V = 0 \quad \text{in } \Omega, \tag{6.4-3}$$

$$V(\mathbf{x}) = f(\mathbf{x}) \quad \text{(given) on } \partial\Omega. \tag{6.4-4}$$

The third classical problem is that of finding the so-called *Green's function* $G(\mathbf{x}, \mathbf{y})$ for the region Ω, which is of the form

$$G(\mathbf{x}, \mathbf{y}) = \frac{1}{|\mathbf{x} - \mathbf{y}|^{n-2}} + g(\mathbf{x}, \mathbf{y}), \tag{6.4-5}$$

where, for each fixed $\mathbf{y}$ in Ω, g is a solution of the particular Dirichlet problem

$$\nabla^2 g = 0 \quad \text{in } \Omega \tag{6.4-6}$$

(∇^2 denotes the Laplacian with respect to the components of $\mathbf{x}$) and

$$g(\mathbf{x}, \mathbf{y}) = -\frac{1}{|\mathbf{x} - \mathbf{y}|^{n-2}} \quad \text{for } \mathbf{x} \text{ in } \partial\Omega. \tag{6.4-7}$$

Hence, for fixed $\mathbf{y}$ in Ω, $G(\mathbf{x}, \mathbf{y})$ satisfies Laplace's equation, except at $\mathbf{x} = \mathbf{y}$ where it has a singularity as indicated in (6.4-7), and vanishes for $\mathbf{x}$ on $\partial\Omega$.

We state without proof that if Ω is a region with a reasonable boundary $\partial\Omega$ (see below), then all three classical problems are solvable. Their solutions are related as follows: If the Dirichlet problem (6.4-3, 4) has a solution for any given continuous $f(\mathbf{x})$ on $\partial\Omega$, then the problem (6.4-6, 7) has a solution, and consequently the Green's function is given by (6.4-5). For given $\mathbf{y}$, $G(\mathbf{x}, \mathbf{y})$ is the solution of a special case of the Poisson problem (6.4-1, 2) in which a point charge of unit amount is located at the point $\mathbf{x} = \mathbf{y}$, so that $\rho(\mathbf{x})$ is equal to the n-dimensional delta function of $\mathbf{x} - \mathbf{y}$. Then the solution of the general Poisson problem (6.4-1, 2) is given by the integral

$$V(\mathbf{x}) = \int_\Omega G(\mathbf{x}, \mathbf{y})\rho(\mathbf{y})d^n\mathbf{y}. \qquad (6.4\text{-}8)$$

Lastly, if $f(\mathbf{x})$ is any reasonable function given on $\partial\Omega$, say of class C^1, and if $\tilde{f}(\mathbf{x})$ is any function of class C^2 in Ω and class C^1 in $\tilde{\Omega}$ that takes on the given values on $\partial\Omega$, i.e.,

$$\tilde{f}(\mathbf{x}) = f(\mathbf{x}) \quad \text{on } \partial\Omega,$$

then the solution of the Dirichlet problem (6.4-3, 4) is given by $V(\mathbf{x}) = \tilde{f}(\mathbf{x}) + V_1(\mathbf{x})$, where $V_1(\mathbf{x})$ is the solution of the Poisson problem (6.4-1, 2) with $\rho(\mathbf{x})$ given by

$$\rho = \frac{1}{c_n}\nabla^2\tilde{f}. \qquad (6.4\text{-}9)$$

For practical purposes, a sufficient condition for solvability of the classical problems is the so-called *external cone condition*: It must be possible to find numbers $\varepsilon > 0$ and $h > 0$ such that for each point $\mathbf{x}$ of $\partial\Omega$ there is a circular cone of angle ε and height h with vertex at $\mathbf{x}$ and lying exterior to Ω; see Figure 6-1. This condition ensures that the boundary $\partial\Omega$ has no infinitely sharp reentrant edges, corners, or spikes. See Courant and Hilbert 1962 for details in the 3-dimensional case.

Figure 6.1 External cone condition.

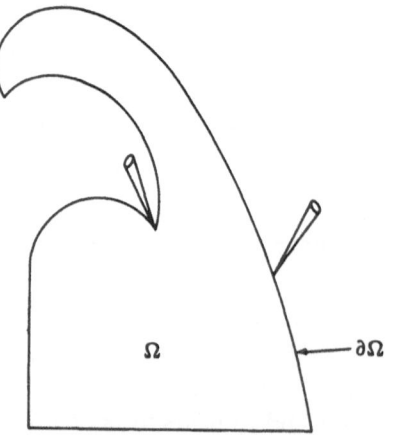

Ω $\partial\Omega$

The classical *Neumann problem* is similar to the Dirichlet problem but involves the normal component of the gradient of $V(\mathbf{x})$ on the boundary, rather than $V(\mathbf{x})$ itself. It is:

$$\nabla^2 V = 0 \quad \text{in } \Omega, \tag{6.4-10}$$

$$\mathbf{n} \cdot \nabla V = h(\mathbf{x}) \quad \text{(given) on } \partial\Omega, \tag{6.4-11}$$

where $\mathbf{n} = \mathbf{n}(\mathbf{x})$ is the outward unit vector normal to $\partial\Omega$ at the point $\mathbf{x}$ of $\partial\Omega$. An application of Gauss's theorem to the vector field ∇V shows that a necessary condition for this problem to have a solution is that

$$\int_{\partial\Omega} h(\mathbf{x}) d\mathscr{A} = 0. \tag{6.4-12}$$

If this condition is satisfied, then the external cone condition is sufficient for the existence of the solution. The corresponding *Poisson–Neumann problem* is:

$$\nabla^2 V = \rho \quad \text{(given) in } \Omega, \tag{6.4-13}$$

$$\mathbf{n} \cdot \nabla V = 0 \quad \text{on } \partial\Omega. \tag{6.4-14}$$

In this case the necessary conditions for the existence of the solution is

$$\int_{\Omega} \rho(\mathbf{x}) d^n\mathbf{x} = 0. \tag{6.4-15}$$

For a sort of analogue of the Green's function, see Exercise 7 below; see also Garabedian 1964 under "Neumann function" for the related operator ∇^2-const.

The solutions of the Neumann problems (6.4-10, 11) and (6.4-13, 14) are not unique because an arbitrary constant can be added to $V(\mathbf{x})$.

We state without proof that the above equations (6.4-1, 2, 8) hold also when ρ is an arbitrary distribution with support in Ω, in the following sense: We write (6.4-5) as

$$G(\mathbf{x}, \mathbf{y}) = k_{n-2}(\mathbf{x} - \mathbf{y}) + g_{\mathbf{x}}(\mathbf{y}), \tag{6.4-16}$$

where $k_{n-2}(\mathbf{x}) = |\mathbf{x}|^{-(n-2)}$ and $g_{\mathbf{x}}(\mathbf{y}) = g(\mathbf{x}, \mathbf{y})$. Then (6.4-8) is interpreted as

$$V = k_{n-2} * \rho + \langle \rho, g_{\mathbf{x}} \rangle \quad \text{in } \Omega, \tag{6.4-17}$$

in accordance with the following observations: (1) In n dimensions, the singularity of $k_{n-2}(\mathbf{x})$ is integrable, hence k_{n-2} can be identified with the distribution defined by

$$\langle k_{n-2}, \varphi \rangle = \int k_{n-2}(\mathbf{x}) \varphi(\mathbf{x}) d^n\mathbf{x};$$

(2) the definition of the functional $\langle \rho, \cdot \rangle$ can be continuously extended to any function $g(\mathbf{x})$ that is of class C^∞ on the support of ρ but does not itself

necessarily have bounded support; (3) Poisson's equation holds in the distribution sense; and (4) $V(\mathbf{x})$ is a continuous function outside the support of ρ and satisfies the boundary condition (6.4-2).

If the support of ρ is not restricted to Ω but only to $\tilde{\Omega}$, the only change is that the boundary condition (6.4-2) is satisfied only in the weak sense of Section 5.12.

EXERCISES

1. Let Ω be the ball $|\mathbf{x}| < a$ in $\mathbb{R}^n$. Show that the Green's function is

$$G(\mathbf{x}, \mathbf{y}) = \frac{1}{|\mathbf{x} - \mathbf{y}|^{n-2}} - \frac{1}{|(|\mathbf{y}|/a)\mathbf{x} - (a/|\mathbf{y}|)\mathbf{y}|^{n-2}} \qquad (6.4\text{-}18)$$

and verify that G is symmetric, $G(\mathbf{x}, \mathbf{y}) = G(\mathbf{y}, \mathbf{x})$, as in the general case, according to Exercise 3 below.

2. Derive Green's identity

$$\int_{\Omega} (u\nabla^2 v - v\nabla^2 u)d^n\mathbf{x} = \int_{\partial\Omega} (u\nabla v - v\nabla u) \cdot \mathbf{n}\, d\mathscr{A}, \qquad (6.4\text{-}19)$$

by applying Gauss's theorem to the vector field $u\nabla v - v\nabla u$; $\mathbf{n} = \mathbf{n}(\mathbf{x})$ is the outward normal unit vector for $\mathbf{x}$ in $\partial\Omega$.

3. Show that the Green's function of a region Ω is symmetric, $G(\mathbf{x}, \mathbf{y}) = G(\mathbf{y}, \mathbf{x})$ by taking $u(\mathbf{x}) = G(\mathbf{x}, \mathbf{y})$ and $v(\mathbf{x}) = G(\mathbf{x}, \mathbf{w})$ for fixed $\mathbf{y}$ and $\mathbf{w}$ in Ω ($\mathbf{y} \neq \mathbf{w}$), by applying Green's identity to the region

$$\Omega' = \{\mathbf{x} \in \Omega : |\mathbf{x} - \mathbf{y}| > \varepsilon, |\mathbf{x} - \mathbf{w}| > \varepsilon\}$$

(see Figure 6-2), and by then letting $\varepsilon \to 0$.

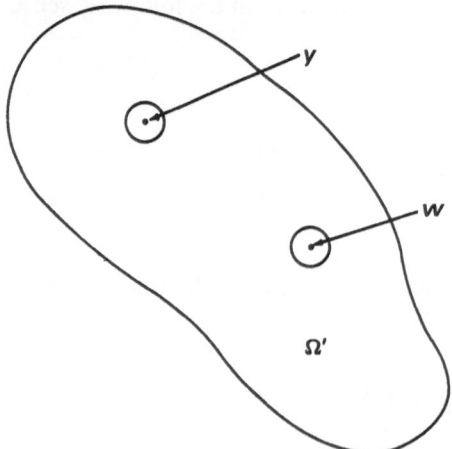

Figure 6.2 Symmetry of the Green's function.

4.   Show that the solution of the Dirichlet problem (6.4-3, 4) is given by the *Poisson integral formula*

$$V(\mathbf{x}) = \frac{1}{c_n} \int_{\partial\Omega} f(\mathbf{y})\mathbf{n}(\mathbf{y}) \cdot \nabla_{\mathbf{y}} G(\mathbf{y}, \mathbf{x}) d\mathscr{A}(\mathbf{y}), \tag{6.4-20}$$

on the assumption that the various processes involved can be justified.

5.   Show that the processes involved in the preceding exercise cannot be justified if $\partial\Omega$ has a reentrant edge or corner by considering a problem in which a charge faces such a projection, as in Figure 6-3 and showing that the field strength $\nabla_{\mathbf{y}} G(\mathbf{y}, \mathbf{x})$ is infinite at the corner, $\mathbf{y} = \mathbf{y}_0$.

6.   Show that if Ω is the ball $|\mathbf{x}| < a$ in $\mathbb{R}^3$, Poisson's integral formula is

$$V(\mathbf{x}) = \frac{a^2 - |\mathbf{x}|^2}{4\pi a} \int_{\partial\Omega} \frac{f(\mathbf{y}) d\mathscr{A}(\mathbf{y})}{|\mathbf{x} - \mathbf{y}|^3} \tag{6.4-21}$$

or, in polar coordinates,

$$V(\mathbf{x}) = \frac{a^3 - ar^2}{4\pi} \iint \frac{f(\mathbf{y})\sin\theta \, d\theta \, d\varphi}{(a^2 + r^2 - 2ar\cos\theta)^{3/2}}, \tag{6.4-22}$$

where $r = |\mathbf{x}| < a$ and θ is the angle between $\mathbf{x}$ and $\mathbf{y}$ (i.e., the polar axis for the variable point $\mathbf{y}$ on the sphere $|\mathbf{y}| = a$ is taken in the $\mathbf{x}$ direction).

7.   Consider the Poisson–Neumann problem (6.4-13, 14) in which the charge distribution $\rho(\mathbf{x})$ consists of a positive point charge at $\mathbf{x} = \mathbf{y}$ and a negative one at $\mathbf{x} = \mathbf{y}'$. Since the solution $V(\mathbf{x})$ contains an arbitrary additive constant, we consider the difference $V(\mathbf{x}) - V(\mathbf{x}')$ for two points $\mathbf{x}$ and $\mathbf{x}'$ and denote it by

$$G(\mathbf{x}, \mathbf{x}', \mathbf{y}, \mathbf{y}'). \tag{6.4-23}$$

Find a characterization of this function analogous to (6.4-5) for the Green's function.

The function $G(\mathbf{x}, \mathbf{x}', \mathbf{y}, \mathbf{y}')$ is interpreted as an *electrical resistance* by supposing that the region Ω is filled with a homogeneous material of unit resistivity and is insulated at its boundary $\partial\Omega$, that unit current is led in at point $\mathbf{y}$ and out at point $\mathbf{y}'$, and that the potential difference between the points $\mathbf{y}$ and $\mathbf{y}'$ is measured. Electrical resistance is thus conceptually a four-point function. For that reason, in classical electrical-measurements practice, precision standard resistors of low resistance were made with separate

Figure 6.3 Singularity of the field.

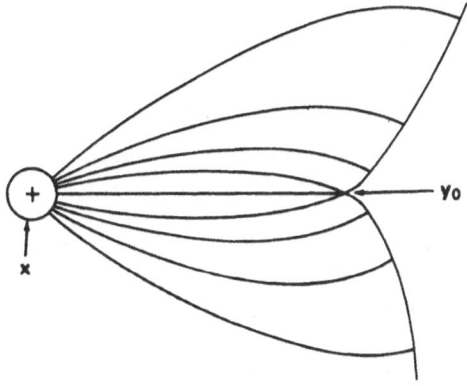

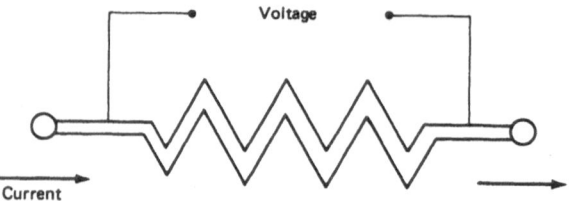

Figure 6.4 Idealized resistor.

current and voltage terminals, as in Figure 6-4. If we set $x = y$ or $x' = y'$, G becomes infinite; the interpretation of this is that if a finite current is led into a body of finite resistivity at a mathematical point, the resulting "contact" resistance is infinite (it diverges logarithmically as the radius of the "point" tends to zero).

EXERCISE

8. Show that

$$G(x, x', y, y') = G(y, y', x, x').$$

(This is one of many so-called *reciprocity relations* in electromagnetic theory.)

6.5 Schwartz's Nuclear Theorem; The Direct Product $f(x)g(y)$

The convolution product of ordinary functions is commutative; $f * g = g * f$, but it is not obvious from the definition (6.2-4) whether the same is true for distributions. (The commutativity was not used in the preceding discussion.) Let us assume that the distributions f and g on $\mathbb{R}^n$ *both* have bounded support (that assumption can be somewhat relaxed). Then the question whether $f * g = g * f$ becomes: Given the linear functionals $\langle f, \cdot \rangle$ and $\langle g, \cdot \rangle$; is it then true that

$$\langle f(x), \langle g(y), \varphi(x + y) \rangle \rangle = \langle g(y), \langle f(x), \varphi(x + y) \rangle \rangle \quad \text{for all } \varphi \text{ in } C_0^\infty(\mathbb{R}^n)?$$

$$(6.5\text{-}1)$$

The question makes sense, because in each member the quantity after the first comma is a test function, owing to the assumed boundedness of the supports of f and g. We first generalize the question a little: Is

$$\langle f(x), \langle g(y), \psi(x, y) \rangle \rangle = \langle g(y), \langle f(x), \psi(x, y) \rangle \rangle \quad \text{for all } \psi \text{ in } C_0^\infty(\mathbb{R}^{2n})?$$

$$(6.5\text{-}2)$$

[It might seem that this does not cover the preceding equation, because the support of $\varphi(x + y)$ necessarily extends to infinity in $\mathbb{R}^{2n}$ in the 45° directions $x + y = $ constant. However, to make the two equations agree, it is only necessary that $\psi(x, y)$ agree with $\varphi(x + y)$ in a certain rectangular region in $\mathbb{R}^{2n}$ determined by the supports of $f(x)$ and $g(y)$; outside that region, ψ can go to zero.]

The question in this more general form concerns the so-called direct product of two distributions. The left member of (6.5-2), as a linear functional defined for all ψ in $C_0^\infty(\mathbb{R}^{2n})$, defines a distribution on $\mathbb{R}^{2n}$, which we denote by $f(x)g(y)$ and call the *direct product* of f and g. Similarly the right member of (6.5-2) defines $g(y)f(x)$. Thus the question at hand is whether the direct product is commutative.

[We have already used the direct product in certain simple cases, like $\delta(x)\delta(y)$, where the equality of the two members of (6.5-2) is immediately evident.]

The same question arises when f is a distribution on $\mathbb{R}^n$ and g is one on $\mathbb{R}^m$; then $f(x)g(y)$ and $g(y)f(x)$ are distributions on $\mathbb{R}^{n+m}$. We may also ask about associativity: Is

$$f(x)[g(y)h(z)] = [f(x)g(y)]h(z)? \tag{6.5-3}$$

All these questions are answered (affirmatively) by Schwartz's *nuclear theorem*. We note first that equation (6.5-2) is evidently valid in the special case $\psi(x, y) = \psi(x)\chi(y)$, for then both members are simply

$$\langle f, \psi \rangle \langle g, \chi \rangle; \tag{6.5-4}$$

this is a *bilinear* functional of ψ and χ. We state Schwartz's theorem without proof:

Nuclear Theorem.   *Let $B[\psi, \chi]$ be a bilinear functional defined for test functions ψ and χ on $\mathbb{R}^n$ and $\mathbb{R}^m$ and continuous in each argument with respect to the convergence mode $\overset{\mathscr{D}}{\rightarrow}$. Then there is a unique linear functional $L(\varphi)$ defined for test functions $\varphi(x, y)$ on $\mathbb{R}^{n+m}$ and also continuous with respect to $\overset{\mathscr{D}}{\rightarrow}$, such that*

$$L[\psi(x)\chi(y)] = B[\psi, \chi] \qquad \forall \psi, \chi. \tag{6.5-5}$$

The same holds with C_0^∞ replaced by $\mathscr{S}$ throughout and $\overset{\mathscr{D}}{\rightarrow}$ by $\overset{\mathscr{S}}{\rightarrow}$ and also for other kinds of continuity of the functionals.

If $B[\psi, \chi]$ is taken as (6.5-4), we derive the commutativity of the direct product from the uniqueness of L in the theorem, because, as noted above, the bilinear functionals corresponding to the two sides of (6.5-2) are the same.

By repeated use of the nuclear theorem, we conclude that a multilinear functional $M[\psi_1, \ldots, \psi_k]$ determines a unique linear functional $L[\varphi]$ such that

$$L[\psi_1(x_1) \cdots \psi_k(x_k)] = M[\psi_1, \ldots, \psi_k] \qquad \forall \psi_1, \ldots, \psi_k. \tag{6.5-6}$$

The trilinear case shows the associativity of the direct product.

Going back to (6.5-1), we then derive immediately the commutativity and associativity of the convolution product for distributions with compact support on $\mathbb{R}^n$:

$$f * g = g * f, \qquad f * (g * h) = (f * g) * h. \tag{6.5-7}$$

Some authors write the direct product as $f(\mathbf{x}) \times g(\mathbf{y})$, but that is surely unnecessary, since it is not done for ordinary functions.

For further discussion and generalization of the nuclear theorem, see Gel'fand and Vilenkin (1964), volume 4 of *Generalized Functions*, where, however, it is called the "kernel" theorem (because the Russian words for "kernel" and "nucleus" are the same).

For distributions with noncompact support, the convolution, when it exists, may not be associative. That is already true for functions; consider

$$f(x) \equiv 1$$
$$g(x) = xe^{-x^2}$$
$$h(x) = \begin{cases} 1 & \text{for } x > 0, \\ 0 & \text{for } x < 0. \end{cases} \qquad (6.5\text{-}8)$$

Calculation gives immediately $f * (g * h) = \text{const.} \neq 0$, $(f * g) * h \equiv 0$. However, associativity holds if two of f, g, h have compact support or, more generally, if the supports are suitably related.

In order to define the convolution in terms of the direct product as

$$\langle f * g, \varphi \rangle = \langle f(\mathbf{x})g(\mathbf{y}), \varphi(\mathbf{x} + \mathbf{y}) \rangle \qquad (6.5\text{-}9)$$

for all φ in $C_0^\infty(\mathbb{R}^n)$, it must be possible to find a test function $\psi(\mathbf{x}, \mathbf{y})$ in $C_0^\infty(\mathbb{R}^{2n})$ such that $\varphi(\mathbf{x} + \mathbf{y}) = \psi(\mathbf{x}, \mathbf{y})$ for all points $\mathbf{x}, \mathbf{y}$ in the support of the direct product, that is, in the Cartesian product

$$\text{supp}(f) \times \text{supp}(g) \qquad (6.5\text{-}10)$$

Hence the requirement is that if $\mathfrak{R}$ is any bounded region in $\mathbb{R}^n$, to be thought of as the support of φ, then the intersection of the set (6.5-10) with the set given by $\mathbf{x} + \mathbf{y} \in \mathfrak{R}$ (a "slab" at 45° with the axes) should be a bounded set. Then we can choose $\psi(\mathbf{x}, \mathbf{y}) = \varphi(\mathbf{x} + \mathbf{y})$ in that set and let $\psi \to 0$ smoothly outside it.

For distributions on $\mathbb{R}$ (see Exercise 1 in Section 6.3), the above condition is satisfied if f and g both have supports bounded from below on $\mathbb{R}$ (or both from above). The generalization to distributions on $\mathbb{R}^n$ is that f and g should both have supports (which may extend to infinity) lying in a circular cone in $\mathbb{R}^n$ of half angle $< \pi/2$. If $f, g,$ and h all have supports in such a cone, then $f * (g * h) = (f * g) * h$.

6.6 The Variational Method for the Eigenfunctions of the Laplacian

On the basis of many classical examples where separation of variables can be used, it is natural to expect that for the problem $\nabla^2 u + \lambda u = 0$ in Ω with the boundary condition $u = 0$ on $\partial\Omega$ there is always a complete orthonormal set $\{u_j\}_1^\infty$ of eigenfunctions, with corresponding eigenvalues $\{\lambda_j\}$, such that $\lambda_j \leq \lambda_{j+1}$ and $\lambda_j \to \infty$ as $j \to \infty$.

The classical variational method for this problem, which is independent of separation of variables, depends on the successive minima of the *Dirichlet integral*

$$D(u) = \int_\Omega |\nabla u|^2 \, d^n\mathbf{x} = \int_\Omega \nabla u \cdot \nabla u \, d^n\mathbf{x}, \qquad (6.6\text{-}1)$$

under various constraints on the function u. For example, the fundamental eigenfunction $u_1(\mathbf{x})$ (the one with the lowest eigenvalue λ_1) is given by minimizing $D(u)$ under the constraints that (1)

$$\int_\Omega u^2 \, d^n\mathbf{x} = 1 \qquad (6.6\text{-}2)$$

and (2)

$$u = 0 \quad \text{on } \partial\Omega, \qquad (6.6\text{-}3)$$

assuming that such a minimizing function $u_1(\mathbf{x})$ exists and is sufficiently differentiable. Namely, let $u_1 + \delta u$ be any nearby function that also vanishes on $\partial\Omega$; to lowest order, (6.6-1) and (6.6-2) give

$$\int \nabla u_1 \cdot \nabla(\delta u) d^n\mathbf{x} = 0, \qquad (6.6\text{-}4)$$

$$\int u_1 \delta u \, d^n\mathbf{x} = 0, \qquad (6.6\text{-}5)$$

and integration by parts in (6.6-4) gives

$$-\int (\nabla^2 u_1)\delta u \, d^n\mathbf{x} = 0,$$

because $\delta u = 0$ on $\partial\Omega$. Hence,

$$\int_\Omega (\nabla^2 u_1 + \lambda u_1)\delta u \, d^n\mathbf{x} = 0 \qquad (6.6\text{-}6)$$

for any value of the so-called Lagrange multiplier λ. Equation (6.6-5) says that δu must be orthogonal to u_1, but otherwise it is an arbitrary smooth function that vanishes on $\partial\Omega$. However, if λ is so chosen that $\nabla^2 u_1 + \lambda u_1$ is orthogonal to u_1, i.e. if λ is determined by

$$\int_\Omega (\nabla^2 u_1 + \lambda u_1)u_1 \, d^n\mathbf{x} = 0, \qquad (6.6\text{-}7)$$

then (6.6-6) holds whether δu is orthogonal to u_1 or not. Hence,

$$\nabla^2 u_1 + \lambda u_1 = 0 \quad \text{in } \Omega, \qquad (6.6\text{-}8)$$

which is the desired eigenfunction equation if we set $\lambda_1 = \lambda$; it is the *Euler-Lagrange equation* of the variational problem.

The higher eigenfunctions are similarly obtained. Namely, after the eigen-functions $u_1, \ldots, u_{j-1}$ have been found, u_j is the function u that minimizes the Dirichlet integral (6.6-1) subject to the conditions:

$$\int_\Omega u^2 \, d^n\mathbf{x} = 1,$$

$$\int_\Omega u_k u \, d^n\mathbf{x} = 0, \qquad k = 1, \ldots, j-1,$$

$$u(\mathbf{x}) = 0 \quad \text{on } \partial\Omega.$$

The variational calculation is carried out in detail in Section 6.8, where the existence of the minimizing functions is proved.

Similarly, the solution $V(\mathbf{x})$ of the Dirichlet problem (6.4-3, 4), when it exists, is the function u that minimizes the integral (6.6-1) subject just to the condition that $u(\mathbf{x}) = f(\mathbf{x})$ on the boundary $\partial\Omega$.

The existence of a function $u(\mathbf{x})$ that minimizes (6.6-1) under various conditions is known as *Dirichlet's principle* and was simply taken as obvious until toward the end of the nineteenth century, when counterexamples were found for certain special shapes of the boundary $\partial\Omega$. Under suitable restrictions on $\partial\Omega$, such as the external cone condition described in Section 6.4, the existence of the minimizing function was proved by various mathematicians, starting with Hilbert in 1899. See Courant 1950.

If the cone condition is violated, there may be no minimizing function. For example, if a sufficiently sharp needle, say one whose diameter is $= ae^{-b/z}$, where z is the distance from the point, sticks into the region, there is in general no solution of the Dirichlet problem. A simplified example of that case, in which the needle is strictly one-dimensional, is the subject of the following exercise.

EXERCISE

1. Consider the region Ω shown in Figure 6-5, in which a straight line segment along the axis of a very long closed cylinder is part of the boundary $\partial\Omega$. Suppose that the boundary function $f(\mathbf{x})$ is $= 0$ on the line segment and $= 1$ on the rest of the boundary. Suppose that the "trial function" $u(\mathbf{x})$ in the Dirichlet integral (6.6-1) is taken as a function of the radial coordinate r only (except near the ends), namely

$$u = \begin{cases} 0 & \text{for } 0 \le r \le \varepsilon, \\ \log\dfrac{r}{\varepsilon} \Big/ \log\dfrac{a}{\varepsilon} & \text{for } \varepsilon \le r \le a, \end{cases}$$

where a is the radius of the cylinder and ε is a parameter in $(0, a)$. Show that the integral (6.6-1) $\to 0$ as $\varepsilon \to 0$, neglecting end effects. Conclude that if there were a minimizing function u, it would satisfy $\nabla u = 0$, hence would be a constant, hence could not satisfy the boundary condition both on the axis and on the cylinder.

Figure 6.5 Dirichlet problem with no solution.

6.7 A Compactness Theorem for the Sobolev Space W^1

The classical Arzelà or Ascoli–Arzelà theorem says that any uniformly bounded and equicontinuous sequence of functions on a compact region in $\mathbb{R}^n$ contains a convergent, in fact uniformly convergent, subsequence (see Courant and Hilbert 1953 or Dunford and Schwartz 1957). In particular, the theorem applies if the functions have first derivatives that are also bounded by a common bound K, for then they are equicontinuous. The theorem is used for proving the existence of the solution of certain variational problems. For the variational problem of the Laplacian described in the preceding Section, a similar theorem, known as Rellich's Lemma, is needed, in which the function and its first derivatives are bounded not pointwise but in the L^2 norm, and the subsequence converges not pointwise but in L^2.

Functions u are said to constitute an equicontinuous family, if the differences $u(x + y) - u(x)$ are bounded for given y, by an amount $\varepsilon(y)$, which is the same for all functions in the family and all x and which $\to 0$ as $y \to 0$. In the present theorem, we have an L^2 bound on $u(x + y) - u(x)$, rather than one that is uniform in x, as shown by the first lemma below.

Let $W^1 = W^1(\mathbb{R}^n)$ be the Sobolev space discussed in Section 5.11, namely the Hilbert space consisting of all u in $L^2(\mathbb{R}^n)$ that have finite values of the norm $\|u\|_1$ given by

$$\|u\|_1^2 = \|u\|^2 + \|\nabla u\|^2, \tag{6.7-1}$$

where $\|u\|$ is the L^2 norm and

$$\|\nabla u\|^2 \stackrel{\text{def}}{=} \sum_{k=1}^{n} \left\| \frac{\partial u}{\partial x_k} \right\|^2. \tag{6.7-2}$$

Let K be > 0. We denote by $\mathcal{K}$ the set of u in W^1 for which $\|u\|_1 \leq K$. For any such u, $\|u\| \leq K$ and $\|\nabla u\| \leq K$.

Lemma 1. *Let u be in $\mathcal{K}$. Then, for any $\mathbf{y}$ and any $\delta > 0$,*

$$\|T_y u - u\| < K|2\mathbf{y}|^{1/2} \tag{6.7-3}$$

where T_y is the translation operator: $(T_y f)(\mathbf{x}) = f(\mathbf{x} - \mathbf{y})$.

PROOF   Let $u_\delta = J_\delta u$, where J_δ is the mollifier described in Section 2.6; since $\|J_\delta f\| \leq \|f\|$ for any f in L^2 (Section 5.10), and since $\nabla J_\delta f = J_\delta \nabla f$ (Section 2.6), we see that u_δ is also in the class $\mathcal{K}$. Then,

$$\|T_y u_\delta - u_\delta\|^2 = \|T_y u_\delta\|^2 - 2 \operatorname{Re}(u_\delta, T_y u_\delta) + \|u_\delta\|^2;$$

the first and third terms on the right are independent of $\mathbf{y}$ (they are in fact equal). Now

$$\frac{d}{ds}(u_\delta, T_{sy}u_\delta) = \frac{d}{ds}\int \overline{u_\delta(\mathbf{x})}u_\delta(\mathbf{x} - s\mathbf{y})d^n\mathbf{x}$$

$$= -\int \overline{u_\delta(\mathbf{x})}\mathbf{y} \cdot \nabla u_\delta(\mathbf{x} - s\mathbf{y})d^n\mathbf{x}.$$

Therefore, by the Schwarz inequality,

$$\left|\frac{d}{ds} 2 \operatorname{Re}(u_\delta, T_{sy}u_\delta)\right| \le |2\mathbf{y}| \,\|u_\delta\|\, \|\nabla u_\delta\|$$

$$\le |2\mathbf{y}|K^2;$$

hence

$$\|T_y u_\delta - u_\delta\|^2 = \int_0^1 \frac{d}{ds} \|T_{sy}u_\delta - u_\delta\|^2 \, ds \le |2\mathbf{y}|K^2,$$

and (6.7-3) follows because

$$T_y u_\delta - u_\delta = J_\delta(T_y u - u) \overset{L_2}{\to} T_y u - u, \quad \text{as } \delta \to 0.$$

It was shown in Section 5.10 that $u_\delta \to u$ in L^2, as $\delta \to 0$, for any fixed u. Here we need a little more.

Lemma 2. *The convergence $u_\delta \to u$ is uniform in the class $\mathcal{K}$. That is, for any $\varepsilon > 0$, there is a $\delta_0 > 0$, independent of u, such that*

$$\|u_\delta - u\| < \varepsilon \quad \text{for all } u \text{ in } \mathcal{K}, \text{ for } \delta \le \delta_0.$$

In fact,

$$\|u_\delta - u\|^2 \le 2\delta K^2 \quad \text{for } u \in \mathcal{K}. \tag{6.7-4}$$

PROOF.   We show first that (6.7-4) holds if u is a C^∞ function ψ in $\mathcal{K}$. Namely,

$$|\psi_\delta(\mathbf{x}) - \psi(\mathbf{x})| \le \int |\psi(\mathbf{x} + \delta\mathbf{y}) - \psi(\mathbf{x})|\rho(\mathbf{y})d^n\mathbf{y}.$$

The square of this can be expressed as the product of two integrals, say with respect to $\mathbf{y}_1$ and $\mathbf{y}_2$. We integrate the result with respect to $\mathbf{x}$ first; the $\mathbf{x}$ integration is

$$\int |\psi(\mathbf{x} + \delta\mathbf{y}_1) - \psi(\mathbf{x})| \cdot |\psi(\mathbf{x} + \delta\mathbf{y}_2) - \psi(\mathbf{x})|d^n\mathbf{x}.$$

We use the Schwarz inequality and Lemma 1. Since the support of ρ has unit radius, it is only necessary to consider $|\mathbf{y}_1| \le 1$ and $|\mathbf{y}_2| \le 1$. Hence the above expression is $\le 2\delta K^2$ by Lemma 1. Then, the functions $\rho(\mathbf{y}_1)$ and $\rho(\mathbf{y}_2)$ both integrate to unity, hence (6.7-4) holds for $u = \psi$. Lastly, for given u in $\mathcal{K}$, we set $\psi = J_{\delta_1}u$. Then ψ is also in $\mathcal{K}$, because, according to (5.10-1),

$$\|\psi\| = \|J_{\delta_1}u\| \le \|u\|$$

and

$$\|\nabla\psi\| = \|\nabla J_{\delta_1}u\| = \|J_{\delta_1}\nabla u\| \le \|\nabla u\|.$$

Furthermore, $\psi_\delta - \psi = J_\delta(u_\delta - u)$, because $J_\delta J_{\delta_1} = J_{\delta_1}J_\delta$. Hence, $\psi_\delta - \psi \to u_\delta - u$ as $\delta_1 \to 0$, and (6.7-4) follows.

We now call $\mathscr{K}(\Omega)$ the set of elements of $\mathscr{K}$ with supports in a bounded region Ω of $\mathbb{R}^n$.

Theorem (Rellich's Lemma). *The set $\mathscr{K}(\Omega)$ is conditionally compact in the L^2 norm. That is, any sequence $\{u_k\}$ of elements in $\mathscr{K}(\Omega)$ contains a subsequence $\{\tilde{u}_k\}$ that that converges in L^2.*

Note. The word "conditionally" indicates that the limit of the sequence $\{\tilde{u}_k\}$ is not necessarily in $\mathscr{K}$ or even in W^1.

PROOF. We first show that the elements of $\mathscr{K}(\Omega)$, when suitably mollified, are equicontinuous. For any u in $\mathscr{K}(\Omega)$, $u_\delta = J_\delta u$ is defined in $\mathbb{R}^n$ and satisfies the equation

$$u_\delta(\mathbf{x} - \mathbf{y}) - u_\delta(\mathbf{x}) = J_\delta(T_y u - u) = \langle T_y u - u, \rho_{\mathbf{x}\,\delta} \rangle$$

(see Section 2.6). Therefore,

$$|u_\delta(\mathbf{x} - \mathbf{y}) - u_\delta(\mathbf{x})| \le \|T_y u - u\| \, \|\rho_{\mathbf{x}\,\delta}\|,$$

but

$$\|\rho_{\mathbf{x}\,\delta}\|^2 = \int \left| \rho\left(\frac{\mathbf{z}}{\delta}\right)\left(\frac{1}{\delta}\right)^n \right|^2 d^n\mathbf{z}$$

$$= \frac{\text{const.}}{\delta^n},$$

hence

$$|u_\delta(\mathbf{x} - \mathbf{y}) - u_\delta(\mathbf{x})| \le K |2\mathbf{y}|^{1/2} \frac{\text{const.}}{\delta^{n/2}}.$$

That is, the functions $J_\delta u = u_\delta(\mathbf{x})$, $u \in \mathscr{K}(\Omega)$, are equicontinuous for any $\delta > 0$. It is easy to show that they are also uniformly bounded for any $\delta > 0$. Lastly, they differ from zero only in a bounded region $\Omega(\delta)$ that extends beyond Ω to a distance δ. Hence the Arzelà theorem can be applied to them. Let $\{u_k\}$ be any sequence of elements in $\mathscr{K}(\Omega)$. We must show that there is a subsequence that converges in L^2. We proceed inductively:

(1) Take $\delta = 1$, $J_\delta = J_1$, and let $\{u_k^1\}$ be a subsequence of $\{u_k\}$ such that $\{J_1 u_k^1\}$ converges.

(2) Take $\delta = \frac{1}{2}$, $J_\delta = J_{1/2}$ and let $\{u_k^2\}$ be a subsequence of $\{u_k^1\}$ such that $\{J_{1/2} u_k^2\}$ converges.

$\vdots$

(q) Take $\delta = 1/q$, $J_\delta = J_{1/q}$ and let $\{u_k^q\}$ be a subsequence of $\{u_k^{q-1}\}$ such that $\{J_{1/q} u_k^q\}$ converges. Then the diagonal subsequence $\{u_k^k\}_{k=1}^\infty$ is such that $\{J_\delta u_k^k\}$ converges for any $\delta = 1, \frac{1}{2}, \ldots$, and we wish to show that $\{u_k^k\}$ itself converges. Given $\varepsilon > 0$, we choose $\delta = 1/q$ such that $\|J_\delta u - u\|$ is $< \varepsilon$ for all u in $\mathscr{K}$, by Lemma 2, and for that δ we choose L such that

$$\|J_\delta(u_k^k - u_l^l)\| < \varepsilon, \quad \text{for } k, l > L.$$

Then

$$\|u_k^k - u_l^l\| < 2\varepsilon, \quad \text{for } k, l > L,$$

and it follows that the subsequence $\{u_k^k\}$ of $\{u_k\}$ converges in the L^2 norm, as required.

EXERCISE

1.   Show that if u is in $W^1(\mathbb{R}^n)$, then $J_\delta u \to u$ in the norm $\| \quad \|_1$ and that the same is true of the restriction of $J_\delta u$ to Ω, if u is in $W^1(\Omega)$.

6.8 Existence of the Eigenfunctions

We now establish the existence and other properties of the solutions of the variational problem of Section 6.6. As in the preceding two sections, Ω is a bounded region of $\mathbb{R}^n$ bounded by a piecewise smooth hypersurface $\partial\Omega$, $L^2(\Omega)$ is the basic Hilbert space and W^1 the corresponding Sobolev space. W^1_0 is the subspace of W^1 of elements that vanish on $\partial\Omega$. According to the definition (6.7-2), the Dirichlet integral $D(u)$ is imply $\|\nabla u\|^2$.

We proceed inductively. We assume that the first $j-1$ eigenfunctions u_k and the corresponding eigenvalues λ_k have been found $(k = 1,\ldots,j-1)$ and that each u_k is in W^1_0 and satisfies the equation $\nabla^2 u_k + \lambda_k u_k = 0$ in Ω. As elements of $L^2(\mathbb{R}^n)$, the u_k are regarded as having support in $\tilde{\Omega}$, i.e. as $\equiv 0$ outside Ω. The construction is such that each new eigenfunction is normalized and orthogonal to the preceding ones, so we assume that $u_1,\ldots,u_{j-1}$ form an orthonormal set. According to Section 6.6, the new eigenfunction is required to minimize $\|\nabla u\|^2$ subject to the conditions just mentioned; hence we define $\mathfrak{M}$ as the corresponding subspace of W^1_0:

$$\mathfrak{M} = \{u \in W^1_0 : (u_k, u) = 0 \ (k = 1,\ldots,j-1)\}. \qquad (6.8\text{-}1)$$

We call

$$\lambda = \inf\{\|\nabla u\|^2 : u \in \mathfrak{M}, \|u\| = 1\} \qquad (6.8\text{-}2)$$

and we wish to show that the infimum is actually attained.

Theorem. *There is an element $\tilde{u}$ in $\mathfrak{M}$ such that $\|\tilde{u}\| = 1$, $\|\nabla\tilde{u}\|^2 = \lambda$, and $\nabla^2\tilde{u} + \lambda\tilde{u} = 0$ in Ω.*

PROOF.   By definition of the infimum, there is a sequence of elements u in $\mathfrak{M}$ with $\|u\| = 1$ on which $\|\nabla u\|^2$ converges (from above) to λ. If $K > \sqrt{1 + \lambda}$, the members of that sequence are all in the set $\mathscr{K}$ described in the preceding section, from some point on. Hence, by the compactness theorem, there is a subsequence, which we shall call $\{v^{(l)}\}_{l=1}^\infty$, that converges in L^2; that is, each $v^{(l)}$ is in $\mathfrak{M}$, and, as $l \to \infty$, $\|\nabla v^{(l)}\|^2 \to \lambda$, $v^{(l)} \to \tilde{u} \in L^2(\Omega)$. Owing to the continuity of the inner product $(\cdot,\cdot)$ in $L^2(\Omega)$, $\tilde{u}$ satisfies the equations

$$\|\tilde{u}\| = 1, \qquad (u_k, \tilde{u}) = 0, \qquad (k = 1,\ldots,j-1), \qquad (6.8\text{-}3)$$

because each $v^{(l)}$ satisfies them. Let w be an arbitrary element of $\mathfrak{M}$. Since λ may be characterized as the infimum of $\|\nabla u\|^2/\|u\|^2$ for u in $\mathfrak{M}$, we have

$$\|\nabla v^{(l)} + \varepsilon\nabla w\|^2 - \lambda\|v^{(l)} + \varepsilon w\|^2 \geq 0, \qquad (6.8\text{-}4)$$

for any ε and any $l = 1, 2, \ldots$. That is,

$$[\|\nabla v^{(l)}\|^2 - \lambda\|v^{(l)}\|^2] + 2\,\mathrm{Re}[\varepsilon(\nabla v^{(l)}, \nabla w) - \lambda\varepsilon(v^{(l)}, w)]$$
$$+ |\varepsilon|^2[\|\nabla w\|^2 - \lambda\|w\|^2] \geq 0. \quad (6.8\text{-}5)$$

As $l \to \infty$, the limit inferior of the left member of this inequality is ≥ 0; furthermore, the first square bracket $\to 0$, and $(v^{(l)}, w) \to (\tilde{u}, w)$. Therefore,

$$\lim \inf 2 \operatorname{Re} \varepsilon(\nabla v^{(l)}, \nabla w) - 2\lambda \operatorname{Re} \varepsilon(\tilde{u}, w) + |\varepsilon|^2 [\|\nabla w\|^2 - \lambda \|w\|^2] \geq 0.$$

Here, we let ε be in succession $= \varepsilon_0, -\varepsilon_0, i\varepsilon_0, -i\varepsilon_0$, where $\varepsilon_0 > 0$, thus getting upper and lower limits on the real and imaginary parts of $(\nabla v^{(l)}, \nabla w)$. In the limit $\varepsilon_0 \to 0$, we have

$$(\nabla v^{(l)}, \nabla w) \to \lambda(\tilde{u}, w), \quad \text{as } l \to \infty. \tag{6.8-6}$$

We use this result in two ways. First, for any φ in C_0^∞ (then, φ is in W_0^1), the function

$$w = \varphi - \sum_{k=1}^{j-1} (u_k, \varphi)u_k$$

is in $\mathfrak{M}$. Also,

$$\nabla^2 w = \nabla^2 \varphi + \sum_{k=1}^{j-1} \lambda_k(u_k, \varphi)u_k.$$

Since w and its first partial derivatives are all in $L^2(\Omega)$ and $v^{(l)}$ is in W_0^1, the integration-by-parts formula (5.12-5) can be used to transform the left member of (6.8-6) into $-(v^{(l)}, \nabla^2 w)$; here is where we use the boundary condition that $v^{(l)} = 0$ on $\partial\Omega$ in the sense that $v^{(l)} \in W_0^1$. Therefore, since $v^{(l)}$ and $\tilde{u}$ are orthogonal to $u_1, \ldots, u_{j-1}$, (6.8-6) gives

$$-(v^{(l)}, \nabla^2 \varphi) \to \lambda(\tilde{u}, \varphi),$$

hence $(\tilde{u}, \nabla^2 \varphi + \lambda\varphi) = 0$, hence, by the definition of distribution derivative,

$$\nabla^2 \tilde{u} + \lambda \tilde{u} = 0 \quad \text{in } \Omega, \tag{6.8-7}$$

which is one of the required results. Second, we take w in (6.8-6) as one of the $v^{(k)}$; then

$$(\nabla v^{(l)}, \nabla v^{(k)}) \to \lambda(\tilde{u}, v^{(k)}), \quad \text{as } l \to \infty,$$

that is, if we let k also $\to \infty$,

$$(\nabla v^{(l)}, \nabla v^{(k)}) \to \lambda\|\tilde{u}\|^2 = \lambda, \quad \text{as } k, l \to \infty,$$

from which it follows that $\{\nabla v^{(l)}\}$ is a Cauchy sequence in L^2. (I.e., for each q, $\{(\partial/\partial x_q)v^{(l)}\}_{l=1}^\infty$ is a Cauchy sequence.) By definition of distribution derivative, the limit of $\nabla v^{(l)}$ is $\nabla \tilde{u}$, and we conclude that $\nabla \tilde{u}$ is in L^2, i.e., $\tilde{u}$ is in the Sobolev space W^1, and $v^{(l)} \to \tilde{u}$ in the norm $\|\cdot\|_1$ of W^1 and $\|\nabla \tilde{u}\|^2 = \lambda$. Since each $v^{(l)}$ is in W_0^1, which is a closed manifold in W^1, $\tilde{u} \in W_0^1$. That is, $\tilde{u}$ satisfies the requirements of the theorem.

It is easy to see that the eigenfunctions that are obtained in this way form a complete set in $L^2(\Omega)$. First, they are orthonormal by construction. Second, after any set $\{u_1, u_2, \ldots\}$ of them has been obtained, if the dimension of $\mathfrak{M}$, which is orthogonal to those functions, is > 0, then still further eigenfunctions can be obtained by the construction, while if dim $\mathfrak{M} = 0$ there are no functions in W_0^1 orthogonal to all the eigenfunctions. With respect to the L^2 norm, W_0^1 is dense in $L^2(\Omega)$, hence the eigenfunction system is complete.

6.9 A Problem from Hydrodynamical Stability; Irrotational and Solenoidal Vector Fields

In Sattinger 1970 the following problem appears: Let Ω be a simply connected region of $\mathbb{R}^3$ with a piecewise smooth boundary $\partial\Omega$. We seek a smooth vector field $\mathbf{u}(\mathbf{x})$ in $\tilde{\Omega} = \Omega \cup \partial\Omega$ that satisfies:

$$\nabla^2 \mathbf{u} + \lambda \mathbf{u} = \nabla p \quad \text{in } \Omega \tag{6.9-1}$$

$$\nabla \cdot \mathbf{u} = 0 \quad \text{in } \Omega \tag{6.9-2}$$

$$\mathbf{u} = 0 \quad \text{on } \partial\Omega, \tag{6.9-3}$$

for some number λ and some scalar field $p(\mathbf{x})$. Then, λ is called an *eigenvalue* of the above problem and $\mathbf{u}(\mathbf{x})$ an *eigenfunction*.

In the full stability problem, the Laplace operator in (6.9-1) is modified by the addition of lower order terms consisting of first derivatives multiplied by functions that describe the basic flow whose stability is to be determined. In Sattinger's method, the solutions are obtained by suitably perturbing the solutions of the above problem.

It seems reasonable to expect the above problem to have a complete set of eigenfunctions, by comparison with the problem of electromagnetic vibrations of the cavity Ω, where the boundary $\partial\Omega$ is a perfect conductor and $\mathbf{u}(\mathbf{x})$ is the electric field. That problem is known to be self-adjoint and to have a complete orthonormal set of eigenfunctions. It differs from the present problem first in that ∇p is $=0$, which restricts the freedom of choice of $\mathbf{u}(\mathbf{x})$, and second in that the only boundary condition is the vanishing of the tangential component of $\mathbf{u}$ on $\partial\Omega$, and that increases the freedom of choice of $\mathbf{u}(\mathbf{x})$. We have thus traded off one function on $\partial\Omega$ (p satisfies Laplace's equation, hence is completely determined by its values on $\partial\Omega$) for another such function, the normal component of $\mathbf{u}(\mathbf{x})$ on $\partial\Omega$.

We first describe the calculation formally and then show how the steps can be justified by distribution theory. The Dirichlet integral is generalized by introducing, in addition to the inner product involving vectors, namely,

$$(\mathbf{u}, \mathbf{v}) = \int_\Omega \bar{\mathbf{u}} \cdot \mathbf{v} \, d^3\mathbf{x}, \tag{6.9-4}$$

also one involving dyadics

$$(\nabla\mathbf{u}, \nabla\mathbf{v}) = \int_\Omega \nabla\bar{\mathbf{u}} : \nabla\mathbf{v} \, d^3\mathbf{x}, \tag{6.9-5}$$

where the colon indicates the dyadic scalar product

$$\nabla\bar{\mathbf{u}} : \nabla\mathbf{v} = \sum_{(k,\,j)}^{3} (\partial_j \bar{u}_k)(\partial_j v_k). \tag{6.9-6}$$

The generalized Dirichlet integral is

$$D(\mathbf{u}) = \|\nabla\mathbf{u}\|^2 = (\nabla\mathbf{u}, \nabla\mathbf{u}). \tag{6.9-7}$$

The lowest eigenvalue λ is the minimum of $D(\mathbf{u})$ subject to the auxiliary and boundary conditions (6.9-2, 3) and the restriction $\|\mathbf{u}\| = 1$, where $\|\mathbf{u}\|$ is the norm obtained from (6.9-4). The construction of the jth eigenfunction is such as to make it orthogonal to the preceding eigenfunctions $\mathbf{u}_1, \ldots, \mathbf{u}_{j-1}$ and normalized, hence we assume that the preceding ones are orthonormal. We let $\mathfrak{M}$ denote the appropriate space of vector fields

$$\mathfrak{M} = \{\mathbf{u}(\mathbf{x}) : (\mathbf{u}_k, \mathbf{u}) = 0 \ (k = 1, \ldots, j-1), \nabla \cdot \mathbf{u} = 0 \text{ in } \Omega, \mathbf{u} = 0 \text{ on } \partial\Omega\}.$$

(6.9-8)

(This space will be defined more precisely later.) We assume that the minimum of $D(\mathbf{u})$ for $\mathbf{u} \in \mathfrak{M}$ and $\|\mathbf{u}\| = 1$ is obtained for $\mathbf{u} = \tilde{\mathbf{u}}$. Then, if we write $\mathbf{u} = \tilde{\mathbf{u}} + \mathbf{w}$ and regard $\mathbf{w} \in \mathfrak{M}$ as a small variation, we see by the same variational method as in Section 6.6 that

$$\int_\Omega [\nabla\tilde{\mathbf{u}} : \nabla\mathbf{w} - \lambda\tilde{\mathbf{u}} \cdot \mathbf{w}]d^3x = 0 \qquad \forall \mathbf{w} \in \mathfrak{M},$$

where λ is a Lagrange multiplier. More specifically,

$$\sum_{j=1}^{3} \int_\Omega [\nabla\tilde{u}_j \cdot \nabla w_j - \lambda\tilde{u}_j w_j]d^3x = 0.$$

Since each component of $\mathbf{w}$ is $=0$ on $\partial\Omega$, we can integrate by parts, and we find

$$\int_\Omega [\nabla^2\tilde{\mathbf{u}} + \lambda\tilde{\mathbf{u}}] \cdot \mathbf{w} \, d^3x = 0 \qquad \forall \mathbf{w} \in \mathfrak{M}. \qquad (6.9\text{-}9)$$

In this last equation the restriction that $\mathbf{w}$ be orthogonal to the preceding eigenfunctions can be dropped, for if $\mathbf{u}_k$ is one of them, we have

$$\int [\nabla^2\tilde{\mathbf{u}} + \lambda\tilde{\mathbf{u}}] \cdot \mathbf{u}_k \, d^3x = \int \tilde{\mathbf{u}} \cdot [\nabla^2\mathbf{u}_k + \lambda\mathbf{u}_k]d^3x$$

$$= (\lambda - \lambda_k) \int \tilde{\mathbf{u}} \cdot \mathbf{u}_k \, d^3x,$$

which is zero because $\tilde{\mathbf{u}}$ is orthogonal to the preceding eigenfunctions. Therefore (6.9-9) holds for arbitrary divergence-free $\mathbf{w}$ that vanishes on the boundary. In particular, if $\mathbf{w} = \nabla \times \boldsymbol{\varphi}$, where $\boldsymbol{\varphi}$ is an arbitrary vector field with support in Ω, then an integration by parts shows that $\nabla \times (\nabla^2\tilde{\mathbf{u}} + \lambda\tilde{\mathbf{u}})$ is orthogonal to every such $\boldsymbol{\varphi}$ and hence is $=0$, and we conclude that $\nabla^2\tilde{\mathbf{u}} + \lambda\tilde{\mathbf{u}}$ is a gradient, i.e., satisfies (6.9-1). (Recall that Ω was assumed to be simply connected.)

Conversely, if $\tilde{\mathbf{u}}$ satisfies (6.9-1) and is divergence-free and vanishes on the boundary, then we can take $\mathbf{w} = \tilde{\mathbf{u}}$ in (6.9-9), and then an integration by parts gives

$$D(\tilde{\mathbf{u}}) = \|\nabla\tilde{\mathbf{u}}\|^2 = \lambda\|\tilde{\mathbf{u}}\|^2.$$

To justify the various steps of the foregoing calculation, we first interpret (6.9-4, 5) in terms of the usual inner product in $L^2(\Omega)$ as

$$(\mathbf{u}, \mathbf{v}) = \sum_{j=1}^{3} (u_j, v_j) \qquad (6.9\text{-}10)$$

$$(\nabla\mathbf{u}, \nabla\mathbf{v}) = \sum_{j,k}^{3} (\partial_j u_k, \partial_j v_k). \qquad (6.9\text{-}11)$$

We call $\mathfrak{H}$ the Hilbert space of all vector fields in Ω with finite $\|\mathbf{u}\|$, and $\mathfrak{H}^1$ the corresponding Sobolev space of fields with finite values of the norm $\|\mathbf{u}\|_1$ defined by

$$\|\mathbf{u}\|_1^2 = \|\mathbf{u}\|^2 + \|\nabla\mathbf{u}\|^2. \qquad (6.9\text{-}12)$$

An element $\mathbf{u}$ of $\mathfrak{H}$ is in $\mathfrak{H}^1$ if all nine partial derivatives $\partial_j u_k$ are in $L^2(\Omega)$.

According to vector analysis any sufficiently smooth vector field can be decomposed as the sum of a divergence-free or *solenoidal* and a curl-free or *irrotational* part. Because of the condition (6.9-2), we wish to find subspaces of $\mathfrak{H}$ and $\mathfrak{H}^1$ consisting of divergence-free fields. However, the classical decomposition is not unique: one can add to one part and subtract from the other the gradient of any harmonic function; if $\nabla^2\psi = 0$, then $\nabla\psi$ is both divergence-free and curl-free. A boundary condition on $\partial\Omega$ (or at infinity) is needed to make the decomposition unique, and we choose somewhat arbitrarily the condition that the normal component of the solenoidal part vanish on $\partial\Omega$, since that condition is implied by (6.9-3). Hence, we wish to define a subspace $\mathfrak{H}_\sigma^1$ of $\mathfrak{H}^1$ ("σ" denotes "solenoidal") with the property that a smooth $\mathbf{u}(\mathbf{x})$ is in $\mathfrak{H}_\sigma^1$ if and only if

$$\begin{aligned} \nabla \cdot \mathbf{u} &= 0 \quad \text{in } \Omega, \\ \mathbf{u} \cdot \mathbf{n} &= 0 \quad \text{on } \partial\Omega. \end{aligned} \qquad (6.9\text{-}13)$$

For a smooth field $\mathbf{u}$, the conditions (6.9-13) are equivalent to the condition that

$$\int_\Omega \mathbf{u} \cdot \nabla\varphi \, d^3\mathbf{x} = 0 \quad \text{for every } \varphi \text{ in } C^\infty(\tilde\Omega), \qquad (6.9\text{-}14)$$

as is easily seen by integrating by parts. (By first considering φ that vanish on $\partial\Omega$ we see that if (6.9-14) holds, then $\nabla \cdot \mathbf{u} = 0$ in Ω, and by then considering general φ we see that $\mathbf{u} \cdot \mathbf{n} = 0$ on $\partial\Omega$). Hence, we define

$$\mathfrak{H}_\sigma = \{\mathbf{u} \in \mathfrak{H} : (\nabla\varphi, \mathbf{u}) = 0 \; \forall\varphi \in C^\infty(\tilde\Omega)\}. \qquad (6.9\text{-}15)$$

$\mathfrak{H}_\sigma$ is a *closed* linear manifold in $\mathfrak{H}$, hence a subspace, because it is an orthogonal complement. $\mathfrak{H}_\sigma^1$ is the corresponding subspace of $\mathfrak{H}^1$.

We interpret the subspace (6.9-8) as

$$\mathfrak{M} = \{\mathbf{u} \in \mathfrak{H}_\sigma^1 : (\mathbf{u}_k, \mathbf{u}) = 0 \; (k = 1, \ldots, j-1), \text{ each component of } \mathbf{u} \in W_0^1\} \qquad (6.9\text{-}16)$$

and we call

$$\lambda = \inf\{\|\nabla\mathbf{u}\|^2 : \mathbf{u} \in \mathfrak{M}, \|\mathbf{u}\| = 1\}.$$

Theorem.   *The infimum is attained; that is, there is an element ũ in $\mathfrak{M}$ such that $\|\nabla ũ\|^2 = \lambda$, $\|ũ\| = 1$; furthermore*

$$\nabla^2 ũ + \lambda ũ = \nabla p \quad in \ \Omega \tag{6.9-17}$$

for some scalar field p.

PROOF. (similar to the proof in the preceding section).   Consider a sequence of elements **u** in $\mathfrak{M}$ with $\|u\| = 1$ and such that $\|\nabla u\|^2 \to \lambda$. Let K be $> \sqrt{1 + \lambda}$; then from some point on in the sequence each **u** is in the set $\mathscr{K}$ referred to in the compactness theorem of Section 6.7. Hence there is a subsequence, which we call $\{v^{(l)}\}_1^\infty$, that converges in L^2. Owing to the continuity of the inner product, the limit ũ of the subsequence satisfies the equations

$$\|ũ\| = 1, \quad (u_k, ũ) = 0 \quad (k = 1, \ldots, j - 1), \tag{6.9-18}$$

because each $v^{(l)}$ satisfies them. By the same argument as in the preceding section (see equations (6.8-5 and 6)), we conclude that if **w** is an arbitrary element of $\mathfrak{M}$,

$$(\nabla v^{(l)}, \nabla w) \to \lambda(ũ, w), \quad as \ l \to \infty. \tag{6.9-19}$$

This result is used in three ways. First, by taking $w = v^{(m)}$ we see that $\{\nabla v^{(l)}\}$ is a Cauchy sequence and it follows as before that its limit is $\nabla ũ$, hence ũ is in $\mathfrak{H}^1$, and $v^{(l)} \to ũ$ in $\mathfrak{H}^1$, but $\mathfrak{M}$ is a subspace of $\mathfrak{H}^1$ (i.e. a *closed* linear manifold), hence ũ is in $\mathfrak{M}$. From (6.9-19), then

$$(\nabla ũ, \nabla w) = \lambda(ũ, w) \tag{6.9-20}$$

for $w \in \mathfrak{M}$, and in particular

$$\|\nabla ũ\|^2 = \lambda\|ũ\|^2 = \lambda. \tag{6.9-21}$$

Second, let **w** in (6.9-20) be u_k ($1 \leq k \leq j - 1$); since $(ũ, u_k) = 0$, we see that $(\nabla ũ, \nabla u_k) = 0$. Third, let φ be any vector field in $C_0^\infty(\Omega)$. Then

$$w = \nabla \times \varphi - \sum_{k=1}^{j-1} (u_k, \nabla \times \varphi)u_k$$

is in $\mathfrak{M}$, and we see from (6.9-13) that

$$(ũ, \nabla^2 \nabla \times \varphi + \lambda \nabla \times \varphi) = 0,$$

hence, by definition of distribution derivative,

$$\nabla \times (\nabla^2 ũ + \lambda ũ) = 0 \quad in \ \Omega \tag{6.9-22}$$

and it follows from the lemma below that

$$\nabla^2 ũ + \lambda ũ = \nabla p \quad in \ \Omega$$

for some scalar field p.

The completeness of the system of eigenfunctions is proved as in the preceding section.

Lemma.   *Let Ω be a simply connected region in $\mathbb{R}^3$ with boundary $\partial\Omega$ consisting of piecewise smooth closed surfaces and satisfying the external cone condition. Let **v** be a vector field, each component of which is a distribution in $L^2(\Omega)$, and suppose that $\nabla \times v = 0$ in Ω. Then there is a (scalar) distribution p in Ω such that $v = \nabla p$ in Ω.*

PROOF.   We show that for each $\delta > 0$ there is a distribution $p = p^\delta$ such that $\mathbf{v} = \nabla p$ in a region Ω_δ consisting of points in Ω at distance $> \delta$ from $\partial\Omega$. Furthermore, by careful choice of the arbitrary additive constants in the p^δ it will appear that if $0 < \delta' < \delta$, then $p^\delta = p^{\delta'}$ in Ω_δ. Then the required distribution p is obtained from the p^δ by the piecing together principle of Section 3.5. Let ψ_0 be a fixed test function with support in some Ω_{δ_0} and such that $\int_\Omega \psi_0 \, d^3\mathbf{x} = 1$, and consider values of $\delta \le \delta_0$. For any test function ψ with support in Ω_0, let $\chi(\mathbf{x})$ be the solution of the Poisson–Neumann problem

$$\nabla^2\chi = \psi - c\psi_0 \quad \text{in } \Omega_\delta$$
$$\mathbf{n} \cdot \nabla\chi = 0 \quad \text{on } \partial\Omega_\delta \qquad\qquad (6.9\text{-}23)$$
$$\chi(0) = 0,$$

where

$$c = \int_\Omega \psi \, d^3\mathbf{x}$$

and where the cartesian coordinate system is so chosen that the origin 0 is some point of Ω_{δ_0}. The problem has a solution because the integral of $\psi - c\psi_0$ is zero. We then define

$$\langle p, \psi \rangle = -\sum_{j=1}^{3} (\bar{v}_j, \partial_j \chi)_{\Omega_\delta} = -\lim_{\varepsilon \to 0} \int_{\Omega_\delta} J_\varepsilon \mathbf{v} \cdot \nabla\chi \, d^n\mathbf{x}. \qquad (6.9\text{-}24)$$

The limit exists because $J_\varepsilon \mathbf{v} \to \mathbf{v}$ in $L^2(\Omega_\delta)$. For $0 < \varepsilon < \delta$, $\nabla \times J_\varepsilon \mathbf{v} = 0$ in Ω_δ, hence by classical vector analysis there is a potential $q^\varepsilon(\mathbf{x})$ such that $J_\varepsilon \mathbf{v} = \nabla q^\varepsilon$ in Ω_δ. We fix the additive constant in q^ε by the requirement $\int q^\varepsilon \psi_0 d^n\mathbf{x} = 0$. Then

$$\langle p, \psi \rangle = -\lim_{\varepsilon \to 0} \int_{\Omega_\delta} \nabla q^\varepsilon \cdot \nabla\chi \, d^3\mathbf{x}$$

$$= \lim \int_{\Omega_\delta} q^\varepsilon \nabla^2\chi \, d^3\mathbf{x}$$

$$= \lim \int_{\Omega_\delta} q^\varepsilon(\psi - c\psi_0) d^3\mathbf{x} = \lim\langle q^\varepsilon, \psi \rangle.$$

Therefore $q^\varepsilon \to p$ as $\varepsilon \to 0$ (convergence of distributions). Therefore

$$\langle \nabla p, \psi \rangle = -\langle p, \nabla\psi \rangle = -\lim_{\varepsilon \to 0} \langle q^\varepsilon, \nabla\psi \rangle$$

$$= \lim\langle \nabla q^\varepsilon, \psi \rangle = \lim\langle J_\varepsilon \mathbf{v}, \psi \rangle$$

$$= \langle \mathbf{v}, \psi \rangle.$$

Conclusion.   For each $\delta \,\exists\, a$ distribution $p = p^\delta$ such that

$$\nabla p = \mathbf{v} \quad \text{in } \Omega_\delta, \qquad \langle p, \psi_0 \rangle = 0.$$

It follows that for $0 < \delta' < \delta$,

$$p^{\delta'} = p^\delta \quad \text{in } \Omega_\delta, \text{ as stated.}$$

6.10 The Cauchy–Riemann Equations; Harmonic Distributions

In complex analysis, it is known that if a function $f(z)$ of a complex variable
has a derivative $f'(z)$ for all z in a region Ω, then Re f and Im f satisfy the
Cauchy–Riemann equations in Ω. The converse is not true unless the partial
derivatives of Re f and Im f with respect to Re z and Im z are continuous,
as is shown by the example

$$f(z) = e^{-1/z^4} \quad \text{for } z \neq 0$$
$$f(0) = 0 \tag{6.10-1}$$

in which the Cauchy Riemann equations hold in the whole z plane. As
pointed out by P. D. Lax (private communication), if the Cauchy–Riemann
equations hold in the *distribution* sense, then $f'(z)$ exists; then, the continuity
of the partial derivatives is automatic and does not have to be assumed in
advance. It is not even necessary to assume that Re f and Im f are functions.
If u and v are any distributions on $\mathbb{R}^2$ that satisfy the Cauchy–Riemann
equations in Ω, then they are in fact real analytic functions and satisfy those
equations in the ordinary sense.

An equivalent statement is that if a distribution on $\mathbb{R}^2$ satisfies Laplace's
equation in Ω, i.e., is a harmonic distribution in Ω, then it is a harmonic
function $u(x, y)$ in Ω. In this form it is valid in $\mathbb{R}^n$.

Theorem. *A distribution f in $\mathbb{R}^n$ that is harmonic in a region Ω is equal in Ω
to a harmonic function.*

PROOF. First let f_δ be the result of smoothing f over a distance δ by a spherically symmetric mollifier $\rho_\delta(\mathbf{x}) = \rho(\mathbf{x}/\delta)(1/\delta)^n$, as described in Section 2.6. That is,

$$f_\delta = f * \rho_\delta = \langle f(\mathbf{y}), \rho_\delta(\mathbf{x} - \mathbf{y}) \rangle. \tag{6.10-2}$$

Then $f_\delta = f_\delta(\mathbf{x})$ is of class C^∞ for any $\delta > 0$, and $\to f$ in the sense of convergence of
distributions as $\delta \to 0$. By assumption, $\nabla^2 f$ is equal in Ω to the zero distribution. Let Ω_δ
be the slightly smaller region

$$\Omega_\delta = \{\mathbf{x} \in \Omega: \text{dist}(\mathbf{x}, \partial\Omega) > \delta\}. \tag{6.10-3}$$

If $\mathbf{x} \in \Omega_\delta$, then $\rho_\delta(\mathbf{x} - \mathbf{y})$ as a function of $\mathbf{y}$ has support in Ω. Therefore, by the rule for
differentiating a convolution

$$\nabla^2 f_\delta = (\nabla^2 f) * \rho_\delta = 0 \quad \text{in } \Omega_\delta. \tag{6.10-4}$$

That is, f_δ is a harmonic function in Ω_δ. Second, suppose $u(\mathbf{x})$ is a harmonic function in a
region Ω'. We assert that then

$$u_\delta(\mathbf{x}) \equiv u(\mathbf{x}) \quad \text{in } \Omega'_\delta. \tag{6.10-5}$$

The mean-value theorem of classical potential theory says that for $\mathbf{x}$ in Ω'_δ, $u(\mathbf{x})$ is equal to
the average of its values on any sphere

$$\{\mathbf{y}: |\mathbf{x} - \mathbf{y}| = \text{const.} \leq \delta\}. \tag{6.10-6}$$

Since the mollifier ρ has spherical symmetry, (6.10-5) follows. Third, by combining these two results, we show that f_δ is independent of δ, for small enough δ. Namely, let δ and δ' be any positive numbers such that $\delta + \delta' \leq \delta_0$ for some $\delta_0 > 0$. Then,

$$f * \rho_\delta * \rho_{\delta'} = f * \rho_{\delta'} * \rho_\delta.$$

By (6.10-5), the left number is $= f * \rho_\delta = f_\delta$ and the right is $= f * \rho_{\delta'} = f_{\delta'}$, i.e., $f_\delta = f_{\delta'}$. Hence f_δ is a harmonic function independent of δ so its limit f as $\delta \to 0$ is that same harmonic function. That holds in Ω_{δ_0}, but δ_0 was arbitrary > 0, hence it holds in Ω, as was to be proved.

CHAPTER 7

Linear Operators in a Hilbert Space

Linear operators or transformations in a Hilbert space; domain; range; bound; Extension Theorem; Banach algebras; adjoints; symmetric, self-adjoint, and unitary operators; integral and differential operators; symmetric operators with no self-adjoint extension and ones with many; simple Sturm–Liouville operators; closed and closable operators; the graph of an operator; radial momentum operators.

Prerequisites: Chapters 1–5.

7.1 Linear Operators

The idea of a linear operator or transformation in a Hilbert space $\mathfrak{H}$ (or a Banach space) is a direct generalization of the idea of a linear transformation in a finite-dimensional space. One point, however, needs emphasis (mainly because it is sometimes ignored, especially in books on quantum mechanics), namely, an operator A cannot be regarded as fully specified until its domain of definition (i.e., the set of those x in $\mathfrak{H}$ for which Ax is meaningful) has been specified; operators with different domains of definition have to be regarded as different operators. It is customary to require the domain of definition to be a linear set (manifold) in $\mathfrak{H}$, for the obvious reason that if A is linear and Ax is defined in a set S, then Ay can be uniquely defined, by linearity, when y is any finite linear combination of elements of S. However, further extensions are not generally unique, except in special circumstances.

The formal definitions are these: A linear *operator* or transformation A is a linear mapping of a linear subset $\mathfrak{D}(A)$ of $\mathfrak{H}$, the *domain* of A, onto a subset $\mathfrak{R}(A)$, the *range* of A. The domain and range are linear manifolds. An operator A' is an *extension* of A (in symbols, $A \subset A'$) if, first, $\mathfrak{D}(A) \subset \mathfrak{D}(A')$, and second, $Au = A'u$ for every u in $\mathfrak{D}(A)$. A is *bounded* if there is a K such that $\|Au\| \leq K\|u\|$ for every u in $\mathfrak{D}(A)$; the *bound* $\|A\|$ is the smallest such K. According to the Extension Theorem, proved below, a bounded linear operator A has a unique bounded extension $\tilde{A}$ whose domain is the closure of $\mathfrak{D}(A)$, and then $\|\tilde{A}\| = \|A\|$; in particular, if $\mathfrak{D}(A)$ is dense, then $\mathfrak{D}(\tilde{A}) = \mathfrak{H}$. If $\mathfrak{D}(B) \subset \mathfrak{R}(A)$, then BA is defined; in this case, if A and B are bounded, then $\|BA\| \leq \|B\| \|A\|$. If $u = 0$ is the only solution of the equation $Au = 0$, then A has an *inverse* A^{-1}, whose domain and range are the range and domain, respectively, of A; furthermore, $(BA)^{-1} = A^{-1}B^{-1}$. *Note*: In this book, the symbol $\subset$ is taken to include equality; if $A \subset A'$, but $A \neq A'$, then A' is a *proper* extension of A.

The foregoing definitions are valid in any Banach space. However, in a Hilbert space, there are expressions for the bound $\|A\|$ in terms of the inner product, namely,

$$\|A\| = \sup_{\substack{u \neq 0 \\ v \neq 0}} \frac{|(Au, v)|}{\|u\| \|v\|} = \sup_{\substack{u \neq 0 \\ v \neq 0}} \frac{\mathrm{Re}(Au, v)}{\|u\| \|v\|}. \tag{7.1-1}$$

PROOF.   By the Schwarz inequality and the definition of bound,

$$\mathrm{Re}(Au, v) \leq |(Au, v)| \leq \|Au\| \|v\| \leq \|A\| \|u\| \|v\|,$$

for all u and v. On the other hand, u can be chosen so that $\|Au\|/\|u\|$ is arbitrarily close to $\|A\|$; hence, if v is taken as Au, then (Au, v) is real and $= \|Au\|^2$, so that $(Au, v)/\|u\| \|v\|$ is equal to $\|Au\|/\|u\|$, hence is arbitrarily close to $\|A\|$. Equations (7.1-1) now follow.

Extension Theorem.   *If A is a bounded operator, then A has a unique extension $\tilde{A}$ to the closure of $\mathfrak{D}(A)$ such that $\|\tilde{A}\| = \|A\|$.*

PROOF.   The existence of $\tilde{A}$ is proved first. Let $\mathfrak{D}$ be the closure of $\mathfrak{D}(A)$; let u be any vector in $\mathfrak{D}$ and let $\{u_n\}$ be a sequence in $\mathfrak{D}(A)$ that converges to u. Au_n is defined for all n, and $\{Au_n\}$ is a Cauchy sequence, because $\|Au_n - Au_m\| \leq \|A\| \|u_n - u_m\|$ which $\to 0$ as $n, m \to \infty$, since $\{u_n\}$ is a Cauchy sequence. If, for every such u, $\tilde{A}u$ is defined as the limit of Au_n, as $n \to \infty$, then it is clear first that $\|\tilde{A}\| \leq \|A\|$, because $\|\tilde{A}u\| = \lim\|Au_n\| \leq \lim\|A\| \|u_n\| = \|A\| \|u\|$, and second that $\tilde{A}u = Au$ if u is in $\mathfrak{D}(A)$, so that $\tilde{A}$ is an extension of A, and $\|\tilde{A}\| = \|A\|$.

To prove uniqueness, suppose that $\tilde{A}'$ is any other bounded extension of A to $\mathfrak{D}$. If $\{u_n\}$ and u are as above, then

$$\|\tilde{A}'u - Au_n\| = \|\tilde{A}'u - \tilde{A}'u_n\| \leq \|\tilde{A}'\| \|u - u_n\|,$$

hence $\tilde{A}'u$ is also $= \lim Au_n$, i.e., $\tilde{A}' = \tilde{A}$.

An application of this theorem to integral operators is given in Section 7.4.

An important class of operators for quantum statistics and for applications to other parts of physics and mathematics is the class $\mathfrak{B}(\mathfrak{H})$ of bounded operators defined in all $\mathfrak{H}$. Their importance arises from the fact that $\mathfrak{B}(\mathfrak{H})$ is an *algebra*: not only is it a linear space, containing $c_1 A_1 + c_2 A_2$ for all A_1, A_2 in it and all c_1, c_2 in $\mathbb{C}$, but it contains the product BA of any A and B in it. Furthermore, $\mathfrak{B}(\mathfrak{H})$ is a complete normed linear space (i.e., a Banach space—see Chapter 15 and the note in Section 1.2) with the norm of A taken as the bound $\|A\|$, because this satisfies the usual requirements of a norm, including the triangle inequality $\|A + B\| \leq \|A\| + \|B\|$, and in addition the inequality $\|BA\| \leq \|B\| \|A\|$. Such an algebra is called a *Banach algebra*. For further discussion of bounded operators, see Section 9 and 14.6. Although many of the observables in quantum mechanics are unbounded operators, the same information can be provided, in principle, by bounded operators (bounded observables—i.e., ones whose possible measured values are restricted to bounded sets of real numbers) and this is worth while for some purposes—see Sections 14.5 and 14.6.

Note. Such algebras were formerly called "rings of operators" and are frequently called "normed rings" in the Russian literature.

7.2 Adjoints; Self-Adjoint and Unitary Operators

Observables are represented in quantum mechanics by self-adjoint operators in a Hilbert space $\mathfrak{H}$. These operators are the analogues of Hermitian matrices, but the infinite dimensionality of $\mathfrak{H}$ introduces one fundamental difference. If an $n \times n$ matrix A is such that $(Au, v) = (u, Av)$ for all u, v in V^n (in which case A is called *Hermitian*), then A has a complete orthonormal set of eigenvectors. The same is true (in a sense) for a *bounded* operator A in $\mathfrak{H}$; namely, if $(Au, v) = (u, Av)$ for all u, v in $\mathfrak{H}$, then A has a complete set of eigenvectors, in the sense of the spectral decomposition theorem in Chapter 9. However, most of the operators of quantum mechanics are unbounded and hence are not defined in all $\mathfrak{H}$ but only on a domain $\mathfrak{D}(A)$. If $(Au, v) = (u, Av)$ for all u and v in a domain $\mathfrak{D}(A)$ dense in $\mathfrak{H}$ (in this case A is called *Hermitian* by the physicists and *symmetric* by the mathematicians), then A may or may not have a complete set of eigenvectors (in the sense mentioned); if it does, it is called an *observable* by the physicists and a *self-adjoint* operator by the mathematicians. (The actual definition is given below).

Confusion has sometimes arisen because operators that are merely symmetric have been referred to as self-adjoint in some books on quantum mechanics and on ordinary differential equations.

The key to the self-adjointness of a symmetric operator is to have an adequate domain $\mathfrak{D}(A)$ (which is of course usually at one's disposal when A is being defined). If the domain is maximal in a certain sense, and A is symmetric on this domain, then A is self-adjoint. However, there are symmetric operators, such as the radial momentum operators discussed in Section 7.8 below, which cannot be made self-adjoint by any choice of the domain. On the other hand, there are operators (even ones with domain dense in $\mathfrak{H}$), which can be made into many different self-adjoint operators by suitably extending the domain; an example is given in Section 7.5; see also Section 8.6 on deficiency indices.

Proof of self-adjointness has generally been regarded as too intricate a matter, mathematically, to be discussed in books on quantum mechanics (see e.g., Messiah, 1958, p. 188). It is my opinion, however, that this need not be the case. The operators in question are mostly differential operators; if distribution theory is used, the main complications disappear, and the remaining difficulties have mostly to do with boundary conditions, hence are of a physical nature.

Theorem. *If A is a bounded operator defined in all $\mathfrak{H}$ (i.e., with $\mathfrak{D}(A) = \mathfrak{H}$), then there is a unique bounded linear operator A^*, the adjoint of A, such that $\mathfrak{D}(A^*)$ is also $= \mathfrak{H}$, and $(A^*u, v) = (u, Av)$ for all u and v in $\mathfrak{H}$; also, $\|A^*\| = \|A\|$; furthermore, if A and B are two such operators, then $(AB)^* = B^*A^*$, and $(A^*)^*$ (usually written simply as A^{**}) is equal to A.*

PROOF.   For any given u, (u, Av) is a bounded linear functional $l(v)$, and we take A^*u to be the unique element, whose existence is assured by the Riesz–Fréchet representation theorem (see Section 1.8), such that $l(v) = (A^*u, v)$ for all v; A^*u depends linearly on u, hence A^* is a linear operator in $\mathfrak{H}$. To find the bound of A^*, first write

$$|(A^*u, v)| = |(u, Av)| \leq \|u\| \|Av\| \leq \|u\| \|A\| \|v\|,$$

and then set $v = A^*u$;

$$\|A^*u\|^2 \leq \|u\| \|A\| \|A^*u\|; \quad \text{i.e.,} \quad \|A^*u\| \leq \|A\| \|u\|, \qquad \forall u;$$

therefore, $\|A^*\| \leq \|A\|$. The same argument, with A and A^* interchanged, shows that $\|A\| \leq \|A^*\|$; hence, $\|A^*\| = \|A\|$. The proofs of the last two assertions of the theorem are left as an exercise.

A more general definition of adjoint is now given, for which A does not need to be bounded or defined in all $\mathfrak{H}$, but is only required to be a linear operator with domain $\mathfrak{D}(A)$ *dense* in $\mathfrak{H}$. A linear subset $\mathfrak{D}^*$ of $\mathfrak{H}$ is first defined as follows: v is in $\mathfrak{D}^*$ if and only if there is a w in $\mathfrak{H}$ such that

$$(w, u) = (v, Au) \quad \text{for all } u \text{ in } \mathfrak{D}(A). \tag{7.2-1}$$

This $w = w(v)$ is unique, for any given v. *Proof:* If w_1 and w_2 were two such, we should have $(w_1 - w_2, u) = 0$ for all u in $\mathfrak{D}(A)$ and therefore for all u in $\mathfrak{H}$ by the continuity of the inner product, because $\mathfrak{D}(A)$ is dense; this would be true in particular for $u = w_1 - w_2$, so that $\|w_1 - w_2\|^2 = 0$, or $w_1 = w_2$. Clearly, w depends linearly on v; we therefore define: $\mathfrak{D}(A^*) = \mathfrak{D}^*$, $A^*v = w(v)$ for any $v \in \mathfrak{D}^*$. Hence, to find A^*, one must find all pairs $\{v, w\}$ that satisfy (7.2-1).

Warning. A^{**} may not exist, for $\mathfrak{D}(A^*)$ may not be dense; even if A^{**} exists, it may not be equal to A.

EXERCISE

1.   Show that if A^{**} exists, then $A \subseteq A^{**}$.

The *sum* $A + B$ of two operators is defined by the equation $(A + B)v = Av + Bv$, for all v in the domain

$$\mathfrak{D}(A + B) \overset{\text{def}}{=} \mathfrak{D}(A) \cap \mathfrak{D}(B).$$

EXERCISE

2.   Suppose that $A + B$ is densely defined in $\mathfrak{H}$, so that A^*, B^*, and $(A + B)^*$ exist. Show that $(A + B)^* \supset A^* + B^*$. Show by an example that $(A + B)^*$ may be a proper extension of $A^* + B^*$, i.e. $(A + B)^* \neq A^* + B^*$. Show however, that if B is bounded and defined in all $\mathfrak{H}$, then $(A + B)^*$ is always $= A^* + B^*$. *Hint*: According to the general definition, in order to find $(A + B)^*$, one must find all pairs v, w such that

$$(v, (A + B)u) = (w, u), \quad \text{for all } u \text{ in } \mathfrak{D}(A) \cap \mathfrak{D}(B).$$

It should be pointed out that operator addition, as defined above, has certain drawbacks. The domain of $A + B$ may be empty, except for the zero vector. Furthermore, addition is not generally associative, and in particular, $(A + B) - B$ is not necessarily $= A$. In some contexts, addition is restricted to operators all having the same domain, for example the domain C_0^∞ or $\mathscr{S}$, when one is dealing with operators in L^2.

With either definition of the adjoint, if $A^* = A$, then A is called *self-adjoint*, (abbreviated s.a.). If $A^* = A$, then $(Au, v) = (u, Av)$ for all u, v in $\mathfrak{D}(A)$; however, this condition is not sufficient for A to be self-adjoint. If merely $(Au, v) = (u, Av)$ for all u, v in $\mathfrak{D}(A)$, and if $\mathfrak{D}(A)$ is dense in $\mathfrak{H}$, i.e., if $A \subset A^*$, then A is called (*Hermitian*) *symmetric*; $\mathfrak{D}(A^*)$ may be larger than $\mathfrak{D}(A)$, in which case A^* is a proper extension of A. If A^* is in turn also symmetric (it need not be), then A^* is self-adjoint, i.e., $A^{**} = A^*$, but more generally, A^{**} may not exist, or A^{**} may be $\subset A^*$. (See examples in Section 7.5 below.) An alternative definition of self-adjointness is given in Section 8.6.

If A has a unique self-adjoint extension $\tilde{A}$, then A is called *essentially self-adjoint*. In practice, one often deals with A rather than $\tilde{A}$; A is often easier to describe than $\tilde{A}$. In Chapter 11 the Laplacian is discussed; first, an operator, there called A_0, is defined, whose domain is $C_0^\infty(\mathbb{R}^n)$, and which is defined by $A_0 \varphi = \nabla^2 \varphi$ for all φ in that domain. It is proved that A_0 is essentially self-adjoint. The domain of $\tilde{A}_0$ consists of certain distributions in $L^2(\mathbb{R}^n)$, but is not easy to describe.

If U is a linear operator with an inverse, and if $\mathfrak{D}(U) = \mathfrak{D}(U^{-1}) = \mathfrak{H}$, then the following conditions on U are equivalent:

$$(Uv, Uv) = (v, v) \quad \text{for all } v \text{ in } \mathfrak{H}. \tag{7.2-2}$$

$$(Uv, Uw) = (v, w) \quad \text{for all } v, w \text{ in } \mathfrak{H}. \tag{7.2-3}$$

$$U^{-1} = U^* \quad \text{(i.e., } UU^* = U^*U = I\text{)}. \tag{7.2-4}$$

(the implication (7.2-2) $\rightarrow$ (7.2-3) uses the polarization procedure, equation (1.11-1)—otherwise the proof is obvious); such an operator is called *unitary*; clearly $\|U\| = 1$. [Note that the more general definition of adjoint is not needed here, since U and U^{-1} are bounded and defined in all $\mathfrak{H}$.]

7.3 Examples in l^2

(See Section 1.4.)

1. Let $\xi = (x_1, x_2, \ldots)$ be any element of l^2, and define $A\xi = (x_2, x_3, \ldots)$ (x_1 is omitted, and the $n + 1$st coordinate replaces the nth, for $n = 1$, 2, $\ldots$); $\mathfrak{D}(A) = l^2$. For any $\eta = (y_1, y_2, \ldots)$ it is seen that $A^*\eta = (0, y_1, y_2, \ldots)$. Although $\mathfrak{D}(A^*) = l^2 = \mathfrak{H}$ and $\|A^*\eta\| = \|\eta\|$ for all η, A^* is not unitary, for its inverse is not defined in all $\mathfrak{H}$. $\|A\xi\|$ can be $< \|\xi\|$.

2. If $\xi = (x_1, x_2, \ldots)$, then put $A\xi = (x_2, x_4, x_1, x_6, x_3, x_8 \cdots x_{2n+2}, x_{2n-1} \ldots)$. A is unitary.

3. Let M be any $n \times n$ matrix. Let $\xi = (x_1, x_2, \ldots)$ and $A\xi = (w_1, w_2, \ldots)$, where

$$w_j = \sum_{k=1}^{n} M_{jk}x_k, \quad \text{for } j = 1, 2, \ldots, n,$$

$$w_j = x_j, \quad \text{for } j > n,$$

and $\mathfrak{D}(A) = l^2 = \mathfrak{H}$. A is self-adjoint if M is hermitian; A is unitary if M is unitary.

4. $\mathfrak{D}(A)$ is the set of all $\xi = (x_1, x_2, \ldots)$ for which only finitely many of the x_j are $\neq 0$. If $\xi = (x_1, x_2, \ldots)$ is in $\mathfrak{D}(A)$, then $A\xi = (x_1, 2x_2, 3x_3, \ldots, nx_n, \ldots)$. A is symmetric. $\mathfrak{D}(A)$ is dense in l^2, and A^* is an extension of A with domain given by

$$\mathfrak{D}(A) = \left\{ \xi = (x_1, x_2, \ldots): \sum_{j=1}^{\infty} j^2 |x_j|^2 < \infty \right\};$$

A^* is self-adjoint: $A^{**} = A^*$.

7.4 Integral Operators in $L^2(a, b)$

Let $K(x, y)$ be a continuous function of x and y, for $a \leq x, y \leq b$. Let $\mathfrak{D}(A) = C[a, b]$, i.e. $\mathfrak{D}(A)$ is the set of functions $\varphi(x)$ continuous for $x \in [a, b]$; for any such φ, define $A\varphi$ by

$$A\varphi = [A\varphi](x) = \int_a^b K(x, y)\varphi(y)dy. \tag{7.4-1}$$

By the Schwarz inequality, for any fixed x,

$$\left| \int_a^b K(x, y)\varphi(y)dy \right|^2 \leq \int_a^b |K(x, y)|^2 \, dy \int_a^b |\varphi(y)|^2 \, dy;$$

integrating with respect to x gives

$$\|A\varphi\|^2 \leq \int_a^b \int_a^b |K(x, y)|^2 \, dx \, dy \|\varphi\|^2. \tag{7.4-2}$$

Therefore, A is a bounded operator with domain dense in $\mathfrak{H} = L^2(a, b)$. According to the Extension Theorem, A has a unique bounded linear extension $\tilde{A}$ with $\mathfrak{D}(\tilde{A}) = \mathfrak{H}$. If the kernel $K(x, y)$ satisfies the equation $K(y, x) = \overline{K(x, y)}$, then A is symmetric, and $\tilde{A}$ is self-adjoint.

Integral operators of this kind can be generalized by allowing the interval (a, b) or the kernel $K(x, y)$ to be unbounded or by allowing the number of independent variables to be greater than one. If the integral in (7.4-2) is finite, then the operator is said to be of *Hilbert–Schmidt type* (see Section 12.4). However, an integral operator does not need to be of this type to be bounded. The Fourier transform operator F given in $L^2(\mathbb{R}^n)$ by the equations

$$\mathfrak{D}(F) = C_0^\infty(\mathbb{R}^n)$$

$$(F\varphi)(\mathbf{x}) = (2n)^{-n/2} \int \cdots \int e^{i\mathbf{x} \cdot \mathbf{y}} \varphi(\mathbf{y}) dy_1 \cdots dy_n$$

is densely defined and has bound $\|F\|$ equal to 1, but is obviously not of Hilbert–Schmidt type. Its extension $\tilde{F}$ to all L^2 is a unitary mapping of all L^2 onto itself, according to Section 5.10.

7.5 Differential Operators Via Distribution Theory

First, consider an operator T defined in $\mathfrak{H} = L^2(a, b)$ by the equations

$$\mathfrak{D}(T) = \{f \in L^2 : f' \in L^2 \quad \text{and} \quad f(a) = f(b) = 0\}$$
$$Tf = -if', \tag{7.5-1}$$

where f' is understood as the distribution derivative of f. T is symmetric, because integration by parts is permitted for distributions in L^2, and the integrated part vanishes on account of the boundary conditions, that is,

$$(Tf, g) = \int_a^b \overline{-if'} g \, dx = i\bar{f}g \Big|_a^b - i \int_a^b \bar{f}g' \, dx = (f, Tg). \tag{7.5-2}$$

To find the adjoint T^*, one looks for pairs $\{g, h\}$ of elements of L^2 such that

$$(Tf, g) = (f, h), \quad \text{for all } f \text{ in } \mathfrak{D}(T)$$
$$\text{i.e. } i(f', g) = (f, h), \quad \text{for all } f \text{ in } \mathfrak{D}(T). \tag{7.5-3}$$

The set of all resulting elements g is $\mathfrak{D}(T^*)$, and, for any such pair $\{g, h\}$, $T^*g = h$. In particular,

$$i(\bar{\varphi}', g) = (\bar{\varphi}, h), \quad \text{for all } \varphi \text{ in } C_0^\infty(a, b).$$

I.e., for all test functions, because if φ is in C_0^∞, then φ and $\bar{\varphi}$ are both in L^2. In the notation of Chapter 2, this is $i\langle g, \varphi' \rangle = \langle h, \varphi \rangle$, but $\langle g, \varphi' \rangle = \langle -g', \varphi \rangle$, hence

$$\langle -ig', \varphi \rangle = \langle h, \varphi \rangle, \quad \text{for all } \varphi \text{ in } C_0^\infty;$$

that is, h and $-ig'$ are the same distribution, i.e., the same element of $L^2(a, b)$. Consequently, a necessary condition for a pair $\{g, h\}$ to satisfy (7.5-3) is that $h = -ig'$, but this condition is also sufficient, because of (7.5-2). Since $f(a) = f(b) = 0$, no boundary condition on g is needed.

Conclusion.
$$\mathfrak{D}(T^*) = \{g \in L^2 : g' \in L^2\},$$
$$T^*g = -ig'. \tag{7.5-4}$$

T^* is a nonsymmetric extension of T, and neither T nor T^* is self-adjoint.

If $\mathfrak{D}(T)$ had been further restricted to functions in the class C^1 or C^k ($k > 0$) or C^∞ or even to analytic functions in $[a, b]$ or even to polynomials (the boundary conditions being retained in each case), then $\mathfrak{D}(T)$ would still be dense in $L^2(a, b)$ and T^* would still be given by (7.5-4). T^{**} would still be the old T, given by (7.5-1), so that now $T \subset T^{**}$.

Further examples in $L^2(a, b)$ are (1)

$$\mathfrak{D}(T_1) = \{f \in L^2 : f' \in L^2, f(a) = 0\}, \qquad T_1 f = -if';$$

then,

$$\mathfrak{D}(T_1^*) = \{f \in L^2 : f' \in L^2, f(b) = 0\}, \qquad T_1 f = -if',$$

and (2) (periodic boundary condition)

$$\mathfrak{D}(A) = \{f \in L^2 : f' \in L^2, f(a) = f(b)\}, \qquad Af = -if',$$

$$A \text{ is self-adjoint}; \qquad A^* = A.$$

Note. In the classical theory, where f and f' were considered to be ordinary functions (although f' was generally undefined on a set of measure zero), it was necessary to restrict $\mathfrak{D}(A)$ to functions of the class

$$\{f(x) \in L^2 : f(x) \text{ is absolutely continuous}, f'(x) \in L^2\},$$

where "$f'(x) \in L^2$" means that $f'(x)$, which exists almost everywhere, is Lebesgue measurable and that the Lebesgue integral $\int_a^b |f'(x)|^2\, dx$ is finite. In the present treatment, it is unnecessary to use Lebesgue theory and unnecessary to introduce the notion of absolute continuity. That is, $f(x)$ is automatically absolutely continuous if f', as a distribution, is in L^2.

The operator T^* given by (7.5-4) is characterized by having the largest possible domain for an operator in $L^2(a, b)$ whose action is $-i(d/dx)$, in the distribution theory sense. The operators T_1, T_1^*, and A are intermediate between the operator T of (7.5-1) and T^*. In particular, A is a self-adjoint operator such that $T \subset A \subset T^*$. However, A is not the only self-adjoint extension of T; there are uncountably many others, defined as follows: Let θ be any real number in $[0, 2\pi)$, and define A_θ by the equations

$$\mathfrak{D}(A_\theta) = \{f \in L^2 : f' \in L^2, f(b) = e^{i\theta}f(a)\}$$
$$A_\theta f = -if'. \tag{7.5-5}$$

EXERCISES

1. (a) Show that A_θ is self-adjoint, for each θ, and that $T \subset A_\theta \subset T^*$. (b) Find the eigenvalues and eigenfunctions of A_θ. (c) Show that the eigenfunctions form a complete set. [Hint: Take $(a, b) = (0, 2\pi)$; then, if $f(x)$ is an arbitrary smooth function, expand $e^{-i\theta x/2\pi} f(x)$ in an ordinary Fourier series. (d) Show that the boundary condition cannot be replaced by the condition $f(b) = \alpha f(a)$ with $|\alpha| \neq 1$.

2. Show that if $p(x)$ and $q(x)$ are real C^∞ functions and $p(x) > 0$, for $a \leq x \leq b$, then the regular Sturm–Liouville operator A given by

$$\mathfrak{D}(A) = \{f \in L^2(a, b): -(pf')' + qf \in L^2(a, b), f(a) = f(b) = 0\}$$
$$Af = -(pf')' + qf,$$

(7.5-6)

where the prime denotes differentiation, is self-adjoint.

A simple Sturm–Liouville operator with one singular end point (and with $p(x) \equiv 1$) is now considered. Let $q(x)$ be real, continuous, and ≥ 0, for $0 \leq x < \infty$. In the Hilbert space $\mathfrak{H} = L^2 = L^2(0, \infty)$, an operator A is defined as follows:

$$\mathfrak{D}(A) = \{f(x) \in L^2: -f''(x) + q(x)f(x) \in L^2, f(0) = 0\};$$
$$\text{for } f \text{ in } \mathfrak{D}(A), \ Af = -f'' + qf.$$

(7.5-7)

(More general operators of this type are discussed in Chapter 10.) Since f and Af are a fortiori in $L^2(0, b)$, for any finite b, it follows that qf and therefor f'' are in $L^2(0, b)$; hence, $f(x)$ and $f'(x)$ are continuous $(0 \leq x < \infty)$. Consequently, the boundary conditions $f(0) = 0$ in the definition of $\mathfrak{D}(A)$ is meaningful. [It doesn't follow that f' or f'' or qf is in $L^2(0, \infty)$; however, it will be seen that f' and $\sqrt{q} f$ are in $L^2(0, \infty)$.]

Since f'' is in $L^2(0, b)$, integration-by-parts may be used as follows:

$$(f, -f'' + qf) = \int_0^\infty \bar{f}(x)[-f''(x) + q(x)f(x)] \, dx$$

$$= \lim_{b \to \infty} \int_0^b \bar{f}(x)[-f''(x) + q(x)f(x)] \, dx$$

$$= \lim_{b \to \infty} \left\{ -\bar{f}(x)f'(x) \Big|_0^b + \int_0^b (|f'(x)|^2 + q(x)|f(x)|^2) dx \right\}.$$

The integral here, by its very nature, can only tend to a positive value or $+\infty$. In the latter case, Re $\bar{f}f'$ would $\to +\infty$; i.e., $d|f|^2/dx$ would $\to \infty$, which would contradict the quadratic integrability of $f(x)$. Therefore,

$$\int_0^\infty |f'(x)|^2 \, dx < \infty \quad \text{and} \quad \int_0^\infty q(x)|f(x)|^2 \, dx < \infty,$$

that is, f' and $\sqrt{q} f$ are in $L^2(0, \infty)$. It is noted for future reference that the boundary condition $f(0) = 0$ has not been used in this argument.

The adjoint operator $A*$ will now be found, and it will be shown that $A* = A$. We seek all possible pairs of elements $\{g, h\}$ in $L^2(0, \infty)$ such that

$$(-f'' + qf, g) = (f, h), \quad \text{for all } f \text{ in } \mathfrak{D}(A); \qquad (7.5\text{-}8)$$

The set of all resulting elements g is $\mathfrak{D}(A*)$, and if $\{g, h\}$ is any pair, then $A*g = h$. In particular, if $\{g, h\}$ is such a pair, then

$$(-\bar{\varphi}'' + q\bar{\varphi}, g) = (\bar{\varphi}, h) \quad \text{for all } \varphi \text{ in } C_0^\infty(0, \infty); \qquad (7.5\text{-}9)$$

that is,

$$\langle g, -\varphi'' + q\varphi \rangle = \langle h, \varphi \rangle \quad \text{for all } \varphi.$$

Now $\langle g, -\varphi'' \rangle = \langle -g'', \varphi \rangle$ by the definition of differentiation of distributions, and $\langle g, q\varphi \rangle = \langle qg, \varphi \rangle$. [Note: it is not asserted that the distribution qg is in $L^2(0, \infty)$]. Therefore, $\langle -g'' + qg, \varphi \rangle = \langle h, \varphi \rangle$ for all test functions; i.e.

$$h = -g'' + qg; \qquad (7.5\text{-}10)$$

this condition is necessary for $\{g, h\}$ to be a pair of the required kind, but not sufficient, because (7.5-9) does not imply (7.5-8).

To find a further condition, suppose that g and h satisfy the above equation (7.5-10). By the above argument for f, applied to g, it follows that g and g' are continuous and that g' and $\sqrt{q}g$ are also in $L^2(0, \infty)$. For any f in $\mathfrak{D}(A)$,

$$(-f'' + qf, g) = \lim_{b \to \infty} \left[-\bar{f}'g \Big|_0^b + \int_0^b (\bar{f}'g' + q\bar{f}g)dx \right];$$

the integral here tends to a finite limit, because f', g', $\sqrt{q}f$, and $\sqrt{q}g$ are all in $L^2(0, \infty)$. Therefore, the limit of $\bar{f}'(b)g(b)$ exists, but then this limit must be zero because $\int_0^\infty \bar{f}'(x)g(x)dx$ is finite. Integration by parts once more gives

$$(-f'' + qf, g) = \bar{f}'(0)g(0) + \lim_{b \to \infty} \left[\bar{f}g' \Big|_0^b + \int_0^b \bar{f}(-g'' + qg)dx \right];$$

again the integral tends to a finite limit, hence $\bar{f}(b)g'(b) \to 0$ as $b \to \infty$, by the same argument as before. Also $\bar{f}(0)g'(0) = 0$ because $f(0) = 0$, hence

$$(-f'' + qf, g) = \bar{f}'(0)g(0) + (f, -g'' + qg)$$
$$= \bar{f}'(0)g(0) + (f, h).$$

In order that $(Af, g) = (f, h)$ for all f in $\mathfrak{D}(A)$, it is therefore necessary and sufficient that $g(0) = 0$ in addition to (7.5-10). Therefore, $\mathfrak{D}(A*) = \mathfrak{D}(A)$, and A is self-adjoint.

A similar Sturm–Liouville operator $Af = -f'' + qf$ can be defined on $\mathbb{R}$ with singular end points at $\pm\infty$. In this case, the boundary condition $f(0) = 0$ is not needed, and $A* = A$ for $q \geq 0$.

3. Show that if $q(x) = x^2$ (A is then the oscillator Hamiltonian), then f'', xf', and $x^2 f$ are also in $L^2(\mathbb{R})$. Hint: consider the inner product of $-f'' + x^2 f$ with itself.

In Chapter 10 the spectra of these and more general Sturm–Liouville operators will be discussed, and certain eigenfunction expansion theorems will be derived.

4. Consider the operator T defined in $\mathfrak{H} = L^2(a, b)$ by the equations

$$\mathfrak{D}(T) = \{f \in L^2 : f'' \in L^2, f(a) = f(b) = f'(a) = f'(b) = 0\},$$
$$Tf = f''.$$

Show that T is a symmetric operator. Find T^*. Show that T is not essentially self-adjoint by finding a two-(complex)-parameter family of different self-adjoint extensions of T.

Lastly, the operation of multiplication by a given function, although not a differential operator, often appears in connection with differential equations, and is appropriately discussed at this point. Let $\xi(\mathbf{x})$ be any real continuous function on $\mathbb{R}^n$. According to Section 5.4, the product ξf is defined, as a distribution, for any f in L^2; an operator A is defined by the equations

$$\mathfrak{D}(A) = \{f \in L^2 : \xi f \in L^2\}, \quad \text{where } L^2 = L^2(\mathbb{R}^n),$$
$$Af = \xi f. \tag{7.5-11}$$

It will be shown that A is self-adjoint. Consider all pairs g, h such that

$$(Af, g) = (\xi f, g) = (f, h), \quad \text{for all } f \text{ in } L^2; \tag{7.5-12}$$

for any such pair, then

$$(\xi \bar{\varphi}, g) = (\bar{\varphi}, h), \quad \text{for all test functions } \varphi;$$

i.e.,

$$\langle g, \xi \varphi \rangle = \langle h, \varphi \rangle \quad \text{for all } \varphi.$$

According to Section 5.4, the left member is the functional that defines the product of ξ by g; therefore

$$\langle \xi g, \varphi \rangle = \langle h, \varphi \rangle, \quad \text{for all } \varphi;$$

i.e., g must be in $\mathfrak{D}(A)$ and $h(= A^*g) = \xi g = Ag$. This last is also sufficient for the validity of (7.5-12), for if $\{\varphi_n\}$ is a Cauchy sequence converging in L^2 to f, then $(\xi g, \varphi_n) = (h, \varphi_n)$; (7.5-12) follows in the limit by the continuity of $(\cdot, \cdot)$. Therefore $A^* = A$. If the given function $\xi(x)$ is bounded, then A is a bounded operator and $\mathfrak{D}(A)$ is all of L^2; otherwise A is unbounded.

7.6 Closed Operators

Let A be a linear operator whose domain $\mathfrak{D}(A)$ is not closed, and let ξ be a point of the closure $\bar{\mathfrak{D}}(A)$ which is not in $\mathfrak{D}(A)$ itself; we ask whether we can define $A\xi$ in some reasonable way. This is always possible if A is bounded, according to the Extension Theorem of Section 7.1. Namely, if $\{u_n\}$ is a sequence of points in $\mathfrak{D}(A)$ such that $u_n \to \xi$, as $n \to \infty$, then $\{Au_n\}$ is a Cauchy sequence, and $A\xi$ can be defined as the limit of Au_n, as $n \to \infty$.

If A is unbounded, the procedure just described fails, because the sequence $\{Au_n\}$ may have no limit, even though $\{u_n\}$ does; e.g., $\|Au_n\|$ may $\to \infty$. [Even if $\{Au_n\}$ has a limit, it may happen that for another sequence $\{u'_n\}$ converging to the same ξ, $\{Au'_n\}$ has a different limit from that of $\{Au_n\}$—see Example 3 below.]

Suppose, however, that for every sequence $\{u_n\}$ with the property that

(i) $u_n \in \mathfrak{D}(A)$, for all n,
(ii) $\lim u_n$ exists (as $n \to \infty$), and
(iii) $\lim Au_n$ exists (as $n \to \infty$),

it happens that $\lim u_n$ is also in $\mathfrak{D}(A)$, and $A(\lim u_n) = \lim(Au_n)$; then, A is a *closed operator*.

An unbounded operator which is not closed may or may not have a closed extension; if it does, it is called *closable*, and the smallest such extension is called the *closure* of A, and is denoted by $\tilde{A}$. Clearly A is closable if and only if it has the property that whenever two sequences $\{u_n\}$ and $\{v_n\}$ converges to the same limit ξ in $\mathfrak{H}$ and $\{Au_n\}$ and $\{Av_n\}$ also both converge, then these latter sequences also have the same limit, say η. If A has this property, its closure is obtained by defining $\tilde{A}\xi = \eta$ for each such pair ξ, η. Clearly, $\tilde{A}$ is the smallest (with respect to domain) or *minimal* closed extension of A; namely, if A_1 is any other closed extension of A, then A_1 is an extension of $\tilde{A}$. If A itself is closed, then $\tilde{A} = A$.

EXAMPLE 1

In Example 4 of Section 7.3, the operator A is closable, and its closure is its adjoint, $\tilde{A} = A^*$. Note that $\mathfrak{D}(\tilde{A}) \neq \mathfrak{H}$.

EXAMPLE 2

As in Section 7.5, let $\mathfrak{H}$ be $L^2(a, b)$ and let A be "d/dx" with any one of the boundary conditions considered in that section. Various classes of functions (and possibly distributions) can be considered for $\mathfrak{D}(A)$. First, if $\mathfrak{D}(A)$ is restricted to the functions $C^1(a, b)$ (continuous functions with continuous derivatives), then A is not closed, but is is closable and becomes closed if its domain is enlarged to include all continuous f such that the derivative f', *as a distribution*, is in $L^2(a, b)$. *Proof*: It will be shown that, whenever $\{f_n(x)\}$ is such that $\|f_n - f\| \to 0$ and $\|f'_n - g\| \to 0$ where $f_n(x)$ and $f'_n(x)$ are continuous in $[a, b]$, it then follows that $g = f'$, hence that f is in the enlarged domain, and $g = Af$, hence that the extension of A is closed. To show this, integrate by parts:

$$(-\varphi', f_n) = (\varphi, f'_n) \quad \text{for all } \varphi \text{ in } C_0^\infty(a, b)$$

(recall that $\varphi(a) = \varphi(b) = 0$ for any test function); let $n \to \infty$:

$$(-\varphi', f) = (\varphi, g) \quad \text{for all } \varphi;$$

hence, by the definition of the derivative of a distribution, it follows that $g = f'$.

To exhibit an operator that is not closable, we proceed in the spirit of the Lebesgue theory:

EXAMPLE 3

$\mathfrak{H} = L^2(a, b)$ again. Whenever $f(x)$ is continuous on $[a, b]$ (then $f(x)$ is in L^2) and is such that $f'(x)$ is defined (in the strict classical sense) almost everywhere on $[a, b]$ and is equal almost everywhere to a continuous function $g(x)$ on $[a, b]$ (then $g(x)$ is in L^2), then we say that $f(x) \in \mathfrak{D}(A)$ and that $Af(x) = g(x)$. This defines a linear operator A (it is left as an exercise to show that $g(x)$ is uniquely determined by $f(x)$ and that $g(x)$ depends linearly on $f(x)$). It has a larger domain than any of the corresponding operators in the preceding section.

A is not closable, because we can construct a sequence of functions $f_n(x)$ in $\mathfrak{D}(A)$ which converges in norm (and pointwise uniformly, for that matter) to the function $f(x) \equiv x$, while $Af_n(x) \equiv 0$ for each n. [Hence, if $\{g_n\}$ is another sequence defined simply by $g_n(x) \equiv x$ for every n, then $\lim f_n = \lim g_n$, while $\lim Af_n \neq \lim Ag_n$. To get a closed operator from this A, one must *remove* functions from $\mathfrak{D}(A)$, rather than add them.] To construct $f_n(x)$, we take $[a, b] = [0, 1]$ and we divide the unit square in the x, y plane into n^2 small squares or cells by horizontal and vertical lines. In each of the diagonal cells (see Figure 7-1), we define $y = f_n(x)$ to be a scaled-down replica of the Cantor function defined in Section 13.1. Then $f'_n(x) = 0$ almost everywhere, $f_n(x)$ is continuous, and $f_n(x) \to x$, as $n \to \infty$. Therefore, A is not closable.

Note.   As is seen in these examples, even if A is closed, its domain $\mathfrak{D}(A)$ is not necessarily closed; in fact, by the famous *closed-graph theorem*, if A is closed and has a closed domain, then A is bounded. In particular, if A is closed and defined in all $\mathfrak{H}$, then A is bounded. For a more general statement of the closed-graph theorem and its proof, see Kato (1966).

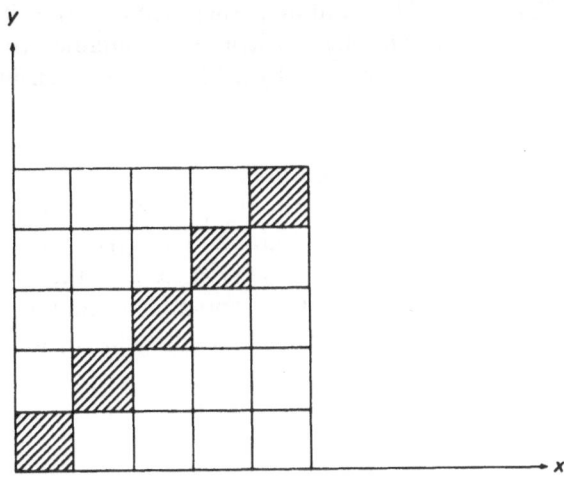

Figure 7.1 Construction of the sequence $\{f(x)\}$.

1. Show that for any A, if A^* exists (i.e., if $\mathfrak{D}(A)$ is dense in $\mathfrak{H}$), then A^* is closed. Hence, a symmetric operator is closable, and a self-adjoint operator is closed.

2. Determine which of the following operators are closable. For each closable one, find the domain of the closure.

(a) $\mathfrak{H} = l^2$, $\mathfrak{D}(A) =$ set of sequences $\xi = \{x_1, x_2, \ldots\}$ for which only finitely many x_n are $\neq 0$. For such ξ, define $A\xi$ as $A\xi = \{2x_1, 4x_2, 8x_3, \ldots, 2^n x_n, \ldots\}$.

(b) Same $\mathfrak{H}$, same $\mathfrak{D}(A)$; $A\xi = \{\sum_{n=1}^{\infty} x_n, 0, 0, 0, \ldots\}$.

(c) Same $\mathfrak{H}$, same $\mathfrak{D}(A)$; $A\xi = \{\sum_{n=1}^{\infty} (1/n)x_n, 0, 0, 0, \ldots\}$.

(d) $\mathfrak{H} = L^2(0, 1)$, $\mathfrak{D}(A) =$ set of continuous functions in $[0, 1]$; for $f(x)$ in $\mathfrak{D}(A)$, Af is the function $[Af](x) = f(\frac{1}{2})\sin \pi x$.

(e) Same $\mathfrak{H}$, same $\mathfrak{D}(A)$; $[Af](x) = \int_0^1 f(x')dx' \sin \pi x$.

7.7 The Graph of an Operator; Range and Nullspace

Closure and other properties of an operator can be interpreted geometrically in terms of its graph, which is defined as follows: The set of all ordered pairs $\{u, v\}$ of elements in $\mathfrak{H}$ is another Hilbert space, called $\mathfrak{H} \times \mathfrak{H}$, in which each such pair is a point or an element, and in which the inner product is defined by

$$(\{u_1, v_1\}, \{u_2, v_2\}) = (u_1, u_2) + (v_1, v_2),$$

the terms on the right being the original scalar products in $\mathfrak{H}$. The *graph* of an operator, denoted by $\Gamma(A)$, is the subset of $\mathfrak{H} \times \mathfrak{H}$ consisting of all pairs of the form $\{u, Au\}$, for u in $\mathfrak{D}(A)$. It is a linear set or manifold in $\mathfrak{H} \times \mathfrak{H}$ (if A is a linear operator). It is a *closed* linear manifold or a subspace of $\mathfrak{H} \times \mathfrak{H}$ if and only if A is a *closed* linear operator. A linear manifold in $\mathfrak{H} \times \mathfrak{H}$ is the graph of some linear operator A if and only if it contains no elements of the form $\{0, v\}, v \neq 0$. [If it contained such elements, then it would contain $\{u, w + \alpha v\}$, α an arbitrary number, whenever it contained $\{u, w\}$, so that Au could never be uniquely determined by u.] If A' is an extension of A, then $\Gamma(A')$ contains $\Gamma(A)$.

As a special case, if $\mathfrak{H}$ is the one-dimensional real space $\mathbb{R}$, an operator (linear or not) is a function $f(x)$ that maps $\mathbb{R}$ into itself: $\mathbb{R} \times \mathbb{R}$ is the Euclidean x, y plane, and the graph of $f(x)$ is the set of points $(x, f(x))$ in this plane. [This justifies the term "graph" for $\Gamma(A)$.] A *linear* operator in $\mathbb{R}$ is a homogeneous linear function $f(x) = ax$ and the graph is then a straight line through the origin, that is, a one-dimensional subspace. (In a finite number of dimensions, every linear manifold is closed, so the question of closure does not come up.)

1. The linear manifold $\hat{\Gamma}(A)$, defined as the set of elements $\{Av, -v\}$ in $\mathfrak{H} \times \mathfrak{H}$, for v in $\mathfrak{D}(A)$, is called the *rotated graph* of A, because it corresponds, in the one-dimensional case, to the result of rotating the graph in the x, y plane through $90°$ about the origin.

Show, from the definition of adjoint, that if A is densely defined, then A^* is the operator whose graph is the orthogonal complement (in $\mathfrak{H} \times \mathfrak{H}$) of $\Gamma(A)$, that is,

$$\Gamma(A^*) = \Gamma(A)^\perp, \qquad (7.7\text{-}1)$$

(while if A is not densely defined, then $\Gamma(A)^\perp$ is not the graph of an operator at all). This gives a second solution of Exercise 1 of the preceding section, because the orthogonal complement of a set is always a *closed* linear manifold, whether the set is closed or not. [If A is itself a closed operator, then $\Gamma(A)$ and $\Gamma(A^*)$ are both closed linear manifolds, and the space $\mathfrak{H} \times \mathfrak{H}$ is their orthogonal direct sum.]

2. Show that if A is a linear operator with dense domain, then its nullspace $\mathfrak{N}(A) = \{u \in \mathfrak{H}: Au = 0\}$ is the orthogonal complement of the range of A^*:

$$\mathfrak{N}(A) = \mathfrak{R}(A^*)^\perp. \qquad (7.7\text{-}2)$$

[N.B., (1) Here, "$\perp$" denotes orthogonal complement in the original space $\mathfrak{H}$. N.B., (2) $\mathfrak{N}(A)$ is automatically a closed linear manifold, but $\mathfrak{R}(A^*)$ is not, so that when the above statement is turned around, it becomes $\overline{\mathfrak{R}}(A^*) = \mathfrak{N}(A)^\perp$.]

3. Show that if A is closable, and A^* exists, then A^{**} exists and is the closure $\bar{A}$ of A.

Comment. For A closed, the representation of $\mathfrak{H} \times \mathfrak{H}$ as an orthogonal direct sum (see Section 1.6) is

$$\mathfrak{H} \times \mathfrak{H} = \Gamma(A) \oplus \Gamma(A^*).$$

The signs $\times$ and $\oplus$ have similar though slightly different meanings. If we define subspaces $\mathfrak{H}_1$ and $\mathfrak{H}_2$ of $\mathfrak{H} \times \mathfrak{H}$ as the sets of all pairs $\{u, 0\}$ and of all pairs $\{0, u\}$, respectively, then each of $\mathfrak{H}_1$ and $\mathfrak{H}_2$ is isomorphic to $\mathfrak{H}$, while

$$\mathfrak{H} \times \mathfrak{H} = \mathfrak{H}_1 \oplus \mathfrak{H}_2.$$

The difference is that $\oplus$ is used to connect *subsets* of a *given* space, and $\times$ is used to construct new spaces out of old ones. According to this usage, the sign $\oplus$ would have no meaning if the same symbol appeared on both sides of it.

7.8 The Radial Momentum Operators

According to the general rules of quantum mechanics, the momentum operator corresponding to a coordinate x is $-i(\partial/\partial x)$, in units such that $\hbar = 1$. If x is taken as the radial coordinate r, then its range is $0 < r < \infty$.

It is noted first that the operator $-i(\partial/\partial r)$ cannot be made self-adjoint in the Hilbert space $L^2(0, \infty)$ by any choice of its domain. For example, if an operator A is defined by the equations

$$\mathfrak{D}(A) = \{f \in L^2(0, \infty): f' \in L^2\}, \qquad Af = -if',$$

where the prime denotes distribution derivative, then A^* is given by the equations

$$\mathfrak{D}(A^*) = \{f \in L^2: f' \in L^2, f(0) = 0\}, \qquad A^*f = -if',$$

as the reader can easily verify. Furthermore, the adjoint of $A*$ is A, so neither A nor $A*$ is self-adjoint. $A*$ is symmetric, because $A* \subset A$, but has no self-adjoint extension (see Section 8.4).

A more appropriate Hilbert space is the L^2 space of distributions f, g in which the inner product is given by

$$(f, g) = \int_0^\infty \overline{f(r)}g(r)r^k \, dr, \tag{7.8-1}$$

when r is regarded as the radial coordinate of a polar coordinate system in a space of $k + 1$ dimensions. This Hilbert space is of the type denoted by L_σ^2 in Section 5.9, with

$$\sigma(r) = \begin{cases} \dfrac{r^{k+1}}{k+1} & \text{if } r \geq 0, \\ 0 & \text{if } r < 0. \end{cases}$$

To make the operator even formally self-adjoint in this Hilbert space, it is necessary to replace $-i(\partial/\partial r)$ by $-i(\partial/\partial r) + (k/2r)$. This suggests consideration of the operator A_k defined by the equation

$$\mathfrak{D}(A_k) = \left\{ f \in L_\sigma^2 : f' + \frac{k}{2r}f \in L_\sigma^2 \right\}, \qquad A_k f = -i\left(f' + \frac{k}{2r}f \right)$$

$$= -ir^{-k/2}(r^{k/2}f)'. \tag{7.8-2}$$

[The following curious fact is noted: In three dimensions ($k = 2$), A_k^2 is equal to minus the r-dependent part of the Laplacian, $-r^{-2}(\partial/\partial r)r^2(\partial/\partial r)$, whereas, for $k \neq 2$, an additional term appears:

$$A_k^2 = -r^{-k}\frac{\partial}{\partial r}r^k\frac{\partial}{\partial r} - \frac{k}{2}\left(\frac{k}{2} - 1\right)\left(\frac{1}{r}\right)^2.]$$

To find A_k^*, let f and g be in the domain of A_k. Integration by parts gives

$$\int_a^b \overline{A_k f}gr^k \, dr = i\int_a^b (r^{k/2}\overline{f})'(r^{k/2}g)dr$$

$$= i[r^k\overline{f}g]_a^b - i\int_a^b (r^{k/2}\overline{f})(r^{k/2}g)' \, dr. \tag{7.8-3}$$

Therefore,

$$(A_k f, g) = i\lim_{\substack{b \to \infty \\ a \to 0}} [r^k\overline{f(r)}g(r)]_a^b + (f, A_k g) \tag{7.8-4}$$

if the limits exist. From (7.8-3), with $g = f$, and the Schwarz inequality, it follows that

$$[r^k |f(r)|^2]_a^b = 2 \operatorname{Im} \int_a^b \overline{A_k f} f r^k \, dr$$

$$\leq 2 \left\{ \int_a^b |A_k f|^2 r^k \, dr \int_a^b |f|^2 r^k \, dr \right\}^{1/2}.$$

Each of the integrals in the curly brackets is finite, in the limit $a \to 0$ and $b \to \infty$, because $A_k f$ and f are both in L_σ^2. Hence, as in Section 5.6, if a and b both $\to \infty$ or both $\to 0$, the curly bracket $\to 0$, hence $r^k |f(r)|^2$ has definite limits, as $r \to \infty$ and as $r \to 0$. The limit as $r \to \infty$ is zero, for otherwise f could not be in L_σ^2, but the limit as $r \to 0$ need not be zero. [For example, if $f(r) = r^{-k/2} e^{-r}$, then f and $A_k f$ are both in L_σ^2, but $r^k |f(r)|^2 \to 1$ as $r \to 0$.] Therefore, (7.8-4) shows that the adjoint A_k^* of A_k is given by

$$\mathfrak{D}(A_k^*) = \left\{ g \in L_\sigma^2 : g' + \frac{k}{2r} g \in L_\sigma^2, \text{ and } \lim_{r \to 0} r^{k/2} |g(r)| = 0 \right\},$$

$$A_k^* g = -i \left(g' + \frac{k}{2r} g \right).$$

Hence, again, $A_k \neq A_k^*$, while $A_k^* \subset A_k$.

The non-self-adjointness of A_k and A_k^* is not mere mathematical pendantry. For every complex α with Im $\alpha > 0$, the function $f(r) = r^{-k/2} e^{i\alpha r}$ is an eigenfunction of A_k with eigenvalue α, whereas the eigenvalues of a *self-adjoint* operator are all real. On the other hand, A_k^* has not even a continuous spectrum. Symmetric operators, which, like A_k^*, have no self-adjoint extension, are characterized by their so-called deficiency indices, defined in Section 8.6.

7.9 Positive Operators; Numerical Range

Considerable information about an operator A can be obtained from the values taken on by (v, Av). If A is densely defined, then A is symmetric if and only if (v, Av) is real for all v in $\mathfrak{D}(A)$, because polarization (see Section 1.11) of the equation $(v, Av) = (Av, v)$ gives $(u, Av) = (Au, v)$. A symmetric operator is called *positive* if $(v, Av) > 0$ for all nonzero v in $\mathfrak{D}(A)$ and is called *nonnegative* if $(v, Av) \geq 0$ for all v in $\mathfrak{D}(A)$; in symbols, $A > 0$ or $A \geq 0$. The terms *positive definite* and *positive semidefinite* are also used. Negative and nonpositive operators are similarly defined. One writes $A \geq B$ if $A - B \geq 0$ and $A > B$ if $A - B > 0$. If A is any bounded operator, then the self-adjoint operators A^*A and AA^* are nonnegative, because $(v, A^*Av) = (Av, Av) \geq 0$ and $(v, AA^*v) = (A^*v, A^*v) \geq 0$.

If A is unbounded, then A^*A need not be self-adjoint, but von Neumann proved (see Kato (1966)) that, if A is closed and has a dense domain, then A^*A is self-adjoint. See Note below. By exercise 3 of Section 7.7, A^* also has dense domain, since A is closed, so that $A^{**} = A$; hence AA^* is also defined and self-adjoint. Clearly, AA^* and A^*A are nonnegative.

The *numerical range* or *field of values* of A is the set of complex numbers (v, Av) obtained as v ranges over all elements in $\mathfrak{D}(A)$ such that $\|v\| = 1$. Clearly, the eigenvalues, if any, of A lie in the numerical range. The continuous spectrum (see next chapter) lies in the closure of the numerical range, and so does the entire spectrum, if A is bounded (Kato). Any A such that $\mathfrak{D}(A)$ is dense is closable if the numerical range is not the whole complex plane. (Kato)

Note.   For the purpose of von Neumann's theorem, an operator product AB is defined on a domain $\mathfrak{D}(AB) = \{v \in \mathfrak{D}(B): Bv \in \mathfrak{D}(A)\}$, and then $(AB)v$ is defined as $A(Bv)$. Hence, $\mathfrak{D}(A^*A)$ may be smaller than $\mathfrak{D}(A)$, but von Neumann proved that it is at least a so-called *core* of A; that is, if A_1 is the restriction of A to $\mathfrak{D}(A^*A)$, then the closure of A_1 is $= A$.

CHAPTER 8

Spectrum and Resolvent

Continuous, point, and residual spectrum; eigenvectors and approximate eigenvectors; resolvent; analyticity of the resolvent; the Cayley transform; von Neumann's theory of the extension of symmetric operators; the deficiency indices of a symmetric operator; second definition of self-adjoint operator.

Prerequisites: Chapters 1–5 and 7.

8.1 Definitions

The eigenvalues of an $n \times n$ matrix M constitute a (finite) point set in the complex plane, called the *spectrum of* M. If A is any linear operator in a Hilbert space $\mathfrak{H}$, the complex plane $\mathbb{C}$ is similarly decomposed into two parts: the spectrum of A, denoted by $\sigma(A)$, and the resolvent set, denoted by $\rho(A)$. The spectrum of A is further decomposed into the point spectrum $P\sigma(A)$, the continuous spectrum $C\sigma(A)$, and the residual spectrum $R\sigma(A)$.

The classification is based on the existence and properties of the operator $(A - \lambda)^{-1}$, which is an abbreviation for $(A - \lambda I)^{-1}$, where I is the identity operator. It is recalled that a linear operator

$$T: \mathfrak{D}(T) \to \mathfrak{R}(T)$$

has an inverse

$$T^{-1}: \mathfrak{D}(T) \leftarrow \mathfrak{R}(T)$$

if and only if the transformation $u \to Tu$ is one-to-one, that is, if $Tu_1 = Tu_2$ implies $u_1 = u_2$, or $Tu = 0$ implies $u = 0$; that is, if 0 is not an eigenvalue of T.

The *point spectrum* $P\sigma(A)$ is the set of eigenvalues of A; that is:

$$P\sigma(A) = \{\lambda \in \mathbb{C}: Au = \lambda u \text{ for some nonzero } u \text{ in } \mathfrak{H}\}, \qquad (8.1\text{-}1)$$

or alternatively

$$P\sigma(A) = \{\lambda \in \mathbb{C}: A - \lambda \text{ has no inverse}\}. \qquad (8.1\text{-}2)$$

143

A number λ is in $C\sigma(A)$ (or possibly in $R\sigma(A)$) if there is no $u \neq 0$ such that $Au - \lambda u = 0$, but, given any $\varepsilon > 0$, there is an "approximate eigenvector" $u = u(\varepsilon)$ with norm $\|u\| = 1$ such that $\|Au - \lambda u\| < \varepsilon$. In this case, $(A - \lambda)^{-1}$ exists, but is unbounded. A further distinction is made according to whether the domain $\mathfrak{D}((A - \lambda)^{-1})$, which is $\mathfrak{R}(A - \lambda)$, is dense in $\mathfrak{H}$.

The *continuous spectrum* $C\sigma(A)$ of A is defined as

$$C\sigma(A) = \{\lambda \in \mathbb{C}: A - \lambda \text{ has an unbounded inverse with}$$

$$\text{domain dense in } \mathfrak{H}\}, \qquad (8.1\text{-}3)$$

and the *residual spectrum* $R\sigma(A)$ is defined as

$$R\sigma(A) = \{\lambda \in \mathbb{C}: A - \lambda \text{ has an inverse (bounded or unbounded)}$$

$$\text{with domain not dense in } \mathfrak{H}\}. \qquad (8.1\text{-}4)$$

[For most operators of interest (including all self-adjoint, unitary, and, more generally, normal ones), the residual spectrum is empty, and the continuous spectrum may be characterized as consisting of those λ's for which an approximate eigenvector of any desired accuracy can be found, but no true eigenvector.]

Finally, the *resolvent set* $\rho(A)$ is the remainder of the complex plane:

$$\rho(A) = \{\lambda \in \mathbb{C}: A - \lambda \text{ has a bounded inverse with domain dense in } \mathfrak{H}\}.$$

$$(8.1\text{-}5)$$

If $\lambda \in \rho(A)$, there are not even approximate corresponding eigenvectors, because

$$\|Au - \lambda u\| \geq (\|(A - \lambda)^{-1}\|)^{-1}\|u\|.$$

[This inequality is simply a reformulation of the inequality $\|(A - \lambda)^{-1}v\| \leq \|(A - \lambda)^{-1}\| \, \|v\|$, with v set equal to $Au - \lambda u$.]

For λ in the resolvent set $\rho(A)$, the operator $(A - \lambda)^{-1}$ is called the *resolvent* of A; it is denoted by R_λ or $R_\lambda(A)$. It is a family of operators depending on the complex parameter λ, for λ in $\rho(A)$; that is, it is an operator-valued function of the complex variable λ, defined on $\rho(A)$. Its properties play an important role in the analysis of self-adjoint and related operators. According to the definition of $\rho(A)$, R_λ is, for each λ, a bounded operator with domain dense in $\mathfrak{H}$.

8.2 Examples and Exercises

The reader should verify the statements given below, as far as possible. To show that a given λ is in $P\sigma(A)$, find an eigenvector; to show that λ is in $C\sigma(A)$, find a sequence of approximate eigenvectors; to show that λ is in

$\rho(A)$, solve the equation $Au - \lambda u = v$, for arbitrary v, and show that the values of $\|u\|$ are bounded, if $\|v\| = 1$.

1. Let $\mathfrak{H}$ be $L^2(\mathbb{R})$, and let A be the self-adjoint operator $i(d/dx)$ with domain consisting of all f in L^2 such that f' (as a distribution) is also in L^2. A has a purely continuous spectrum consisting of the entire real axis. *Hint*: Approximate eigenfunctions can be found (for λ real) in the form of wave packets $\alpha(x)\exp\{i\lambda x\}$, where $\alpha(x)$ is suitably chosen.

2. Let $\mathfrak{H}$ be $L^2(\mathbb{R})$, and let A be the operator of multiplication by x, defined by

$$\mathfrak{D}(A) = \{f(x) \in L^2 : xf(x) \in L^2\}, \qquad Af(x) = xf(x).$$

    A has a purely continuous spectrum consisting of the entire real axis.

3. The operator $-(d/dx)^2$, with a suitably chosen domain in $L^2(\mathbb{R})$—see Section 7.5—has a purely continuous spectrum consisting of the nonnegative real axis. (The negative real axis belongs to the resolvent set.)

4. The operator $-(d/dx)^2 + x^2$, with a suitably chosen domain in $L^2(\mathbb{R})$, has a pure point spectrum consisting of the positive odd integers, each of which is a simple eigenvalue. *Hint*: the eigenfunctions (Hermite functions) form a complete set.

5. Let M be a given $n \times n$ matrix. Let A be the operator in $\mathfrak{H} = l^2$ represented by the infinite matrix

$$A_{np+j\,np+k} = M_{jk} \qquad (j, k = 1, \ldots, n, p = 0, 1, 2, \ldots)$$
$$A_{r\,s} = 0 \text{ otherwise}$$

    (see Figure 8-1). A has a pure point spectrum, consisting of finitely many eigenvalues, each with infinite multiplicity.

Figure 8.1 An infinite matrix.

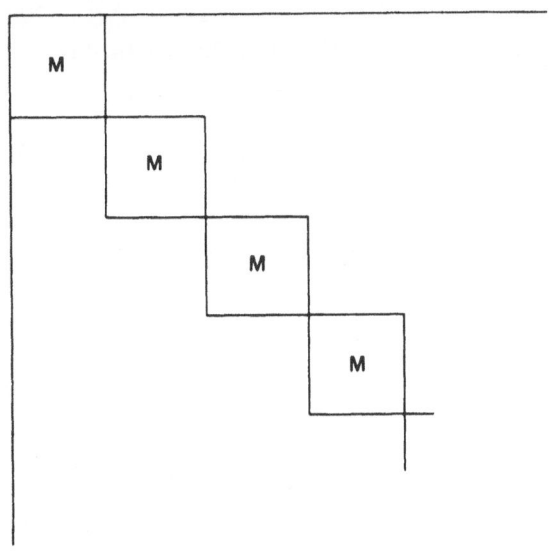

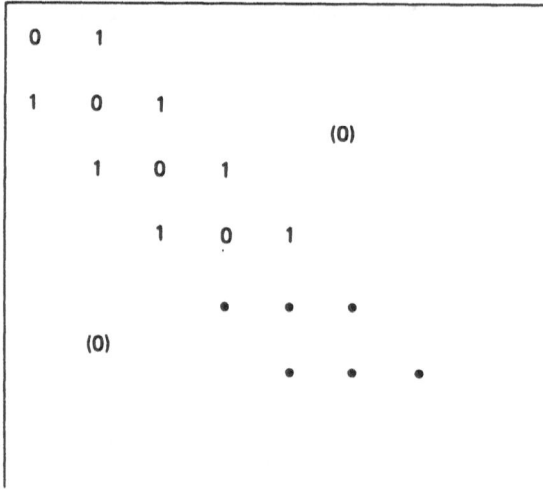

Figure 8.2 Another infinite matrix.

6. In $\mathfrak{H} = l^2$, let A be the self-adjoint operator whose matrix is

$$A_{jk} = \begin{array}{ll} 1 & \text{if } j = k + 1, \\ 1 & \text{if } j = k - 1, \\ 0 & \text{otherwise.} \end{array}$$

The interval $-2 \leq \lambda \leq 2$ is the continuous spectrum, and the remainder of the real axis is in the resolvent set.

7. In $\mathfrak{H} = l^2$, let A be the unitary operator, discussed in Section 7.3, that maps

$$\xi = (x_1, x_2, x_3, x_4, x_5, \ldots, x_{2n}, x_{2n+1}, \ldots)$$

onto

$$A\xi = (x_2, x_4, x_1, x_6, x_3, \ldots, x_{2n+2}, x_{2n-1}, \ldots);$$

the point spectrum is empty and the continuous spectrum is the entire unit circle in the λ plane.

8. In the Hilbert space $\mathfrak{H} = l^2$, *annihilation* and *creation operators* denoted by a and a^* are defined as follows: They have a common domain

$$\mathfrak{D}_1 = \mathfrak{D}(a) = \mathfrak{D}(a^*) = \{\xi = (x_0, x_1, x_2, \ldots): \sum n|x_n|^2 < \infty\}.$$
$$(8.2\text{-}1)$$

Then, $a\xi$ and $a^*\xi$ are given by

$$a(x_0, x_1, x_2, \ldots) = (x_1, \sqrt{2}x_2, \sqrt{3}x_3, \ldots),$$
$$a^*(x_0, x_1, x_2, \ldots) = (0, x_0, \sqrt{2}x_1, \sqrt{3}x_2, \ldots).$$

The physical interpretation for a simple model is that the vector

$$\varphi_n = (0, 0, \ldots, x_n = 1, 0, 0, \ldots)$$

represents a state of a physical system in which n particles are present; in particular, φ_0 represents the vacuum state. The action of the operators a and a^* on these states is given by

$$a\varphi_n = \sqrt{n}\,\varphi_{n-1}, \qquad a^*\varphi_n = \sqrt{n+1}\,\varphi_{n+1}.$$

Show that a^* is the adjoint of a according to the definition in Section 7.2—see Equation (7.2-1). Show that

$$a^*a - aa^* = -1,$$

in the sense that for all ξ in a certain domain $\mathfrak{D}_2(\subset \mathfrak{D}_1)$

$$a^*a\xi - aa^*\xi = -\xi,$$

and find $\mathfrak{D}_2$. The operator $N = a^*a$ with domain $\mathfrak{D}_2$ is called the *particle-number operator*; its action on the states φ_n is given by $N\varphi_n = n\varphi_n$. Show that the point spectrum of a is the entire complex plane, by finding the solution ξ_λ of the equation

$$a\xi_\lambda = \lambda\xi_\lambda$$

for any $\lambda \in \mathbb{C}$, and verify that $\xi_\lambda \in l^2$. Show that the point spectrum of a^* is empty by showing that for any $\lambda \in \mathbb{C}$ the equation $a^*\xi = \lambda\xi$ implies $\xi = 0$. Lastly, show from the range-nullspace relation (7.7-2) that the residual spectrum of a^* is the entire complex plane.

8.3 Spectra of Symmetric, Self-Adjoint, and Unitary Operators

First suppose that A is symmetric, i.e., that $\mathfrak{D}(A)$ is dense in $\mathfrak{H}$, and $(u, Aw) = (Au, w)$ for all u and w in $\mathfrak{D}(A)$ (then (u, Au) is always real), and consider the equation

$$Au - \lambda u = v. \tag{8.3-1}$$

The imaginary part of the resulting equation $(u, Au) - \lambda(u, u) = (u, v)$ is

$$-\operatorname{Im}\lambda\|u\|^2 = \operatorname{Im}(u, v). \tag{8.3-2}$$

This shows that for λ nonreal, v cannot vanish unless u vanishes, hence, λ is not an eigenvalue of A, and $(A - \lambda)$ has an inverse. By the Schwarz inequality,

$$|\operatorname{Im}\lambda|\,\|u\|^2 = |\operatorname{Im}(u, v)| \le |(u, v)| \le \|u\|\,\|v\|,$$

but $u = (A - \lambda)^{-1}v$, and therefore

$$\|(A - \lambda)^{-1}v\| \le \frac{1}{|\operatorname{Im}\lambda|}\,\|v\|. \tag{8.3-3}$$

It follows that for any symmetric operator A, any nonreal λ is either in the resolvent set or in the residual spectrum.

Now, the further assumption is made that A is self-adjoint. It is recalled from Section 7.8 that if T is an operator with dense domain, then the closure

$\Re(T)$ of its range is equal to $\Re(T^*)^\perp$, the orthogonal complement of the nullspace of T^*. Therefore, if λ is nonreal,

$$\Re(A - \lambda) = \Re(A^* - \bar{\lambda})^\perp,$$

but $A^* = A$, and $\bar{\lambda}$ is not an eigenvalue; hence, the nullspace is empty (except for the zero element), and its orthogonal complement is all of $\mathfrak{H}$. It is concluded that the range of $A - \lambda$, which is the domain of $(A - \lambda)^{-1}$, is dense in $\mathfrak{H}$ and therefore that *any nonreal λ is in the resolvent set $\rho(A)$*, because (8.3-3) shows that $(A - \lambda)^{-1}$ is bounded. $(A - \lambda)^{-1}$ is also closed, because A is closed (hence $A - \lambda$ is closed), and the graph of $(A - \lambda)^{-1}$ is the rotated graph of $-A + \lambda$, hence is a closed graph. Therefore, for λ in $\rho(A)$, $R_\lambda = (A - \lambda)^{-1}$ is defined in all $\mathfrak{H}$. In ract, for any λ not an eigenvalue, $(A - \lambda)^{-1}$ exists and has dense domain; hence there is no residual spectrum. These conclusions are now summarized:

Theorem 1. *Let A be self-adjoint. The spectrum $\sigma(A)$ lies on the real axis; the upper and lower half planes are in the resolvent set $\rho(A)$; the residual spectrum is empty; for λ in $\rho(A)$, the domain of $R_\lambda = R_\lambda(A)$ is all of $\mathfrak{H}$; for λ nonreal,*

$$\|R_\lambda\| \le \frac{1}{|\text{Im } \lambda|}. \tag{8.3-4}$$

The arguments leading to this theorem can be mimicked rather closely for a unitary operator U. If

$$Uv - \lambda v = w,$$

so that $v = R_\lambda w$, then,

$$\|v\|^2 = \|Uv\|^2 = |\lambda|^2\|v\|^2 + 2\,\text{Re}(\lambda v, w) + \|w\|^2.$$

The remainder of the argument is left as an exercise. [One has to solve a quadratic equation to get limits on $\|v\|/\|w\|$.] The result is the following:

Theorem 2. *Let U be unitary. The spectrum $\sigma(U)$ lies on the unit circle $|\lambda| = 1$; the interior and exterior of the unit circle are in the resolvent set $\rho(U)$; the residual spectrum is empty; for λ in $\rho(U)$, the domain of $R_\lambda = R_\lambda(U)$ is all of $\mathfrak{H}$; for $|\lambda| \ne 1$,*

$$\|R_\lambda\| \le \frac{1}{|1 - |\lambda||} \tag{8.3-5}$$

The inequalities (8.3-4) and (8.3-5) become equalities if the denominator is replaced by the distance from the point λ to the spectrum of the operator, and they can be generalized to normal operators. An operator T is called *normal* if it commutes with its adjoint (in the strict sense; not only must $T^*Tx = TT^*x$ for all x for which both sides are defined, but T^*T and TT^* must have the same domain, so that, whenever T^*Tx is defined, TT^*x is also defined, and conversely). Self-adjoint and unitary operators are normal. For λ in the resolvent set $\rho(T)$, define

$$d(\lambda, \sigma(T)) = \inf\{|\lambda - \mu| : \mu \in \sigma(T)\}$$

("min" could have been written for "inf", because the spectrum $\sigma(T)$ is a closed set in the complex plane, according to Section 8.5, below.)

Theorem 3.   *If T is a normal operator and R_λ is its resolvent, then, for $\lambda \in \rho(T)$,*

$$\|R_\lambda\| = \frac{1}{d(\lambda,\, \sigma(T))}.\tag{8.3-6}$$

For proof, see Kato, *Perturbation Theory for Linear Operators* (1966).

It is proved in the Section 8.5 that, for any operator A, if $\lambda \in \rho(A)$, then the disc consisting of complex numbers μ such that $|\mu - \lambda| < \|R_\lambda\|^{-1}$ is also in $\rho(A)$, hence

$$\|R_\lambda\| \geq \frac{1}{d(\lambda,\, \sigma(A))},\tag{8.3-7}$$

whether A is normal or not.

8.4 Modification of the Spectrum When an Operator is Extended

If a linear operator A is replaced by A', an extension of A, then the sets $P\sigma(A)$, $C\sigma(A)$, $R\sigma(A)$, and $\rho(A)$ in the complex plane are modified. For example, although the point spectrum $P\sigma(A)$ cannot decrease, it may increase, because an eigenvector of A' may not be in $\mathfrak{D}(A)$. The residual spectrum $R\sigma(A)$ cannot increase (because $\mathfrak{D}((A - \lambda)^{-1})$ is contained in $\mathfrak{D}((A' - \lambda)^{-1})$, and if the latter is not dense, then neither is the former) but it can decrease. The various possibilities are the ones illustrated in Figure 8-3, as is easily verified. The symbolism $S_1 \to S_2$ means that a given point λ in the complex plane, which was previously in the set S_1, may find itself in the set S_2 after A is replaced by A'.

The action of the arrow at the bottom of the diagram is illustrated by the unfortunate operator A of radial momentum in quantum mechanics. (See Section 7.8.) Namely, let $\mathfrak{H} = L^2(0, \infty)$; A is defined by the equations

$$\mathfrak{D}(A) = \{f \in L^2 : f' \in L^2,\, f(0) = 0\}$$

$$Af = -if'.$$

[Since f' is in L^2, f is a continuous function in $[0, \infty)$; hence the boundary condition $f(0) = 0$ is meaningful.] The Adjoint A^* is an extension of A; its domain includes elements of L^2 that do not satisfy the boundary condition; namely,

$$\mathfrak{D}(A^*) = \{f \in L^2 : f' \in L^2\}$$

$$A^*f = -if'.$$

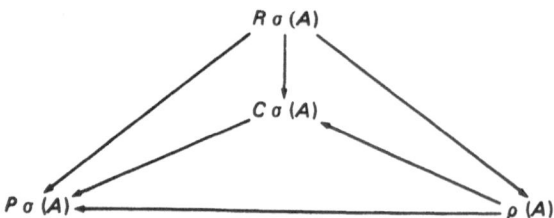

Figure 8.3 Modification of the spectrum by extension.

It is readily seen that A is symmetric, i.e., that $(Af, g) = (f, Ag)$ for all f and g in $\mathfrak{D}(A)$. However, it is not self-adjoint and has no self-adjoint extension; A^* is not symmetric, because $(A^*f, g) - (f, A^*g) = -if(0)g(0)$. Let λ be in the upper half plane; if λ is an eigenvalue of A or A^*, then the eigenfunction is the solution of the equation

$$-if'(x) = \lambda f(x);$$

hence,

$$f(x) = \text{constant } e^{i\lambda x}.$$

This is an eigenfunction of A^* for $\text{Im } \lambda > 0$, but not of A, for it is not in $\mathfrak{D}(A)$ unless the constant is zero, in which case $f(x) \equiv 0$. However, λ is in the resolvent set $\rho(A)$ of A, as can be seen by solving the equation

$$-if'(x) - \lambda f(x) = g(x)$$

for given g in L^2. The solution is

$$f(x) = i \int_0^x e^{i\lambda(x-y)} g(y) dy$$

and is unique, because of the boundary condition $f(0) = 0$. The function f is continuous; furthermore, if g has bounded support, and $\text{Im } \lambda > 0$, then $f(x)$ decreases exponentially as $x \to \infty$, hence is in L^2. The functions of bounded support are dense in L^2, hence $(A - \lambda)^{-1}$ exists and has dense domain. The inequality (8.3-3) applies, because A is symmetric, hence $(A - \lambda)^{-1}$ is bounded. The conclusion is that the upper half plane is in the resolvent set $\rho(A)$ of A but in the point spectrum $P\sigma(A^*)$ of A^*.

EXERCISE

1. Analyze the remainder of the complex plane ($\text{Im } \lambda \leq 0$) with respect to the spectra of A and A^*.

Note. The radial momentum operator A has no self-adjoint extension. If B were such an extension, then A^* would be an extension of $B^* = B$, hence the relations

$$\mathfrak{D}(A) \subset \mathfrak{D}(B) = \mathfrak{D}(B^*) \subset \mathfrak{D}(A^*)$$

would hold, where the inclusions are proper. However, $\mathcal{D}(A^*)$ is the span of $\mathcal{D}(A)$ together with a one-dimensional set, consisting, for example, of the functions const. e^{-x}. Therefore, there is no room between $\mathcal{D}(A)$ and $\mathcal{D}(A^*)$ for $\mathcal{D}(B)$; that is, if $\mathcal{D}(B)$ contained a single element not in $\mathcal{D}(A)$, then it would contain all of $\mathcal{D}(A^*)$; hence, B would $= A^*$, but A^* is not self-adjoint, not even symmetric. For further discussion, see Section 8.6, on deficiency indices.

An important case of extension of an operator is the mere replacement of a closable operator A by its closure $\tilde{A}$. Then, most of the arrows in the above diagram disappear. It is readily verified that in this case the resolvent set is unaltered (however, the domain of the operator R_λ generally increases from a dense set in $\mathfrak{H}$ to all of $\mathfrak{H}$) and that the diagram reduces to

$$C\sigma(A) \to P\sigma(A) \leftarrow R\sigma(A).$$

If the closure $\tilde{A}$ of a closable operator A is self-adjoint, then A is called *essentially self-adjoint*. In this case, $\tilde{A}$ is the only self-adjoint extension of A, because any other self-adjoint extension A_1, being closed, would have to be an extension of $\tilde{A}$, and if it were a proper extension, then A_1^* would have to be a proper restriction of $\tilde{A}^*$, which would be a contradiction, because $A_1^* = A_1$ and $\tilde{A}^* = \tilde{A}$. The replacement of an essentially self-adjoint operator by its closure leaves all parts of the spectrum unchanged; the only effect is that $\mathcal{D}(R_\lambda)$ increases to all $\mathfrak{H}$, for λ in $\rho(A)$.

8.5 Analytic Properties of the Resolvent

As in the finite-dimensional case, the resolvent R_λ of any closed operator A satisfies the *resolvent equation*

$$\frac{R_\lambda - R_\mu}{\lambda - \mu} = R_\lambda R_\mu, \tag{8.5-1}$$

for λ and μ in $\rho(A)$; this equation can be verified by multiplying both sides on the left by $(A - \lambda)$, which is permissible because (1) if either side is applied to an arbitrary v in $\mathfrak{H}$, the result is in $\mathcal{D}(A)$, and (2) $(A - \lambda)$ has an inverse. [Note: For any v in $\mathcal{D}(A)$, $R_\lambda(A - \lambda)v = v = (A - \lambda)R_\lambda v$; however, R_λ and $(A - \lambda)$ do not commute, in the strict sense, because the domain of $(A - \lambda)R_\lambda$ is larger (namely, is all $\mathfrak{H}$) than that of $R_\lambda(A - \lambda)$. That is,

$$R_\lambda(A - \lambda) \subset (A - \lambda)R_\lambda = I. \tag{8.5-2}$$

However, $A - \lambda$ commutes with $A - \mu$, and R_λ with R_μ.]

Now let A be any closed operator and λ any fixed complex number in $\rho(A)$. We shall show that the open disk with λ as center and radius $\|R_\lambda\|^{-1}$ is also in $\rho(A)$. Hence $\rho(A)$ is an open set, and the inequality (8.3-7) holds. Also, we shall find a series expansion of R_μ for μ in the disk. Namely, let μ be such that

$$\alpha \overset{\text{def}}{=} |\mu - \lambda|\,\|R_\lambda\| < 1. \tag{8.5-3}$$

A sequence of operators $B_n = B_n(\mu)$ is defined by the equation

$$B_n = R_\lambda + (\mu - \lambda)R_\lambda^2 + \cdots + (\mu - \lambda)^n R_\lambda^{n+1}; \qquad (8.5\text{-}4)$$

it will be shown that $B_n \to R_\mu$, as $n \to \infty$. For any u in $\mathfrak{H}$, and for $n > l$,

$$\|B_n u - B_l u\| \leq (\alpha^{l+1} + \alpha^{l+2} + \cdots + \alpha^n)\|R_\lambda u\|; \qquad (8.5\text{-}5)$$

therefore $\{B_n u\}$ is a Cauchy sequence and has a limit, as $n \to \infty$, which is linear in u and is taken as defining an operator B:

$$Bu = \lim_{n \to \infty} B_n u \quad (\text{all } u \text{ in } \mathfrak{H}).$$

Furthermore, by summing the geometric series, it is seen that

$$\|Bu\| = \lim \|B_n u\| \leq \frac{1}{1 - \alpha} \|R_\lambda\| \|u\|, \qquad (8.5\text{-}6)$$

so that B is a bounded operator. If the left member of the defining equation (8.5-4) of B_n is multiplied by $(A - \mu)$ on the left, and the right member is multiplied on the left by $(A - \lambda) - (\mu - \lambda)$, which is the same thing, a telescoping sum results on the right, and the result is

$$(A - \mu)B_n = I - (\mu - \lambda)^{n+1} R_\lambda^{n+1}.$$

Therefore, for any u in $\mathfrak{H}$,

$$(A - \mu)B_n u \to u, \quad \text{as } n \to \infty;$$

we have seen that

$$B_n u \to Bu, \quad \text{as } n \to \infty,$$

and since $(A - \mu)$ is a closed operator, it follows that Bu is in $\mathfrak{D}(A)$, that $\mu \in \rho(A)$, and that

$$B = (A - \mu)^{-1} = R_\mu,$$

as required. Furthermore, from (8.5-6) and (8.5-3), it follows that

$$\|R_\mu\| \leq \frac{\|R_\lambda\|}{1 - |\mu - \lambda| \|R_\lambda\|}. \qquad (8.5\text{-}7)$$

[From (8.5-5) it follows that $\|B_n - B_l\| \to 0$, as n and $l \to \infty$, hence that $\|B - B_l\| \to 0$, as $l \to \infty$. One says that B_n converges *in norm* to B. Various modes of convergence of operators will be discussed in the next chapter.]

From (8.5-4), it follows that, for λ in $\rho(A)$ and $|\mu - \lambda| < \|R_\lambda\|^{-1}$, and for any u and v in $\mathfrak{H}$, $(u, R_\mu v)$ is given by the convergent power series

$$(u, R_\mu v) = \sum_{l=0}^{\infty} (\mu - \lambda)^l (u, R_\lambda^{l+1} v);$$

hence $(u, R_\mu v)$ is an analytic function of μ. In particular,

$$\frac{d}{d\mu}(u, R_\mu v)\bigg|_{\mu = \lambda} = (u, R_\lambda^2 v),$$

and it is interesting to note that the resolvent equation (8.5-1) gives the same result, because

$$\lim_{\mu \to \lambda} \frac{(u, R_\mu v) - (u, R_\lambda v)}{\mu - \lambda} = \lim_{\mu \to \lambda}(u, R_\mu R_\lambda v),$$

and the right member is $(u, R_\lambda^2 v)$, if $(u, R_\mu R_\lambda v)$ is assumed continuous in μ.

In Section 9.7, it will be seen that we can regard R_λ itself as an operator-valued analytic function of λ; for example, $dR_\lambda/d\lambda$ is R_λ^2.

Examples of explicit formulas for the resolvent $R_\lambda(A)$ of certain operators A will be found in Chapters 10 and 11.

8.6 Extensions of a Symmetric Operator; Deficiency Indices; The Cayley Transform; Second Definition of Self-Adjointness

A rather complete theory of the possible symmetric extensions of a symmetric operator (i.e. of a densely defined operator A such that $(Au, v) = (u, Av)$ for all u and v in $\mathfrak{D}(A)$) was given by von Neumann in 1929. The examples given in Chapter 7 show that such an operator may have no self-adjoint extension, it may be self-adjoint or more generally have a unique self-adjoint extension (in which case it is called "essentially self-adjoint"), or it may have many different self-adjoint extensions. The "deficiency indices," which will now be defined, play an important role in this connection. First, if $\mathfrak{M}$ is any linear manifold, its *codimension*, denoted by codim $\mathfrak{M}$, is defined as the dimension of the orthogonal complement $\mathfrak{M}^\perp$ of $\mathfrak{M}$. It is a nonnegative integer or a transfinite cardinal (which can only be $\aleph_0$ in a separable Hilbert space). If A is a symmetric operator, its *deficiency indices* are the numbers (m, n), where $m = \text{codim } \mathfrak{R}(A + i)$ and $n = \text{codim } \mathfrak{R}(A - i)$; clearly, m and n describe the extent to which the ranges of $A \pm i$ (more precisely, the closures of these ranges) fail to fill out the Hilbert space. It was shown in Section 8.3 that $A - \lambda$ has a bounded inverse for Im $\lambda \neq 0$; hence the number i (or $-i$) is in the resolvent set $\rho(A)$ if $m = 0$ (or $n = 0$) and is in the residual spectrum $R\sigma(A)$ if $m > 0$ (or $n > 0$).

The following lemma shows that the choice of the numbers $\pm i$ in the definition of the deficiency indices is arbitrary; they could be replaced by any numbers λ_1 and λ_2, of which the first is in the upper half plane and the second in the lower.

Lemma. *If T is any densely defined linear operator, then* codim $\mathfrak{R}(T - \lambda)$ *is constant, as a function of λ, in any connected region of the λ plane in which $(T - \lambda)$ has a bounded inverse. (For proof, see Akhiezer and Glazman 1963, Section 78.)*

Corollary. *If T is symmetric, then* codim $\mathfrak{R}(T - \lambda)$ *has constant values m and n in the upper and lower half planes. If, furthermore, $T - \lambda$ has a bounded inverse for some real λ, then $m = n$.*

Now suppose that T is a symmetric operator with deficiency indices (m, n), and suppose that we wish to find all self-adjoint extensions A of T, if any exist. If such extensions exist, then T is at least closable and can be assumed to be closed, without loss of generality. [*Proof*: If A is a self-adjoint extension of T (hence is closed), if $\{u_n\}$ and $\{Tu_n\}$ are Cauchy sequences with $u_n \in \mathfrak{D}(T)$ $(n = 1, 2, \ldots)$ and with limits v and w, then $u_n \in \mathfrak{D}(A)$ and $Tu_n = Au_n$ $(n = 1, 2, \ldots)$ hence $w = Av$, so that $\lim Tu_n$ depends only on $\lim u_n$; hence T is closable.] It is furthermore assumed provisionally that $m = n$ and even that $m = n < \infty$. (It will be seen that T has no self-adjoint extension if $m \neq n$.) Obviously $T + i$ is also a closed operator, and it is asserted that the range of $T + i$ is a closed linear manifold. To see this, let $\{u_n\}$ be a Cauchy sequence of elements in $\mathfrak{R}(T + i)$; $(T + i)^{-1}$ is bounded, hence the elements $w_n = (T + i)^{-1} u_n$ form a Cauchy sequence, and $u_n = (T + i)w_n$; $T + i$ is closed, hence $\lim w_n \in \mathfrak{D}(T + i)$ and $\lim u_n \in \mathfrak{R}(T + i)$, so that $\mathfrak{R}(T + i)$ is closed.

The *Cayley transform* V of a symmetric operator T is defined in complete analogy with the Cayley transform of a Hermitian matrix, namely,

$$\mathfrak{D}(V) = \mathfrak{R}(T + i)$$
$$Vw = (T - i)(T + i)^{-1}w, \quad \text{for all } w \in \mathfrak{D}(V). \tag{8.6-1}$$

If $(T + i)^{-1}w = u$, i.e., $w = Tu + iu$, then $Vw = Tu - iu$; hence, $\|Vw\|^2 = (Tu - iu, Tu - iu) = \|Tu\|^2 + \|u\|^2 = \|w\|^2$, because $(Tu, u) = (u, Tu)$, i.e., $(Tu, iu) = -(iu, Tu)$. That is, the transformation $w \to Vw$ is a one-to-one isometric mapping of $\mathfrak{R}(T + i)$ onto $\mathfrak{R}(T - i)$, or simply an *isometry* for short. [It is noted that if $\|Vw\| = \|w\|$ for all w in a domain $\mathfrak{D}(V)$, then $(Vu, Vw) = (u, w)$ for all u and w in $\mathfrak{D}(V)$ by the polarization procedure that was used for equations (7.2-2 and 3).] If T is self-adjoint, then $\mathfrak{R}(T + i) = \mathfrak{R}(T - i) = \mathfrak{H}$, and V is unitary; and conversely, if V is unitary, then T is self-adjoint.

The following lemma gives the reverse of the Cayley transform:

Lemma. *Let T be a symmetric operator with Cayley transform V. Then, the operator $V - 1$ is invertible, and T is given in terms of V by the equations*

$$\mathfrak{D}(T) = \mathfrak{R}(V - 1), \qquad T = -i(V + 1)(V - 1)^{-1}. \tag{8.6-2}$$

Conversely, if V is any isometry such that $\mathfrak{R}(V - 1)$ is dense in $\mathfrak{H}$, then $V - 1$ is invertible, the operator T defined by (8.6-2) is symmetric, and V is its Cayley transform.

PROOF.   With w and u as above, for any $u \in \mathfrak{D}(T)$, we have

$$w = (T + i)u$$
$$Vw = (T - i)u \tag{8.6-3}$$
$$(V - 1)w = -2iu.$$

Hence, $(V - 1)w = 0 \Rightarrow u = 0 \Rightarrow w = 0$; hence, $V - 1$ has an inverse. It is easy to see from these same equations that u is in $\mathfrak{D}(T)$ if it is in $\mathfrak{R}(V - 1)$, and that then $Tu =$

$-i(V + 1)(V - 1)^{-1}u$, as claimed. Conversely, let V be any isometry such that $\Re(V - 1)$ is dense in $\mathfrak{H}$. It will be shown that $V - 1$ has an inverse. Namely, suppose that $(V - 1)w = 0$ for some w. Then, for all z in $\mathfrak{D}(V)$, since $(w, z) = (Vw, Vz)$, we have

$$0 = (Vw - w, z) = (Vw, z) - (w, z) = -(Vw, Vz - z).$$

Since elements of the form $Vz - z$ are dense in $\mathfrak{H}$, it follows that $Vw = 0$, hence $w = 0$. Therefore, $V - 1$ has an inverse. Now let T be defined in terms of V by (8.6-2). For any u and v in $\mathfrak{D}(T)$, we can then write $u = (V - 1)x$, $v = (V - 1)y$, and then $Tu = -i(V + 1)x$ and $Tv = -i(V + 1)y$. Therefore

$$(u, Tv) = ((Vx - x), -i(Vy + y))$$

$$(Tu, v) = (-i(Vx + x), (Vy - y)),$$

and these are seen to be equal, because $(Vx, Vy) = (x, y)$, hence T is symmetric. A final calculation then shows that the Cayley transform of T coincides with V.

The first theorem on the extensions of a symmetric operator can now be given.

Theorem (von Neumann). *Let T be a closed symmetric operator with deficiency indices (m, m) $(m < \infty)$, and let V be its Cayley transform. Then (1) if A is any self-adjoint extension of T, and U is its Cayley transform, then U is an extension of V which maps $\Re(T^* - i)$ isometrically onto $\Re(T^* + i)$. (2) Any isometric mapping V' of $\Re(T^* - i)$ onto $\Re(T^* + i)$ determines a unique self-adjoint extension A of T, given by equations (8.6-4) below. Since V' can be represented by an $m \times m$ unitary matrix, there is an m^2-parameter family of self-adjoint extensions of T. (There are m^2 real parameters.)*

PROOF.   According to the range-nullspace theorem, Exercise 2 in Section 7.7, the Hilbert space can be represented as an orthogonal direct sum in two ways, as follows:

$$\mathfrak{H} = \Re(T + i) \oplus \Re(T^* - i)$$

$$\mathfrak{H} = \Re(T - i) \oplus \Re(T^* + i)$$

[**Note.** $\Re(T \pm i)$ are closed linear manifolds, because $(T \pm i)^{-1}$ are bounded closed operators.] (1) Since A is an extension of T, U is an extension of V. Hence, by definition of V, U maps $\Re(T + i)$ isometrically onto $\Re(T - i)$. Hence, since U preserves orthogonality, it also maps $\Re(T^* - i)$ isometrically onto $\Re(T^* + i)$, as stated. (2) Let V' be *any* isometric mapping of $\Re(T^* - i)$ onto $\Re(T^* + i)$, and define a unitary operator U by writing

$$Uv = Vv, \quad \text{for } v \in \Re(T + i)$$

$$Uv = V'v, \quad \text{for } v \in \Re(T^* - i)$$

and then extending the definition of U to all $\mathfrak{H}$ by linearity. Since $V - 1$ has a dense range, $U - 1$ also has a dense range. Therefore, according to the foregoing lemma, an operator A can be defined by the equations

$$\mathfrak{D}(A) = \Re(U - 1)$$

$$Av = -i(U + 1)(U - 1)^{-1}v,$$

(8.6-4)

and then U is the Cayley transform of A. Since U is unitary, A is self-adjoint.

An example is provided by the operator T discussed in Section 7.5, given in $L^2(-1, 1)$ by the equations

$$\mathfrak{D}(T) = \{f \in L^2 : f' \in L^2, f(-1) = 0, f(1) = 0\}$$

$$Tf = -if'.$$

It was found that T^* is the same as T except that it has no boundary conditions at all. The self-adjoint extensions of T will now be found. For any λ in $\mathbb{C}$, the equation $(T^* - \lambda)f = 0$ has the solution $f(x) = e^{i\lambda x}$, which is always in $L^2(-1, 1)$; hence, the deficiency indices of T are $(1, 1)$. The nullspaces $\mathfrak{N}(T^* + i)$, $\mathfrak{N}(T^* - i)$ consist of multiples of the functions e^x, e^{-x}, respectively; hence the general isometric mapping of $\mathfrak{N}(T^* - i)$ onto $\mathfrak{N}(T^* + i)$ is $V' = V'_\alpha : e^{-x} \to e^{i\alpha}e^x$, where α is a fixed real number. The domain of the self-adjoint operator A corresponding to this choice of V' is the range of $U - 1$, according to (8.6-4), where U is the unitary operator given as above in terms of V' and the Cayley transform $V = (T - i)(T + i)^{-1}$ of T. The domain of V is the range of $T + i$. The decomposition of $\mathfrak{H}$ shows that the ranges of $T + i$ and $T - i$ consist of all f in L^2 such that $\int_{-1}^{1} e^{-x}f(x)\,dx = 0$ and $\int_{-1}^{1} e^x f(x)dx = 0$, respectively, as can also be shown directly by computing all f of the form $(T + i)g$ with g in $\mathfrak{D}(T)$. Direct calculation shows that for any f_1 in $\mathfrak{R}(T + i)$, Vf_1 or $(T - i)(T + i)^{-1}f_1$ is given by the equation

$$(Vf_1)(x) = 2e^x \int_{-1}^{x} e^{-y}f_1(y)dy + f_1(x).$$

Therefore, if an arbitrary f in $L^2(-1, 1)$ is written as $f_1(x) + ce^{-x}$, where $f_1 \in \mathfrak{R}(T + i)$ and $ce^{-x} \in \mathfrak{N}(T^* - i)$, then

$$((U - 1)f)(x) = 2e^x \int_{-1}^{x} e^{-y}f_1(y)dy + c(e^{x+i\alpha} - e^{-x}),$$

or, in abbreviation,

$$g(x) = g_1(x) + g_2(x).$$

Now $g_1(x)$ vanishes at $x = \pm 1$ and is in fact a completely general element of $\mathfrak{D}(T)$. The effect of the term $g_2(x) = c(e^{x+i\alpha} - e^{-x})$ is to extend $\mathfrak{D}(T)$ to $\mathfrak{D}(A)$ by relaxing the boundary condition $g(\pm 1) = 0$ to the condition

$$g(1) = \frac{e^{1+i\alpha} - e^{-1}}{e^{-1+i\alpha} - e^{1}}\, g(-1) = e^{i\theta}g(-1),$$

where the constant $e^{i\theta}$ is given by

$$e^{i\theta} = \frac{e^{1+(1/2)i\alpha} - e^{-1-(1/2)i\alpha}}{e^{-1+(1/2)i\alpha} - e^{1-(1/2)i\alpha}},$$

and hence has unit modulus. It is concluded that the self-adjoint extensions of T are precisely the ones found earlier and given by (7.5-5), one for each θ in $[0, 2\pi)$; they were called A_θ in Section 7.5.

1.   Verify that, in this example, equation (8.6-4) gives $A_\theta f = -if$, for f in $\mathfrak{D}(A_\theta)$. *Hint*: It suffices to consider $f(x) = e^{x+i\alpha} - e^{-x}$.

2.   Give a similar discussion of the operator T given by the equations

$$\mathfrak{D}(T) = \{f : f^{(m)} \in L^2, f^{(p)}(-1) = f^{(p)}(1) = 0 \ (p = 0, 1, \ldots, m-1)\}$$

$$Tf = (-i)^m f^{(m)},$$

where, for any integer $p \geq 0$, $f^{(p)}$ denotes the pth distribution derivative of f.

The method used in the proof of the foregoing theorem makes the general situation evident. Let T be any closed symmetric operator, and call its deficiency indices (m, n). Let F and G be closed linear manifolds contained in $\mathfrak{R}(T^* - i)$ and $\mathfrak{R}(T^* + i)$, respectively, both having dimension $m_0 \leq \min(m, n)$. Let V' be any isometry of F onto G. The linear transformation U given by

$$Uw = Vw, \quad \text{for } w \text{ in } \mathfrak{R}(T + i)$$

$$Uw = V'w, \quad \text{for } w \text{ in } F$$

is an isometry of $\mathfrak{R}(T + i) \oplus F$ onto $\mathfrak{R}(T - i) \oplus G$. Then, the operator A given by (8.6-4) is a symmetric extension of T, and all symmetric extensions are obtainable in this way. A is maximal (i.e. cannot be further extended) if and only if its deficiency indices are either $(0, l)$ or $(l, 0)$. A is self-adjoint if and only if its deficiency indices are $(0, 0)$, which requires that the deficiency indices of T be of the form (m, m). If the deficiency indices of T are (∞, ∞), then T has both self-adjoint and non-self-adjoint maximal extensions.

The radial momentum operators (Section 7.8) have deficiency indices $(1, 0)$, hence are maximal but not self-adjoint.

A second definition of self-adjointness, equivalent to the one given in the preceding chapter, but more convenient for some purposes, is now given: A symmetric operator (or closed symmetric operator) is *essentially self-adjoint* (or is *self-adjoint* respectively) if $+i$ and $-i$ are in its resolvent set.

Hence, to show that a given operator A with dense domain such that $(Au, v) = (u, Av)$ for all u and v in $\mathfrak{D}(A)$ is essentially self-adjoint, i.e. is an observable, it suffices to show that the equations $Af \pm if = g$ have a unique solution f for every g in a set dense in $\mathfrak{H}$. For examples, see Sections 8.2, and Chapters 10 and 11.

CHAPTER 9

Spectral Decomposition of Self-Adjoint and Unitary Operators

Applications of complex variable methods to matrix theory;
projectors; resolution of the identity; canonical form of a matrix;
Jordan form; nilpotent part of a matrix;
generalized eigenvector and eigenspace; Schur's theorem on
triangularization; functions and distributions as boundary values of
analytic functions; the Laplace transform; canonical representation
of self-adjoint and unitary operators; weak, strong, and uniform
convergence of bounded operators; spectrum of A as the t-set on
which E_t is not constant; functions of operators; bounded
observables; the polar decomposition of an operator.

Prerequisites: Mainly, Chapters 1, 7, 8.

The main subject of this chapter is the analogue, for a self-adjoint operator in a Hilbert Space, of the problem of diagonalizing a Hermitian matrix and thereby expressing the matrix in terms of its eigenvalues and eigenvectors.

9.1 Spectral Decomposition of a Hermitian Matrix

Let A be an $n \times n$ Hermitian matrix. The transformation $\mathbf{x} \to A\mathbf{x}$ in $\mathbb{C}^n$ can be described geometrically in terms of the eigenvalues $\lambda_1, \ldots, \lambda_n$ and an orthonormal set of eigenvectors $\mathbf{v}_1, \ldots, \mathbf{v}_n$ of A. The eigenvectors determine the invariant directions, and the eigenvalues determine the corresponding dilations and contractions produced by the transformation. However, this description is highly nonunique: any one of the vectors $\mathbf{v}_k$ can be multiplied by any constant α having modulus $|\alpha|$ equal to 1; if $\mathbf{v}_{k_1}, \ldots, \mathbf{v}_{k_p}$ all belong to the same eigenvalue, they can be subjected to an arbitrary unitary

158

transformation in the p-dimensional subspace (eigenspace) spanned by those vectors. On the other hand, specification of the *distinct* eigenvalues $\lambda_1, \ldots, \lambda_q$ (say, in increasing order) and of the corresponding eigen*spaces* $\mathfrak{E}_1, \ldots, \mathfrak{E}_q$ provides a unique description of the transformation—a description that can be generalized to the infinite-dimensional case.

Since A is Hermitian, the eigenspaces are mutually orthogonal and span $\mathbb{C}^n$. Any vector $\mathbf{x}$ has a unique decomposition as $\mathbf{x}_1 + \cdots + \mathbf{x}_q$, where $\mathbf{x}_j$ is a vector lying in the subspace $\mathfrak{E}_j$; that is, $\mathbb{C}^n$ is the orthogonal direct sum

$$\mathbb{C}^n = \mathfrak{E}_1 \oplus \mathfrak{E}_2 \oplus \cdots \oplus \mathfrak{E}_q. \tag{9.1-1}$$

For fixed j and varying $\mathbf{x}$, the mapping $\mathbf{x} \to \mathbf{x}_j$ is a linear transformation of $\mathbb{C}^n$ onto $\mathfrak{E}_j$, called a *projection*, and the matrix P_j of that transformation is called a *projector*, specifically the *projector onto the jth eigenspace*; $P_j\mathbf{x} = \mathbf{x}_j$. The decomposition of $\mathbf{x}_j$ itself is $0 + \cdots + \mathbf{x}_j + 0 + \cdots + 0$, because $\mathbf{x}_j$ already belongs to $\mathfrak{E}_j$; hence $P_j\mathbf{x}_j = \mathbf{x}_j$, while $P_k\mathbf{x}_j = 0$ for $k \neq 0$. That is, $P_j^2\mathbf{x} = P_j\mathbf{x}$ for every $\mathbf{x}$, while $P_k P_j\mathbf{x} = 0$; hence,

$$P_j^2 = P_j \tag{9.1-2}$$

$$P_k P_j = 0 \qquad (k \neq j). \tag{9.1-3}$$

Since every vector in $\mathfrak{E}_j$ is an eigenvector corresponding to the eigenvalue λ_j, we have $A\mathbf{x} = A \sum_{(j)} \mathbf{x}_j = \sum_{(j)} \lambda_j \mathbf{x}_j$, hence the original matrix A is given by

$$A = \sum_{j=1}^{q} \lambda_j P_j. \tag{9.1-4}$$

Because of the unique decomposition $\mathbf{x} = \mathbf{x}_1 + \cdots + \mathbf{x}_q$ referred to above, for arbitrary $\mathbf{x}$, where $\mathbf{x}_j = P_j\mathbf{x}$, it is seen that

$$I = \sum_{j=1}^{q} P_j, \tag{9.1-5}$$

and if $f(\lambda)$ is any polynomial (or any function of a suitable kind whose domain of definition includes all the eigenvalues λ_i), then

$$f(A) = \sum_{j=1}^{q} f(\lambda_j) P_j. \tag{9.1-6}$$

Because of equation (9.1-5), the projectors P_j are said to constitute a *resolution of the identity*.

The main objective of the present chapter is to generalize the above equations to any self-adjoint operator in a Hilbert space. In the general case, there may be a continuum of "eigenvalues" (i.e., a continuous spectrum), hence an integration must replace the summation. There may of course also be discrete eigenvalues, so the integral clearly has to be a Stieltjes integral, to allow for both.

9.2 Projectors in a Hilbert Space $\mathfrak{H}$

If an operator P is bounded, defined in all $\mathfrak{H}$, and idempotent (i.e., $P^2 = P$), then it is called a *projector*; $I - P$ is clearly also a projector, because $(I - P)^2 = I - 2P + P^2 = I - P$. Let $\mathfrak{M}$ and $\mathfrak{N}$ be the ranges of P and of $I - P$, respectively. If u is in $\mathfrak{M}$, then $Pu = u$, while, if u is in $\mathfrak{N}$, then $Pu = 0$.

PROOF. $\mathfrak{M}$ is the range of P, so that if u is in $\mathfrak{M}$, then $u = Pv$ for some v in $\mathfrak{H}$; then $Pu = P^2v = Pv = u$. If u is in $\mathfrak{N}$, then, by the same argument, $(I - P)u = u$, so that $Pu = 0$.

EXERCISE

1. Show that the linear manifolds $\mathfrak{M}$ and $\mathfrak{N}$ are closed.

Note. $\mathfrak{M}$ and $\mathfrak{N}$ are not necessarily orthogonal—see below.

Assertion.   Any u in $\mathfrak{H}$ can be uniquely decomposed as $u = u_1 + u_2$, where $u_1 \in \mathfrak{M}$ and $u_2 \in \mathfrak{N}$ (u_1 and u_2 are not necessarily orthogonal). See Figure 9-1. In fact, $u_1 = Pu$ and $u_2 = (I - P)u$; then $u_1 + u_2 = u$. To prove uniqueness, Suppose that $u_1' + u_2'$ were any other such decomposition of u; then $u_1 - u_1'$ would be in both $\mathfrak{M}$ and $\mathfrak{N}$ (it would be equal to $u_2' - u_2$), hence (1) $u_1 - u_1' = P(u_1 - u_1')$, and similarly (2)

$$u_1 - u_1' = (I - P)(u_1 - u_1') = (I - P)P(u_1 - u_1') = (P - P^2)(u_1 - u_1') = 0.$$

Therefore, u_1' would be equal to u_1, and u_2' to u_2. This proves the assertion. The only vector common to $\mathfrak{M}$ and $\mathfrak{N}$ is the zero element of $\mathfrak{H}$. The elements u_1 and u_2 are called the *projections* of u on (or in) $\mathfrak{M}$ and $\mathfrak{N}$, respectively.

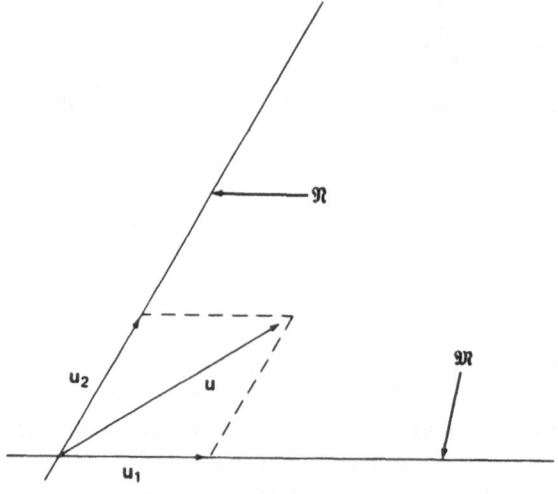

Figure 9.1 Projections of a vector.

Conversely, if $\mathfrak{M}$ and $\mathfrak{N}$ are any two closed linear manifolds that span $\mathfrak{H}$ and are such that an arbitrary element u has a unique decomposition $u_1 + u_2$, where $u_1 \in \mathfrak{M}$ and $u_2 \in \mathfrak{N}$, then a corresponding operator P can be defined by the equation $Pu = u_1$, and it is easily verified that the operator P so defined is a projector with range $\mathfrak{M}$ and nullspace $\mathfrak{N}$. (The *nullspace* of any operator A is the set of all u in $\mathfrak{H}$ such that $Au = 0$).

In general, $\mathfrak{M}$ and $\mathfrak{N}$ are not orthogonal, but if the projector P is also self-adjoint, then, for any u in $\mathfrak{M}$ and any v in $\mathfrak{N}$,

$$(u, v) = (Pu, (I - P)v) = (u, P(I - P)v) = (u, (P - P^2)v) = 0;$$

in this case, $\mathfrak{M}$ and $\mathfrak{N}$ are *orthogonal*—in symbols, $\mathfrak{M} \perp \mathfrak{N}$, and P is called an *orthogonal* projector.

EXERCISE

2. If P is an orthogonal projector, then $\|P\| = 1$.

9.3 Construction of the Spectral Projectors for a Matrix

In the finite-dimensional case, the projectors P_j can be constructed either directly from a complete orthonormal set of eigenvectors or by a function-theoretic method, but only the latter method is suitable for extending to the infinite-dimensional case, and it will now be described.

Let A be an arbitrary $n \times n$ matrix (not necessarily Hermitian). For λ not equal to any eigenvalue, the matrix $(A - \lambda I)$ has an inverse, namely, the resolvent $R_\lambda = (A - \lambda I)^{-1}$. According to the usual formula (Cramer's rule) for the inverse of a matrix in terms of its determinant and its minors, each matrix element of R_λ is a rational function of λ with poles at the eigenvalues $\lambda_1, \ldots, \lambda_q$ of A. For $|\lambda| \geq a$, where a is any constant greater than all the $|\lambda_i|$, the resolvent can be expanded as

$$R_\lambda = -\frac{1}{\lambda}\left(I + \frac{1}{\lambda}A + \frac{1}{\lambda^2}A^2 + \cdots\right). \tag{9.3-1}$$

Therefore,

$$-\frac{1}{2\pi i}\oint_{|\lambda|=a} R_\lambda \, d\lambda = I, \tag{9.3-2}$$

$$-\frac{1}{2\pi i}\oint_{|\lambda|=a} \lambda R_\lambda \, d\lambda = A, \tag{9.3-3}$$

and generally

$$-\frac{1}{2\pi i}\oint_{|\lambda|=a} \lambda^m R_\lambda \, d\lambda = A^m, \qquad m = 0, 1, 2, \ldots.$$

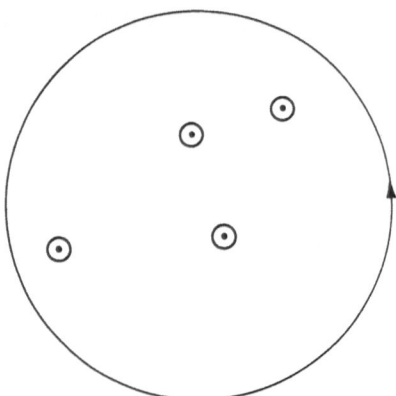

Figure 9.2 Deformation of the contour of integration.

In each case, the integral on the left means the matrix whose j, k element is the integral of $\lambda^m (R_\lambda)_{j\,k}$. Furthermore, if $f(\lambda)$ is any function defined by a power series for $|\lambda| \le a$, then

$$-\frac{1}{2\pi i} \oint_{|\lambda|=a} f(\lambda) R_\lambda \, d\lambda = f(A), \tag{9.3-4}$$

as can be seen by multiplying the power series for $f(\lambda)$ by (9.3-1) before taking the residue. [This last equation, when written as

$$\frac{1}{2\pi i} \oint_{|\lambda|=a} f(\lambda)(\lambda I - A)^{-1} \, d\lambda = f(A), \tag{9.3-5}$$

is a generalization of the Cauchy integral formula

$$\frac{1}{2\pi i} \oint f(\lambda)(\lambda - z)^{-1} \, d\lambda = f(z).]$$

Now shrink the contour as much as possible, namely until it has become a set of small contours, each encircling one of the eigenvalues, as in Figure 9-2. Then (9.3-2) gives

$$\sum_{j=1}^{q} P_j = I, \tag{9.3-6}$$

where

$$P_j = \frac{1}{2\pi i} \oint^{(\lambda_j-)} R_\lambda \, d\lambda. \tag{9.3-7}$$

[The symbol (λ_j-) indicates that the contour encircles λ_j once negatively (clockwise), but does not encircle any other singularity of the integrand.]

To show that P_j is a projector, P_j^2 must be computed;

$$P_j^2 = \left(\frac{1}{2\pi i}\right)^2 \oint^{(\lambda_j-)} \oint^{(\lambda_j-)} R_\lambda R_\mu \, d\mu \, d\lambda. \tag{9.3-8}$$

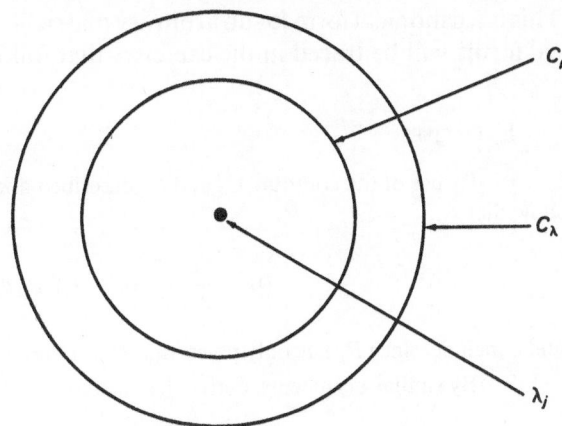

Figure 9.3 The contours C_μ and C_λ.

It is not necessary to use exactly the same contour for both integrations; it is in fact convenient to suppose that the contour for $\int d\mu$ lies inside the one for $\int d\lambda$, as shown in Figure 9-3. These contours will be called C_μ and C_λ respectively. The resolvent R_λ satisfies the resolvent equation (8.5-1); therefore

$$P_j^2 = \left(\frac{1}{2\pi i}\right)^2 \int_{C_\lambda} \int_{C_\mu} \left[\frac{R_\lambda}{\lambda - \mu} - \frac{R_\mu}{\lambda - \mu}\right] d\mu \, d\lambda$$

If the first term is integrated with respect to μ first, the result is zero, because λ lies outside C_μ; if the second term is integrated with respect to λ first, the result of the λ integration is $2\pi i R_\mu$ (recall that C_λ is traced negatively), hence

$$P_j^2 = \frac{1}{2\pi i} \int_{C_\mu} R_\mu \, d\mu = P_j. \tag{9.3-9}$$

Hence, P_j is a projector. A similar argument shows that if $j \neq k$, then $P_j P_k = 0$, for in this case each contour lies outside the other. Therefore,

$$P_j P_k = P_k P_j = \delta_{jk} P_j. \tag{9.3-10}$$

To express A in terms of these projectors, by means of equation (9.3-3), matrices D_j are first defined by writing

$$\frac{1}{2\pi i} \oint^{(\lambda_j-)} \lambda R_\lambda \, d\lambda = \lambda_j P_j + \frac{1}{2\pi i} \oint^{(\lambda_j-)} (\lambda - \lambda_j) R_\lambda \, d\lambda$$

$$\stackrel{\text{def}}{=} \lambda_j P_j + D_j;$$

Then

$$A = \sum_{(j)} [\lambda_j P_j + D_j]. \tag{9.3-11}$$

This is a canonical form for an arbitrary matrix A. Its relation to other canonical forms will be traced in the exercises that follow.

EXERCISES

1.   By use of the contours C_λ and C_μ described above, and by an induction on m, show that

$$D_j^m = \frac{1}{2\pi i} \oint_{C_\mu} (\mu - \lambda_j)^m R_\mu \, d\mu,$$

and conclude, since R_μ has only poles, that D_j is nilpotent. ($D_j^m = 0$ for some m).

2.   By similar arguments, derive the further equations

$$D_j P_k = P_k D_j = \delta_{jk} D_j,$$

$$A P_k = P_k A = \lambda_k P_k + D_k,$$

and conclude (1) that D_j maps the range of P_j into itself (this range is the generalized eigenspace $\mathfrak{E}_j$ corresponding to the eigenvalue λ_j), hence A maps $\mathfrak{E}_j$ into itself, and (2) that if $\mathbf{v}$ is any vector $\neq 0$ in $\mathfrak{E}_j$, then $(A - \lambda_j)^m \mathbf{v} = 0$ for some positive integer m. If m is the smallest such integer, then $\mathbf{v}$ is a *generalized eigenvector* of A of *order m*.

From equation (9.3-11) and the nilpotent character of the matrices, it can be shown (we omit details) that any $n \times n$ matrix A can be put into the *Jordan canonical form* by a similarity transformation; that is, there is a nonsingular matrix T such that

$$T^{-1}AT = J = \begin{bmatrix} J_1 & & (0) \\ & J_2 & \\ & & \ddots \\ (0) & & \end{bmatrix};$$

the only nonzero elements of J are located in square *Jordan blocks* $J_1, J_2, \ldots$ straddling the main diagonal; each Jordan block is of the form

$$J_r = \begin{bmatrix} \lambda & 1 & & & (0) \\ & \lambda & 1 & & \\ & & \ddots & \ddots & \\ & & & \ddots & 1 \\ (0) & & & & \lambda \end{bmatrix},$$

where λ is one of the eigenvalues of A. The columns of T are ordinary and generalized eigenvectors of A, in a suitable order. Different Jordan blocks do not necessarily contain different eigenvalues, i.e., several blocks may correspond to subspaces of one eigenspace $\mathfrak{E}_j$.

3.   Show, using equation (8.3-3) that if A is Hermitian, then R_λ has only simple poles, hence $D_j = 0$ for each j, and A is given by

$$A = \sum_{j=1}^{q} \lambda_j P_j;$$

in this case, the P_j are Hermitian, T can be taken to be unitary, and the Jordan form is diagonal.

4.   By applying the Gram–Schmidt orthonormalization procedure to the columns T in (9.3-12), prove *Schur's theorem*: any matrix A can be put into upper triangular form (a form in which all elements below the main diagonal are zero) by a unitary transformation.

If A is an operator (not necessarily self-adjoint) in $\mathfrak{H}$, and *if* formulas like (9.3-6, 10, 11) hold, with summation suitably replaced by Stieltjes integration, then A is called a *spectral operator*. In the infinite-dimensional case, it is not easy to say when that is true. (It is not in general possible to shrink the contour onto a set of discrete points, or even necessarily to find a contour that encloses the spectrum). In these terms, the main result of this chapter is that every self-adjoint operator is a spectral operator. A spectral operator whose nilpotent part vanishes identically is called an *operator of scalar type*. That is the case for self-adjoint operators in $\mathfrak{H}$ as well as for Hermitian matrices.

5.   Let A be any nonsingular $n \times n$ matrix, and $R_\lambda = (A - \lambda I)^{-1}$ its resolvent. Let C be a simple closed curve in the λ plane that encircles all the eigenvalues of A counterclockwise but does not encircle the origin. On C, the multivalued function $\log \lambda$ splits into independent single-valued continuous branches; let $\log \lambda$ denote one of them, and define a matrix $\log A$ by the equation

$$\log A = -\frac{1}{2\pi i} \oint_C R_\lambda \log \lambda \, d\lambda.$$

Show that

$$(\log A)^n = -\frac{1}{2\pi i} \oint_C R_\lambda (\log \lambda)^n \, d\lambda \qquad (n = 2, 3, \ldots)$$

and that $\exp(\log A) = A$. Show that if a different branch of $\log \lambda$ had been used, $\log A$ would be altered by the addition of an integer multiple of $2\pi i I$.

9.4 Connection with Analytic Functions

Let A be a self-adjoint operator and $R_\lambda = (A - \lambda)^{-1}$ its resolvent. As in the finite-dimensional case, the spectral projectors are of the form $(2\pi i)^{-1} \int R_\lambda d\lambda$, integrated around a closed contour that encircles a part of the spectrum of A

(the spectrum lies on the real axis). For λ in the resolvent set, R_λ is an analytic operator-valued function of λ (see Exercise 4 in Section 9.9), which is most conveniently dealt with by studying the ordinary analytic function

$$\varphi(\lambda) = \varphi(\lambda; v) = (v, R_\lambda v) = (v, u), \qquad (9.4\text{-}1)$$

where v is an arbitrary element of $\mathfrak{H}$. The polarization procedure (used below) shows that R_λ is completely determined for given λ when $\varphi(\lambda; v)$ is known for all v.

According to Section 8.5, $\varphi(\lambda)$ is analytic in the upper and lower half planes. As in Section 8.3, the imaginary part of the equation

$$(v, u) = (Au, u) - \bar{\lambda}(u, u)$$

gives

$$\text{Im}(v, u) = \text{Im } \lambda \|u\|^2,$$

so that $\text{Im } \varphi(\lambda)$ has the same sign as $\text{Im } \lambda$. Furthermore, the Schwarz inequality and the bound (8.3-4) on the resolvent imply that

$$|(v, u)| \leq \frac{1}{|\text{Im } \lambda|} \|v\|^2.$$

Therefore, $\varphi(\lambda)$ has the following properties:

(i) $\varphi(\lambda)$ is analytic
(ii) $|\varphi(\lambda)| \leq \|v\|^2/\text{Im } \lambda$ } for $\text{Im } \lambda > 0$ (9.4-1)
(iii) $\text{Im } \varphi(\lambda) > 0$

[Similar statements hold in the lower half plane $\text{Im } \lambda < 0$.]

Theorem.   *A function having the above properties can be expressed as*

$$\varphi(\lambda) = \int_{-\infty}^{\infty} \frac{1}{t - \lambda} \, d\sigma(t) (\text{Im } \lambda > 0), \qquad (9.4\text{-}2)$$

where $\sigma(t)$ is a nondecreasing function with finite limits as $t \to \pm\infty$; furthermore, if we set $\sigma(-\infty) = 0$, $\sigma(t)$ is determined from $\varphi(\lambda)$ by the equation

$$\sigma(t) = \lim_{\varepsilon \downarrow 0} \frac{1}{\pi} \int_{-\infty}^{t} \text{Im } \varphi(s + i\varepsilon) ds, \qquad (9.4\text{-}3)$$

which holds at the points of continuity of $\sigma(t)$. At the jumps, we arbitrarily impose the normalization condition

$$\sigma(t) = \lim_{\delta \downarrow 0} \sigma(t + \delta) \qquad (9.4\text{-}4)$$

of continuity on the right.

For the proof, see Akhiezer and Glazman 1963, Chapter 59.

Property (ii) above shows that the total variation $\sigma(\infty) - \sigma(-\infty)$ is proportional to $\|v\|^2$. It turns out in fact that

$$\sigma(\infty) - \sigma(-\infty) = \|v\|^2. \qquad (9.4\text{-}5)$$

9.5 Functions and Distributions as Boundary Values of Analytic Functions

Equation (9.4-2 and 3), which are used in the spectral theory in Section 9.6 below, can be expressed in terms of distributions, as follows: Let f denote the derivative σ' (in the distribution theory sense) of the function $\sigma(t)$, then

$$\frac{1}{\pi} \operatorname{Im} \varphi(t + i\varepsilon) \to f(t), \quad \text{as } \varepsilon \to 0 \tag{9.5-1}$$

in the sense of convergence of distributions, and

$$\varphi(\lambda) = \left\langle f(t), \frac{1}{t - \lambda} \right\rangle, \quad \text{for Im } \lambda > 0. \tag{9.5-2}$$

Although the function $\psi_\lambda(t) = 1/(t - \lambda)$ is not a test function (it is in C^∞ but is neither in C_0^∞ nor in $\mathscr{S}$), the above equation has nevertheless a valid interpretation, because $\sigma(t)$ has bounded total variation (it is nondecreasing and has finite limits as $t \to \pm\infty$). Namely if we call

$$\sigma_T(t) = \begin{cases} \sigma(-T) & \text{if } t < -T, \\ \sigma(t) & \text{if } -T \le t < T, \\ \sigma(T) & \text{if } t \ge T, \end{cases}$$

then the distribution $f_T = \sigma'_T$ has bounded support, hence $\langle f_T, \psi \rangle$ is well defined for any ψ in C^∞, and the right member of (9.5-2) is the limit of $\langle f_T, \psi_\lambda \rangle$ as $T \to \infty$.

Equation (9.5-1) says that in a sense the distribution $f(t)$ is the set of boundary values or the *trace* on the real axis of the harmonic function $(1/\pi) \operatorname{Im} \varphi(\lambda)$. The trace of $\operatorname{Re} \varphi(\lambda)$ is also a distribution, but a slightly more complicated one, because $\operatorname{Re} \varphi(\lambda)$ was not assumed to be ≥ 0 in the half plane; it is in fact the *second* distribution derivative of a continuous function.

EXERCISE

1. Starting from

$$\operatorname{Re} \varphi(x + iy) = \int_{-\infty}^{\infty} \frac{t - x}{(t - x)^2 + y^2} \, d\sigma(t), \tag{9.5-3}$$

which follows from (9.4-2), show that

$$\operatorname{Re} \varphi(x + iy) \to \left(\frac{d}{dx}\right)^2 \int_{-\infty}^{\infty} (t - x)\log \frac{|t - x|}{\sqrt{t^2 + 1}} \, d\sigma(t) \tag{9.5-4}$$

as $y \to 0 \, (y > 0)$, where both $\to$ and d/dx are to be understood in the distribution sense. First verify that the integral in (9.5-4) converges and is a continuous function of x.

Boundary values of analytic functions have been extensively studied; see Johnson 1968 and the literature cited there for some recent developments. Among the earlier results is the following theorem: If $\varphi(z)$ is analytic in the upper half plane (Im $z > 0$) and if, for some $p \geq 1$,

$$M \overset{\text{def}}{=} \sup_{y > 0} \int_{-\infty}^{\infty} |\varphi(x + iy)|^p \, dx < \infty, \qquad (9.5\text{-}5)$$

in which case $\varphi(z)$ is said to belong to the *Hardy class* H_p, the boundary values of $\varphi(z)$ on the real axis constitute an element of the space $L^p(\mathbb{R})$. That is, there is a function or distribution f in $L^p(\mathbb{R})$ such that $\varphi(x + iy)$, regarded as a function of x for fixed $y > 0$ converges in the L^p norm to $f(x)$ as $y \to 0$; namely

$$\int_{-\infty}^{\infty} |\varphi(x + iy) - f(x)|^p \, dx \to 0, \quad \text{as } y \to 0.$$

The L^p norm of f is $= M^{1/p}$. See Hille 1962, volume 2, Chapter 19.

In order to give further examples, we map the half plane Im $z > 0$ onto the unit disk $|w| < 1$ by the mapping $w = (z - i)/(z + i)$ and thereby avoid minor complications that arise for $z \to \infty$—see Johnson 1968 and Baernstein 1972 for details. If $\varphi(w)$ is analytic for $|w| < 1$, we call

$$\varphi_r(\theta) = \varphi(re^{i\theta}) \qquad 0 \leq r < 1. \qquad (9.5\text{-}6)$$

Corresponding to the Hardy class H_p and the space $L^p(\mathbb{R})$, we have the class $\tilde{H}_p$ of functions $\varphi(w)$ such that

$$\|\varphi\| \overset{\text{def}}{=} \sup_{r < 1} \left\{ \int_0^{2\pi} |\varphi_r(\theta)|^p \, d\theta \right\}^{1/p} < \infty \qquad (9.5\text{-}7)$$

and the space $L^p(S^1)$ (S^1 is the unit circle or one-dimensional sphere) of functions and distributions with the norm

$$\|f\| = \left\{ \int_0^{2\pi} |f(\theta)|^p \, d\theta \right\}^{1/p}. \qquad (9.5\text{-}8)$$

For $p = 2, \tilde{H}_p = \tilde{H}_2$ is the Hilbert space discussed in Section 1.10.

We state without proof (see Johnson 1968) that if $\varphi(w)$ is analytic for $|w| < 1$ and satisfies the inequality

$$|\varphi(w)| < \frac{C}{(1 - |w|)^k} \qquad (9.5\text{-}9)$$

for some C and some integer k, then the boundary values of $\varphi(w)$ on the circle $|w| = 1$ are a distribution. A distribution on S^1 is a continuous linear functional on the space $C^\infty(S^1)$ of C^∞ functions $\psi(\theta)$ periodic (2π) in θ. [Convergence of a sequence in $C^\infty(S^1)$ is the same as in $C_0^\infty(\mathbb{R})$ except that there is no requirement on the supports of the members of the sequence, because S^1 is compact.] That is, there is a distribution f on S^1 such that

$$\varphi_r(\theta) \to f(\theta), \quad \text{as } r \to 1$$

in the sense of convergence of distributions. Furthermore, f is the $(k + 1)st$ distribution derivative of a continuous function on S^1, where k is the integer that appears in (9.5-9)

The following example shows that a function $\varphi(w)$ analytic in the unit disk need not satisfy an inequality of the kind (9.5-9): Let

$$\varphi(w) = \sum_{n=0}^{\infty} a^{\sqrt{n}}w^n \qquad (a = \text{const.} > 1).$$

The radius of convergence of this series is $= 1$, and

$$\varphi_r(0) = \sum a^{\sqrt{n}}r^n \qquad (0 \le r < 1)$$

For r close to 1, the largest term of this sum is

$$\max a^{\sqrt{n}}r^n \approx \exp\left\{\frac{(\log a)^2}{4|1 - r|}\right\},$$

which is not bounded, as $r \to 1$, by $\text{const.}/|1 - r|^k$ for any k.

When $\varphi(w)$ does not satisfy an inequality of the type (9.9-5) its boundary values on the circle $|w| = 1$ are not a distribution but an *analytic functional*. Let $\mathscr{A}$ be the class of test functions $\psi(w)$ analytic on the unit circle. That is, each $\psi(w)$ in $\mathscr{A}$ is analytic in some annulus $1 - \varepsilon < |w| < 1 + \varepsilon$ that contains the unit circle. A sequence $\{\psi_n\}$ in $\mathscr{A}$ *converges to* ψ if there is an annulus $1 - \varepsilon_0 < |w| < 1 + \varepsilon_0$ in which the $\psi_n(w)$ are all analytic and in which they converge uniformly to $\psi(z)$. An *analytic functional* f on the unit circle is a continuous linear functional $\langle f, \cdot \rangle$ on $\mathscr{A}$; this generalizes the concept of a distribution on the unit circle. A sequence $\{f_n\}$ of analytic functionals *converges to* f if $\langle f_n, \psi \rangle \to \langle f, \psi \rangle$ for every ψ in $\mathscr{A}$. The main result of this theory is that if $\varphi(w)$ is analytic in the disk $|w| < 1$, there is an analytic functional f such that

$$\varphi_r(\theta) \to f(\theta) \quad \text{as } r \to 1$$

in the sense of convergence of analytic functionals as defined above. For more details, see Johnson 1968.

At present it seems unclear to what extent analytic functionals can replace distributions for purposes of analysis and physical applications. The theory of local properties is presumably more difficult, because the support of a test function ψ in $\mathscr{A}$ necessarily contains the entire unit circle unless $\psi \equiv 0$.

In connection with the converse of the statement that the boundary values of an analytic function $\varphi(w)$ are a distribution or an analytic functional f, we must bear in mind that the real and imaginary parts of f are not independent, because they are respectively the traces on $|w| = 1$ of Re φ and Im φ, which satisfy the Cauchy–Riemann equations in $|w| < 1$, i.e., are *conjugate harmonic functions*. In the special case where $\varphi(w)$ is continuous for $|w| \le 1$, we have $f(\theta) = \varphi(e^{i\theta}) = \varphi_1(\theta)$ and φ is given in terms of f by the Poisson integral

$$\varphi_r(\theta) = \int_0^{2\pi} P_r(\theta - t)f(t)dt, \tag{9.5-10}$$

where

$$P_r(t) = \frac{1 - r^2}{2\pi(1 - 2r\cos t + r^2)}. \tag{9.5-11}$$

Equation (9.5-10) is a convolution, and we write

$$\varphi_r = P_r * f; \tag{9.5-12}$$

in this form it holds generally for any function $\varphi(w)$ analytic in $|w| < 1$, where f is its trace on $|w| = 1$ (an analytic functional), and where the convolution of analytic functionals is defined in the same way as that of distributions. Since the kernel P_r is real, the real and imaginary parts of $\varphi(w)$ and of f are separated in equations (9.5-10 and 12), and in fact (9.5-12) establishes a one-to-one correspondence between real harmonic functions $\varphi(r, \theta)$ and real analytic functionals f, and also, for that matter, between complex harmonic functions (not assumed analytic, i.e., whose real and imaginary parts are not assumed to be conjugate functions) and complex analytic functionals. If f is a distribution, then $\varphi(w)$ satisfies (9.5-9) for some C and some k.

A final example is provided by the Laplace transform of a tempered distribution on $\mathbb{R}$ with support in $[0, \infty)$. It is recalled that if $f(t)$ is a continuous function defined for $t \geq 0$ and with growth bounded according to an inequality $|f(t)| < \text{const.}\ e^{\alpha t}$ as $t \to \infty$, then its Laplace transform

$$F(z) = \int_0^\infty e^{-zt} f(t)\, dt$$

is analytic in the half plane $\text{Re}\ z > \alpha$. If $f(t)$ is of slow growth at infinity, we can take $\alpha = 0$.

Let $f = f(t)$ be any tempered distribution on $\mathbb{R}$ with support in $[0, \infty)$. Let $\chi(t)$ be a C^∞ function that is $= 1$ for $t \geq 0$ and $= 0$ for $t \leq -1$; see Figure 9-4. For any z with $\text{Re}\ z > 0$, the function $\varphi_z(t) \overset{\text{def}}{=} \chi(t)e^{-zt}$ is in $\mathscr{S}$. Furthermore, $(1/\alpha)(\varphi_{z+\alpha} - \varphi_z) \overset{\mathscr{S}}{\to} d\varphi_z/dz$ (as $\alpha \to 0$); therefore, the function

$$F(z) \overset{\text{def}}{=} \langle f, \varphi_z \rangle \tag{9.5-13}$$

is analytic in the right half plane $\text{Re}\ z > 0$. It is called the Laplace transform of f. Since $f = 0$ in $(-\infty, 0)$, $F(z)$ does not depend on the behavior of the

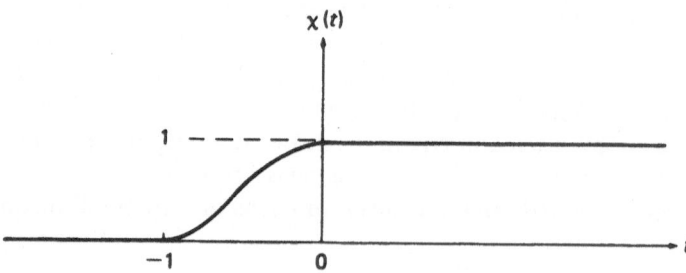

Figure 9.4 The function $\chi(t)$.

function $\chi(t)$ in $(-1, 0)$. Now let $\psi(y)$ be any test function in $\mathscr{S}$. For any $x > 0$,

$$\int_a^b F(x + iy)\psi(y)dy = \langle f, \Psi \rangle,$$

where

$$\Psi(t) = \chi(t) \int_a^b e^{-(x + iy)t}\psi(y)dy.$$

Clearly,

$$\Psi(t) \xrightarrow{\mathscr{S}} \sqrt{2\pi} \chi(t)\hat{\psi}(t)$$

as $x \to 0$, $b \to \infty$, $a \to -\infty$. Therefore, as $x \to 0$,

$$\frac{1}{\sqrt{2\pi}} \int_{-\infty}^{\infty} F(x + iy)\psi(y)dy \to \langle f, \chi\hat{\psi} \rangle = \langle f\chi, \hat{\psi} \rangle = \langle f, \hat{\psi} \rangle = \langle \hat{f}, \psi \rangle.$$

That is, the boundary values of the function $(2\pi)^{-1/2}F(x + iy)$ on the imaginary axis, obtained as $x \downarrow 0$, are the distribution $\hat{f}$, which is the Fourier transform of the given distribution f.

9.6 The Resolution of the Identity for a Self-Adjoint Operator

In this section we show that the theorem of Section 9.4 leads to the existence of a family of operators $E_t(-\infty < t < \infty)$ called the *resolution of the identity* of A, and we rephrase equations (9.4-2 and 3) of that theorem in terms of the operators E_t.

To acknowledge the dependence on the element v of $\mathfrak{H}$, equation (9.4-3) is rewritten as

$$\sigma(t) = \sigma(t; v) = \frac{1}{\pi} \lim_{\varepsilon \downarrow 0} \int_{-\infty}^t \text{Im}(v, R_{s+i\varepsilon}v)ds$$

$$= \frac{1}{2\pi i} \lim_{\varepsilon \downarrow 0} \int_{-\infty}^t [(v, R_{s+i\varepsilon}v) - (v, R_{s-i\varepsilon}v)]ds, \qquad (9.6\text{-}1)$$

where the relation $R_{s-i\varepsilon} = R_{s+i\varepsilon}^*$ has been used. The integral is equal to the integral of $(v, R_\lambda v)$ on the contour in the λ plane consisting of two straight pieces C_1 and C_2 parallel to the real axis at distance ε above and below it, as shown in Figure 9-5. We replace C_1 and C_2 by the equivalent contours C_1' and C_2' shown in Figure 9-6 and then let $\varepsilon \to 0$. The result is

$$\sigma(t; v) = \frac{1}{2\pi i} \int_{C(t)} (v, R_\lambda v)d\lambda, \qquad (9.6\text{-}2)$$

where $C(t)$ is a contour that comes from $-\infty + ia$ in the upper half plane, crosses the real axis at $\lambda = t$ and then returns to $-\infty - ia$ in the lower half plane, and where the Cauchy principal value of the integral, i.e., the limit as

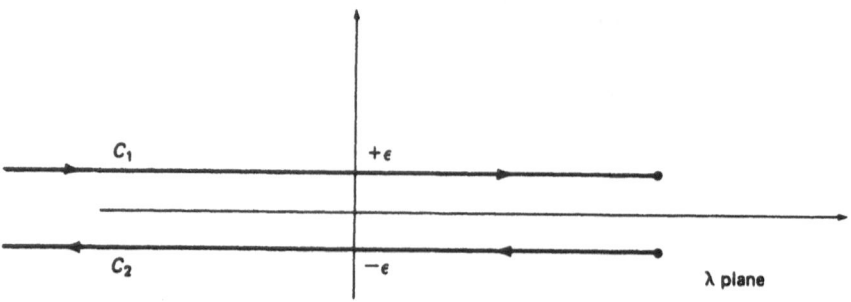

Figure 9.5 The original contours.

$\varepsilon \downarrow 0$, is understood. It is easily seen from equations (9.4-2 and 3) that the limit exists at any point of continuity of $\sigma(t)$. Values of t at which σ has jumps (they are at most denumerable) are taken care of, and in such a way that the normalization condition (9.4-4) is satisfied, by evaluating the integral (9.6-2) at $t' = t + \delta > t$ and then letting t' approach t through a sequence of values where σ is continuous. This procedure will be indicated by writing $C(t+)$ for the contour of integration.

A function $\sigma(t; u, v)$, depending on the real variable t and two elements u of v of $\mathfrak{H}$, is defined by polarization of $\sigma(t; v)$ as

$$\sigma(t; u, v) = \frac{1}{4} \sum_{k=0}^{3} i^{-k} \sigma(t; u + i^k v). \tag{9.6-3}$$

Since polarization of $(v, R_\lambda v)$ gives $(u, R_\lambda v)$,

$$\sigma(t; u, v) = \frac{1}{2\pi i} \int_{C(t+)} (u, R_\lambda v) d\lambda. \tag{9.6-4}$$

For fixed t and v, the right member is an antilinear functional of u, hence is equal to (u, w) for some unique fixed w in $\mathfrak{H}$, according to the Riesz–Fréchet

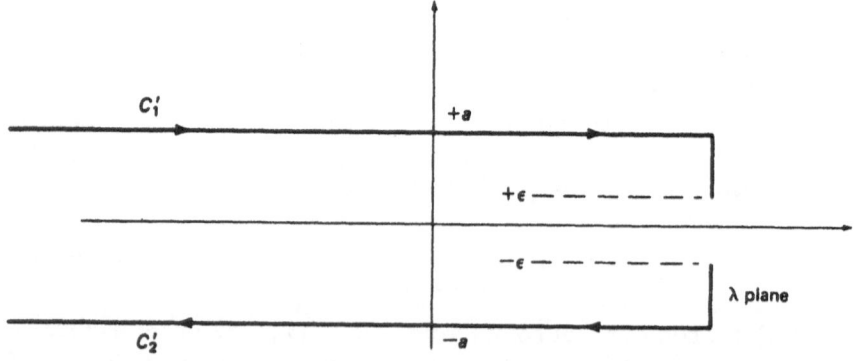

Figure 9.6 The modified contours.

representation theorem; clearly, the resulting w depends linearly on v, so $w = E_t v$, where E_t is a linear operator for each real t; that is, $\sigma(t; u, v) = (u, E_t v)$; hence,

$$(u, E_t v) = \frac{1}{2\pi i} \int_{C(t+)} (u, R_\lambda v) d\lambda. \tag{9.6-5}$$

The family of operators $\{E_t\}$ $(-\infty < t < \infty)$ is called the *resolution of the identity* for the self-adjoint operator A.

Note. In the finite-dimensional case, where A is a Hermitian matrix, and $R_\lambda = (A - \lambda I)^{-1}$, equation (9.6-5) gives $E_t = \sum P_j$, where the summation is over those values of j such that $\lambda_j \leq t$, i.e., such that λ_j is enclosed by the countour $C(t+)$. Then, E_t is a (matrix-valued) step function, and the equation $A = \sum \lambda_j P_j$ can be written as a Stieltjes integral $A = \int t \, dE_t$; this is precisely the formula that will be obtained for a general self-adjoint operator A in $\mathfrak{H}$ (equation (9.7-2) below). Equation (9.6-5) is often written simply as

$$E_t = \frac{1}{2\pi i} \int_{C(t+)} R_\lambda \, d\lambda, \tag{9.6-6}$$

which follows immediately in the finite-dimensional case and follows in the infinite-dimensional case if the integral is understood in the appropriate sense of convergence of operators—See Section 9.9.

The inverse of this relation follows from the theorem of Section 9.4, which says that

$$(v, R_\lambda v) = \int_{-\infty}^{\infty} \frac{1}{t - \lambda} d(v, E_t v),$$

for Im $\lambda \neq 0$. Polarization of this equation gives

$$(u, R_\lambda v) = \int_{-\infty}^{\infty} \frac{1}{t - \lambda} d(u, E_t v), \quad \text{for all } u \text{ and } v \tag{9.6-7}$$

or

$$R_\lambda = \int_{-\infty}^{\infty} \frac{1}{t - \lambda} dE_t, \tag{9.6-8}$$

where this integral is also understood in the appropriate sense of convergence of operators.

9.7 The Properties of the Operators E_t

In Appendix A, at the end of this Chapter, it is proved that the family $\{E_t\}$ of operators has the following properties:

1. For each real value of t, E_t is a bounded, self-adjoint, and idempotent operator, hence is an orthogonal projector.

2. If $s < t$, then $E_t E_s = E_s E_t = E_s$. It follows that if $\mathfrak{M}_t$ denotes the range of E_t, for each t, then $\mathfrak{M}_s \subset \mathfrak{M}_t$, when $s < t$. It follows furthermore that $(E_t - E_s)^2 = E_t - E_s$, so that $E_t - E_s$ is also an orthogonal projector; it is in fact the projector onto the orthogonal complement of $\mathfrak{M}_s$ in $\mathfrak{M}_t$, namely, onto the manifold, denoted by $\mathfrak{M}_t \ominus \mathfrak{M}_s$, consisting of all v in $\mathfrak{M}_t$ that are orthogonal to all u in $\mathfrak{M}_s$. The projector $E_t - E_s$ is regarded as associated with the interval $\Delta = (s, t]$ of the real axis and is denoted by $E(\Delta)$; its range is the manifold $\mathfrak{M}(\Delta) = \mathfrak{M}_t \ominus \mathfrak{M}_s$. If Δ_1 and Δ_2 are disjoint intervals, then $E(\Delta_1)E(\Delta_2) = 0$, so that $\mathfrak{M}(\Delta_1) \perp \mathfrak{M}(\Delta_2)$. It will be seen that the projectors $E(\Delta)$ are somewhat analogous to the P_j of the finite-dimensional case, and the $\mathfrak{M}(\Delta)$ are somewhat analogous to the corresponding eigenspaces $\mathfrak{E}_j$, at least when the intervals Δ are very small.

3. $E_{-\infty} = 0$ and $E_\infty = I$, in the sense that for every v in $\mathfrak{H}$, $E_t v \to 0$, as $t \to -\infty$, and $E_t v \to v$, as $t \to +\infty$. As t increases from $-\infty$ to $+\infty$, the manifold $\mathfrak{M}_t$ grows continually larger, starting with the zero manifold, and finally opens out to fill all of $\mathfrak{H}$. If

$$-\infty = t_0 < t_1 < \cdots < t_N = +\infty \qquad (9.7\text{-}1)$$

is any partition of the real line into intervals $\Delta_j = (t_{j-1}, t_j]$, then

$$I = E(\Delta_1) + \cdots + E(\Delta_N) \qquad (9.7\text{-}2)$$

and

$$\mathfrak{H} = \mathfrak{M}(\Delta_1) \oplus \cdots \oplus \mathfrak{M}(\Delta_N). \qquad (9.7\text{-}3)$$

This last is similar to the expression of $\mathbb{C}^n$ as an orthogonal direct sum of the eigenspaces $\mathfrak{E}_j$ of a Hermitian matrix, except that the decomposition (9.7-3) can usually be refined indefinitely by simply repeatedly refining the partition (9.7-1) of $\mathbb{R}$.

4. The projector-valued function E_t is continuous on the right, in the sense that, for every v in $\mathfrak{H}$, $E_{t+\varepsilon} v \to E_t v$, as $\varepsilon \downarrow 0$; that is, the manifold $\mathfrak{M}_{t+\varepsilon} \ominus \mathfrak{M}_t$ shrinks to zero. For given t, E_t may or may not be continuous also on the left. If it is not, then $\mathfrak{M}_t \ominus \mathfrak{M}_{t-\varepsilon}$ shrinks, it will be seen, to an eigenspace (in the strict sense) of A corresponding to the eigenvalue t, as $\varepsilon \downarrow 0$. The continuity of E_t on the right is not really essential, however, because all statements can be so phrased as to be independent of it. For example, the eigenspace just referred to can be described as the limit to which the manifold $\mathfrak{M}_{t+\varepsilon} \ominus \mathfrak{M}_{t-\varepsilon}$ shrinks, as $\varepsilon \downarrow 0$.

9.8 The Canonical Representation of a Self-Adjoint Operator

It has been seen that every self-adjoint operator A determines, via its resolvent and equation (9.6-5) or (9.6-6), a unique resolution of the identity, i.e., a unique family $\{E_t\}$ of projectors that satisfy the conditions (1)–(4) of the preceding section. Conversely, $\{E_t\}$ determines A uniquely, by equations (9.8-5, 6, 7), given below, and there results a one-to-one correspondence between the set of all self-adjoint operators A and the set of all resolutions of the identity $\{E_t\}$.

The idea behind the formula that expresses A in terms of $\{E_t\}$ comes from the formula $A = \sum \lambda_j P_j$ for a Hermitian matrix (see Section 9.1). If Δ_j ($j = 1, \ldots, N$) are the intervals of a partition of $\mathbb{R}$, as in (9.7-1), then, for each j, the projector $E(\Delta_j) = E_{t_j} - E_{t_{j-1}}$ is roughly analogous to one of the P_j, and the corresponding eigenvalue is roughly some number λ_j in the interval Δ_j. Hence, one expects that

$$A \approx \sum \lambda_j (E_{t_j} - E_{t_{j-1}}); \tag{9.8-1}$$

hence, one expects that, in the limit as the partition of $\mathbb{R}$ is indefinitely refined, A will be given in some sense by a Stieltjes integral

$$A = \int_{-\infty}^{\infty} t \, dE_t, \tag{9.8-2}$$

although the sense in which the Riemann–Stieltjes sum converges is not yet clear (hence, the meaning of the integral itself is unclear), because the right member of (9.8-1) is a bounded operator defined in all $\mathfrak{H}$, while A is generally unbounded and defined only on a domain $\mathfrak{D}(A) \neq \mathfrak{H}$.

If (9.8-2) has any validity, then presumably

$$(u, Av) = \int_{-\infty}^{\infty} t \, d(u, E_t v) \tag{9.8-3}$$

for all u and for all v in $\mathfrak{D}(A)$. This is an ordinary Stieltjes integral, but it may fail to converge because of the infinite range of t. For each $n = 1, 2, \ldots$, however, an operator A_n (bounded and defined in all $\mathfrak{H}$) can be defined by the equation

$$(u, A_n v) = \int_{-n}^{n} t \, d(u, E_t v) \tag{9.8-4}$$

For given v, $A_n v$ may or may not have a limit, as $n \to \infty$, and it is reasonable to suppose that if it does, then v is in $\mathfrak{D}(A)$, and $Av = \lim A_n v$. That is indeed correct, according to Appendix B of this Chapter, devoted to a proof of the theorem below. For that to be the case it is necessary (and also sufficient, it turns out) that $\|A_n v\|$ have a limit, and this is the case if and only if $\int t^2 \, d(v, E_t v)$ converges. The result of these considerations, which are given in detail in the appendix, is the following theorem:

Theorem.   *Each resolution of the identity $\{E_t\}$ (each family that satisfies the conditions (1)–(4) of Section 9.7) determines a unique self-adjoint operator A, and conversely. A is determined from $\{E_t\}$ as follows:*

$$\mathfrak{D}(A) = \left\{ v \colon \int_{-\infty}^{\infty} t^2 \, d(v, E_t v) < \infty \right\}; \tag{9.8-5}$$

then

$$\|Av\|^2 = \int_{-\infty}^{\infty} t^2 \, d(v, E_t v) \tag{9.8-6}$$

and, for such v and for all u in $\mathfrak{H}$,

$$(u, Av) = \int_{-\infty}^{\infty} t \, d(u, E_t v). \qquad (9.8\text{-}7)$$

This construction is often indicated by the symbolic equation (9.8-2) and is called the spectral decomposition *of A.*

EXERCISE

1.   Show that, if A is bounded, then E_t is constant for $t < -\|A\|$ and for $t > \|A\|$, namely $E_t = 0$ for $t < -\|A\|$ and $E_t = I$ for $t > \|A\|$. *Hint*: If the contrary is assumed, a vector v can be found such that the right member of (9.8-6) exceeds $\|A\|^2\|v\|^2$.

9.9   Modes of Convergence of Bounded Operators; Connection Between the Continuity Properties of E_t and the Spectrum of A

Let $B, B_n (n = 1, 2, \ldots)$ be bounded operators. If $(u, B_n v) \to (u, Bv)$, as $n \to \infty$, for all u and all v in $\mathfrak{H}$, the sequence $\{B_n\}$ is said to converge *weakly* to B. If $\|B_n v - Bv\| \to 0$, as $n \to \infty$, for all v in $\mathfrak{H}$, the sequence is said to converge *strongly* to B. If $\|B_n - B\| \to 0$, as $n \to \infty$, the sequence is said to converge *uniformly* or *in norm* to B. Clearly, convergence in norm implies strong convergence (and to the same limit), because $\|B_n v - Bv\| \le \|B_n - B\| \|v\|$, and strong convergence implies weak, because

$$|(u, B_n v) - (u, Bv)| \le \|u\| \, \|B_n v - Bv\|.$$

Also, to say that the operators B_n converge strongly or weakly to the operator B is to say that the vectors $B_n v$ converge strongly or weakly to the vector Bv, as defined in Section 1.9, for every v in $\mathfrak{H}$.

   [These concepts apply in any Banach space $\mathfrak{B}$, except that weak convergence is defined as follows: First, a linear functional $l(v)$, defined for all v in $\mathfrak{B}$, is called *bounded*, as in Section 1.8, if there is a K such that $|l(v)| \le K\|v\|$, for all v; then B_n is said to converge *weakly* to B if $l(B_n v) \to l(Bv)$ for every v in $\mathfrak{B}$ and every bounded linear function $l(\cdot)$. In a Hilbert space, this agrees with the definition given above, because, according to the Riesz–Fréchet representation theorem (Section 1.8), a bounded linear functional $l(v)$ can always be written as (u, v). The notions of strong convergence and convergence in norm of operators in a Banach space will be encountered in connection with well-posed initial-value problems in Chapters 15 and 16].

EXAMPLES IN $L^2(\mathbb{R})$

(1)   Let B_n be a translation operator

$$(B_n f)(x) = f(x + 2n).$$

Then,

$$(g, B_n f) = \int_{-\infty}^{\infty} \overline{g(x)} f(x + 2n) dx.$$

This integral is broken into two parts, $\int_{-\infty}^{-n}$ and $\int_{-n}^{\infty}$; in the second, the new variable $y = x + 2n$ is introduced:

$$(g, B_n f) = \int_{-\infty}^{-n} \overline{g(x)} f(x + 2n) dx + \int_{n}^{\infty} \overline{g(y - 2n)} f(y) dy$$

$$= I_1 + I_2.$$

The Schwarz inequality yields

$$|I_1|^2 \le \int_{-\infty}^{-n} |g(x)|^2 \, dx \|f\|^2,$$

$$|I_2|^2 \le \|g\|^2 \int_{n}^{\infty} |f(y)|^2 \, dy,$$

and the integrals appearing here tend to zero, as $n \to \infty$, because $f(\cdot)$ and $g(\cdot)$ are quadratically integrable. Therefore B_n converges weakly to the zero operator. However, B_n does not converge strongly to anything, because if it did the limit would have to be zero, whereas, for any f, $\|B_n f\| = \|f\|$, which does not tend to zero.

EXAMPLE 2

Let B_n be a truncation operator

$$(B_n f)(x) = \begin{cases} f(x) & \text{if } |x| < n, \\ 0 & \text{if } |x| > n. \end{cases}$$

(Note that B_n is a projector, because $B_n^2 = B_n$.) Obviously $\|B_n f - f\| \to 0$, as $n \to \infty$, hence B_n converges strongly to the identity operator I. However, B_n does not converge in norm to I, because $\|B_n - I\| = 1$ for every n, as can be seen by applying $B_n - I$ to a function whose suppose lies outside the interval $(-n, n)$.

EXAMPLE 3

Let B and B_n be Hilbert–Schmidt integral operators

$$(B_n f)(x) = \int_{-\infty}^{\infty} K_n(x, y) f(y) dy,$$

$$(B f)(x) = \int_{-\infty}^{\infty} K(x, y) f(y) dy,$$

where the kernels are such that the quantity

$$M_n \overset{\text{def}}{=} \int_{-\infty}^{\infty} \int_{-\infty}^{\infty} |K_n(x, y) - K(x, y)|^2 \, dx \, dy \to 0, \quad \text{as } n \to \infty$$

(i.e., $K_n \to K$ in $L^2(\mathbb{R}^2)$). A simple application of the Schwartz inequality shows that

$$\|(B_n - B)f\|^2 \leq M_n\|f\|^2, \quad \text{for all } f,$$

so that $\|B_n - B\| \leq M_n$; therefore B_n converges in norm to B.

Corresponding to each of the three types of convergence there is a type of continuity. The one-parameter family of bounded operators $B(t)$ is said to be *weakly* or *strongly continuous* or *continuous in norm* at t if $B(t + \delta)$ converges weakly, strongly, or in norm, to $B(t)$, as $\delta \to 0$. Right and left one-sided continuity of each type can be similarly defined.

If $B(t)$ is the resolution of the identity E_t given by (9.6-5 or 6), then it is automatically weakly continuous on the right. Furthermore, for a resolution of the identity, weak continuity (on the right, on the left, or both) automatically implies strong continuity (on the right, on the left, or both), because if

$$(u, (E_{t+\delta} - E_t)v) \to 0, \quad \text{as } \delta \to 0,$$

for every u and every v, then this is true in particular for $u = v$. Since $E_{t+\delta} - E_t$ is a projector, the above quantity is then equal to

$$(v, (E_{t+\delta} - E_t)^2 v),$$

and since the projector is self-adjoint, the quantity is also equal to

$$((E_{t+\delta} - E_t)v, (E_{t+\delta} - E_t)v) = \|(E_{t+\delta} - E_t)v\|^2.$$

Therefore, in this particular case, strong and weak continuity are equivalent. Consequently the cases to be considered, for any real t_0, are:

(a) E_t discontinuous (i.e., on the left) at t_0,
(b) E_t strongly continuous (but not in norm) at t_0,
(c) E_t continuous in norm at t_0.

It will be seen that in case (a) t_0 is in the point spectrum of A; in case (b) it is in the continuous spectrum and in case (c) in the resolvent set. [Subcases might also be considered, such as continuity in norm on one side and merely strong continuity on the other, but the implications of such behavior for the spectrum will be evident, when the above cases have been analyzed.]

The continuity of E_t on the right (both weak and strong, but not generally in norm) was achieved in Section 9.6 by defining E_t by

$$(u, E_t v) = \frac{1}{2\pi i} \lim_{\delta \downarrow 0} \int_{C(t+\delta)} (u, R_\lambda v)d\lambda,$$

where $C(s)$ is a contour that comes from $-\infty$ above the real axis in the λ plane, cuts the real axis at s, and then returns to $-\infty$ below the real axis. It is evident that an operator, which will be called E_{t-}, can be similarly defined by letting δ approach zero from below rather than from above. E_{t-} is a family of projectors with all the properties of E_t except that it is continuous on the left rather than on the right. By the methods of Section 9.6, it is seen that the operator

$$P_t \overset{\text{def}}{=} E_t - E_{t-} \tag{9.9-1}$$

is a projector; at a point of continuity of E_t, P_t is the zero projector (the zero operator, which maps all $\mathfrak{H}$ onto the zero element), but at a point of discontinuity of E_t, P_t is not zero. The manifold onto which P_t projects, i.e., its range $\mathfrak{R}(P_t)$, consists of those vectors in the range of E_t that are $\perp$ the range of E_s for any $s < t$. This is expressed by the equations

$$P_t E_s = E_s P_t = \begin{cases} P_t & \text{if } s \geq t, \\ 0 & \text{if } s < t, \end{cases} \tag{9.9-2}$$

which are easy consequences of the methods of Appendix A.

Suppose now that for a given real number t_0 the above case (a) holds, so that $P_{t_0} \neq 0$. For $v\,(\neq 0)$ in $\mathfrak{R}(P_{t_0})$,

$$(u, Av) = \int_{-\infty}^{\infty} t\, d(u, E_t v), \quad \text{for all } u. \tag{9.9-3}$$

According to (9.9-2), since $v = P_t v$, the function $(u, E_t v)$ is constant except for a jump of magnitude (u, v) at $t = t_0$. Therefore,

$$(u, Av) = t_0(u, v) \quad \text{for all } u \text{ in } \mathfrak{H},$$

that is, $Av = t_0 v$; hence t_0 is in the point spectrum of A, and v is an eigenvector; the subspace

$$\mathfrak{E}_{t_0} \overset{\text{def}}{=} \mathfrak{R}(P_{t_0}) \tag{9.9-4}$$

is the eigenspace that corresponds to the eigenvalue t_0. See Comment below.

Case (b) is considered next. In Section 9.8 it was seen that if v is in the domain of the self-adjoint operator A, then

$$\|Av\|^2 = (Av, Av) = \int_{-\infty}^{\infty} t^2\, d(v, E_t v). \tag{9.9-5}$$

Now suppose that E_t is strongly continuous at $t = t_0$, but not continuous in norm. Then, for each $\delta > 0$, there is a nonzero element $v = v_\delta$ in the range of $E_{t_0 + \delta} - E_{t_0 - \delta}$; for this v, the function $(v, E_t v)$ is constant for $|t - t_0| > \delta$. Therefore,

$$(Av - t_0 v, Av - t_0 v) = \int_{t_0 - \delta}^{t_0 + \delta} (t - t_0)^2\, d(v, E_t v). \tag{9.9-6}$$

[The equations $\int t\, d(v, E_t v) = (v, Av)$ and $\int d(v, E_t v) = (v, v)$ have also been used, as well as 9.9-5.] In the interval $(t_0 - \delta, t_0 + \delta)$, $(v, E_t v)$ increases from zero to $\|v\|^2$; therefore

$$\|Av - t_0 v\|^2 \leq \delta^2 \|v\|^2.$$

It follows that the vectors v_δ are approximate eigenvectors, in the sense of Section 8.1, and that t_0 is a point of the continuous spectrum.

To investigate case (c), let t_0 be a point of continuity in norm of E_t. Then, $\|E_t - E_{t_0}\| \to 0$ as $t \to 0$. But $\|E_t - E_{t_0}\|$ is always either $= 1$ or $= 0$, because $E_t - E_{t_0}$ is either an orthogonal projector or the zero operator, if $t \geq t_0$, and the same holds for $E_{t_0} - E_t$, if $t \leq t_0$. (See Exercise 2 in Section 9.2). Therefore, there is an interval $[t_0 - \varepsilon, t_0 + \varepsilon]$, $\varepsilon > 0$, in which $E_t = E_{t_0}$. In this interval, $(u, E_t v)$ is constant, for any u and v in $\mathfrak{H}$. In place of (9.9-6), we have

$$(Av - t_0 v, Av - t_0 v) = \left(\int_{-\infty}^{t_0 - \varepsilon} + \int_{t_0 + \varepsilon}^{\infty} \right)(t - t_0)^2 \, d(v, E_t v).$$

Therefore,

$$\|Av - t_0 v\|^2 \geq \varepsilon^2 \left(\int_{-\infty}^{t_0 - \varepsilon} + \int_{t_0 + \varepsilon}^{\infty} \right) d(v, E_t v)$$

$$= \varepsilon^3 \int_{-\infty}^{\infty} d(v, E_t v) = \varepsilon^2 \|v\|^2,$$

for any v in $\mathfrak{H}$. It follows that $(A - t_0 I)^{-1}$ is bounded; in fact, $\|(A - t_0 I)^{-1}\| \leq 1/\varepsilon$; hence t_0 is in the resolvent set of A.

Comment. It was established above that any vector v in $\mathfrak{R}(P_{t_0})$ is an eigenvector with t_0 as eigenvalue. It is noted that all eigenvectors are obtained in that way. Let v be any vector $\neq 0$ such that $Av = t_0 v$ for some t_0. Then $0 = \|Av - t_0 v\|^2 = \int_{-\infty}^{\infty} (t - t_0)^2 \, d(v, E_t v)$.

Now, $(v, E_t v)$ is nondecreasing and $(t - t_0)^2$ is positive except at $t = t_0$. Therefore, since the integral is zero, $(v, E_t v)$ is constant except for a jump at $t = t_0$. Since $E_t \to 0$ and I, respectively, for $t \to -\infty$ and $+\infty$, the magnitude of the jump is (v, v), but it is also $= (v, P_{t_0} v)$, by definition of P_{t_0}. Hence $(v, (I - P_{t_0})v) = 0$, but $I - P_{t_0}$ is a projector, hence is $= (I - P_{t_0})^2$ and is self-adjoint, so that

$$(v - P_{t_0} v, v - P_{t_0} v) = 0,$$

$$\therefore \quad v = P_{t_0} v,$$

hence $v \in \mathfrak{R}(P_{t_0})$ as claimed.

Summary. A discontinuity of E_t at t_0 implies $t_0 \in P\sigma(A)$. Strong continuity but not continuity in norm implies that $t_0 \in C\sigma(A)$. Continuity in norm implies that $t_0 \in \rho(A)$.

EXERCISES

1. Let $A_n (n = 1, 2, \ldots)$ be the operators introduced in Section 9.8 as approximations to a self-adjoint operator A. Show that if $v \in \mathfrak{D}(A)$, then $A_n v \to Av$ strongly. *Hint*: Use Exercise 5 of Section 1.9. Conclude that if A is bounded, then $A_n \to A$ strongly. It follows from the exercise in Section 9.8, of course, that if A is bounded, then $A_n = A$

for large enough n, so that then $A_n \rightarrow A$ also in norm. If A is unbounded, then A_n does not converge to A in any of the three senses introduced in this section.

2. If A is bounded, in what sense does the Riemann–Stieltjes sum in (9.8-1) converge to A? *Hint*: The sum can be written as $\int f(t)dE_t$, where $f(t)$ is a step function.

3. If A is unbounded, in what sense does $E_n - E_{-n}$ converge to I, as $n \rightarrow \infty$?

4. By use of equations (8.5-7) and the resolvent equation (8.5-1), show that R_λ is differentiable with respect to λ in the complex variable sense, at any point λ in the resolvent set, with convergence of the difference quotient $(1/\alpha)(R_{\lambda+\alpha} - R_\lambda)$ in norm to the derivative, which is equal to R_λ^2, in agreement with formal differentiation of $R_\lambda = (A - \lambda)^{-1}$. We say that R_λ is an analytic or holomorphic operator-valued function of the complex variable λ in $\rho(A)$.

5. Suppose that, for each λ in a region Ω, $F(\lambda)$ is a bounded operator, and suppose that $F(\lambda)$ is differentiable in Ω in the sense of Exercise 4. Show that Cauchy's theorem and Cauchy's integral formula hold for $F(\lambda)$, after deciding what is meant by an integral $\oint F(\lambda)d\lambda$.

6. If R_λ is the resolvent of a self-adjoint operator A and λ_0 is an isolated eigenvalue of A, what kind of singularity does R_λ have at $\lambda = \lambda_0$, and what is the residue there?

9.10 Unitary Operators; Functions of Operators; Bounded Observables; Polar Decomposition

The Cayley transform $U = (A - i)(A + i)^{-1}$ of a self-adjoint operator A was introduced in Section 8.6 in connection with the extensions of a symmetric operator. Let $\{E_t\}$ be the resolution of the identity that corresponds to A. Then, it is asserted,

$$U = \int_{-\infty}^{\infty} \frac{t - i}{t + i} \, dE_t. \tag{9.10-1}$$

This equation can be interpreted either in terms of strong convergence (in fact convergence in norm) of the corresponding Riemann–Stieltjes sum (see Exercise 1, below) or as an abbreviation for

$$(u, Uv) = \int_{-\infty}^{\infty} \frac{t - i}{t + i} \, d(u, E_t v). \tag{9.10-2}$$

To establish this equation, note first that, for any v and w in $\mathfrak{D}(A)$,

$$((A + i)w, v) = (w, (A - i)v) = \int_{-\infty}^{\infty} (t - i)d(w, E_t v).$$

Now let u be any vector in $\mathfrak{H}$ and set

$$w = \int_{-\infty}^{\infty} \frac{1}{s + i} \, dE_s u;$$

then

$$((A + i)w, v) = \int_{-\infty}^{\infty} (t - i)d_t \int_{-\infty}^{\infty} \frac{1}{s - i} \, d_s(E_s u, E_t v).$$

Equation (9.B-5) of the appendix gives a way to reduce such double integrals to single ones and the result is

$$((A + i)w, v) = \int_{-\infty}^{\infty} \frac{t - i}{t - i} \, d(u, E_t v) = (u, v).$$

Therefore $(A + i)w = u$, or $w = (A + i)^{-1}u$, i.e.,

$$(A + i)^{-1}u = \int_{-\infty}^{\infty} \frac{1}{s + i} \, dE_s u. \tag{9.10-3}$$

One further application of (9.B-5) then establishes (9.10-2).

Equations (9.10-2) and (9.10-3) suggest the following definition: If $f(t)$ is any continuous or piecewise continuous function, an operator $f(A)$ is defined as

$$f(A) = \int_{-\infty}^{\infty} f(t) dE_t. \tag{9.10-4}$$

The interpretation of this equation is similar to that of the equation $A = \int_{-\infty}^{\infty} t \, dE_t$, namely

$$\mathfrak{D}(f(A)) = \left\{ v : \left(v, \int_{-\infty}^{\infty} |f(t)|^2 \, dE_t v \right) < \infty \right\},$$

$$(u, f(A)v) = \int_{-\infty}^{\infty} f(t) d(u, E_t v). \tag{9.10-5}$$

If $f(t)$ is a bounded function, for all t, as in (9.10-2) and (9.10-3), then $f(A)$ is a bounded operator; if $f(t)$ is a real function, then $f(A)$ is self-adjoint; if $|f(t)| = 1$, then $f(A)$ is unitary.

As an example, another unitary operator can be obtained from A by taking $f(t) = e^{i\alpha t}$, α real. [The unitary operator obtained by the Cayley transform has the advantages over this one that it can be defined without using the spectral decomposition of A and that the equation $U = (A - i)(A + i)^{-1}$ can be solved for A.]

As a further example, let $f(t) = \tanh t$. Then $f(A)$ is bounded and self-adjoint. If A is regarded as an observable in quantum mechanics, then $f(A)$ is an equivalent bounded observable. The apparatus for measuring $f(A)$ is the same as that for measuring A but has attached to it a simple computer for computing $\tanh t$ from the measured value t of A. The observable $f(A)$ gives the same information as A, because t can always be obtained from $f(t)$. This point of view will be found useful in Chapter 14.

If the mapping $t \to f(t)$ is one-to-one, for all real t, as in all the above examples except $f(t) = e^{i\alpha t}$, then (9.10-4) can be rewritten as follows: Let $\mathscr{C}$ be the curve in the complex plane given by $z = f(t)$ ($-\infty < t < \infty$), and let g be the function inverse to f, so that $t = g(z)$. A family of orthogonal projectors is then defined on $\mathscr{C}$ by writing $F_z = E_{g(z)}$, where $\{E_t\}$ is the resolution of the identity for A. Then (9.10-4) becomes

$$f(A) = \int_{\mathscr{C}} z \, dF_z.$$

In particular, the choice $f(t) = (t - i)/(t + i)$ shows that any unitary operator of which 1 is not an eigenvalue is characterized by a resolution of the idendity $\{F_z\}$ defined on the unit circle $\mathscr{C}$: $|z| = 1$. In this case, F_z is usually written as F_θ, where $z = e^{i\theta}$ (hence, $t = -\cot \frac{1}{2}\theta$); the canonical representation of a unitary operator is therefore taken as

$$U = \int_0^{2\pi} e^{i\theta} \, dF_\theta,$$ (9.10-7)

where the family $\{F_\theta\}$ satisfies the requirements 1–4 of Section 9.7, with the interval $(-\infty, \infty)$ replaced by $(0, 2\pi)$.

EXERCISE

1. Show that if $f(t)$ is bounded, then the Riemann–Stieltjes sum for (9.10-4) converges strongly to $f(A)$. Under what circumstances does it converge in norm to $f(A)$?

Fractional powers of a nonnegative self-adjoint operator can be defined after a short preliminary consideration.

Theorem. *If A is a self-adjoint operator, and E_t is its resolution of the identity, if v is a unit vector ($\|v\| = 1$) in the range of the projector $E_b - E_a$, where $a < b$, then*

$$a \leq (v, Av) \leq b.$$ (9.10-8)

PROOF. The nondecreasing function $(v, E_t v)$ is $=0$ for $t < a$ and is $=1$ for $t > b$. Therefore,

$$b - (v, Av) = \int_{a-0}^{b+0} (b - t)(v, E_t v) \geq 0;$$

the other inequality is proved similarly.

This theorem shows that the spectrum of A lies in the closure of the numerical range, for if t_0 is such that E_t is nonconstant in every interval $(t_0 - \varepsilon, t_0 + \varepsilon)$, then t_0 can be approximated to within any $\varepsilon > 0$ by (v, Av), where v is a unit vector.

The theorem shows also that if A is nonnegative, then $E_t = 0$ for all $t < 0$, for otherwise a vector v could be found such that $(v, Av) < 0$. Therefore, if $A \geq 0$, the function $f(A) = A^{1/2}$ and more generally $A^\alpha (\alpha > 0)$ can be defined according to

$$A^\alpha = \int_0^\infty t^\alpha \, dE_t,$$ (9.10-9)

where the positive root t^α is meant. In particular, if T is any closed operator with domain dense in $\mathfrak{H}$, so that T^*T is defined and self-adjoint by the theorem of von Neumann mentioned in Section 7.9, then $(T^*T)^{1/2}$ is a self-adjoint operator (also ≥ 0) which can often serve as a sort of absolute value of T. However, it is different from $(TT^*)^{1/2}$, unless T is a normal operator.

The so-called polar decomposition of a general operator is now discussed, first for a bounded operator A defined in all $\mathfrak{H}$. Call $R = (A^*A)^{1/2}$; R is nonnegative. Call $\hat{R}$ the restriction of R to $\mathfrak{R}(R) = \mathfrak{N}(R)^{\perp} \overset{\text{def}}{=} \mathfrak{D}(\hat{R})$, where $\mathfrak{N}$ stands for nullspace. (This step is of course unnecessary if R is positive, not merely nonnegative.) Since any w in $\mathfrak{H}$ can be written as $u + v$, where $u \in \mathfrak{D}(\hat{R})$ and $v \in \mathfrak{N}(R)$, and hence $Rw = \hat{R}u = Ru$, it is seen that R and $\hat{R}$ have the same range, namely $\mathfrak{D}(\hat{R})$. Therefore $\hat{R}$ maps $\mathfrak{D}(\hat{R})$ one-to-one onto itself. Call

$$\hat{V} = A\hat{R}^{-1}, \qquad \mathfrak{D}(\hat{V}) = \mathfrak{R}(\hat{R}) = \mathfrak{D}(\hat{R}).$$

Then, for any v in $\mathfrak{H}$,

$$\hat{V}Rv = A\hat{R}^{-1}Rv = Av.$$

Note. $\mathfrak{R}(R) = \mathfrak{N}(A)$. Now $\hat{V}$ is an isometric mapping of its domain onto its range (which is the range of A), because if v is any vector in $\mathfrak{R}(\hat{R})$ and if $\hat{R}^{-1}v$ is called w, then $\|v\|^2 = \|Rw\|^2 = (Rw, Rw) = (w, R^2w) = (w, A^*Aw) = (Aw, Aw) = \|Aw\|^2 = \|A\hat{R}^{-1}v\|^2 = \|\hat{V}v\|^2$. Hence $\hat{V}$ is isometric. An operator V is now defined as the extension of $\hat{V}$ to all of $\mathfrak{H}$ obtained by defining $Vw = 0$ for $w \perp \mathfrak{R}(R)$. Such an operator V is called *partially isometric*. Clearly, $\|V\| = 1$, except when A is the zero operator.

Conclusion. Any bounded operator A can be written as $A = VR$, where R is ≥ 0 and V is partially isometric. The decomposition is unique if one requires that $Vw = 0$ for $w \perp \mathfrak{R}(R)$. If $Av \neq 0$ for $v \neq 0$, then R is >0 and V is unitary. The conclusion holds also if A is unbounded but is closed and defined densely in $\mathfrak{H}$; see Kato (1966), also Section 7.9. VR is called the *polar decomposition* of A. If V is unitary, it can always be written as $\exp\{i\Theta\}$, where Θ is self-adjoint. Since self-adjoint operators correspond to real numbers, the equation $A = VR$ is then the analogue of the equation $z = e^{i\theta}r$ for an arbitrary complex number z.

EXERCISES

2.   Show that if V is partially isometric, then V^* is also partially isometric.

3.   Show that any bounded operator A can also be written as $R_1 V_1$, where $R_1 = (AA^*)^{1/2}$, and where V_1 is partially isometric.

Appendix A to Chapter 9—The Properties of the Operators E_t

It will first be proved that E_t is a bounded operator, for each t. According to the theorem of Section 9.4, $0 \leq \sigma(t) \leq C\|v\|^2$ for all t, where C is a constant, that is,

$$\sigma(t) = \sigma(t; v) = (v, E_t v) \leq C\|v\|^2;$$

it follows by polarization and by the device in Section 1.11 (see equation (1.11-3)) that

$$|(u, E_t v)| \leq 4C\|u\| \, \|v\|;$$

setting $u = E_t v$ and cancelling one factor yields

$$\|E_t v\| \leq 4C\|v\|;$$

i.e., E_t is a bounded operator (as follows also from the closed-graph theorem, since $E_t v$ was defined for all v in $\mathfrak{H}$); it will be seen below that the bound $4C$ can be replaced by 1.

Since $\sigma(-\infty) = 0$, it is clear that $E_{-\infty}$ is the zero operator, and it will now be shown that $E_{+\infty}$ is the identity operator I. In equation (9.6-7), take $u = (A - \bar{\lambda}I)w$, where w is any element in $\mathfrak{D}(A)$. Then

$$(Aw - \bar{\lambda}w, R_\lambda v) = (w, (A - \lambda I)R_\lambda v) = (w, v)$$

$$= \int_{-\infty}^{\infty} \frac{1}{t - \lambda} \, d(Aw - \bar{\lambda}w, E_t v)$$

$$= \int_{-\infty}^{\infty} \frac{1}{t - \lambda} \, d(Aw, E_t v) + \int_{-\infty}^{\infty} \frac{-\lambda}{t - \lambda} \, d(w, E_t v).$$

Now, $(Aw, E_t v)$ and $(w, E_t v)$ were obtained by polarizations of $\sigma(t; v)$, which has finite total variation, as a function of t; hence, they also have finite total variation. Therefore, as $\lambda \to i\infty$, the first integral goes to zero and the second to $\int_{-\infty}^{\infty} d(w, E_t v) = (w, E_\infty v)$; hence $(w, v) = (w, E_\infty v)$, for every w in $\mathfrak{D}(A)$, but $\mathfrak{D}(A)$ is dense in $\mathfrak{H}$, so $v = E_\infty v$; in other words $E_\infty = I$, as was to be proved.

For each t, the operator E_t is self-adjoint, because $\sigma(t; v)$ is real, and hence the function $\sigma(t; u, v)$ obtained by polarization satisfies the equation $\sigma(t; v, u) = \overline{\sigma(t; u, v)}$, or

$$(v, E_t u) = \overline{(u, E_t v)} = (E_t v, u);$$

hence, since E_t is bounded and defined in all $\mathfrak{H}$, $E_t^* = E_t$.

It will now be shown that for any two real s, t,

$$E_t E_s = E_s E_t = E_{\min(s\ t)}. \tag{9.A-1}$$

First, suppose that $s \neq t$ and even, without loss of generality, that $s < t$; then,

$$(u, E_t E_s v) = (E_t u, E_s v) = \frac{1}{2\pi i} \int_{C(s+)} d\lambda (E_t u, R_\lambda v)$$

$$= \frac{1}{2\pi i} \int_{C(s+)} d\lambda (u, E_t R_\lambda v)$$

$$= \frac{1}{(2\pi i)^2} \int_{C(s+)} d\lambda \int_{C(t+)} d\mu (u, R_\mu R_\lambda v)$$

$$= \frac{1}{(2\pi i)^2} \int_{C(s+)} d\lambda \int_{C(t+)} d\mu \left(u, \frac{R_\mu - R_\lambda}{\mu - \lambda} v\right), \tag{9.A-2}$$

where the resolvent equation has been used.

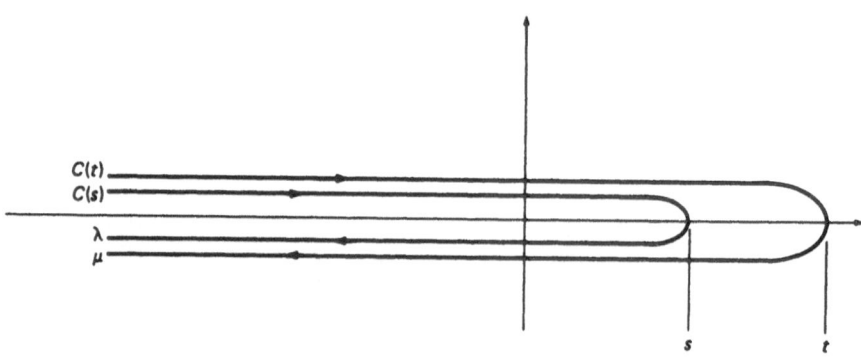

Figure 9.7 The contours $C(s)$ and $C(t)$.

It is now supposed that the contour $C(s)$ lies inside the contour $C(t)$ as in Figure 9-7. [Compare with the treatment of the equation (9.3-8) in the finite-dimensional case.] The last integral above is written as the difference of two integrals, the first containing R_μ and the second R_λ. In the first, integration first with respect to λ gives

$$\int_{C(s+)} \frac{1}{\mu - \lambda} \, d\lambda = 0,$$

because μ lies outside the contour $C(s)$. In the second, integration first with respect to μ gives

$$\int_{C(t+)} \frac{1}{\mu - \lambda} \, d\mu = -2\pi i,$$

because λ lies inside the contour $C(t)$ (note that $C(t)$ is traced clockwise). The result is

$$(u, E_t E_s v) = \frac{1}{2\pi i} \int_{C(s+)} d\lambda (u, R_\lambda v) = (u, E_s v),$$

so that $E_t E_s = E_s$. It is evident from (9.A-2) that E_s and E_t commute. So far, the case $s = t$ has been excluded. By construction, $(u, E_t w)$, as a function of t, is continuous on the right, for any u and any w, hence for $w = E_s v$; therefore, if t converges to s from the right, the equation $(u, E_t E_s w) = (u, E_s w)$ shows that $E_s^2 = E_s$, as was claimed. Hence, E_s is a projector. For $s < t$, $E_t - E_s$ is also a projector, because

$$(E_t - E_s)^2 = E_t^2 - 2E_s E_t + E_s^2 = E_t - 2E_s + E_s = E_t - E_s.$$

Summary. $\{E_t\}$ is a one-parameter family of self-adjoint projectors such that

$$E_t E_s = E_s E_t = E_{\min(s\ t)}; \tag{9.A-3}$$

and for any u, v in $\mathfrak{H}$,

$$\lim_{t \to -\infty} (u, E_t v) = 0, \qquad \lim_{t \to +\infty} (u, E_t v) = (u, v), \qquad (9.\text{A-}4)$$

$$\lim_{\varepsilon \downarrow 0} (u, E_{t+\varepsilon} v) = (u, E_t v). \qquad (9.\text{A-}5)$$

Equations (9.A-4) were paraphrased by saying that $E_{-\infty} = 0$, $E_{+\infty} = I$. Equation (9.5-A) describes a kind of continuity, on the right, of E_t; the continuity properties of E_t are further discussed in Section 9.9.

Appendix B to Chapter 9—The Canonical Representation of a Self-Adjoint Operator

It will be shown that any family $\{E_t\}$ of self-adjoint projectors having the properties 1–4 in Section 9.7 determines a self-adjoint operator $\hat{A}$, and that if $\{E_t\}$ is obtained from an operator A by (9.6-5), where $R_\lambda = R_\lambda(A) = (A - \lambda)^{-1}$, then $\hat{A} = A$, so that there is a one-to-one correspondence between the class of all such families and that of the self-adjoint operators.

Operators A_n $(n = 1, 2, \ldots)$ are first defined by the equation

$$(u, A_n v) = \int_{-n}^{n} t \, d(u, E_t v). \qquad (9.\text{B-}1)$$

It is evident that each A_n is a linear, bounded, self-adjoint operator, and the question arises whether, for given v in $\mathfrak{H}$, $A_n v$ tends to a limit, as $n \to \infty$. A necessary (and, as will be seen, also sufficient) condition is that the norms $\|A_n v\|$ be bounded, as $n \to \infty$. The complex conjugate of the above equation is

$$(A_n v, u) = \int_{-n}^{n} t \, d(E_t v, u).$$

In this equation, set $u = A_n v$ and use (9.B-1) again, with t replaced by s:

$$\|A_n v\|^2 = \int_{-n}^{n} t \, d_t \int_{-n}^{n} s \, d_s(E_t v, E_s v). \qquad (9.\text{B-}2)$$

By (9.A-1), the function $(E_t v, E_s v) = (E_s E_t v, v)$ is independent of s, for $s \geq t$. Therefore,

$$\int_{-n}^{n} s \, d_s(E_t v, E_s v) = \int_{-n}^{t} s \, d(E_s v, v) \qquad (9.\text{B-}3)$$

and when d_t is applied to this, the result is simply $t \, d(E_t v, v)$. Consequently, the double integral reduces to a single one, and

$$\|A_n v\|^2 = \int_{-n}^{n} t^2 \, d(v, E_t v). \qquad (9.\text{B-}4)$$

[Since the integral is $\leq n^2 \|v\|^2$, it is seen that $\|A_n\| \leq n$, but this result will not be used.] For later reference, the relation between double and single integrals is generalized; namely,

$$\int_{-\infty}^{\infty} f(t) \, d_t \int_{-\infty}^{\infty} g(s) d_s(E_t u, E_s v) = \int_{-\infty}^{\infty} f(t)g(t) d(u, E_t v), \qquad (9.\text{B-}5)$$

whenever $f(\cdot)$ and $g(\cdot)$ are such that the integrals converge.

An operator $\hat{A}$ is now defined by setting

$$\mathfrak{D}(\hat{A}) = \left\{ v \in \mathfrak{H} : \int_{-\infty}^{\infty} t^2 \, d(v, E_t v) < \infty \right\}, \qquad (9.\text{B-}6)$$

and, for v in $\mathfrak{D}(\hat{A})$, $\hat{A}v$ is defined as $= \lim A_n v$, as $n \to \infty$. By the same argument as above, an expression similar to (9.B-4) is obtained for $\|A_n v - A_m v\|^2$, but integrated only over values of t such that $n < |t| < m$. Hence, $\{A_n v\}$ is a Cauchy sequence in $\mathfrak{H}$ if the integral in (9.B-6) is finite. It will be shown, in succession, (1) that $\hat{A}$ is densely defined (hence has an adjoint), (2) that $\hat{A}$ is self-adjoint, and (3) that $R_\lambda(\hat{A} - \lambda)w = w$, for all w in $\mathfrak{D}(\hat{A})$. From the last follows that R_λ is also the resolvent of $\hat{A}$; that is, $(\hat{A} - \lambda)^{-1} = (A - \lambda)^{-1}$; hence, $\hat{A} = A$.

To show that $\mathfrak{D}(\hat{A})$ is dense, let v be an arbitrary element of $\mathfrak{H}$. Then the sequence $\{v_k\}$, where $v_k = (E_k - E_{-k})v$, converges to v, as $k \to \infty$, while $A_n v_k = A_k v_k$, for all $n \geq k$, so that v_k is in $\mathfrak{D}(\hat{A})$.

To find $\hat{A}^*$, consider all pairs of elements u, w in $\mathfrak{H}$ such that

$$(u, \hat{A}v) = \int_{-\infty}^{\infty} t \, d(u, E_t v) = (w, v)$$

for all v in $\mathfrak{D}(\hat{A})$. Taking complex conjugates,

$$(v, w) = \int_{-\infty}^{\infty} t \, d(E_t v, u) = \int_{-\infty}^{\infty} t \, d(v, E_t u);$$

The problem of finding u and w to satisfy this equation is precisely the same problem that was encountered above of determining $\mathfrak{D}(\hat{A})$ and $\hat{A}u$, and it is concluded that $\hat{A}^* = \hat{A}$.

Lastly, for any u, v, according to (9.6-7),

$$(u, R_\lambda v) = \int_{-\infty}^{\infty} \frac{1}{t - \lambda} d(u, E_t v) = \int_{-\infty}^{\infty} \frac{1}{t - \lambda} d(E_t u, v), \qquad (9.\text{B-}7)$$

for any nonreal λ. Next, set $v = \hat{A}w - \lambda w$, where w is any element in $\mathfrak{D}(\hat{A})$, given by (9.B-6). By definition of $\hat{A}$,

$$(E_t u, \hat{A}w - \lambda w) = \int_{-\infty}^{\infty} (s - \lambda) d_s(E_t u, E_s w);$$

substitution into (9.B-7) gives

$$(u, R_\lambda(\hat{A} - \lambda I)w) = \int_{-\infty}^{\infty} \frac{1}{t - \lambda} d_t \int_{-\infty}^{\infty} (s - \lambda)d_s(E_t u, E_s w);$$

use of the relation (9.B-5) between double and single integrals gives

$$(u, R_\lambda(\hat{A} - \lambda I)w) = \int_{-\infty}^{\infty} \frac{t - \lambda}{t - \lambda} d(u, E_s w) = (u, w).$$

Therefore $R_\lambda = (\hat{A} - \lambda I)^{-1}$, as was to be proved; hence, letting $n \to \infty$ in (9.B-1) gives the desired representation of A, for $v \in \mathfrak{D}(A)$.

CHAPTER 10

Ordinary Differential Operators

The operators $(-i d/dx)$ and $-(d/dx)^2$ on $\mathbb{R}$; regular and singular Sturm–Liouville operators; limit-point and limit-circle endpoints; method of Frobenius; formulas for the resolvent and the spectral projectors; eigenfunction expansions; the radial equations for the hydrogen atom in the nonrelativistic and relativistic cases.

Prerequisites: Chapters 1–9

The basic theory of second-order ordinary differential operators, largely due to Hermann Weyl, is summarized in this chapter.

10.1 Resolvent and Spectral Family for the Operator $- id/dx$

Let T_0 denote the operator defined for all distributions f on $\mathbb{R}$ by the equation

$$T_0 f = -if'. \tag{10.1-1}$$

In the Hilbert space $L^2(\mathbb{R})$ we define an operator A:

$$\mathfrak{D}(A) = \{ f \in L^2 : f' \in L^2 \},$$
$$Af = T_0 f = -if'. \tag{10.1-2}$$

The considerations of Section 7.5 show that A is self-adjoint. We note that any f in $\mathfrak{D}(A)$ automatically satisfies the boundary condition $f(x) \to 0$ as $x \to \pm\infty$, according to Section 5.6.

The point spectrum $P\sigma(A)$ is empty, for if λ were an eigenvalue, the eigenfunction would be $Ce^{i\lambda x}$ ($C = \text{const.} \neq 0$), which is not in L^2 for any λ. On the other hand, an approximate eigenfunction in the form of a wave packet $f(x) = \beta^{1/4} e^{-\beta x^2} e^{i\lambda x}$ ($\beta > 0$) can be constructed for any real λ. As $\beta \to 0$, $\|f\|$ is constant, while $\|Af - \lambda f\| \to 0$. Therefore, the continuous spectrum $C\sigma(A)$ occupies the entire real axis in the λ plane.

To find the resolvent, we suppose that Im $\lambda \neq 0$, and we seek a solution f of the equation

$$Af - \lambda f = g, \qquad \text{that is,} \quad -if' - \lambda f = g, \tag{10.1-3}$$

where g is an arbitrary element of $L^2(\mathbb{R})$. The solution is

$$f(x) = (R_\lambda g)(x) = \begin{cases} i \int_{-\infty}^{x} e^{i\lambda(x-x')} g(x')dx', & \text{for Im } \lambda > 0, \\[3mm] -i \int_{x}^{\infty} e^{i\lambda(x-x')} g(x')dx', & \text{for Im } \lambda < 0, \end{cases} \tag{10.1-4}$$

Hence, the resolvent R_λ is an integral operator.

EXERCISES

1. Show that $\|f\| \leq |\text{Im } \lambda|^{-1} \|g\|$, i.e., that $\|R_\lambda\| \leq |\text{Im } \lambda|^{-1}$. This gives a second proof that A is self-adjoint; namely, for Im $\lambda \neq 0$, $(A - \lambda)^{-1}$ is defined in all L^2 and bounded, hence the upper and lower half planes are in the resolvent set. (See Section 8.6.)

2. Show that if g has bounded support, then $\|f\|$, i.e., $\|R_\lambda g\|$ is $\leq$ Const. $|\text{Im } \lambda|^{-1/2}$, where the constant depends on g but not on λ. *Suggestion*: Compute the Fourier transform, and recall that $\|\hat{f}\| = \|f\|$.

3. By integrating R_λ on a suitable path in the λ plane (see Section 9.6) show that, for $s < t$, $E_t - E_s$ is the integral operator given by

$$((E_t - E_s)g)(x) = \int_{-\infty}^{\infty} \frac{e^{it(x-x')} - e^{is(x-x')}}{2\pi i(x-x')} g(x')dx'. \tag{10.1-5}$$

10.2 Resolvent and Spectral Family for the Operator $- (d/dx)^2$

Here the operator T_0 is given by $T_0 f = -f''$, where f is any distribution on $\mathbb{R}$. A self-adjoint operator A is defined by

$$\begin{aligned} \mathfrak{D}(A) &= \{f \in L^2 : f'' \in L^2\}, \\ Af &= T_0 f = -f'' \end{aligned} \tag{10.2-1}$$

By reasoning similar to that of the preceding section it is seen that the point spectrum of A is empty and that the continuous spectrum occupies the nonnegative real axis. For any λ not on the nonnegative real axis, the equation $Af - \lambda f = g$ can be solved to give the resolvent, and we find

$$f(x) = (R_\lambda g)(x) = \frac{1}{2k} \int_{-\infty}^{\infty} e^{-k|x-x'|} g(x')dx', \tag{10.2-2}$$

where

$$k = \sqrt{-\lambda}, \qquad \text{Re } k > 0.$$

1. By integrating the resolvent on a suitable contour in the λ plane, show that the spectral projector E_t is given by

$$(E_t g)(x) = \begin{cases} \displaystyle\int_{-\infty}^{\infty} \frac{\sin\sqrt{t}\,|x - x'|}{\pi |x - x'|} g(x')dx' & \text{if } t > 0, \\ 0 & \text{if } t \le 0. \end{cases} \qquad (10.2\text{-}3)$$

2. Verify directly that

$$E_t E_s = E_{\min(t,\, s)}.$$

10.3 The Fourier Transform Method

Let T_0 be either of the operators discussed in the last two sections or any ordinary differential operator that has constant coefficients and is of the form

$$T_0 = p\left(-i\frac{d}{dx}\right), \qquad (10.3\text{-}1)$$

where $p(k)$ is a real polynomial for real k. We define an operator in $L^2(\mathbb{R})$ as follows:

$$\mathfrak{D}(A) = \{f \in L^2 : T_0 f \in L^2\}$$

$$Af = T_0 f = p\left(-i\frac{d}{dx}\right)f. \qquad (10.3\text{-}2)$$

If $\hat{f}(k)$ is the Fourier transform of any tempered distribution $f(x)$ on $\mathbb{R}$, then $k\hat{f}(k)$ is the Fourier transform of $(-id/dx)f(x)$, and $p(k)\hat{f}(k)$ is the Fourier transform of $p(-id/dx)f(x)$. Therefore, if $\mathscr{F}$ denotes the unitary operator of Fourier transform in $L^2(\mathbb{R})$, so that $\hat{f} = \mathscr{F}f$, and if $\hat{A}$ denotes $\mathscr{F}A\mathscr{F}^*$, equations (10.3-2) take the form

$$\mathfrak{D}(\hat{A}) = \{\hat{f} \in L^2 : p(k)\hat{f} \in L^2\}$$

$$\hat{A}\hat{f} = p(k)\hat{f}. \qquad (10.3\text{-}3)$$

Clearly $\hat{A}$ is self-adjoint, hence A, which is $\mathscr{F}^*\hat{A}\mathscr{F}$ is also self-adjoint.

The Fourier transform of the equation $Af - \lambda f = g$ is $\hat{A}\hat{f} - \lambda\hat{f} = \hat{g}$, hence if R_λ is the resolvent of A, then the operator $\hat{R}_\lambda = \mathscr{F}R_\lambda\mathscr{F}^*$ is the resolvent of $\hat{A}$, and clearly,

$$(\hat{R}_\lambda \hat{g})(k) = \frac{1}{p(k) - \lambda}\hat{g}(k). \qquad (10.3\text{-}4)$$

If $C(t+)$ is the contour described in Section 9.6, then

$$\frac{1}{2\pi i}\int_{C(t+)}\frac{1}{p(k)-\lambda}d_\lambda = \begin{cases} 1 & \text{if } p(k) \le t, \\ 0 & \text{if } p(k) > t, \end{cases} \qquad (10.3\text{-}5)$$

hence the projector $\hat{E}_t$ is given by

$$(\hat{E}_t\hat{g})(k) = \begin{cases} \hat{g}(k) & \text{for all } k \text{ such that } p(k) \le t, \\ 0 & \text{for all } k \text{ such that } p(k) > t. \end{cases} \qquad (10.3\text{-}6)$$

More precisely, this equation determines $\hat{E}_t\hat{g}$ whenever $\hat{g}$ is a continuous function, but the continuous functions are dense in L^2 and the resulting operator is bounded, so $\hat{E}_t$ is thereby determined in all L^2, according to the Extension Theorem at the beginning of Chapter 7. The spectral projectors for A are now given by

$$E_t = \mathscr{F}^*\hat{E}_t\mathscr{F}.$$

As a by-product, if f is in the domain $\mathfrak{D}(A)$ given by (10.3-2), then all derivatives $f', \ldots, f^{(m)}$ are in $L^2(\mathbb{R})$, where m is the degree of the polynomial $p(\cdot)$. To prove that, we note first that since $p(k)\hat{f}$ is in L^2, $|p(k)|\,\hat{f}$ is also in L^2, hence

$$c_1\hat{f} + c_2|p(k)|\,\hat{f}$$

is in L^2. We now choose c_1 and c_2 so that

$$|k^r| < c_1 + c_2|p(k)|$$

for all k and for $r = 1, \ldots, m$. Since $\hat{f}$ is in $L^2(\mathbb{R})$, hence a fortiori in $L^2(-K, K)$, K finite, the distribution $k^r\hat{f}$ is in $L^2(-K, K)$ for any r, and the Lemma in Section 5.5 shows that

$$\|k^r\hat{f}\| \le \|c_1\hat{f} + c_2|p(k)|\,\hat{f}\|$$

for every K. The right member remains bounded as $K \to \infty$, hence $k^r\hat{f}$ is in L^2, hence so is $f^{(r)}$, as stated.

The Fourier transform method is used in the next chapter in the study of the Laplacian and of a certain integro-differential operator.

We close this section with some remarks about the Fourier transform operator $\mathscr{F}$ itself. In n dimensions,

$$(\mathscr{F}f)(\mathbf{x}) = \hat{f}(\mathbf{x}) = (2\pi)^{-n/2}\int_{\mathbb{R}^n}\cdots\int e^{-i\mathbf{x}\cdot\mathbf{x}'}f(\mathbf{x}')dx_1'\cdots dx_n'. \qquad (10.3\text{-}7)$$

Its resolvent is

$$f(\mathbf{x}) = (R_\lambda g)(\mathbf{x}) = \frac{1}{1-\lambda^4}[\lambda^2\hat{g}(\mathbf{x}) + \hat{g}(-\mathbf{x}) + \lambda^3 g(\mathbf{x}) + \lambda g(-\mathbf{x})],$$

$$\text{for } \lambda^4 \ne 1, \qquad (10.3\text{-}8)$$

for it is easily seen that then $\mathscr{F}f - \lambda f = g$, if it is recalled that $\hat{\hat{g}}(\mathbf{x}) = g(-\mathbf{x})$.

EXERCISE

1.   Equation (10.3-8) shows that the entire λ plane, minus the points $\lambda = \pm 1, \pm i$, is the resolvent set of the Fourier transform operator $\mathscr{F}$. Therefore, the corresponding resolution of the identity F_z, defined on the unit circle $|z| = 1$ (see 9.10-7) is a jump function with jumps at $z = \pm 1, \pm i$; the jumps are projectors P_1, P_{-1}, P_i, P_{-i}. Find these projectors, and verify that

$$\mathscr{F} = \sum_{(z = 1, i, -1, -i)} zP_z,$$

$$I = \sum_{(z = 1, i, -1, -i)} P_z.$$

Show that this operator $\mathscr{F}$ is not the Cayley transform of a self-adjoint operator A, but $\sqrt{i}\mathscr{F}$ is, and find A. Find the eigenfunctions of $\mathscr{F}$ in the one-dimensional case by first showing that a Gaussian of a suitable width is invariant under Fourier transformation and then showing that if $f(x)$ is an eigenfunction then so is $f'(x) \pm xf(x)$ and using the properties of the Hermite functions.

10.4 Regular Sturm–Liouville Operator

Suppose that $p(x)$ and $q(x)$ are real functions of class C^1 and C, respectively, for $a \le x \le b$, and that $p(x) > 0$. Let T_0 denote the operator

$$T_0 f = -(pf')' + qf; \tag{10.4-1}$$

According to Section 5.4, if f is in L^2, then pf' and qf are well defined, as distributions, hence $T_0 f$ is well defined, (but is not necessarily in L^2). Consider the boundary conditions

$$\alpha f(a) + \beta f'(a) = 0,$$
$$\gamma f(b) + \delta f'(b) = 0, \tag{10.4-2}$$

where $\alpha, \beta, \gamma, \delta$ are real constants, of which α and β are not both zero, and γ and δ are not both zero. An operator A of Sturm–Liouville type in the Hilbert space $L^2 = L^2(a, b)$ is now defined as follows:

$$\mathfrak{D}(A) = \{f \in L^2 : T_0 f \in L^2, f \text{ satisfies } (10.4\text{-}2)\} \tag{10.4-3}$$

$$Af = T_0 f \quad \text{for } f \in \mathfrak{D}(A) \tag{10.4-4}$$

[Note that since f is in L^2, qf is in L^2, hence $(pf')'$ is in L^2, hence f' is a continuous function, hence the boundary conditions (10.4-2) make sense.] It is readily seen by the method of Section 7.5 that A is self-adjoint. It will be shown that A has a pure point spectrum with eigenvalues λ_j such that $|\lambda_j| \to \infty$ as $j \to \infty$ (actually, $\lambda_j \to +\infty$). The resolvent of A will be shown to be a compact integral operator whose kernel is the Green's function of $A - \lambda$; according to Section 8.6, the existence of the resolvent for all nonreal λ gives another proof that A is self-adjoint, because it is obviously symmetric.

The symmetry of A is due to the *formal self-adjointness* of T_0, by which is meant that whenever the integrations by parts are justified,

$$\int_a^b f T_0 g \, dx = \int_a^b g T_0 f \, dx + \text{boundary terms.} \qquad (10.4\text{-}5)$$

Sometimes one introduces a third coefficient function $r(x)$, assumed continuous and positive in $[a, b]$, and writes the eigenvalue equation in the more general form

$$-(pf')' + qf = \lambda r f. \qquad (10.4\text{-}6)$$

That is equivalent to introducing the operator S_0 defined by

$$S_0 f = \frac{1}{r} [-(pf')' + qf], \qquad (10.4\text{-}7)$$

which is formally self-adjoint in the Hilbert space $L_\sigma^2(a, b)$, where the measure σ is given by $d\sigma(x) = r(x)dx$, so that the inner product is

$$(f, g) = \int_a^b \overline{f(x)} g(x) r(x) dx. \qquad (10.4\text{-}8)$$

Clearly, a quite general second-order operator can be written in the form (10.4-7) by suitable choice of $P(x)$, $q(x)$, and $r(x)$, so that the essence of Sturm–Liouville theory is the choice of an inner product with respect to which the operator is formally self-adjoint. The choice of a Hilbert space is of course a matter of physics, and the above choice reflects the importance of self-adjointness in many physical applications. Although the form (10.4-7) is often convenient for calculation, the simpler form (10.4-1), which will be used in the remainder of this chapter, suffices for the development of the theory.

10.5 Existence and Uniqueness of the Solution; the Integral Equation; The Eigenfunctions

Although all quantities in the formulation of the problem are real, and there is no mention of analyticity, and $p'(x)$ and $q(x)$ need not be differentiable, nevertheless analytic function theory plays an important role in the analysis of the properties of the operator. The following lemma shows that the solution of the one-point boundary problem of the operator $T_0 - \lambda$ depends analytically on λ.

Lemma. *For specified values of $\varphi(a)$ and $\varphi'(a)$ and for any real or complex λ, the differential equation*

$$T_0 \varphi = -(p\varphi')' + q\varphi = \lambda\varphi \qquad (10.5\text{-}1)$$

has a unique solution $\varphi(x, \lambda)$, for $a \le x \le b$, which, for given x, is an entire function of λ. [Note: This φ is not generally in $\mathfrak{D}(A)$.]

PROOF.   If φ' is called ψ, the differential equation is equivalent to the following coupled integral equations for φ and ψ:

$$p(x)\psi(x) = p(a)\varphi'(a) + \int_a^x [q(x') - \lambda]\varphi(x')dx',$$

$$\varphi(x) = \varphi(a) + \int_a^x \psi(x')dx'. \tag{10.5-2}$$

These equations are solved by Picard's iterative method, according to which φ and ψ are replaced by φ_s and ψ_s $(s = 0, 1, 2, \ldots)$ in the integrals and by φ_{s+1} and ψ_{s+1} on the left sides of the equations. Furthermore, $\varphi_0(x)$ and $\psi_0(x)$ are set $= 0$; hence $\varphi_1(x)$ and $\psi_1(x)$ are the constants $\varphi(a)$ and $\varphi'(a)$, and it is proved that $\varphi_s(x)$ and $\psi_s(x)$ converge, as $s \to \infty$, and that the limit functions satisfy the integral equations. Namely, let K be any compact set in the λ plane, and call

$$M = \max\left\{\sup \frac{|q(x') - \lambda|}{p(x)}, 1\right\},$$

where the supremum is for all x and x' in $[a, b]$ and all λ in K. Call $\Delta_s\varphi = \varphi_{s+1} - \varphi_s$, $\Delta_s\psi = \psi_{s+1} - \psi_s$. Then,

$$|\Delta_{s+1}\psi(x)| \leq M \int_a^x |\Delta_s\varphi(x')|dx',$$

$$|\Delta_{s+1}\varphi(x)| \leq M \int_a^x |\Delta_s\psi(x')|dx'. \tag{10.5-3}$$

If furthermore $m = \max\{|\varphi(a)|, |\varphi'(a)|\}$, then it follows by an easy induction on s that

$$|\Delta_s\psi(x)|, |\Delta_s\varphi(x)| \leq \frac{1}{s!} M^s(x - a)^s m \tag{10.5-4}$$

for all x in $[a, b]$ and all λ in K. Hence the sums

$$\varphi(a) + \sum_{s=1}^{\infty} \Delta_s\varphi(x)$$

$$\varphi'(a) + \sum_{s=1}^{\infty} \Delta_s\psi(x) \tag{10.5-5}$$

converge uniformly with respect to x and λ, since the partial sums are dominated by partial sums of a power series for an exponential. according to (10.5-4). Therefore, the sums (10.5-5) can be integrated term by term, which shows that their limits satisfy the integral equation, hence that $\varphi(x, \lambda)$, defined by the first of (10.5-5), satisfies the differential equation, as required. Uniqueness is proved, using (10.5-3), by letting $\Delta\varphi$ and $\Delta\psi$, with the subscripts omitted, denote the differences $\varphi - \tilde{\varphi}$ and $\varphi' - \tilde{\varphi}'$ for two solutions of the one-point boundary problem and by showing that the assumption $\Delta\varphi \neq 0$ leads to a contradiction. The analytic dependence on λ now comes as a simple by-product. From (10.5-2) it is seen that the partial sums of (10.5-5) are polynomials in λ, and they converge uniformly with respect to λ in any compact set K in the λ plane. From Weierstrass's theorem on uniform convergence of analytic functions (see Knopp, 1945, Section 19), it follows that $\varphi(x, \lambda)$ is an entire function of λ, for given x.

Now suppose that the given quantities $\varphi(a)$ and $\varphi'(a)$ are fixed (i.e., independent of λ), are not both zero, and are such that the left boundary condition,

$$\alpha\varphi(a) + \beta\varphi'(a) = 0,$$

is satisfied. Then λ is an eigenvalue of A if and only if the right boundary condition,

$$\gamma\varphi(b, \lambda) + \delta\varphi'(b, \lambda) = 0,$$

is also satisfied. The left member of this equation is an entire function of λ and is not identically zero, because A is a symmetric operator and hence has no nonreal eigenvalues. The zeros of an entire function not $\not\equiv 0$ are isolated points, hence the eigenvalues λ_j of A are real numbers with no accumulation point. The uniqueness of $\varphi(x, \lambda)$, for any λ, shows that the eigenspaces are one-dimensional.

10.6 The Resolvent; the Green's Function; Completeness of the Eigenfunctions

We come now to the construction of the Green's function. The roles of the endpoints a and b can be interchanged; therefore, there is another solution $\chi(x) = \chi(x, \lambda)$ of (10.5-1), having given fixed values of $\chi(b)$ and $\chi'(b)$ that satisfy the *right* boundary condition,

$$\gamma\chi(b) + \delta\chi'(b) = 0.$$

It follows from (10.5-1) that the Wronskian $\chi \, \partial\varphi/\partial x - \varphi \, \partial\chi/\partial x$ of the two solutions is a constant times $1/p(x)$, for given λ; hence, a function $h(\lambda)$ is defined, for λ not an eigenvalue, by the equation

$$h(\lambda)p(x)\left[\chi(x, \lambda)\frac{\partial\varphi}{\partial x}(x, \lambda) - \varphi(x, \lambda)\frac{\partial\chi}{\partial x}(x, \lambda)\right] = 1 \qquad (10.6\text{-}1)$$

For λ not equal to any eigenvalue of A, the Green's function has the form

$$G(x, y) = G(x, y; \lambda) = h(\lambda)\begin{cases}\varphi(x, \lambda)\chi(y, \lambda) & \text{if } x \le y \\ \chi(x, \lambda)\varphi(y, \lambda) & \text{if } x \ge y\end{cases}(a \le x, y \le b).$$

$$(10.6\text{-}2)$$

For any fixed y, $G(x, y)$ satisfies the boundary conditions at $x = a$ and $x = b$; also, the operator $T_0 - \lambda$, when applied to $G(x, y)$, gives zero for all $x \ne y$, but not for $x = y$, because λ is not an eigenvalue (in fact $\partial G/\partial x$ has a discontinuity at $x = y$). In fact, $h(\lambda)$ has been chosen so that

$$\left[-\frac{\partial}{\partial x}p(x)\frac{\partial}{\partial x} + q(x) - \lambda\right]G(x, y) = \delta(x - y), \qquad (10.6\text{-}3)$$

i.e., so that $-p(x)(\partial/\partial x)G(x, y)$ has a unit jump at $x = y$. Then, if g is any distribution in L^2, and if

$$f(x) = \int_a^b G(x, y; \lambda)g(y)dy, \qquad (10.6\text{-}4)$$

it is seen that f satisfies the boundary conditions and is in L^2 (it is continuous) and that $T_0 f = \lambda f + g$, hence $T_0 f$ is in L^2, hence f is in $\mathfrak{D}(A)$, and $Af - \lambda f = g$, or $f = R_\lambda g$, where R_λ is the resolvent of A; that is, the resolvent is the integral operator in (10.6-4); it is a bounded operator (in fact compact — see Chapter 12) and defined in all L^2, for any λ that is not an eigenvalue of A.

From this follows a new proof of the self-adjointness of A (only its symmetry has been used so far), since $+i$ and $-i$ are in the resolvent set. It follows also that the continuous spectrum is empty, because any real λ not equal to an eigenvalue is also in the resolvent set. Therefore, the eigenfunctions span all of L^2; i.e., any f in L^2 can be expanded in them, and the expansion converges in the mean to f. Hence, there are infinitely many eigenvalues (since each eigenspace is one-dimensional), and $|\lambda_j| \to \infty$ as $j \to \infty$.

10.7 More General Boundary Conditions

Consider the boundary conditions

$$\alpha_i f(a) + \beta_i f'(a) + \gamma_i f(b) + \delta_i f'(b) = 0 \qquad (i = 1, 2), \quad (10.7\text{-}1)$$

where it is assumed that the matrix of this pair of equations has rank 2, so that the equations are independent. These are called *coupled boundary conditions*. They can be solved for two of the unknowns in terms of the other two, and it will be assumed that they can be solved for $f'(a)$ and $f'(b)$, so that

$$\begin{aligned} f'(a) &= \varepsilon_1 f(a) + \zeta_1 f(b), \\ f'(b) &= \varepsilon_2 f(a) + \zeta_2 f(b); \end{aligned} \qquad (10.7\text{-}2)$$

the discussion of the other cases, which are similar, is left to the reader. In order for the resulting operator A to be symmetric, since the operator of multiplying by $q(x)$ is already symmetric, it is necessary and sufficient that

$$\int_a^b \bar{g}(pf')' \, dx = \int_a^b (p\bar{g}')'f \, dx$$

for all f and g in $\mathfrak{D}(A)$, i.e., for all f and g in L^2 such that $(pf')'$ and $(pg')'$ are in L^2 and such that the above boundary conditions are satisfied by f and g. Integrating by parts yields the condition

$$[p(x)(\bar{g}(x)f'(x) - \bar{g}'(x)f(x))]_a^b = 0.$$

Substituting for f' at $x = a$ and b from the boundary conditions (10.7-2), and for $\bar{g}'$ at $x = a$ and b from the corresponding complex conjugate expression with f replaced by $\bar{g}$, gives an equation with eight terms. The values of f and g at both $x = a$ and $x = b$ can be chosen arbitrarily, provided that f' and g'

are then given at a and b by (10.7-2) and it is easily seen that the eight-term equations referred to is satisfied if and only if

$$\text{Im } \zeta_2 = \text{Im } \varepsilon_1 = 0,$$
$$\varepsilon_2 p(b) + \zeta_1 p(a) = 0. \tag{10.7-3}$$

Then, just as for the preceding problem, the methods of Section 7.5 can be used to show that the operator A defined by

$$\mathfrak{D}(A) = \{f \in L^2 : T_0 f \in L^2, (10.7\text{-}2, 3) \text{ hold}\},$$
$$Af = T_0 f,$$

is self-adjoint.

· These results illustrate von Neuman's theorem on the possible self-adjoint extensions of a symmetric operator. Let T be the operator defined as follows:

$$\mathfrak{D}(T) = \{f \in L^2 : T_0 f \in L^2, f(a) = f(b) = f'(a) = f'(b) = 0\}.$$
$$Tf = T_0 f \quad \text{for } f \in \mathfrak{D}(T). \tag{10.7-4}$$

T is a symmetric operator whose adjoint T^* has no boundary conditions at all. In a sense, T is a minimal operator in $\mathfrak{H}$ (minimal with respect to domain) obtained from T_0, and T^* is a maximal one. For any λ, the equation $T^* f = \lambda f$ has two independent solutions, hence the deficiency indices of T are $(2, 2)$. Therefore, by von Neumann's theorem in Section 8.6, there is a 2-(complex)-parameter family, i.e., a 4-(real)-parameter family of self-adjoint operators A between T and T^* $(T \subset A \subset T^*)$. The foregoing boundary conditions provide such a family; namely, there are four complex constants in the equations (10.7-2), and the equations (10.7-3) impose four real constraints, so that four free real parameters are left.

EXERCISE

1. Find the resolvent of the above operator A; i.e., find the Green's function for $Af - \lambda f = g$.

10.8 Sturm–Liouville Operator with One Singular Endpoint

So far, the range of x has been assumed to be a closed bounded interval $[a, b]$ and the coefficients $p(x)$ and $q(x)$ have been assumed continuous in $[a, b]$. If $[a, b]$ is replaced by an interval of the form $[a, \infty)$ or $[a, b)$ (in the latter case, the coefficients may become infinite as $x \to b$), the right endpoint $(x = b)$ is called *singular*. A Sturm–Liouville problem may have one or two singular endpoints. It often happens that no boundary condition is needed at a singular endpoint—the requirement that the solution be in L^2 takes the place of a boundary condition. That is the so-called limit-point case (see below) and is the case that usually (but not always) occurs in quantum mechanics. The radial equations that result from separation of variables in the Laplace, Schrodinger, and Dirac equations have singular endpoints at $r = 0$ and $r = \infty$. The endpoint at ∞ is of the limit-point type, hence has an automatic or built-in boundary condition, while the endpoint at 0 is sometimes of the limit-point type and sometimes of the limit-circle type, in which case a further boundary or endpoint condition has to be supplied from physical considerations; see Sections 10.15–17 below.

In this section and the next two we consider the case of one singular endpoint; we take the interval to be $[0, \infty)$, but any interval of the form $[a, b)$ is treated in exactly the same way.

Suppose that $p(x) \in C^1$, $q(x) \in C$, $p(x) > 0$, all for $0 \le x < \infty$. If f is in $L^2(0, \infty)$ and if T_0 is the operator defined by

$$T_0 f = -(pf')' + qf, \tag{10.8-1}$$

then $T_0 f$ is well defined as a distribution, because f is in $L^2(0, b)$ for any finite b, hence the arguments of the preceding sections apply.

It is not easy to give an exact analogue of the minimal operator T defined by (10.7-4), because the appropriate boundary condition at $+\infty$ is not yet known. Therefore, we choose a domain that is *sure* to be small enough, namely $C_0^\infty(0, \infty)$, even though the resulting operator is not closed. $C_0^2(0, \infty)$ would be equally satisfactory; in either case, the functions in this domain vanish identically in some neighborhood of 0, and in some neighborhood of ∞. T is defined as follows:

$$\mathfrak{D}(T) = C_0^\infty$$
$$Tf = T_0 f, \quad \text{for } f \text{ in } C_0^\infty. \tag{10.8-2}$$

Integrating twice by parts in (Tf, g) yields (f, Tg), showing that T is symmetric. The method of Section 7.5 shows that the adjoint of T is the operator given by

$$\mathfrak{D}(T^*) = \{f \in L^2 : T_0 f \in L^2\},$$
$$T^*f = T_0 f, \tag{10.8-3}$$

which has no boundary conditions.

10.9 The Boundary Condition at a Singular Endpoint

According to von Neumann's theorem in Section 8.6, the existence and number of self-adjoint extensions of the operator T given by (10.8-1, 2) in the preceding section are determined by the deficiency indices (m, n) of T, which are the codimensions of the ranges of $T \pm i$, that is, the dimensions of the nullspaces of $T^* \mp i$ or the number of linearly independent solutions of the equation $T^*f = \pm if$, where T^* is given by (10.8-3). Now the differential equation $T_0 f = \lambda f$ is of second order and hence has two independent solutions for any λ, and according to the definition (10.8-3) of T^*, a solution f of $T_0 f = \lambda f$ is in the domain of T^* if and only if it is in $L^2(0, \infty)$. It can happen that both solutions of $T_0 f = \lambda f$ (hence all solutions) are in L^2. According to the lemma in Section 8.6, if that happens for one nonreal λ, it also happens for all λ in the same half plane (upper or lower). Furthermore, if $T_0 f = \lambda f$, then $T_0 \bar{f} = \bar{\lambda} \bar{f}$, and $\bar{f}$ is in L^2 if f is, hence if it happens in one half plane it happens in the other, too; in fact, in the present problem, it then happens for all λ, real or complex; i.e., all solutions are in L^2 for all λ (see Coddington and Levinson 1955). We conclude that the deficiency indices of T are $(0, 0)$, $(1, 1)$, or $(2, 2)$. It will be seen below that there is always at least one solution of $T_0 f = \lambda f$ in L^2, hence the case $(0, 0)$ is excluded. (When there are *two* singular endpoints, the case $(0, 0)$ can occur, as in Section 10.2, where the operator $-(d/dx)^2$ on $\mathbb{R}$ was found to be self-adjoint without any boundary conditions.)

Let $f_i(x) = f_i(x; \lambda)$ $(i = 1, 2)$ be solutions of the equation $T_0 f = \lambda f$, that is, of

$$-(pf')' + qf = \lambda f \qquad (10.9\text{-}1)$$

subject to initial conditions as follows:

$$f_1(0) = 1, \qquad p(0)f'_1(0) = 0,$$
$$f_2(0) = 0, \qquad p(0)f'_2(0) = 1. \qquad (10.9\text{-}2)$$

Just as for the regular Sturm–Liouville problem, the Wronskian of two solutions (for the same λ) is a constant times $1/p(x)$; we find in fact that

$$p(x)[f_1(x)f'_2(x) - f_2(x)f'_1(x)] \equiv 1. \qquad (10.9\text{-}3)$$

The general solution for given λ, apart from an arbitrary multiplicative constant, is

$$f(x) = f_1(x) + mf_2(x), \qquad (10.9\text{-}4)$$

where m is a complex number. We shall show that if $\text{Im } \lambda \neq 0$ there is at least one value of m for which $\int_0^\infty |f|^2 \, dx$ is finite. If we multiply (10.9-1) by $\bar{f}(x)$ and integrate from 0 to b (we shall eventually let $b \to \infty$), we find, after integrating by parts,

$$-[\bar{f}pf']_{x=0}^{x=b} + \int_0^b (p|f'|^2 + q|f|^2)dx = \lambda \int_0^b |f|^2 \, dx. \qquad (10.9\text{-}5)$$

The integral on the left is real, and we shall take imaginary parts throughout the equation. We note first, using the initial conditions (10.9-2) that

$$p(0)\text{Im}[\bar{f}(0)f'(0)] = \text{Im } m.$$

Therefore,

$$-p(b)\text{Im}[\bar{f}(b)f'(b)] \overset{\text{def}}{=} F(m; b) = -\text{Im } m + \text{Im } \lambda \int_0^b |f|^2 \, dx. \qquad (10.9\text{-}6)$$

For simplicity we now assume that $\operatorname{Im} \lambda > 0$; clearly the other case is similar. We shall show that for given b $F(m; b)$ is then negative in a certain disk D_b in the m plane and positive outside that disk. Furthermore as b increases, the disks contract; that is, if $b' > b$, then $D_{b'}$ is contained in D_b. Explicitly,

$$F(m; b) = \frac{ip}{2} [(\bar{f}_1 + \bar{m}\bar{f}_2)(f'_1 + mf'_2) - (f_1 + mf_2)(\bar{f}'_1 + \bar{m}\bar{f}'_2)]_{x=b}$$

(10.9-7)

which can be written as

$$F(m; b) = A|m|^2 + B \operatorname{Re} m + C \operatorname{Im} m + D$$ (10.9-8)

where A, B, C, and D are real coefficients, and

$$A = -p(b)\operatorname{Im}[\bar{f}_2(b)f'_2(b)] = \operatorname{Im} \lambda \int_0^b |f_2|^2 \, dx.$$ (10.9-9)

Here, (10.9-5) has been used once more, with f_2 for f. Clearly the locus $F(m; b) = 0$ is a circle in the m plane, and a detailed calculation using (10.9-2) shows that its radius is $1/2A$. Clearly, $F(m; b)$ is >0 for large m (A is >0 by (10.9-9)) hence it is <0 in the disk D_b inside the circle and >0 outside. Lastly, equation (10.9-6) shows that $F(m; b)$ is an increasing function of b, for fixed m, hence the disks contract as b increases.

If the disks D_b contract to a point m_∞ as $b \to \infty$, the problem is said to be *in the limit-point case* at the right endpoint ($x = \infty$). Then $F(m_\infty; b)$ is ≤ 0 for all b, and (10.9-6) shows that the solution $f(x) = f_1(x) + m_\infty f_2(x)$ is quadratically integrable over $(0, \infty)$; in fact,

$$\int_0^\infty |f|^2 \, dx \leq \operatorname{Im} m_\infty / \operatorname{Im} \lambda.$$ (10.9-10)

Since the disk radius $1/2A \to 0$ as $b \to \infty$, equation (10.9-9) shows that $f_2(x)$ is not quadratically integrable, hence neither is any other solution except multiples of $f(x)$. As noted earlier, if this is the case for one λ, then it is the case for all λ. In effect, the condition of quadratic integrability takes the place, in the limit-point case, of a single homogeneous boundary condition at the end point.

If the disks D_b contract to a disk D_∞ of nonzero radius, the problem is said to be *in the limit-circle case* at the right endpoint. Then, by taking any two values of m in D_∞, it is easily seen that every linear combination of f_1 and f_2 is quadratically integrable. In this case, quadratic integrability is not equivalent to a boundary condition. Hence, for Sturm–Liouville problems in the limit-circle case at one end (or both), some additional condition on the solution is generally required, in lieu of a boundary condition, to make the operator self-adjoint. Examples are given in Sections 10.15-17 below.

The following criterion is given in Coddington and Levinson for determining whether a Sturm–Liouville operator T_0 is in the limit-point case at infinity. If there is a function $M(x) > 0$ of class C^1 such that

$$\int^\infty (pM)^{-1/2}\, dx = \infty, \qquad\qquad (10.9\text{-}11)$$

and

$$P^{1/2} M' M^{-3/2} \text{ is bounded,} \qquad 0 \le x < \infty, \qquad\qquad (10.9\text{-}12)$$

and if $q(x) \ge -M(x)$, then T_0 is in the limit-point case at ∞. Two special cases are obtained by taking $M = $ constant and $p = $ constant, respectively.

(a) If $q(x)$ is bounded from below by a (possibly negative) constant, and if $\int^\infty p^{-1/2} dx$ is infinite, then T_0 is in the limit point case.
(b) If $q(x) \ge -kx^2$, where k is a constant, then the operator T_0 defined by $T_0 f = -f'' + qf$ is in the limit-point case.

10.10 Regular Singular Point; Method of Frobenius

A further classification of an endpoint, say at $x = a$, is possible if $p(x)$ and $q(x)$ are analytic near a. We write the equation $T_0 f = \lambda f$ as

$$f''(x) + P(x)f'(x) + Q(x)f(x) = 0, \qquad\qquad (10.10\text{-}1)$$

where $P(x) = p'(x)/p(x)$, $Q(x) = -[q(x) - \lambda]/p(x)$. The endpoint a is a *regular singular point* of (10.10-1) if $P(x)$ and $Q(x)$ are analytic in some neighborhood of a in the complex x plane, except possibly for a simple pole of P and a pole of not more than second order of Q, both at $x = a$.

Note that the functions $p(x)$ and $q(x)$ need not be singular to make $P(x)$ and $Q(x)$ singular; it suffices, for example, that $p(a) = 0$.

We shall see that a regular singular point can be either of the limit-point type or the limit-circle type.

We note also that for a given $p(x)$ and $q(x)$, the point $x = a$ may be a regular singular point of (10.10-1) for some λ and not for others. For example, with $p = q = x^3$, 0 is a regular singular point only for $\lambda = 0$. According to the preceding section, however, the distinction between the limit-point and limit-circle cases is independent of λ, hence any value for which a is a regular singular point can be used for determining which case applies.

The so-called *method of Frobenius* for finding solutions about a regular singular point will now be described.

The Laurent expansions of $P(x)$ and $Q(x)$ about $x = a$ are of the form

$$\left. \begin{array}{l} P(x) = P_0(x - a)^{-1} + P_1 + P_2(x - a) + \cdots \\ Q(x) = Q_0(x - a)^{-2} + Q_1(x - a)^{-1} + Q_2 + \cdots \end{array} \right\} 0 < |x - a| < R$$

$$(10.10\text{-}2)$$

and we assume a solution of (10.10-1) in the form of a power series

$$f(x) = (x - a)^{\alpha} \sum_{n=0}^{\infty} a_n(x - a)^n \qquad (a_0 \neq 0). \tag{10.10-3}$$

We substitute this into (10.10-1) and equate to zero the net coefficients of various powers of $x - a$. That gives first an *indicial equation*

$$f(\alpha) \overset{\text{def}}{=} \alpha^2 - \alpha + P_0\alpha + Q_0 = 0 \tag{10.10-4}$$

for α, and then a sequence of *recurrence relations*

$$F(\alpha + l)a_l + \sum_{j=1}^{l} [(\alpha + l - j)P_j + Q_j]a_{l-j} = 0 \qquad (l = 1, 2, \ldots) \tag{10.10-5}$$

for the coefficients $a_1, a_2, \ldots$, in terms of a_0, which is arbitrary. Let α_1 and α_2 be the roots of the indicial equation (10.10-4) so ordered that $\operatorname{Re} \alpha_1 \geq \operatorname{Re} \alpha_2$.

Theorem.   *For $\alpha = \alpha_1$, the series (10.10-3), with coefficients given by (10.10-5), converges for $|x - a| < R$, and the function $f(x)$ so defined satisfies the differential equation (10.10-1).*

The proof is based on standard analytic function theory and is given in books on differential equations, for example in Jörgens and Rellich 1976.

For $\alpha = \alpha_2$, the same method gives a second solution only if $\alpha_1 - \alpha_2$ is not an integer, but in any case a second solution is given by

$$g(x) = f(x) \int^{x} \frac{dw}{p(w)f(w)^2}, \tag{10.10-6}$$

as is easily seen by direct substitution into the equation $-(pg') + qg = \lambda g$. If the power series for f and p are substituted into the last equation, $g(x)$ is seen to have the form

$$g(x) = \begin{cases} (x - a)^{\alpha_2}\varphi_1(x) & \text{if } \alpha_1 - \alpha_2 \neq \text{integer,} \\ (x - a)^{\alpha_2}\varphi_2(x) + \text{const.} \, f(x)\log(x - a) & \text{if } \alpha_1 - \alpha_2 = \text{integer,} \end{cases}$$

$$\tag{10.10-7}$$

where $\varphi_1(x)$ and $\varphi_2(x)$ are analytic for $|x - a| < R$ and are $\neq 0$ at $x = a$.

If we recall that $\operatorname{Re} \alpha_1 \geq \operatorname{Re} \alpha_2$, we see from (10.10-3 and 7) that f and g are both in $L^2(a, a + c)$ for $c < R$ if $\operatorname{Re} \alpha_2 > -\frac{1}{2}$; that is therefore the criterion for the limit-circle case.

EXERCISES

1. Apply the Frobenius method to Bessel's equation for $x = 0$ and to Legendre's equation for $x = \pm 1$.

2. What does the method give in the special case in which a is a regular point of the differential equation (10.10-1)?

10.11 Self-Adjoint Extension of T in the Limit-Point Case

In this section it is shown that if the endpoint ∞ is in the limit-point case, self-adjoint versions of $T_0 - (d/dx)p(d/dx) + q$ can be obtained by imposing a boundary condition at 0 only. For this, the following feature of the limit-point case is crucial:

It was found that the function $f_2(x)$ was not quadratically integrable in that case. More generally, the discussion can be repeated with f_1 and f_2 redefined so as to satisfy, in place of (10.9-2), the boundary conditions

$$f_1(0) = \cos \alpha \qquad p(0)f'_1(0) = \sin \alpha,$$
$$f_2(0) = -\sin \alpha \qquad p(0)f'_2(0) = \cos \alpha, \qquad (10.11\text{-}1)$$

where α is a real parameter. The entire derivation, starting with (10.9-3) is unchanged, and we conclude that the new function $f_2(x)$ is not quadratically integrable, either, for $\operatorname{Im} \lambda \neq 0$, i.e., *no* solution satisfying a *real* boundary condition

$$f(0)\cos \alpha + p(0)f'(0)\sin \alpha = 0 \qquad (10.11\text{-}2)$$

is quadratically integrable in $(0, \infty)$ for $\operatorname{Im} \lambda \neq 0$.

We now define an extension A_α of T in the limit-point case in which f is not required to vanish identically near $x = 0$ but only to satisfy (10.11-2), and we show that A_α is essentially self-adjoint:

$$\mathfrak{D}(A_\alpha) = \left\{ f \in C^\infty : \begin{array}{l} f(x) = 0 \text{ for all } x > \text{ some } b \\ f(x) \text{ satisfies } (10.11\text{-}2) \end{array} \right\}$$

By the method of Section 7.5 it is easily seen that

$$\mathfrak{D}(A_\alpha^*) = \{ f \in L^2(0, \infty) : T_0 f \in L^2, (10.11\text{-}2) \text{ holds} \}.$$

According to the feature of the limit-point case stated above, the equation $A_\alpha^* f = \lambda f$ has no nonzero solutions f for $\operatorname{Im} \lambda \neq 0$, hence the deficiency indices of A_α are $(0, 0)$; therefore A_α is essentially self-adjoint, and A_α^* is its self-adjoint closure. We thus have a one-parameter family $\{A_\alpha^*\}$ of self-adjoint extensions of T such that $T \subset A_\alpha^* \subset T^*$.

The resolvent $R_\lambda = (A_\alpha - \lambda)^{-1}$ of A_α, which is needed in the next section for the eigenfunction expansion, is given by the Green's function for $A_\alpha - \lambda$, $\operatorname{Im} \lambda \neq 0$. The Green's function contains the quadratically integrable solution of $T_0 f = \lambda f$, which will now be called $f_3(x)$ or $f_3(x, \lambda)$. It is given by

$$f_3(x, \lambda) = f_1(x, \lambda) + m_\infty(\lambda)f_2(x, \lambda), \qquad (10.11\text{-}3)$$

where f_1 and f_2 are the solutions that satisfy the boundary conditions (10.11-1) at $x = 0$. It follows from (10.9-3) that

$$p(x)[f_2(x)f'_3(x) - f_3(x)f'_2(x)] \equiv -1, \qquad (10.11\text{-}4)$$

from which it follows that the Green's function is

$$G(x, y) = G(x, y; \lambda) = \begin{cases} f_2(x)f_3(y), & \text{for } x \le y, \\ f_3(x)f_2(y), & \text{for } x \ge y. \end{cases} \qquad (10.11\text{-}5)$$

For fixed y, $G(x, y)$ is in L^2 and satisfies the boundary condition (10.11-2) at $x = 0$ and the equation

$$\left[-\frac{\partial}{\partial x} p(x) \frac{\partial}{\partial x} + q(x) - \lambda \right] G(x, y) = \delta(x - y). \qquad (10.11\text{-}6)$$

Therefore, for any continuous $g(x)$ in $L^2(0, \infty)$, and for Im $\lambda \ne 0$,

$$(R_\lambda g)(x) = ((A_\alpha - \lambda)^{-1} g)(x)$$

$$= \int_0^\infty G(x, y)g(y)dy$$

$$= f_3(x) \int_0^x f_2(y)g(y)dy + f_2(x) \int_x^\infty f_3(y)g(y)dy. \qquad (10.11\text{-}7)$$

For λ real but not an eigenvalue, the integral operator in (10.11-7) exists, but may be an unbounded operator, because the interval of integration is now $(0, \infty)$. In fact, it is unbounded precisely when λ is a point of the continuous spectrum of A. In any case, however, whenever the equation $Af - \lambda f = g$ has a solution, the solution is given by (10.11-7).

10.12 The Eigenfunction Expansion

The expansion of a given function, more precisely of a given element $g(x)$ of $L^2(0, \infty)$, in the eigenfunctions of the operator A_α contains a summation over the discrete eigenfunctions that come from the point spectrum of A_α and an integration over the continuous-spectrum eigenfunctions (these functions are of course not in L^2). The two parts will be combined by writing the expansion as a Stieltjes integral.

If λ is in $P\sigma(A_\alpha)$, then the function $f_2(x, \lambda)$ that satisfies the boundary condition (10.11-2), which was used in defining the operator A_α, is the eigenfunction (not necessarily normalized), hence is quadratically integral over $0, \infty$. (It was found that that cannot happen for nonreal λ, but of course any λ in the spectrum is real.) If λ is in $C\sigma(A_\alpha)$, then $f_2(x, \lambda)$ is not quadratically integrable, but can be used in the construction of approximate eigenfunctions in the form of wave packets (we shall not use that construction). Therefore, we expect an expansion of the form

$$g(x) = \int_{-\infty}^\infty f_2(x, t)d\gamma(t), \qquad (10.12\text{-}1)$$

where $\gamma(t)$ is some function of bounded variation that has jumps at the eigenvalues of A_α, varies continuously in the continuous spectrum, and is constant in any interval not in the spectrum. An expansion of this type is obtained from the spectral family $\{E_t\}$ of the operator A_α.

Since E_t is obtained by integrating R_λ, given by (10.11-7), on a contour $C(t)$ in the λ plane (see Section 9.6), we need to know the dependence of f_2 and f_3 on λ in more detail. According to Section 10.5, $f_1(x, \lambda)$ and $f_2(x, \lambda)$ are entire functions of λ for each x, while $f_3(x, \lambda)$ is given by (10.11-3). We call $m(b, \lambda) = -f_1(b, \lambda)/f_2(b, \lambda)$. According to (10.9-7), $m(b, \lambda)$ lies on the circle $F(m, b) = 0$ in the m plane. Since that circle contracts to m_∞ as $b \to \infty$, we have

$$m_\infty(\lambda) = \lim_{b \to \infty} \frac{-f_1(b, \lambda)}{f_2(b, \lambda)}. \qquad (10.12\text{-}2)$$

From this it can be shown (see Coddington and Levinson 1955) that $m_\infty(\lambda)$ is analytic in the upper and lower half planes and that Im $m_\infty(\lambda)$ has the same sign as Im λ.

According to Section 9.6 and equation (10.11-7) above, the projector E_t is given by

$$(E_t g)(x) = \lim_{\varepsilon \to 0} \frac{1}{2\pi i} \int_{-\infty}^{t} \int_{0}^{\infty} [G(x, y; s + i\varepsilon) - G(x, y, s - i\varepsilon)]g(y)dy$$

$$(10.12\text{-}3)$$

To simplify the discussion, we assume that $g(y)$ is a continuous function and is $=0$ for all $y >$ some y_0. Later, the interpretation of the final formulas when g is a general element of $L^2(0, \infty)$ will be discussed. When $f_1 + m_\infty f_2$ is substituted for f_3 in the expression for G, the contribution of f_1 gives zero, because f_1 and f_2 are continuous in λ, hence the contributions from $\lambda = s + i\varepsilon$ and $\lambda = s - i\varepsilon$ cancel in the limit $\varepsilon \to 0$. What is left is

$$(E_t g)(x) = \lim_{\varepsilon \to 0} \frac{1}{2\pi i} \int_{-\infty}^{t} \left[f_2(x, \lambda) \int_{0}^{\infty} f_2(y, \lambda)g(y)dy \, m_\infty(\lambda) \right]_{\lambda = s - i\varepsilon}^{\lambda = s + i\varepsilon} ds.$$

The function $m_\infty(s \pm i\varepsilon)$ does not generally have a definite limit as $\varepsilon \to 0$, but its integral with respect to s does (see Coddington and Levinson), and the function

$$\rho(t) = \lim_{\varepsilon \to 0} \frac{1}{2\pi i} \int^{t} [m_\infty(s + i\varepsilon) - m_\infty(s - i\varepsilon)]ds \qquad (10.12\text{-}4)$$

(defined up to an arbitrary additive constant) is real and nondecreasing, because $m_\infty(\bar{\lambda})$ is $= \overline{m_\infty(\lambda)}$ and Im $m_\infty(\lambda)$ has the same sign as Im λ. Therefore

$$(E_t g)(x) = \int_{-\infty}^{t} f_2(x, s)h(s)d\rho(s), \qquad (10.12\text{-}5)$$

where

$$h(s) = \int_{0}^{\infty} f_2(y, s)g(y)dy. \qquad (10.12\text{-}6)$$

Lastly, in the limit $t \to +\infty$, (10.12-5) becomes

$$g(x) = \int_{-\infty}^{\infty} f_2(x, s)h(s)d\rho(s). \qquad (10.12-7)$$

This is the eigenfunction expansion of $g(x)$, and (10.12-6) is the formula for the expansion coefficient $h(s)$. $\rho(s)$ is called the *spectral function* of the operator A_α, and (10.12-4) is called the *formula of Titchmarsh*.

Since E_t has jumps at the eigenvalues of A_α, is continuous in the continuous spectrum, and is constant in any t-interval not in the spectrum, equation (10.12-5) shows that the same is true of $\rho(t)$.

Until now, $g(x)$ was assumed to be a continuous function of bounded support. More generally, it is an element of $L^2(0, \infty)$, and $h(s)$ is an element of the space L_ρ^2 of the kind discussed in Section 5.9. Then, equations (10.12-6 and 7) are to be interpreted as follows:

$$\text{as } T \to \infty, \quad \int_0^T f_2(y, s)g(y)dy \to h(s) \text{ in } L_\rho^2$$

$$\qquad (10.12-8)$$

$$\text{as } T \to \infty, \quad \int_{-T}^T f_2(x, s)h(s)d\rho(s) \to g(s) \text{ in } L^2(0, \infty).$$

In the first of these equations, the integral is a well-defined function of s for each T (in fact analytic) because f_2 and g are both in $L^2(0, T)$, and the convergence to $h(s)$ is in the norm of L_ρ^2. The integral in the second equation has similar meaning; if $\rho_T(s)$ is the same as $\rho(s)$ in $(-T, T)$ and is constant otherwise, then f_2 and h are both in $L_{\rho_T}^2$, and the convergence to $g(x)$ is in the norm of $L^2(0, \infty)$.

The mapping $g \to h$ is similar to a Fourier transformation; corresponding to Parseval's identity, we have, at least formally,

$$(h_1, h_2)_{L_\rho^2} = \int_{-\infty}^{\infty} \overline{h_1(s)}h_2(s)d\rho(s)$$

$$= \int_{-\infty}^{\infty} \overline{h_1(s)} \int_0^{\infty} f_2(y, s)g_2(y)dy \, d\rho(s)$$

$$= \int_0^{\infty} g_2(y) \int_{-\infty}^{\infty} f_2(y, s)\overline{h_1(s)}d\rho(s)dy$$

$$= \int_0^{\infty} g_2(y)\overline{g_1(y)}dy = (g_1, g_2)_{L^2(0, \infty)}.$$

We have arrived somewhat heuristically at the following:

Theorem. *For any $g(x)$ in $L^2(0, \infty)$, there is an element $h(s)$ in L_ρ^2 such that equations (10.12-8) hold; conversely, for an $h(s)$ in L_ρ^2 there is an element $g(x)$ in $L^2(0, \infty)$ such that they hold. The mapping $g \to h$ is an isometric isomorphism (that is, a unitary mapping) of $L^2(0, \infty)$ onto L_ρ^2.*

For justification of the steps and other details, see Coddington and Levinson 1955 or Jörgens and Rellich 1976.

In quantum mechanical terms, $g(x)$ and $h(s)$ may be thought of as wave functions that describe a given state of a physical system in two different representations. In the first representation the observable x is diagonal, and in the second the observable A.

A generalization of this eigenfunction expansion to the case of two singular endpoints is described in Section 10.14 below.

10.13 The Limit-Circle Case

Consider the operator T_0 of Section 10.8 for the interval $[0, \infty)$ on a domain such that f and $T_0 f$ are in L^2 with a boundary condition of the usual kind (10.11-2) at $x = 0$, namely,

$$f(0)\cos \alpha + p(0)f'(0)\sin \alpha = 0 \qquad (\alpha \text{ real}) \qquad (10.13\text{-}1)$$

We assume that T_0 is of the limit-circle type at ∞ and we ask, what are the possible boundary or endpoint conditions at $x = \infty$ that make the operator self-adjoint. Clearly, we cannot write simply $f(\infty)\cos \beta + p(\infty)f'(\infty)\sin \beta = 0$, because the limiting values of f, p, and f' at $x = \infty$ are generally meaningless. In place of $f(\infty)$ and $p(\infty)f'(\infty)$ we need two (generally complex) numbers c_1 and c_2 that characterize the asymptotic behavior of f. It turns out that in the limit-circle case all f such that f and $T_0 f$ are both in L^2 have somewhat similar behavior at ∞ and one can indeed distinguish among them by two such numbers.

Let λ_0 be any given real number, and let $u_1(x)$ and $u_2(x)$ be any linearly independent real solutions of $T_0 u = \lambda_0 u$. For any g in L^2, consider the function

$$h(x) = \frac{1}{W} \int_x^\infty [u_1(y)u_2(x) - u_1(x)u_2(y)]g(y)dy, \qquad (10.13\text{-}2)$$

(which is meaningful because u_1, u_2 and g are all in L^2), where

$$W = p(x)[u_1(x)u_2'(x) - u_1'(x)u_2(x)] = \text{const.} \qquad (10.13\text{-}3)$$

A direct calculation shows that

$$(T_0 - \lambda_0)h = g.$$

Therefore if f is any function such that f and $T_0 f$ are in L^2, and if $g = (T_0 - \lambda_0)f$, then $f - h$ satisfies the homogeneous equation, and we can write

$$f(x) = c_1 u_1(x) + c_2 u_2(x) + \frac{1}{W} \int_x^\infty [u_1(y)u_2(x) - u_1(x)u_2(y)]g(y)dy,$$

$$(10.13\text{-}4)$$

where

$$g = (T_0 - \lambda_0)f. \qquad (10.13\text{-}5)$$

Conversely, if g is any element of L^2 and c_1 and c_2 are any generally complex constants, then the function f given by (10.13-4) is such that f and $T_0 f$ are both in L^2. The constants c_1 and c_2 are called the *limit numbers* of $f(x)$ *with respect to* the given solutions u_1 and u_2. (10.13-4) can also be written as

$$f(x) = u_1(x)[c_1 + \varphi_1(x)] + u_2(x)[c_2 + \varphi_2(x)],$$

where $\varphi_1(x)$ and $\varphi_2(x) \to 0$ as $x \to \infty$. Specification of c_1 and c_2 imposes the same kind of restriction on f for our purposes as specification of $f(a)$ and $f'(a)$ at a finite a; namely, for given λ_0, u_1, and u_2, and for any λ in $\mathbb{C}$, there is exactly one solution of the equation $T_0 f = \lambda f$ having given values of c_1 and c_2. Furthermore, with c_1 and c_2 fixed, the solution is an entire function of λ.

The main result (for proof, see Jörgens and Rellich 1976) is this:

Theorem.   (1) *Let* λ_0, $u_1(x)$, *and* $u_2(x)$ *be given, as above. If A is any self-adjoint extension of the minimal operator T of* (10.8-2), *assumed to be of limit-circle type at* ∞, *with a boundary condition of type* (10.13-1) *at* $x = 0$, *then there is a real number* β *such that the limit numbers c_1 and c_2 of any f in $\mathfrak{D}(A)$ satisfy*

$$c_1 \cos \beta + c_2 \sin \beta = 0. \qquad (10.13\text{-}6)$$

Conversely, for any real α and β the operator $A_{\alpha \beta}$ defined by

$$\mathfrak{D}(A_{\alpha \beta}) = \{f \in L^2 \colon T_0 f \in L^2, (10.13\text{-}1) \text{ and } (10.13\text{-}6) \text{ hold}\},$$

$$A_{\alpha \beta} f = T_0 f, \qquad (10.13\text{-}7)$$

is a self-adjoint extension of T.

   (2) *If λ_0, u_1, and u_2 are replaced by any $\tilde{\lambda}_0$, $\tilde{u}_1$, and $\tilde{u}_2$, where $\tilde{\lambda}_0$ is also real and $\tilde{u}_1(x)$ and $\tilde{u}_2(x)$ are linearly independent real solutions of $T_0 u = \tilde{\lambda}_0 u$, then the value of β is generally different for a given operator, but the same family of self-adjoint extensions of T is obtained.*

Because of (2), it is reasonable to take u_1 and u_2 as solutions of $T_0 u = 0$, for simplicity.

10.14 Case of Two Singular Endpoints

Suppose the operator is singular at both ends of the interval (a, b). (Then, we may have $a = -\infty$ or $b = +\infty$ or both). To classify one end, say $x = a$, we introduce an intermediate point c; $a < c < b$. If all solutions of $T_0 f = \lambda f$ are quadratically integrable over (a, c), i.e., are in $L^2(a, c)$, for every λ, then the left endpoint, $x = a$ is in the limit-circle case; otherwise there is just one solution (except for a multiplicative constant) in $L^2(a, c)$ for each nonreal λ, and a is in the limit-point case. The classification of the right endpoint $x = b$ is similar.

It will be seen that if both ends are in the limit-point case, then no boundary condition is needed at all, and the operator A defined by the equations

$$\mathfrak{D}(A) = \{f \in L^2 \colon T_0 f \in L^2\}, \qquad Af = T_0 f \qquad (10.14\text{-}1)$$

is self-adjoint. If one end is in the limit-circle case, or both, a boundary condition is needed there.

Generally, the spectrum is partly discrete and partly continuous, but Coddington and Levinson 1955 and Jörgens and Rellich 1976 show that it is purely discrete if T_0 is in the limit circle case at both ends and a boundary condition like (10.13-6) is imposed at each. That result incidentally covers the regular Sturm–Liouville problem in the sense that if $p(x)$, $p'(x)$, and $q(x)$ are continuous in a finite closed interval $[a, b]$, then the same problem in the open interval (a, b) (i.e., the problem in which we merely ignore the two values $x = a$ and $x = b$) is in the limit circle case at each end, because all solutions of $T_0 f = \lambda f$ are quadratically integrable in (a, b).

EXERCISE

1. Show that a Sturm–Liouville equation

$$-(p(x)f'(x))' + q(x)f(x) = \lambda f(x)$$

can be transformed into a Sturm–Liouville equation

$$-(\tilde{p}(y)g'(y))' + \tilde{q}(y)g(y) = \lambda g(y)$$

under a transformation

$$x \to y = y(x)$$

by writing $f(x) = \varphi(y)g(y)$, for suitable $\varphi(y)$. Find $\varphi(y)$, $\tilde{p}(y)$, $\tilde{q}(y)$. [Note that the L^2 norm is not preserved; generally $\int |f|^2 dx \neq \int |g|^2 dy$.] Show that if $y(x) = 1/x$, a regular point of the first equation, at $x = 0$, is transformed into a singular endpoint of the second equation of the limit-circle type, at $y = \infty$.

When both endpoints a and b are in the limit-point case, the eigenfunction expansion is similar to that given in Section 10.12, except that the spectrum is of multiplicity 2. That is, there may be two linearly independent eigenfunctions for a given eigenvalue and also two linearly independent continuous spectrum eigenfunctions (not in L^2) at a given point in the continuous spectrum. Hence there are two coefficient functions $h_1(s)$ and $h_2(s)$ in the eigenfunction expansion. In place of the spectral function $\rho(s)$ there is now a 2×2 *spectral matrix* with matrix elements $\rho_{jk}(s)$ ($j, k = 1, 2$). Corresponding to the characterization of $\rho(s)$ as real and nondecreasing, we have that the spectral matrix is Hermitian, in fact real symmetric, and nondecreasing in the sense that for $s' > s$ the matrix $\rho_{jk}(s') - \rho_{jk}(s)$ is positive semidefinite.

Specifically, let $f_1(x, \lambda)$ and $f_2(x, \lambda)$ be solutions of $T_0 f = \lambda f$ satisfying real boundary conditions at some point c in (a, b), and such that

$$p(x)[f_1(x)f'_2(x) - f_2(x)f'_1(x)] \equiv 1.$$

Then, for any λ with Im $\lambda \neq 0$, let f_3 and f_4 be the solutions of $T_0 f = \lambda f$ that are in $L^2(a, c)$ and $L^2(c, b)$, namely

$$f_3(x, \lambda) = f_1(x, \lambda) + m(\lambda)f_2(x, \lambda),$$
$$f_4(x, \lambda) = f_1(x, \lambda) + n(\lambda)f_2(x, \lambda),$$

(10.14-2)

where $m(\lambda)$ and $n(\lambda)$ are determined much as $m_\infty(\lambda)$ was in Section 10.12, namely

$$m(\lambda) = \lim_{x \to a} \frac{-f_1(x, \lambda)}{f_2(x, \lambda)}$$

$$n(\lambda) = \lim_{x \to b} \frac{-f_1(x, \lambda)}{f_2(x, \lambda)}.$$

(10.14-3)

From the analogue of equation (10.9-6), with b in that equation taken first positive, then negative, it is seen that Im $m(\lambda)$ and Im $n(\lambda)$ are nonzero for Im $\lambda \neq 0$ and of opposite sign. Hence, when both endpoints are in the limit-point case, there is no solution of $T_0 f = \lambda f$ in L^2 for nonreal λ, and we can now show that the operator A given by (10.14-1) is self-adjoint. First, if A^* were not $= A$, it would at least be a restriction of A, because the adjoint of the minimal operator T given by

$$\mathcal{D}(T) = C_0^\infty, \qquad Tf = T_0 f$$

is easily seen to be A (see (10.14-1)). Hence A^* is $= T^{**}$, the closure of T, which is $\subset A$. But A has no nonreal eigenvalues, hence the deficiency indices of A^* are $(0, 0)$, and it follows, since A and A^* are closed, that $A^* = A$.

In place of the formulas (10.12-6 and 7) of the eigenfunction expansion of a function $g(y)$ in $L^2(a, b)$, we have (see Coddington and Levinson 1955)

$$h_j(s) = \int_a^b f_j(y, s)g(y)dy \qquad (j = 1, 2)$$

(10.14-4)

$$g(x) = \sum_{j k = 1}^{2} \int_{-\infty}^{\infty} f_j(x, s)h_k(s)d\rho_{j k}(s).$$

(10.14-5)

Lastly, in place of formula (10.12-4) for ρ, we have

$$\rho_{j k}(t) = \lim_{\varepsilon \to 0} \frac{1}{2\pi i} \int^t [M_{j k}(s + i\varepsilon) - M_{j k}(s - i\varepsilon)]ds,$$

(10.14-6)

where

$$M_{j k}(\lambda) = \frac{1}{2[m(\lambda) - n(\lambda)]} \begin{pmatrix} 2 & m(\lambda) + n(\lambda) \\ m(\lambda) + n(\lambda) & 2m(\lambda)n(\lambda) \end{pmatrix}.$$

(10.14-7)

Examples will appear in the next two sections. Often the summation in (10.14-5) is avoided by introducing a new variable in place of $s = \lambda$, or the equivalent. For example, for the operator $-(d/dx)^2$, there are two continuous-spectrum eigenfunctions for each $\lambda > 0$, namely $e^{\pm i\sqrt{\lambda}x}$, but they are rewritten as e^{isx}, where now the variable $s = \sqrt{\lambda}$ ranges from $-\infty$ to ∞.

10.15 Bessel's Equation

Separation of variables in the two-dimensional reduced wave equation $\nabla^2 u + k^2 u = 0$ in plane polar coordinates r, θ, by considering solutions of the form $u(r, \theta) = R(r)\Theta(\theta)$, gives for the factor $R(r)$ the equation

$$\frac{1}{r}\frac{d}{dr} r\frac{dR}{dr} + \left(k^2 - \frac{l^2}{r^2}\right)R = 0, \tag{10.15-1}$$

where l^2 is a separation constant; k^2 plays the role of the eigenvalue parameter λ. In the reduced wave equation problem, only integer values of l appear, but we shall merely require l to be real. In any case, we can assume $l \geq 0$, since only l^2 appears in the equation. If we write x for r, and call $f(x) = \sqrt{r}R(r)$ and $\lambda = k^2$, then

$$-f'' + \frac{l^2 - 1/4}{x^2} f = \lambda f, \tag{10.15-2}$$

which is of the Sturm–Liouville form in the interval $(0, \infty)$, with $p = 1$, $q = (l^2 - \frac{1}{4})/x^2$. According to the criterion in Section 10.9, the right endpoint, at $x = \infty$, is in the limit-point case for all values of l. The left endpoint, at $x = 0$, is a regular singular point, and the indicial equation is $\alpha^2 - \alpha + (-l^2 + \frac{1}{4}) = 0$, hence $\alpha = \frac{1}{2} \pm l$, hence $x = 0$ is in the limit-circle case for $0 \leq l < 1$ and in the limit-point case for $l \geq 1$.

We first suppose that $l \geq 1$ so that no boundary conditions are needed, and the operator A_l defined in $L^2(0, \infty)$ by

$$\mathfrak{D}(A_l) = \{f \in L^2 : T_0 f \in L^2\} \tag{10.15-3}$$

$$A_l f = T_0 f = -f'' + \frac{l^2 - \frac{1}{4}}{x^2} f$$

is self-adjoint. By a slight modification of the discussion of the preceding section, the fundamental solutions f_1 and f_2 of the equation $T_0 f = \lambda f$ can be taken as

$$f_1(r, \lambda) = \sqrt{\frac{\pi r}{2}} J_l(kr) \sim \frac{1}{\sqrt{k}} \cos\left(kr - l\frac{\pi}{2} - \frac{\pi}{4}\right),$$

$$f_2(r, \lambda) = \sqrt{\frac{\pi r}{2}} Y_l(kr) \sim \frac{1}{\sqrt{k}} \sin\left(kr - l\frac{\pi}{2} - \frac{\pi}{4}\right), \tag{10.15-4}$$

where J_l and Y_l are the Bessel and Neumann functions and k is given by

$$k = \sqrt{\lambda}, \qquad \text{Re } k > 0. \tag{10.15-5}$$

The symbol $\sim$ indicates the asymptotic form valid as $r \to \infty$. In contrast with the preceding section, these functions f_1 and f_2 are not entire functions of λ for fixed r, but are analytic in the cut λ plane with a branch cut along the negative real axis. From the asymptotic forms or from the expansions about $r = 0$, we find that the Wronskian has the value

$$f_1 f_2' - f_2 f_1' = 1.$$

As in the preceding section, we denote the solutions quadratically integrable near 0 and ∞ respectively for Im $\lambda \neq 0$ by

$$\begin{aligned} f_3(x, \lambda) &= f_1(x, \lambda) + m(\lambda) f_2(x, \lambda), \\ f_4(x, \lambda) &= f_1(x, \lambda) + n(\lambda) f_2(x, \lambda), \end{aligned} \tag{10.15-6}$$

Then $f_3 f_4' - f_4 f_3' = n(\lambda) - m(\lambda)$, and the Green's function is

$$G(x, y) = G(x, y; \lambda) = \frac{1}{m(\lambda) - n(\lambda)} \begin{cases} f_3(x, \lambda) f_4(y, \lambda) & \text{if } x < y \\ f_4(x, \lambda) f_3(y, \lambda) & \text{if } x > y \end{cases} \tag{10.15-7}$$

We know that $\sqrt{r} J_l(kr)$ is quadratically integrable in any interval $(0, c)$, while $\sqrt{r} Y_l(kr)$ is not, for $l \geq 1$, because of the singularity of Y_l at $r = 0$. Hence $f_3 \equiv f_1$, i.e., $m(\lambda) \equiv 0$. To find $n(\lambda)$, we introduce the Hankel functions

$$H_l^{(1)}(z) = J_l(z) + i Y_l(z) \sim \sqrt{\frac{2}{\pi z}} e^{i(z - l(\pi/2) - (\pi/4))},$$

$$H_l^{(2)}(z) = J_l(z) - i Y_l(z) \sim \sqrt{\frac{2}{\pi z}} e^{-i(z - l(\pi/2) - (\pi/4))}. \tag{10.15-8}$$

To obtain a solution quadratically integrable in (c, ∞), it is clear that we must take f_4 proportional to $\sqrt{r} H_l^{(1)}(kr)$ for Im $k > 0$ and to $\sqrt{r} H_l^{(2)}(kr)$ for Im $k < 0$. For Re $k > 0$, as has been assumed, Im k has the same sign as Im λ, hence,

$$m(\lambda) = 0,$$
$$n(\lambda) = i \text{ sgn Im}(\lambda). \tag{10.15-9}$$

We now discuss the spectrum of A_l. First, $\sqrt{r} J_l(kr)$ is not quadratically integrable over $(0, \infty)$ for any k, hence there are no eigenfunctions (in the strict sense), i.e., the point spectrum is empty. For $\lambda < 0$, we can take $k = i\sqrt{-\lambda}$, where the positive square root is meant, and $f_4 = \sqrt{\pi r/2} H_l^{(1)}(kr)$; then it is easily seen from (10.5-7) that the integral operator $(A_l - \lambda)^{-1} = \int G(x, y) \cdots dy$ is bounded. Therefore the negative real axis in the λ plane is in the resolvent set, and we conclude that the spectrum is purely continuous and lies in $[0, \infty]$.

The eigenfunction expansion is given by equations (10.14-4 to 7) of the preceding section, but with the following modification: As λ crosses the negative real axis, not only does $n(\lambda)$ have a jump, but so do f_1 and f_2 (while $m(\lambda) \equiv 0$), and in just such a way that the Green's function $G(x, y; \lambda)$ is continuous in λ across the axis, because the negative real axis is in the resolvent set, and the Green's function is the kernel of the resolvent $(A_l - \lambda)^{-1}$, which is continuous in the resolvent set. Therefore $G(x, y; s + i\varepsilon) - G(x, y; s - i\varepsilon) \to 0$ as $\varepsilon \to 0$. Hence $\rho_{jk}(t)$ is given by (10.14-6) for $t \geq 0$, but is constant for $t < 0$. Taking the constant as zero, we find

$$(\rho_{j\,k}(t)) = \begin{cases} \begin{pmatrix} 0 & 0 \\ 0 & 0 \end{pmatrix} & \text{for } t < 0, \\[2mm] \begin{pmatrix} t/\pi & 0 \\ 0 & 0 \end{pmatrix} & \text{for } t \geq 0. \end{cases} \tag{10.15-10}$$

Hence the eigenfunction expansion of a given $g(r)$ in $L^2(0, \infty)$ is

$$h(s) = h_1(s) = \int_0^\infty \sqrt{\frac{\pi r}{2}} J_l(\sqrt{sr})g(r)dr, \tag{10.15-11}$$

$$g(r) = \frac{1}{\pi} \int_0^\infty \sqrt{\frac{\pi r}{2}} J_l(\sqrt{sr})h(s)ds. \tag{10.15-12}$$

According to Titchmarsh (1946), this result is due to Hankel.

For $0 \leq l < 1$, the origin is in the limit-circle case, and there are many self-adjoint extensions of the minimal operator corresponding to T_0, depending on a boundary condition at 0, which has to come from physical considerations. We shall not discuss the general case, but merely point out that for $l = 0$, in the problem of the reduced wave equation $\nabla^2 u + k^2 u = 0$ in two dimensions, a reasonable requirement is that ∇^2 be self-adjoint as an operator in $L^2(\mathbb{R}^2)$. Then, as will be seen in the next chapter, although f_1 and f_2 are both bounded, hence quadratically integrable in $(0, c)$ for any $c < \infty$, the solution $R(r) = Y_0(kr)$ has to be excluded because of the logarithmic singularity at $r = 0$. Then, all the above formulas hold, with l set $= 0$. It is noted in passing that in the corresponding problem of the relativistic hydrogen-like atom with angular quantum number $l = 0$, the requirement that the wave function be *finite* at the origin (as $J_0(kr)$ is here) would be unacceptable (the Hamiltonian would not be self-adjoint, and there would be no eigenfunctions at all); see Section 10.17.

Because of the zeros in the matrix (10.15-10), the second solution f_2 of (10.15-4) does not appear in the eigenfunctions expansion, hence the spectrum is simple (of multiplicity $= 1$). That was a consequence of the properties of the function $f_1(x, \lambda)$. In fact, Jörgens and Rellich 1976 have proved the following for any Sturm–Liouville operator A on an x-interval (a, b):

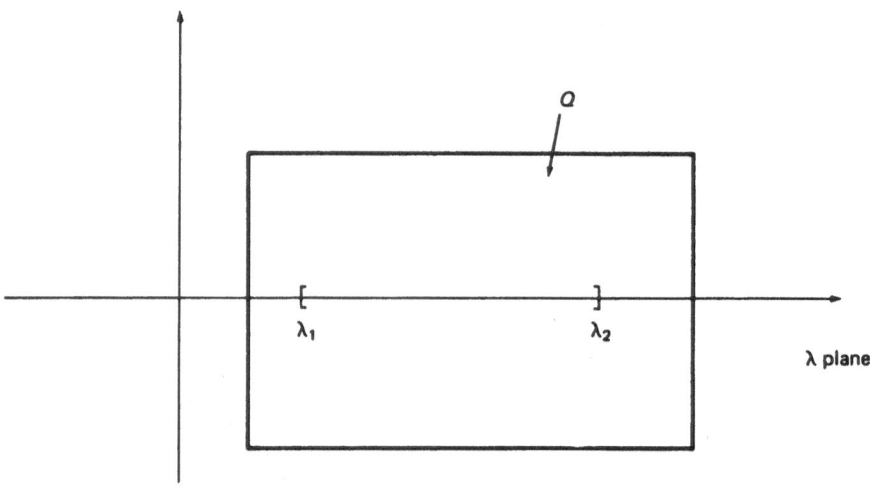

Figure 10.1 Diagram for the theorem of Jörgens and Rellich.

Theorem. *Let Q be an open rectangle in the λ plane symmetric with respect to the real axis (see Figure 10-1) containing an interval $[\lambda_1, \lambda_2]$ that doesn't have any eigenvalues of A in it. Suppose there is a solution $f_1(x, \lambda)$ of $T_0 f = \lambda f$ such that*

1. *f_1 is analytic in λ in Q and real for real λ*
2. *For no λ in Q is $f_1 \equiv 0$ in x*
3. *f_1 is in $L_2(a, c)$ for all λ in Q, for some c, $a < c < b$.*

Then the spectrum of A has multiplicity ≤ 1 in $[\lambda_1, \lambda_2]$.
Clearly, $L^2(a, c)$ in (3.) can be replaced by $L^2(c, b)$.

Note.   Multiplicity $= 0$ would mean that the spectrum is empty in $[\lambda_1, \lambda_2]$. An application will be found in the next section.

EXERCISE

1.   Discuss the eigenvalue problems of equation (10.15-2) on the intervals $(0, a)$, (a, b), and (a, ∞) and the corresponding eigenfunction expansions.

10.16 The Nonrelativistic Hydrogen-Like Atom

The steady-state Schrödinger equation for an electron in the Coulomb field of a fixed point charge Ze at the origin is

$$-\frac{\hbar^2}{2m}\nabla^2 u + \frac{-Ze^2}{r}u = Eu,$$

in the usual notation. There is no dimensionless parameter here, and by suitable choice of the units of length and energy the equation can be written as

$$-\nabla^2 u - \frac{2}{r} u = \lambda u. \qquad (10.16\text{-}1)$$

Solutions in the special form $u = R(r)\Theta(\theta)\Phi(\varphi)$ in spherical polar coordinates r, θ, φ can be obtained by separation of variables. If $rR(r)$ is called $f(r)$, the radial equation takes the form

$$T_0 f \overset{\text{def}}{=} -f'' + \left(\frac{l(l+1)}{r^2} - \frac{2}{r}\right) f = \lambda f, \qquad (10.16\text{-}2)$$

where l is a nonnegative integer. This is a Sturm–Liouville problem in $(0, \infty)$. Like the Bessel equation problem, it is in the limit-point case at ∞ for all l and also at 0 for $l = 1, 2, \ldots$. For $l = 0$, it is in the limit-circle case at 0.

For $l = 1, 2, \ldots$, the operator A_l defined by

$$\mathfrak{D}(A_l) = \{f \in L^2(0, \infty): T_0 f \in L^2\}, \qquad A_l f = T_0 f \qquad (10.16\text{-}3)$$

is self-adjoint. For $l = 0$, a boundary condition is needed at $r = 0$; it can be taken to be of the form (10.13-6) and depends on a parameter β in $[0, \pi)$. The self-adjoint operator with that boundary condition is called $A_{0\,\beta}$.

For $l \geq 1$, the eigenvalues of A_l are given by

$$\lambda_n = -\frac{1}{n^2}, \qquad n = l + 1, l + 2, \ldots, \qquad (10.16\text{-}4)$$

which is the Balmer formula for the energy levels; they are all simple. The corresponding normalized eigenfunctions, which can be found in any book on quantum mechanics, will be called $\varphi_{n,l}(r)$.

The eigenvalues of $A_{0\,\beta}$ were investigated by Jörgens and Rellich 1976 and are found to lie in the intervals

$$-\frac{1}{(n-1)^2} < \lambda_n \leq -\frac{1}{n^2}, \qquad n = 1, 2, \ldots. \qquad (10.16\text{-}5)$$

For one value of β, namely $\beta = 0$ in the formulation of Jörgens and Rellech, λ_n lies in the upper limit of the interval (10.16-5); hence in that case the Balmer formula (10.16-4) holds also for $l = 0$. By considering the full Hamiltonian $-\nabla^2 - 2/r$, it can be shown that $\beta = 0$ gives the physically correct boundary condition (see next chapter).

For $l \geq 1$, the following things are proved in Jörgens and Rellich: (1) The only eigenvalues of A_l are those given by the Balmer formula (10.16-4); (2) there is no continuous spectrum in $\lambda < 0$. That is, every interval of $\lambda < 0$ not containing any eigenvalues is in the resolvent set; (3) the nonnegative real axis $\lambda \geq 0$ is the continuous spectrum; (4) the spectrum is simple.

A solution of $T_0 f = \lambda f$ is the function

$$f(r, \lambda) = r^{l+1} e^{-i\sqrt{\lambda}r} F(l + 1 + i/\sqrt{\lambda}, 2l + 2, 2i\sqrt{\lambda}r), \qquad (10.16\text{-}6)$$

where $\sqrt{\lambda}$ denotes the principal branch of the square root ($|arg\sqrt{\lambda}| < \pi/2$), and F denotes the Kummer confluent hypergeometric function

$$F(a, c, z) = {}_1F_1(a; c; z) = \sum_{n=0}^{\infty} \frac{(a)_n}{(c)_n n!} z^n, \qquad (10.16\text{-}7)$$

where

$$(a)_0 = 1, \qquad (a)_n = a(a + 1) \cdots (a + n - 1) \quad \text{for } n > 0. \qquad (10.16\text{-}8)$$

This solution $f(r, \lambda)$ is analytic in the cut plane $|arg\ \lambda| < \pi$ for fixed r, it is quadratically integrable with respect to r in $(0, c)$ for any λ ($c < \infty$), and, in spite of appearances, it is real for real λ. Therefore the simplicity of the continuous spectrum follows from the theorem in the preceding section.

The eigenfunction expansion of a given $g(r)$ in $L^2(0, \infty)$ is, according to Jörgens and Rellich,

$$g(r) = \sum_{n=l+1}^{\infty} c_n \varphi_{n\ l}(r) + \int_0^{\infty} f(r, s)h(s)d\rho(s), \qquad (10.16\text{-}9)$$

where

$$c_n = \int_0^{\infty} \varphi_{n\ l}(r)g(r)dr \qquad (10.16\text{-}10)$$

$$h(s) = \int_0^{\infty} f(r, s)g(r)dr, \qquad (10.16\text{-}11)$$

while the spectral function $\rho(s)$ is given by

$$d\rho(s) = \frac{1}{2\pi} (2\sqrt{s})^{2l+1} e^{\pi/\sqrt{s}} \left| \frac{\Gamma(l + 1 + i/\sqrt{s})}{\Gamma(2l + 2)} \right|^2 ds. \qquad (10.16\text{-}12)$$

These formulas hold also for $l = 0$ if β is taken $= 0$ in the boundary condition for the operator $A_{0\ \beta}$.

10.17 The Relativistic Hydrogen-Like Atom

In the relativistic case there is a dimensionless parameter $\alpha = \alpha_0 Z$, where

$$\alpha_0 = \frac{e^2}{\hbar c} = (137.03)^{-1}.$$

There is physical reason to believe that the formulation of the problem breaks down for $Z \approx 137$, hence we assume $0 < \alpha < 1$. As indicated in the next chapter, there are two radial functions $f(r)$ and $g(r)$; with $\hbar/mc$ and mc^2 as the units of length and energy, respectively, the equations for f and g are

$$\left(\lambda + 1 + \frac{\alpha}{r}\right)f - g' - \frac{1 + k}{r} g = 0 \qquad (0 < \Gamma < \infty),$$

$$\qquad (10.17\text{-}1)$$

$$\left(\lambda - 1 + \frac{\alpha}{r}\right)g + f' + \frac{1 - k}{r} f = 0 \qquad (0 < r < \infty),$$

where k is an integer $\neq 0$, and λ is the energy. Elimination of one of the functions, say f, leads to a second order equation for the other, namely,

$$g'' + Pg' + Qg = 0, \qquad (10.17\text{-}2)$$

where

$$P(r) = \frac{3}{r} - \frac{\lambda + 1}{(\lambda + 1)r + \alpha},$$

$$\qquad (10.17\text{-}3)$$

$$Q(r) = \left(\lambda + \frac{\alpha}{r}\right)^2 - 1 - \frac{k^2 + k}{r^2} + \frac{(k + 1)\alpha}{r^2[(\lambda + 1)r + \alpha]}.$$

The equation can be put into formally self-adjoint form $-(pg')' + qg = 0$ by defining

$$p(r) = \exp \int P(r)dr, \qquad q(r) = -p(r)Q(r).$$

However, the result is not of the Sturm–Liouville form, owing to the complicated way in which the eigenvalue parameter λ enters; nevertheless, the ideas of the preceding sections can be used, as pointed out by Roos and Sangren 1962.

We introduce a two-component vector $\binom{f}{g}$ and an operator T_0 given by

$$T_0\binom{f}{g} = \frac{1}{r}\binom{(rg)'}{-(rf)'} + \binom{-1 - \alpha/r \quad k/r}{k/r \quad 1 - \alpha/r}\binom{f}{g}. \qquad (10.17\text{-}4)$$

Owing to the symmetry of the matrix that appears here, this operator is seen to be formally self-adjoint with respect to the inner product

$$\left(\binom{f_1}{g_1}, \binom{f_2}{g_2}\right) = \int_a^b (\bar{f_1}f_2 + \bar{g_1}g_2)r^2 \, dr, \qquad (10.17\text{-}5)$$

and the equations (10.17-1) take the form

$$T_0\binom{f}{g} = \lambda\binom{f}{g}. \qquad (10.17\text{-}6)$$

We now assume that f and g satisfy these equations. We multiply the first equation of (10.17-1) by $\bar{f}$ and the second by $\bar{g}$, we add and multiply by r^2, and we integrate from a to b, where $0 < a < b < \infty$; we then integrate the term containing $(r\bar{g})(rf)'$ by parts. Most of the resulting terms are real, and by taking imaginary parts throughout, we find

$$\text{Im } \lambda \int_a^b (|f|^2 + |g|^2)r^2 \, dr + [r^2 \text{ Im } \bar{g}(r)f(r)]_a^b = 0. \qquad (10.17\text{-}7)$$

We now take $a = 1$ and assume that we have two solutions, with subscript 1 and 2, such that

$$f_1(1) = 0 \qquad f_2(1) = 1,$$

$$\qquad (10.17\text{-}8)$$

$$g_1(1) = 1 \qquad g_2(1) = 0.$$

The general solution, up to a multiplicative constant, is

$$\begin{pmatrix} f \\ g \end{pmatrix} = \begin{pmatrix} f_1 \\ g_1 \end{pmatrix} + m \begin{pmatrix} f_2 \\ g_2 \end{pmatrix}, \tag{10.17-9}$$

corresponding to (10.9-4), where m is a complex number, and we put this solution into (10.17-7). The initial conditions (10.17-8) give $\overline{\mathrm{Im}\, g(1)} f(1) = \mathrm{Im}\, m$, hence

$$-\mathrm{Im}\, m + \mathrm{Im}\, \lambda \int_1^b (|f|^2 + |g|^2) r^2 \, dr = F(m; b) \overset{\text{def}}{=} -b^2 \, \mathrm{Im}\, \bar{g}(b) f(b)$$

$$= A|m|^2 + B \,\mathrm{Re}\, m + C \,\mathrm{Im}\, m + D, \tag{10.17-10}$$

as in (10.9-6 to 8), where now

$$A = -b^2 \mathrm{Im}\, \overline{g_2(b)} f_2(b) = \mathrm{Im}\, \lambda \int_1^b (|f_2|^2 + |g_2|^2) r^2 \, dr. \tag{10.17-11}$$

From here on, the theory is the same as in Sections 10.9–10.14 except for replacement of the Hilbert space $L^2(a, b)$ by the Hilbert space $\mathfrak{H}(a, b)$ with the L^2 type of inner product given by (10.17-5). In particular, the properties of the limit-point and limit-circle endpoints are applicable here.

It is easily seen that the endpoint at $r = \infty$ is in the limit-point case. Equations (10.17-1) show that f and g behave asymptotically either both like $e^{\mu r}$ or $e^{-\mu r}$, where $\mu = \sqrt{1 - \lambda^2}$, and only one of these is quadratically integrable in $(1, \infty)$, for $\mathrm{Im}\, \lambda \neq 0$.

The endpoint at $r = 0$ is seen from (10.17-2, 3) to be a regular singular point of the equation for g, when f is eliminated. In the notation of (10.10-2), $P_0 = 3$, $Q_0 = \alpha^2 - k^2 + 1$, hence the indicial equation, with the exponent now called γ instead of α, is

$$(\gamma + 1)^2 + \alpha^2 - k^2 = 0,$$

hence

$$\gamma_1, \gamma_2 = -1 \pm \sqrt{k^2 - \alpha^2}.$$

For small r, the two solutions of (10.17-2) behave according to $g(r) \sim r^{\gamma_1}$ and $g(r) \sim r^{\gamma_2}$.

Now suppose that $\alpha < \sqrt{\tfrac{3}{4}}$, i.e., $Z \leq 118$. The quantum number k is an integer $\neq 0$, hence $k^2 \geq 1$, and we see that $\gamma_1 > -\tfrac{1}{2}$, $\gamma_2 < -\tfrac{3}{2}$. Therefore $|g|^2 r^2$ is integrable in $(0, 1)$ for $\gamma = \gamma_1$ but not for $\gamma = \gamma_2$. Therefore $r = 0$ is in the limit-point case, and the operator A defined by

$$\mathfrak{D}(A) = \left\{ \begin{pmatrix} f \\ g \end{pmatrix} \in \mathfrak{H}(0, \infty) \colon T_0 \begin{pmatrix} f \\ g \end{pmatrix} \in \mathfrak{H} \right\}$$

$$A \begin{pmatrix} f \\ g \end{pmatrix} = T_0 \begin{pmatrix} f \\ g \end{pmatrix}$$

is self-adjoint without any boundary conditions.

For $\sqrt{\frac{3}{4}} < \alpha < 1$ ($119 \leq Z \leq 137$), the endpoint $r = 0$ is still in the limit-point case for $|k| \geq 1$, but is in the limit-circle case for $k = \pm 1$, because then both exponents γ_1 and γ_2 are $\geq -\frac{3}{2}$ so that for all solutions g, $|g|^2 r^2$ is integrable in $(0, 1)$. In this case, for every nonreal λ, the solution that behaves like $e^{-\mu r}$ ($\mu = \sqrt{1 - \lambda^2}$, Re $\mu > 0$) is in $\mathfrak{H}$; hence, the deficiency indices of the minimal operator based on T_0 are $(1, 1)$, and a boundary condition is required at $r = 0$ to obtain a self-adjoint operator. The physically appropriate boundary condition is discussed in the next chapter.

CHAPTER 11

Some Partial Differential Operators of Quantum Mechanics

Schrödinger and Dirac Hamiltonian for a free particle and for hydrogen-like atoms; Schrödinger Hamiltonian of n-electron atoms; self-adjointness and properties of the spectra; resolvent and resolution of the identity for the Laplacian; relatively bounded perturbation of a self-adjoint operator; essential spectrum; absolutely continuous and singular continuous spectra; continuous spectrum in the sense of Hilbert; absolutely continuous and singular-continuous subspaces. Problems of the relativistic hydrogen atom for different values of Z; self-adjointness and spectrum of the Laplacian in a bounded region of space.

Prerequisite: Chapters 1 to 10.

In suitable units, the Hamiltonian H for a free particle is the kinetic energy $-\frac{1}{2}\nabla^2$ in $\mathbb{R}^3$. For a system of N identical particles, it is $-\frac{1}{2}\nabla^2$ in $\mathbb{R}^n$, with $n = 3N$. The Schrödinger Hamiltonian is obtained by adding a potential energy term. Self-adjoint versions of these operators are discussed in this chapter. In the case of just one particle (electron) in a Coulomb field (hydrogen-like atom), the relativistic (Dirac) Hamiltonian is also discussed.

11.1 Self-Adjoint Laplacian in $\mathbb{R}^n$

If u is any distribution on $\mathbb{R}^n$, we denote by $\nabla^2 u$ the distribution given by

$$\nabla^2 u = \left(\frac{\partial^2}{\partial x_1^2} + \cdots + \frac{\partial^2}{\partial x_n^2} \right) u, \qquad (11.1\text{-}1)$$

where the derivatives are understood in the distribution sense. When the operator ∇^2 is restricted to a suitable domain in the Hilbert space $L^2(\mathbb{R}^n)$, it is self-adjoint. Specifically, we shall show that the operator H defined by

$$\mathfrak{D}(H) = \{u \in L^2 : \nabla^2 u \in L^2\}$$
$$Hu = -\tfrac{1}{2}\nabla^2 u \tag{11.1-2}$$

is self-adjoint. That will be done by the Fourier transform method used in Section 10.3.

Let U denote the operator of Fourier transform in L^2, so that if u is in L^2, then $\hat{u} = Uu$ is also in L^2. Let $\hat{H}$ denote the Fourier transform of H; that is,

$$\hat{H} = UHU^*, \qquad H = U^*\hat{H}U, \tag{11.1-3}$$

so that if $v = Hu$, then $\hat{v} = \hat{H}\hat{u}$. Clearly H is self-adjoint if and only if $\hat{H}$ is. But $\hat{H}$ is easily described. We shall show that if $u = u(\mathbf{x})$ is any distribution in $\mathfrak{D}(H)$, and $\hat{u} = \hat{u}(\mathbf{y})$ is its Fourier transform, then the distribution $\tfrac{1}{2}|\mathbf{y}|^2\hat{u}(\mathbf{y})$ is the transform of $Hu = -\tfrac{1}{2}\nabla^2 u$. That is obviously true if u is $= \varphi$, a test function in $\mathscr{S} = \mathscr{S}(\mathbb{R}^n)$, for then

$$\hat{\varphi}(\mathbf{y}) = (2\pi)^{-n/2}\int \cdots \int e^{-i\mathbf{y} \cdot \mathbf{x}}\varphi(\mathbf{x})d^n\mathbf{x},$$

and hence

$$\tfrac{1}{2}|\mathbf{y}|^2\hat{\varphi}(\mathbf{y}) = (2\pi)^{-n/2}\int \cdots \int e^{-i\mathbf{y} \cdot \mathbf{x}}[-\tfrac{1}{2}\nabla^2\varphi(\mathbf{x})]d^n\mathbf{x},$$

after two integrations by parts, which are justified because φ is in $\mathscr{S}$. If u is any element of $\mathfrak{D}(H)$ and $v = Hu = -\tfrac{1}{2}\nabla^2 u$, then, for any φ in $\mathscr{S}$,

$$(\hat{v}, \hat{\varphi}) = (Uv, U\varphi) = (v, \varphi) = (-\tfrac{1}{2}\nabla^2 u, \varphi) = (u, -\tfrac{1}{2}\nabla^2\varphi),$$

where the last step follows from the definition of distribution derivatives. Hence,

$$(\hat{v}, \hat{\varphi}) = (Uu, -\tfrac{1}{2}U\nabla^2\varphi) = (\hat{u}, \tfrac{1}{2}|\mathbf{y}|^2\hat{\varphi})$$

by the previous result. Therefore,

$$(\hat{v}, \hat{\varphi}) = (\tfrac{1}{2}|\mathbf{y}|^2\hat{u}, \hat{\varphi}).$$

The steps can be reversed, of course, and we see that $\hat{u}$ is in $\mathfrak{D}(\hat{H})$ if and only if $|\mathbf{y}|^2\hat{u}$ is in L^2. That is,

$$\mathfrak{D}(\hat{H}) = \{\hat{u} \in L^2 : |\mathbf{y}|^2\hat{u} \in L^2\}$$
$$\hat{H}\hat{u} = \tfrac{1}{2}|\mathbf{y}|^2\hat{u} \tag{11.1-4}$$

Now $\tfrac{1}{2}|\mathbf{y}|^2$ is a real continuous function of $\mathbf{y}$, and it was shown at the end of Section 7.5 that an operator defined in this way by such a function is self-adjoint. We conclude that H is also self-adjoint.

For $n = 2$ and $n = 3$, according to the Remark in Section 5.13, any u in the domain of H is a continuous function and $\to 0$ as $|\mathbf{x}| \to \infty$.

11.2 Resolvent, Spectrum, and Spectral Projectors

If R_λ is the resolvent $(H - \lambda)^{-1}$ of the above operator H, then the operator

$$\hat{R}_\lambda = UR_\lambda U^* \tag{11.2-1}$$

is the resolvent $(\hat{H} - \lambda)^{-1}$ of $\hat{H}$, because

$$(\hat{H} - \lambda)^{-1} = (UHU^* - \lambda)^{-1} = (U(H - \lambda)U^*)^{-1}$$
$$= U^{*-1}(H - \lambda)^{-1}U^{-1} = U(H - \lambda)^{-1}U^*.$$

Equation (11.1-4) shows that $(\hat{H} - \lambda)^{-1}$ is the multiplication operator

$$(\hat{R}_\lambda \hat{v})(\mathbf{y}) = (\tfrac{1}{2}|\mathbf{y}|^2 - \lambda)^{-1}\hat{v}(\mathbf{y}) \tag{11-2-2}$$

with domain the whole Hilbert space $L^2(\mathbb{R}^n)$, for any nonreal λ.

To investigate the spectrum, we note first that if $\lambda < 0$, $\hat{R}_\lambda$ is also a bounded operator defined in all L^2, according to (11.2-2), hence the spectrum is confined to $\lambda \geq 0$. For $\lambda \geq 0$, the inverse $(\hat{H} - \lambda)^{-1}$ exists, but is unbounded. In particular, for any $\hat{v}(\mathbf{y})$ that vanishes smoothly on the sphere $\tfrac{1}{2}|\mathbf{y}|^2 = \lambda$, the equation $(\tfrac{1}{2}|\mathbf{y}|^2 = \lambda)\hat{u} = \hat{v}$ has a unique solution $\hat{u}$ given by the right member of (11.2-2). We see that the point spectrum is empty and the non-negative real axis is the continuous spectrum.

According to Section 9.6, the spectral projector $\hat{E}_t$ is given by

$$\hat{E}_t = \frac{1}{2\pi i}\int_{C(t)} \hat{R}_\lambda \, d\lambda,$$

where $C(t)$ is a contour that comes from $-\infty + ia$ in the upper half plane, crosses the real axis at $\lambda = t$ and goes to $-\infty - ia$ in the lower half plane (a is any number > 0). The integral of $(\tfrac{1}{2}|\mathbf{y}|^2 - \lambda)^{-1}$ on $C(t)$ is $= 2\pi i$ if $\tfrac{1}{2}|\mathbf{y}|^2 < t$ and is $= 0$ if $\tfrac{1}{2}|\mathbf{y}|^2 > t$. Hence,

$$(\hat{E}_t\hat{u})(\mathbf{y}) = \begin{cases} \hat{u}(\mathbf{y}) & \text{for } |\mathbf{y}|^2 < 2t, \\ 0 & \text{for } |\mathbf{y}|^2 > 2t. \end{cases} \tag{11.2-3}$$

Finally, the resolution of the identity E_t of the original operator H, the self-adjoint version of $-\tfrac{1}{2}\nabla^2$, is given by $E_t = U^*\hat{E}_t U$ and we evaluate it first for a function $\varphi \in \mathscr{S}$:

$$(E_t\varphi)(\mathbf{x}) = \begin{cases} (2\pi)^{-n/2}\displaystyle\int\cdots\int_{|\mathbf{y}| < \sqrt{2t}} e^{i\mathbf{y}\cdot\mathbf{x}}\hat{\varphi}(\mathbf{y})d^n\mathbf{y} & \text{if } t > 0, \\ 0 & \text{if } t < 0. \end{cases}$$

$$= (2t)^{n/2}\int\cdots\int_{\mathbb{R}^n} K(\sqrt{2t}\ |\mathbf{x} - \mathbf{x}'|)\varphi(\mathbf{x}')d^n\mathbf{x}', \quad \text{for } t > 0,$$

$$\tag{11.2-4}$$

where

$$K(w) = (2\pi)^{-n}\int\cdots\int_{|\mathbf{y}| < 1} e^{i w y_1} \, d^n\mathbf{y}. \tag{11.2-5}$$

Evidently the same formula (11.2-4) continues to hold if $\varphi(\mathbf{x})$ is replaced by a general element $u(\mathbf{x})$ of L^2, because $\mathscr{S}$ is dense in L^2 and the operator is bounded.

In 1 dimension, $K(w) = \sin w/\pi w$; in three,

$$K(w) = \frac{1}{2\pi^2}\frac{\sin w - w\cos w}{w^3}.$$

EXERCISE

1. Show that in n dimensions, K is given by the formula

$$K(w) = (2\pi w)^{-n/2}J_{n/2}(w).$$

Hints.

1. The integral can be written as

$$\int_{-1}^{1}dy_1 e^{iwy_1}\int_{-\sqrt{1-y_1^2}}^{\sqrt{1-y_1^2}}dy_2\int_{-\sqrt{1-y_1^2-y_2^2}}^{\sqrt{1-y_1^2-y_2^2}}dy_3\cdots;$$

the integrations with respect to $y_2, y_3, \ldots, y_n$ give the volume of an $(n-1)$-dimensional ball of radius $\sqrt{1-y_1^2}$.

2. The formula for the volume of a ball can be obtained from equation (6.3-5).

3. Use the Poisson integral for the Bessel function—see Magnus and Oberhettinger (1943).

We note one difference between the present discussion and that of Section 10.15, where the two-dimensional Laplacian was treated by separation of variables in plane polar coordinate r and θ. (The same difference appears in any number of dimensions.) There, for $l = 0$ (no θ dependence), we found that the radial equation was in the limit-circle case at $r = 0$, hence a boundary condition had to be imposed to make the differential operator self-adjoint, and various self-adjoint operators could be obtained depending on the choice of that boundary condition. Here, no such boundary condition appears. The preferred choice of boundary condition in that treatment was such as to exclude the solution of the differential equation that behaves like $\log r = \log(x^2 + y^2)^{1/2}$ as $r \to 0$. Although $\nabla^2 \log r = 0$ for any $r > 0$, if $\log r$ is regarded as a distribution, its Laplacian is

$$\nabla^2 \log r = 2\pi\delta(x)\delta(y),$$

which is not in $L^2(\mathbb{R}^2)$, hence the prescription (11.1-2) automatically excludes such a solution from the domain of the operator.

EXERCISES

2. Verify that if $\varphi(x, y)$ is any function in $C_0^\infty(\mathbb{R}^2)$, then

$$\int_{\mathbb{R}^2}\int \log r \, \nabla^2\varphi \, dx \, dy = 2\pi\varphi(0, 0).$$

3. The operator of this exercise comes not from quantum theory but from transport theory. Let L^2 denote the Hilbert space $L^2(\mathbb{R} \times [-1, 1])$ consisting of distributions $f(x, \mu)$ with the inner product

$$(f, g) = \int_{-\infty}^{\infty} dx \int_{-1}^{1} d\mu \, \overline{f(x, \mu)} g(x, \mu).$$

Let T be the operator given by

$$\mathfrak{D}(T) = \left\{ f \in L^2 : \mu \frac{\partial f}{\partial x} \in L^2 \right\}$$

$$Tf(x, \mu) = -i\mu \frac{\partial}{\partial x} f(x, \mu).$$

By solving the equation $Tf - \lambda f = g$ compute the resolvent $R_\lambda = (T - \lambda)^{-1}$ for $\text{Im } \lambda \neq 0$. Then let $\hat{T} = UTU^*$, where U is the operator of Fourier transformation with respect to x:

$$(Uf)(k, \mu) = \hat{f}(k, \mu) = (2\pi)^{-1/2} \int_{-\infty}^{\infty} e^{-ikx} f(x, \mu) dx.$$

Show that $\hat{T}$ is self-adjoint, hence so is T. Compute the resolvent $\hat{R}_\lambda = (\hat{T} - \lambda)^{-1}$ and the resolution of the identity $\hat{E}_t$ of $\hat{T}$. By transforming back, find the resolution of the identity E_t of T.

11.3 Schrödinger Operators

One of the aims of recent mathematical work on the theory of Schrödinger and related operators has been to put on a firm basis many of the things that have been regarded as intuitively evident on physical grounds. See Kato (1966) and Jörgens and Weidman (1973). First is the quantum-mechanical requirement that the Hamiltonian of any physical system should be interpretable as a self-adjoint operator. A few remarks about that question are made without proof in the present section, and certain aspects of the continuous spectrum are discussed in the next.

Consider first the Hamiltonian

$$H = H_0 + V, \quad \text{where } H_0 = -\tfrac{1}{2}\nabla^2, \tag{11.3-1}$$

for a single particle moving in a potential $V = V(\mathbf{x})$. The Hilbert space is $\mathfrak{H} = L^2(\mathbb{R}^3)$, and H_0 denotes the self-adjoint operator denoted by H in Section 11.1. If $V(\mathbf{x})$ is a bounded continuous (real) function, then as an operator it is self-adjoint, bounded, and defined in all $\mathfrak{H}$. It follows from Exercise 2 in Section 7.2, that the Hamiltonian H is self-adjoint. One says that V is a bounded perturbation of H_0. If $V(\mathbf{x})$ is unbounded, the question of the self-adjointness and other properties of H is more difficult, but can be answered under reasonable assumptions by the methods of the perturbation theory for operators. If $V(\mathbf{x})$ is in $L^2(\mathbb{R}^3)$, then the product $V(\mathbf{x})\psi(\mathbf{x})$ is well defined as a distribution (in L^1—see Section 5.4), and it can be proved that

that distribution is in L^2 if ψ is in the domain of H_0 given in (11.1-2), so that the operator V is defined on that domain. The following is proved in Kato 1966:

Theorem 1. *If the potential $V(\mathbf{x})$ can be written as the sum of two (real) functions, of which one is continuous and bounded and the other is in $L^2(\mathbb{R}^3)$, then the operator H defined by*

$$\mathfrak{D}(H) = \{\psi \in L^2(\mathbb{R}^3): \nabla^2\psi \in L^2\}$$
$$H\psi = -\tfrac{1}{2}\nabla^2\psi + V\psi \tag{11.3-2}$$

is self-adjoint and bounded from below.

The theorem applies in particular to the hydrogen-like atom, where $V(\mathbf{x}) = -Z/|\mathbf{x}|$; Z is a positive integer, the nuclear charge.

Kato also showed that the same conclusion holds for an n-electron atom of nuclear charge $Z \geq n$, where

$$V = V(\mathbf{x}_1, \ldots, \mathbf{x}_n) = -\sum_{j=1}^{n} \frac{Z}{|\mathbf{x}_j|} + \sum_{j<k=1}^{n} \frac{1}{|\mathbf{x}_j - \mathbf{x}_k|}. \tag{11.3-3}$$

Theorem 2. *The operator H defined by*

$$\mathfrak{D}(H) = \{\psi \in L^2(\mathbb{R}^{3n}): \nabla^2\psi \in L^2\}$$
$$H\psi = -\tfrac{1}{2}\nabla^2\psi + V\psi, \tag{11.3-4}$$

where ∇^2 is the $3n$-dimensional Laplacian and V is given by (11.3-3), is self-adjoint and bounded from below.

The boundedness of H from below means that there is an energy λ_0 (negative in the cases considered here) such that $(\psi, H\psi) \geq \lambda_0(\psi, \psi)$ for all ψ, or equivalently that the entire spectrum lies in $\lambda \geq \lambda_0$. It corresponds to the existence of a ground state, or state of lowest energy.

Jörgens (1967) has proved the self-adjointness of the Hamiltonian for a hydrogen atom in a uniform electric field (Stark effect), or in a uniform magnetic field (Zeeman effect), or in both simultaneously, or in the magnetic field of a current ring (a more physical model of the Zeeman effect, since the total energy of the magnetic field is finite, in this model).

When the problem of the hydrogen-like atom is treated by separation of variables in spherical polar coordinates r, θ, φ, spurious solutions are found for $l = 0$ (zero angular momentum), because the radial equation is in the limit-circle case at $r = 0$, as discussed in Section 10.16. As in the preceding section, the unwanted solutions, which here go like $1/r$ as $r \to 0$, are excluded from the domain in (11.3-2), because

$$\nabla^2 \frac{1}{r} = -4\pi\delta(x)\delta(y)\delta(z), \tag{11.3-5}$$

which is not in L^2.

11.4 Perturbation of the Spectrum; Essential Spectrum; Absolutely Continuous Spectrum

We write $H = H_0 + V$, and think of V as a perturbation. If V is small in some sense, then various aspects of the spectrum are invariant, i.e. are the same for H as for H_0. We suppose that H_0 is self-adjoint and that V is a symmetric operator defined on the domain of H_0; that is, $\mathfrak{D}(V) \supset \mathfrak{D}(H_0)$.

One kind of smallness is relative boundedness. V is called *bounded relative to* H_0 or *H_0-bounded* if there are constants a and b such that

$$\|V\psi\| \le a\|\psi\| + b\|H_0\psi\| \quad \text{for all } \psi \in \mathfrak{D}(H_0). \qquad (11.4\text{-}1)$$

Generally, we can decrease b if we increase a, but we cannot take $b = 0$ for any finite a unless V is a bounded operator. The infimum (greatest lower bound) of the possible value of b is called the *H_0-bound* of V.

Theorem 1. (Rellich). *With H_0 and V as above, if V is H_0-bounded with H_0-bound < 1, then $H = H_0 + V$ is self-adjoint with $\mathfrak{D}(H) = \mathfrak{D}(H_0)$.*

Theorem 2. (Kato). *Under the conditions of Theorem 1, if H_0 is bounded from below (or from above, or both), then $H = H_0 + V$ is bounded from below (or from above, or both), though not necessarily by the same bound(s).*

These theorems underlie the results of Kato quoted in the preceding section. For the Coulomb potential $V(\mathbf{x}) = -Z/|\mathbf{x}|$, although V is not a bounded operator, it is H_0-bounded with H_0-bound $= 0$. (That is, we can let $b \to 0$ in (11.4-1) by letting $a \to \infty$.) Since the spectrum of $-\frac{1}{2}\nabla^2$ lies in $[0, \infty)$, Theorem 2 applies to the Schrödinger problems.

For the one-electron problem it is expected on physical grounds that if $V(\mathbf{x}) \to 0$ as $|\mathbf{x}| \to \infty$, the continuous spectrum of H also occupies $[0, \infty)$, because a particle at very large distance is practically free, and a free particle can have any energy > 0. That expectation is borne out by many computable cases, but the generally correct statement requires a modified spectral notion. Namely, the *essential spectrum* of an operator consists of all points of the spectrum except isolated eigenvalues of finite multiplicity. We have thus added to the continuous spectrum (1) any eigenvalues embedded in or at the edges of the continuous spectrum (2) any limit points of the spectrum, and (3) eigenvalues, if any, of infinite multiplicity. By examining the various cases, it is seen that the points of the essential spectrum can be characterized by approximate eigenvectors (possibly including true eigenvectors) as follows: λ is in the essential spectrum of an operator H if and only if there is a sequence $\{v_j\}_1^\infty$ of linearly independent (or, if you prefer, mutually orthogonal) unit vectors such that $\|Hv_j - \lambda v_j\| \to 0$ as $j \to \infty$.

Now consider the one-electron Hamiltonian $H = H_0 + V$ discussed in the preceding section, where H_0 is the self-adjoint version of the operator $-\frac{1}{2}\nabla^2$ in $\mathbb{R}^3$ and $V(\mathbf{x})$ is the sum of two functions, one bounded and one in $L^2(\mathbb{R}^3)$.

Theorem (Kato). *Under the conditions of theorem 1 of the preceding section, if also $V(\mathbf{x}) \to 0$ as $|\mathbf{x}| \to \infty$, then the essential spectrum of $H = H_0 + V$ is the same as that of $H_0 = -\frac{1}{2}\nabla^2$, namely $[0, \infty)$.*

From the definition of the essential spectrum, it follows that the negative-energy spectrum ($\lambda < 0$) of $H_0 + V$ consists solely of isolated energy levels of finite multiplicity with no accumulation point except possibly at $\lambda = 0$. That is true not only for a hydrogen-like atom but for an electron in any potential $V(\mathbf{x})$ that $\to 0$ as $|\mathbf{x}| \to \infty$. On the other hand, if $V(\mathbf{x})$ is periodic (potential of an electron in a crystal lattice), there can be intervals of continuous spectrum at negative energies, even if the average potential is nonnegative.

For the n-electron atomic Hamiltonian, where the perturbation $V = V(\mathbf{x}_1, \ldots, \mathbf{x}_n)$ is given by (11.3-3), things are more complicated. If an electron is removed to large distances, the remaining ion can be in a bound negative energy state, hence the continuous spectrum is expected to go down to negative values of λ. Furthermore, the Pauli principle requires that the Hamiltonian be restricted to a subspace of the Hilbert space $L^2(\mathbb{R}^{3n})$ consisting of functions antisymmetric with respect to permutation of the electrons. According to Zislin and Sigalov 1965 the essential spectrum of $H_0 + V$, so restricted, is $[\mu, \infty)$, where μ is the lowest (i.e., ground-state) energy of the ion. It is also of interest to restrict the subspace of $L^2(\mathbb{R}^{3n})$ still further by symmetries of the Hamiltonian corresponding to other exactly conserved quantities, for example the total angular momentum, with eventual inclusion of electron spin. For a detailed discussion of these questions, see Jörgens and Weidmann 1973.

Theorems of the above kind fail to give a completely satisfactory characterization of the spectrum, for the following reason: It is possible to define a self-adjoint operator whose eigenvalues are a countable dense set in an interval I (finite or infinite) and whose eigenvectors form a complete set. Clearly that is not what one normally means by a "continuous spectrum," since for example no "continuous spectrum eigenfunctions" are needed for an eigenfunction expansion and the spectral projector E_t is not continuous in t at any point of I. Nevertheless, the entire interval I is essential spectrum. (Some of it is continuous spectrum; see next section.) The theorems referred to do not exclude the possibility of the essential spectrum of a Schrödinger operator being of that kind. Furthermore, a theorem of Weyl and von Neumann says that a purely continuous spectrum (one in which E_t is continuous) can be converted into a spectrum of the kind described by an arbitrarily small relatively compact perturbation (in fact by a perturbation V of Hilbert–Schmidt type with arbitrarily small Hilbert–Schmidt norm—see next chapter).

Even if E_t is continuous, the spectrum may still be lumpy, in a sense. It is recalled that any nondecreasing function $f(t)$ (or any function of locally bounded variation) can be decomposed as

$$f(t) = f_1(t) + f_2(t) + f_3(t), \tag{11.4-2}$$

where f_1 is a pure jump function, f_2 is absolutely continuous, and f_3 is singular continuous; f_2 is equal to the Lebesgue integral of its derivative, and the derivative of f_3 is $=0$ for almost all t. (See Chapter 13: the Cantor function is of type f_3.) In an interval in which f_1 and f_3 are constant, f is absolutely continuous. Now let $\{E_t\}$ be the resolution of the identity of a self-adjoint operator H. For any v in the Hilbert space, $(v, E_t v)$ is a non-decreasing function of t, hence has a decomposition (11.4-2). The jumps of f_1 occur at the eigenvalues of H. The spectrum of H is called *absolutely continuous* in an interval I if $(v, E_t v)$ is absolutely continuous in I for every v in the Hilbert space; otherwise, it is *lumpy*.

It seems reasonable to conjecture that the spectra of the Hamiltonians of atoms and molecules are always absolutely continuous, apart from eigenvalues, in other words that the decomposition of $(v, E_t v)$ is always of the form of the first two terms of (11.4-2); however, that has not been proved, except in a few cases like the hydrogen atom, for which an explicit formula for E_t is known.

One would like to be able to say that for an atom there are no eigenvalues in the continuous spectrum, i.e. above the ionization limit, but that is not true unless electron spin is taken into account. For example, if spin is ignored, there are bound states (so-called quartet states) of the lithium atom (Li) that lie above the ground state of Li^+; if spin-orbit and spin-spin coupling are taken into account, such states are found to be unstable, because of so-called autoionization, a spontaneous transition to Li^+ plus a free electron. Hence, there are no true eigenvalues of the full Hamiltonian for energy λ above the ionization limit. Whether that is always true appears to be an open question.

EXERCISE

1. Show that if T is a symmetric operator with deficiency indices (m, m), $m < \infty$, then all self-adjoint extensions of T have the same essential spectrum.

11.5 Continuous Spectrum in the Sense of Hilbert; Continuous and Absolutely Continuous Subspaces

A self-adjoint operator A is said to have a *pure point spectrum* (*in the sense of Hilbert*) if its eigenvectors span $\mathfrak{H}$. In this case we have no need for a "continuous spectrum"; however, any λ_0 that is a limit point of the point spectrum $P\sigma(A)$, but is not itself in $P\sigma(A)$, is in $C\sigma(A)$ according to the definition given in Section 8.1. To see that, suppose that λ_n ($n = 1, 2, \ldots$) are eigenvalues and $\rightarrow \lambda_0$ (as $n \rightarrow \infty$); for each λ_n, let v_n be a corresponding eigenvector with $\|v_n\| = 1$. Then $\|(A - \lambda_0)v_n\| = |\lambda_n - \lambda_0| \|v_n\| \rightarrow 0$ (as $n \rightarrow \infty$), hence $(A - \lambda_0)^{-1}$ is unbounded; hence, since the residual spectrum is empty, $\lambda_0 \in C\sigma(A)$.

In particular, in the example mentioned in the preceding section of an operator with a pure point spectrum and eigenvalues dense in an interval, every point of the interval that is not an eigenvalue is in the continuous spectrum.

Superfluous points of that kind are avoided by an alternative definition of continuous spectrum for the special case of a self-adjoint operator in a separable Hilbert space attributed to Hilbert by Riesz and Nagy. Since $\mathfrak{H}$ is separable, $P\sigma(A)$ is a finite or countable set $\{\lambda_j\}$ of eigenvalues. For each j, let P_j be the projector onto the jth eigenspace $\mathfrak{E}_j = \mathfrak{N}(A - \lambda_j)$, i.e. the projector whose range is $\mathfrak{E}_j$. Then $\sum_{(j)} P_j$ is a projector (see Exercise 1 below) whose range is the orthogonal direct sum of all the eigenspaces. Subspaces $\mathfrak{H}_p$ and $\mathfrak{H}_c$ of $\mathfrak{H}$ are defined as follows:

$$\mathfrak{H}_p = \mathfrak{R}(\textstyle\sum P_j), \qquad \mathfrak{H}_c = \mathfrak{H}_p^{\perp}; \qquad\qquad (11.5\text{-}1)$$

they are associated with the point and continuous parts of the spectrum of A. $\mathfrak{H}_p$ is invariant under the transformation $v \to Av$, because every v in $\mathfrak{H}_p$ is a linear combination of eigenvectors. $\mathfrak{H}_c$ is also invariant, because if u is in $\mathfrak{H}_c$, that is if $(u, v) = 0$ for every v in $\mathfrak{H}_p$, then $(Au, v) = (u, Av)$ is $= 0$ for every v in $\mathfrak{H}_p$ because Av is also in $\mathfrak{H}_p$, hence Au is in $\mathfrak{H}_c$. Operators A_p and A_c are defined as the restrictions

$$A_p = A\Big|_{\mathfrak{H}_p}, \qquad A_c = A\Big|_{\mathfrak{H}_c}; \qquad\qquad (11.5\text{-}2)$$

they are self-adjoint operators in their respective subspaces; the first has a pure point spectrum (in the sense of Hilbert), and the second a pure continuous spectrum. (If A_c had any eigenvectors, they would also be eigenvectors of A, which gives a contradiction, because the eigenvectors of A all lie in $\mathfrak{H}_p$.) *The continuous spectrum of A, in the sense of Hilbert*, denoted by $HC\sigma(A)$, is defined as $C\sigma(A_c)$. That is, $\lambda \in HC\sigma(A)$ if $(A_c - \lambda)^{-1}$ is an unbounded operator in $\mathfrak{H}_c$. The concept of approximate eigenvector, introduced in Section 8.1, can now be sharpened:

Lemma. *λ_0 is in $HC\sigma(A)$ if and only if there is a sequence $\{u_n\}$ such that $\|u_n\| = 1$, while $\|(A - \lambda_0)u_n\| \to 0$ (as $n \to \infty$) and such that each u_n is orthogonal to every eigenvector of A. Furthermore, $\{u_n\}$ can be chosen as an orthonormal sequence.*

PROOF. The "if" part is obvious, because $(A_c - \lambda_0)^{-1}$ is clearly unbounded under the assumptions made. Therefore, we assume $\lambda_0 \in HC\sigma(A)$, and we shall prove the existence of the sequence $\{u_n\}$ referred to. Let $E_\lambda(A_c)$ be the spectral family of the operator A_c in $\mathfrak{H}_c$. It is strongly continuous with respect to λ, because A_c has no point spectrum. Then there is either an ascending sequence $\{\lambda_n\}$ such that $\lambda_n \uparrow \lambda_0$ or a descending sequence $\{\lambda_n\}$ such that $\lambda_n \downarrow \lambda_0$ and furthermore such that the projectors $E_{\lambda_n}(A_c)$ are all different, for otherwise λ_0 would be in an interval of constancy of $E_\lambda(A_c)$, hence would be in the resolvent set of A_c. Suppose $\lambda_n \uparrow \lambda_0$ (the other case is similar). Let $\{u_n\}$ be a sequence of normalized vectors in $\mathfrak{H}_c$ such that u_n is in the range of the projector

$$P_n \overset{\text{def}}{=} E_{\lambda_{n+1}}(A_c) - E_{\lambda_n}(A_c).$$

Then the u_n are pairwise orthogonal, because $P_n P_m = 0$ for $n \neq m$. Since $E_\lambda(A_c)u_n$ is $= u_n$ for $\lambda > \lambda_{n+1}$ and is $= 0$ for $\lambda < \lambda_n$,

$$\|(A - \lambda_0)u_n\| = \|(A_c - \lambda_0)u_n\| = \left\| \int_{\lambda_n}^{\lambda_{n+1}} (\lambda - \lambda_0) dE_\lambda(A_c)u_n \right\| \leq |\lambda_n - \lambda_0| \|u_n\|$$

$$= |\lambda_n - \lambda_0|.$$

Hence $\|(A - \lambda_0)u_n\| \to 0$ (as $n \to \infty$). Lastly, the u_n are in $\mathfrak{H}_c$, hence are orthogonal to all eigenvectors of A, as claimed.

EXERCISES

1. Show that if $P_j (j = 1, 2, \ldots)$ are mutually orthogonal projectors (i.e., $P_j P_k = P_j \delta_{jk}$ and $P_j^* = P_j$), then $\sum_{j=1}^n P_j$ converges strongly (as $n \to \infty$) to a projector P_0, and the range of P_0 is the orthogonal direct sum of the ranges of the P_j.

2. Show that if A is a self-adjoint operator in a separable Hilbert space, then $HC\sigma(A)$ is a closed set on the real line with no isolated points (i.e., a *perfect* set).

Notes.

1. The spectrum of A (i.e., the complement of $\rho(A)$) is not necessarily the union of $P\sigma(A)$ and $HC\sigma(A)$ but it is the closure of that union.
2. It is possible for a point to be both in $P\sigma(A)$ and in $HC\sigma(A)$.
3. The definition of continuous spectrum in the sense of Hilbert can be extended to normal operators in a separable Hilbert space, but, for non-normal operators and for operators in a general Banach space, the definition of Section 8.1 is still needed.

The subspace $\mathfrak{H}_c$ can be further decomposed. We define $\mathfrak{H}_{ac}$ as the set of all v in $\mathfrak{H}$ such that the function $(v, E_t v) = \|E_t v\|^2$ is absolutely continuous with respect to t in $(-\infty, \infty)$ and $\mathfrak{H}_{sc}$ as the set such that $(v, E_t v)$ is singular continuous. It can be proved (Kato 1966, Section X.1.2) that $\mathfrak{H}_{ac}$ and $\mathfrak{H}_{sc}$ are mutually orthogonal closed linear manifolds (subspaces) and span $\mathfrak{H}_c$. Therefore,

$$\mathfrak{H} = \mathfrak{H}_p \oplus \mathfrak{H}_{ac} \oplus \mathfrak{H}_{sc}. \tag{11.5-3}$$

Furthermore, $\mathfrak{H}_{ac}$ and $\mathfrak{H}_{sc}$ are invariant under the transformation $v \to Av$, and the operators

$$A_{ac} = A\Big|_{\mathfrak{H}_{ac}}, \quad A_{sc} = A\Big|_{\mathfrak{H}_{sc}} \tag{11.5-4}$$

have absolutely continuous and singular continuous spectra respectively. If P_p, P_{ac}, and P_{sc} are the orthogonal projectors corresponding to (11.5-3), then the decompositions

$$E_t = P_p E_t + P_{ac} E_t + P_{sc} E_t \tag{11.5-5}$$

is fully analogous to the decomposition (11.4-2); (11.4-2) applies to a real-valued nondecreasing function $f(t)$ and (11.5-5) to the projector-valued nondecreasing function E_t.

An interesting characterization of the subspace $\mathfrak{H}_{ac}$ in terms of the resolvent $R_\lambda = (A - \lambda)^{-1}$ was given by Gustafson and Johnson 1974. Recall that if λ_0 is in the resolvent set, R_λ is continuous (in fact analytic) at λ_0. If λ_0 is an isolated eigenvalue, R_λ has a pole at λ_0; since A is self-adjoint, λ_0 is real, and the pole is simple. It follows easily that if v is a vector in $\mathfrak{H}_p$, i.e. a linear combination of eigenvectors, then $\|R_\lambda v\|$ becomes infinite like const. $|\operatorname{Im} \lambda|^{-1}$ as $\operatorname{Im} \lambda \to 0$ for some value of $\operatorname{Re} \lambda$ in the point spectrum. Ony may suppose that if v is in $\mathfrak{H}_c$, then $\|R_\lambda v\|$ becomes infinite less rapidly, although just how rapidly might depend on whether v is in $\mathfrak{H}_{ac}$ or not. Consider vectors v for which there is a constant $M(v)$ such that

$$\|R_\lambda v\| \leq M(v)|\operatorname{Im} \lambda|^{-1/2} \quad \text{for all } \lambda. \tag{11.5-6}$$

Gustafson and Johnson showed that any such v is in $\mathfrak{H}_{ac}$, and in fact $\mathfrak{H}_{ac}$ is the closure of the set of all such v.

An example of this behavior appeared in Exercise 2 in Section 10.1 and shows that the spectrum of the operator there considered is absolutely continuous.

11.6 Dirac Hamiltonians

The discussion of the Dirac relativistic Hamiltonians is perforce restricted to the case of one electron in a specified potential, because in relativity the Coulomb interaction between two electrons has to be replaced by the interaction via the electromagnetic field, hence the discussion of the two-or-more-electron case would have to proceed in the framework of quantum electrodynamics.

The operators for the free particle and for hydrogen-like atoms will be discussed briefly. The Hilbert space $\mathfrak{H}$ is the space $(L^2(\mathbb{R}^3))^4$ of 4-component wave functions $\psi = (\psi_1, \psi_2, \psi_3, \psi_4)$, each component of which is a distribution in $L^2(\mathbb{R}^3)$. The Hamiltonian of a free particle is formally

$$-i\hbar c\boldsymbol{\alpha} \cdot \nabla + \alpha_4 mc^2, \tag{11.6-1}$$

where $\boldsymbol{\alpha} = (\alpha_1, \alpha_2, \alpha_3)$, and where the α_i $(i = 1, \ldots, 4)$ are 4×4 Hermitian anticommuting matrices of unit square:

$$\alpha_i\alpha_j + \alpha_j\alpha_i = 0 \quad (i \neq j = 1, \ldots, 4), \tag{11.6-2}$$

$$\alpha_i^2 = 1 \quad (i = 1, \ldots, 4). \tag{11.6-3}$$

See Schiff (1955). On the right of equation (11.6-3), the symbol 1 denotes the 4×4 unit matrix. For any ψ in $\mathfrak{H}$, $\boldsymbol{\alpha} \cdot \nabla\psi$ is well defined as a (4-component) distribution. If the domain of the Hamiltonian H_0 is restricted so that $\boldsymbol{\alpha} \cdot \nabla\psi$ is also in L^2, i.e., by writing

$$\mathfrak{D}(H_0) = \{\psi \in \mathfrak{H}: \boldsymbol{\alpha} \cdot \nabla\psi \in \mathfrak{H}\},$$

$$H_0\psi = (-i\hbar c\boldsymbol{\alpha} \cdot \nabla + \alpha_4 mc^2)\psi, \tag{11.6-4}$$

then H_0 is self-adjoint, as is easily seen by transforming to the momentum representation by a Fourier transformation, whereupon H_0 becomes an operator $\hat{H}_0$ of multiplication by a Hermitian-matrix-valued function, so that an argument like the one used for the Laplacian in Section 11.1 applies.

If the Coulomb potential $-Ze^2/r$ is added, the result is the relativistic hydrogen-like atom Hamiltonian

$$-i\hbar c\alpha \cdot \nabla + \alpha_4 mc^2 - \frac{Ze^2}{r}, \qquad r = |\mathbf{x}|. \tag{11.6-5}$$

As in the nonrelativistic case, it can be shown that ψ/r is in $\mathfrak{H} = (L^2(\mathbb{R}^3))^4$ if ψ is in the domain of the free-particle Hamiltonian, which is now given by (11.6-4). Hence, in analogy with the nonrelativistic case (Theorem 1 of Section 11.3), one might conjecture that the Hamiltonian (11.6-5) is self-adjoint on the domain $\mathfrak{D}(H_0)$.

The question of the self-adjoint versions of the operator (11.6-5) is discussed below. It turns out that the above conjecture is correct for $Z \le 118$, but must be slightly modified for higher values of Z.

First, however, we outline the separation of variables, which leads to the radial equation system already discussed in Section 10.17.

The stationary state equation is

$$[-i\hbar c\alpha \cdot \nabla + \alpha_4 E_0 + V(r)]\psi = E\psi \tag{11.6-6}$$

for an electron in a central force field with potential $V(r)$; E_0 stands for mc^2. Dirac's 4×4 matrix representation of the matrices α_i is given by first expressing the α_i in terms of 2×2 matrices σ_i as follows:

$$\alpha_i = \begin{pmatrix} 0 & \sigma_i \\ \sigma_i & 0 \end{pmatrix}, \qquad \alpha_4 = \begin{pmatrix} I & 0 \\ 0 & -I \end{pmatrix}, \tag{11.6-7}$$

where I is the 2×2 unit matrix and the σ_i are the so-called Pauli spin matrices:

$$\sigma_1 = \begin{pmatrix} 0 & 1 \\ 1 & 0 \end{pmatrix}, \qquad \sigma_2 = \begin{pmatrix} 0 & -i \\ i & 0 \end{pmatrix}, \qquad \sigma_3 = \begin{pmatrix} 1 & 0 \\ 0 & -1 \end{pmatrix}. \tag{11.6-8}$$

Hence,

$$\alpha \cdot \nabla = \begin{pmatrix} 0 & T \\ T & 0 \end{pmatrix}, \qquad T = \begin{pmatrix} \partial_z & \partial_x - i\partial_y \\ \partial_x + i\partial_y & -\partial_z \end{pmatrix}, \tag{11.6-9}$$

where ∂_z is an abbreviation for $\partial/\partial z$, and similarly for ∂_x and ∂_y.

The details of the separation procedure are given in Bethe and Salpeter (1957); the result is the following: One introduces quantum numbers l and j; l is the orbital angular momentum quantum number and is an integer ≥ 0; j is the total angular momentum quantum number and can assume just the two values $l + \frac{1}{2}$ and $l - \frac{1}{2}$, (but only $+\frac{1}{2}$ for $l = 0$). The forms assumed by the four components of ψ are

$$[j = l + \tfrac{1}{2}] \qquad\qquad\qquad [j = l - \tfrac{1}{2}]$$

$$\psi_1 = \sqrt{\frac{l + m + 1}{2l + 1}}\, g Y_l^m \qquad\qquad \psi_1 = \sqrt{\frac{l - m}{2l + 1}}\, g Y_l^m$$

$$\psi_2 = -\sqrt{\frac{l - m}{2l + 1}}\, g Y_l^{m+1} \qquad\qquad \psi_2 = \sqrt{\frac{l + m + 1}{2l + 1}}\, g Y_l^{m+1} \qquad (11.6\text{-}10)$$

$$\psi_3 = -i\sqrt{\frac{l - m + 1}{2l + 3}}\, f Y_{l+1}^m \qquad\qquad \psi_3 = -i\sqrt{\frac{l + m}{2l - 1}}\, f Y_{l-1}^m$$

$$\psi_4 = -i\sqrt{\frac{l + m + 2}{2l + 3}}\, f Y_{l+1}^{m+1} \qquad\qquad \psi_4 = i\sqrt{\frac{l - m - 1}{2l - 1}}\, f Y_{l-1}^{m+1},$$

where f and g are functions of r only, and where $Y_l^m(\theta, \varphi)$ is the normalized tesseral spherical harmonic given by

$$Y_l^m(\theta, \varphi) = \sqrt{\frac{2l + 1}{4\pi}\frac{(l - m)!}{(l + m)!}}\, P_l^m(\cos \theta)e^{im\varphi}$$

$$(11.6\text{-}11)$$

$$(l = 0, 1, \ldots)$$

$$(m = -l, -l + 1, \ldots, l)$$

In each column of the table, m is an integer such that $-j \le m + \frac{1}{2} \le j$; $(m + \frac{1}{2})\hbar$ is the z-component of the total angular momentum. If the functions $\psi_j (j = 1, \ldots, 4)$ from either column of (11.6-11) are substituted into (11.6-6), and if one uses the formulas given in Bethe and Salpeter for the derivatives of a function of the form $h(r)Y_l^m(\theta, \varphi)$ with respect to x, y, and z, one finds a coupled pair of first order ordinary differential equations for $f(r)$ and $g(r)$. The two cases $j = l \pm \frac{1}{2}$ can be combined by introducing a new integer quantum number k given by

$$k = -l - 1 \quad \text{for } j = l + \tfrac{1}{2} \quad (l = 0, 1, \ldots)$$

$$k = l \quad\qquad \text{for } j = l - \tfrac{1}{2} \quad (l = 1, 2, \ldots).$$

Then the coupled equations are

$$\frac{1}{\hbar c}[E + E_0 - V(r)]f(r) - \left[g'(r) + \frac{1 + k}{r}\, g(r)\right] = 0$$

$$(11.6\text{-}12)$$

$$\frac{1}{\hbar c}[E - E_0 - V(r)]g(r) + \left[f'(r) + \frac{1 - k}{r}\, f(r)\right] = 0.$$

If f and g satisfy (11.6-12), for given E, then the functions (11.6-10) are the components of an eigenfunction ψ of H, and E is the corresponding eigenvalue, if and only if ψ is in the domain of H, which has not yet been specified except by way of conjecture, but in any case only if the quantity $\sum_{j=1}^{4} |\psi_j|^2$ has a finite integral over $\mathbb{R}^3$, i.e., only if

$$\int_0^\infty (|f|^2 + |g|^2) r^2 \, dr < \infty. \tag{11.6-13}$$

The system (11.6-12) was discussed in Section 10.17 for the case $V(r) = -Ze^2/r$. When f was eliminated, a second order equation for g was obtained in formally self-adjoint form. (Elimination of g gives the same equation for f.) Although the equation was not in Sturm–Liouville form because of the complicated way in which the eigenvalue parameter $\lambda = E$ occurred, it was found that some of the notions of the Sturm–Liouville theory apply. The interval of r is $(0, \infty)$, and it was found that the endpoint at $r = \infty$ was always in the limit point case. That is, for no real or complex E is more than one independent solution of (11.6-12) such that the integral (11.6-13) converges at the upper limit. In fact, f and g behave asymptotically, as $r \to \infty$, either both like $e^{\mu r}$ or both like $e^{-\mu r}$, where $\mu = (E_0^2 - E^2)^{1/2}/\hbar c$. The endpoint at $r = 0$ is a regular singular point, hence can be analyzed by the method by Frobenius. The result depends on the integer $k = \pm 1, \pm 2, \ldots$ and the dimensionless parameter

$$\alpha = \alpha_0 Z = \frac{e^2}{\hbar c} Z \approx \frac{1}{137.037} Z \tag{11.6-14}$$

(which should not be confused with matrices $\alpha_1, \ldots, \alpha_4$). The exponent γ of the power–series solution

$$g(r) = \sum_{n=0}^{\infty} a_n r^{n+\gamma}$$

was found from the indicial equation to have the values

$$\gamma = -1 \pm \sqrt{k^2 - \alpha^2}.$$

The integral (11.6-13) converges at the lower limit for $\gamma > -\frac{3}{2}$ but not for $\gamma < -\frac{3}{2}$. Hence, we find that the endpoint $r = 0$ is in the limit-circle case if $k^2 = 1$ and $\sqrt{\frac{3}{4}} \le \alpha < 1$, so that we then need a boundary condition at $r = 0$, but is otherwise in the limit-point case. (We do not consider $\alpha \ge 1$.) The need for a further condition in problems like this one was pointed out by Case 1950.

We take as auxiliary condition that

$$\iiint \frac{1}{r} \sum_{j=1}^{4} |\psi_j(\mathbf{x})|^2 \, d^3\mathbf{x} < \infty, \tag{11.6-15}$$

i.e., that we should accept only those states ψ for which the expectation of the potential energy is finite. (Then of course the kinetic energy also has a finite expectation in a stationary state because the total energy E has a precise value.) The integral (11.6-15) converges for $\gamma > -1$ but not for $\gamma < -1$, hence this condition is of just the right kind to select one of the solutions of (11.6-12) in the limit-circle case and to have no effect in the limit-point case.

It seems therefore natural to suppose that, although Sturm–Liouville theory doesn't apply the radial operator in (11.6-12) is always self-adjoint for $0 \leq \alpha < 1$ on a maximal domain subject only to the restriction (11.6-15).

In the nonrelativistic case, the spurious solutions of the radial equation that appeared when the endpoint $r = 0$ was in the limit-circle case were ruled out when the full Hamiltonian was considered, because those solutions were not in the domain of the Laplacian. A similar thing happens here, but, to our embarrassment, it seems to go too far. We ask, for what values of α (i.e., of Z) are the solutions ψ obtained as above in the domain of the unperturbed Hamiltonian H_0 given by (11.6-4). Near the origin ($r = 0$), each component of ψ is r^γ times a tesseral harmonic, according to (11.6-10). We must apply the operator H_0, with the derivatives interpreted in the distribution sense. It is easily seen that for $\gamma > -2$ the differentiations do not introduce any delta-function-like contributions, as in (11.3-5), hence the components of $H_0\psi$ are simply functions of the form $r^{\gamma-1}$ times angular factors near the origin, hence the requirement that $H_0\psi$ be in L^2 is that the integral

$$\int r^{2\gamma-2} r^2 \, dr \tag{11.6-16}$$

converge at $r = 0$, i.e. that γ be $> -\frac{1}{2}$. Unfortunately, for $\alpha \geq \sqrt{\frac{3}{4}}$ (i.e. $Z > 118$), that requirement excludes both solutions of the radial equations (11.6-12) obtained by the method of Frobenius. We conclude that, for $Z > 118$, $H_0 + V$ cannot be self-adjoint on the domain of H_0, but requires a larger domain, subject however to the condition (11.6-15).

We now summarize the information on the full Hamiltonian (11.6-5) as given in Kato 1966, Weidmann 1971, Rejto 1971, and Gustafson and Rejto 1973, after a preliminary remark on the domain of the potential energy operator V.

First, if ψ is in the domain $\mathfrak{D}(H_0)$ of the free-particle Hamiltonian given by (11.6-4), it follows that for each component ψ_j of ψ, $|\nabla\psi_j|^2$ is integrable over $\mathbb{R}^3$. Then the well-known inequality

$$\iiint \frac{1}{r^2} |u|^2 \, d^3\mathbf{x} \leq 4 \iiint |\nabla u|^2 \, d^3\mathbf{x}, \tag{11.6-17}$$

which holds whenever the integral on the right converges, shows that ψ/r is in L^2. Hence the operator V of multiplication by $-Ze^2/r$ is well defined on the domain of H_0. Kato showed that for $\alpha < \frac{1}{2}$, V is H_0-bounded with H_0-bound < 1. It follows from Theorem 1 of Section 11.4 that for $\alpha < \frac{1}{2} (Z \leq 68)$ the operator $H = H_0 + V$ is self-adjoint with $\mathfrak{D}(H)$ taken $= \mathfrak{D}(H_0)$.

In subsequent work, it was shown first that for $\alpha < \sqrt{\frac{3}{4}}$ $(Z \leq 118)$ the minimal operator $H_0 + V$ with domain taken as $C_0^\infty(\mathbb{R}^3)^4$ is essentially self-adjoint, which is all that is needed for many purposes, but then it was shown later that the domain of the self-adjoint version, that is, of the closure of the minimal operator, is the same as the domain of H_0, given by (11.6-4), which can also be characterized as $(W^1)^4$, where W^1 is the Sobolev space $W^1(\mathbb{R}^3)$ described in Section 5.11.

For $\sqrt{\frac{3}{4}} \leq \alpha < 1$, as noted earlier, $H_0 + V$ is not essentially self-adjoint on the domain of H_0, but has deficiency indices $(1, 1)$, hence needs a larger domain. According to K. Gustafson (private communication), the minimal operator $H_0 + V$ becomes essentially self-adjoint if the domain $C_0^\infty(\mathbb{R}^3)^4$ is enlarged by allowing the functions to become infinite like $1/r$ (but no faster) as $r \to 0$. That is in agreement with our findings concerning the radial equations, and suggests the conjecture that for the full range $0 \leq \alpha < 1$, if the Dirac Hamiltonian H for a hydrogen-like atom is defined by

$$\mathfrak{D}(H) = \{\psi \in \mathfrak{H}: T\psi \in \mathfrak{H}, (11.6\text{-}15) \text{ holds}\},$$

$$H\psi = T\varphi,$$

where T is the formal operator given by (11.6-5), then H is self-adjoint; (11.6-15) is the condition that the expectation of the potential energy be finite in the state ψ.

EXERCISES

1. Prove the inequality (11.6-17). *Hint*: First show that if $f(r)$ is real and of class C^1 and vanishes for large r, then

$$\int_0^\infty f(r)^2 \, dr \leq 4 \int_0^\infty r^2 f'(r)^2 \, dr.$$

2. Let A and B be operators in $L^2(0, 1)$ given by

$$\mathfrak{D}(A) = \{f \in L^2: f'''' \in L^2, f(0) = f'(0) = f(1) = f'(1) = 0\}$$

$$Af = f''''$$

$$\mathfrak{D}(B) = \mathfrak{D}(A)$$

$$Bf = -f'''' + f''$$

Show that A and B are self-adjoint and find the deficiency indices of $A + B$.

3. Show that $(\boldsymbol{\alpha} \cdot \mathbf{k} + \alpha_4)^2 = (k^2 + 1)I$, where I is the 4×4 unit matrix.

4. Verify that the 4×4 matrices $\alpha_1, \ldots, \alpha_4$ given by (11.6-7 and 8) satisfy (11.6-2 and 3).

5. Show that for $\sqrt{\frac{3}{4}} < \alpha < 1$ and $k = \pm 1$ the deficiency indices of the radial operator given by (11.6-12) are $(1, 1)$.

11.7 The Laplacian in a Bounded Region

Now suppose that Ω is a bounded region (connected open set) in $\mathbb{R}^3$ whose boundary $\partial\Omega$ consists of finitely many piecewise smooth surfaces and satisfies the external cone condition of Section 6.4 (the reason for choosing $n = 3$

will soon be apparent). An operator A_0 is defined as minus the Laplacian acting on sufficiently smooth functions $f(x)$ that vanish on the boundary:

$$\mathfrak{D}(A_0) = \{f \in C(\tilde{\Omega}): \nabla^2 f \in C(\Omega), \text{ and } f = 0 \text{ on } \partial\Omega\},$$
$$A_0 f = -\nabla^2 f, \quad \text{for } f \in \mathfrak{D}(A_0), \tag{11.7-1}$$

where $\tilde{\Omega}$ denotes the closure of Ω. Green's formula

$$\int_\Omega (f\nabla^2 g - g\nabla^2 f)d\tau = \int_{\partial\Omega} (f\nabla g - g\nabla f) \cdot \mathbf{n}\, d\mathscr{A},$$

where $\mathbf{n} = \mathbf{n}(x)$ is the outward normal to $\partial\Omega$ at x, shows that A_0 is a symmetric operator. It will be shown that A_0 is essentially self-adjoint in the Hilbert space $L^2(\Omega)$ by showing that $+i$ and $-i$ are in the resolvent set (see Section 8.6), and that A_0 has a pure point spectrum.

According to potential theory (see Section 6.4), there is associated with Ω a Green's function $G(x, x')$, and the solution of the problem

$$-\nabla^2 u = 4\pi\rho \quad \text{in } \Omega$$

$$u = 0 \quad \text{on } \partial\Omega \tag{11.7-2}$$

$$u \text{ continuous in } \tilde{\Omega},$$

for given $\rho = \rho(x)$ continuous in $\tilde{\Omega}$, is

$$u(x) = \int_\Omega G(x, x')\rho(x')d\tau'. \tag{11.7-3}$$

For x' in Ω, $G(x, x')$ vanishes for x on $\partial\Omega$, and

$$G(x, x') = \frac{1}{|x - x'|} + g(x, x'),$$

where g is a continuous function. The singularity of G is mild enough so that

$$M \overset{\text{def}}{=} \sup_{(x)} \int_\Omega G(x, x')^2\, d\tau' < \infty, \tag{11.7-4}$$

see Exercise 1 below. [This is not true in more than 3 dimensions; on the other hand, the 2-dimensional case is similar to the present one; in it, G has a logarithmic singularity.]

In the domain $\mathfrak{D}(G_0)$, defined as $C(\Omega)$, the equation

$$(G_0 f)(x) = \int_\Omega G(x, x') f(x')d\tau'$$

defines a bounded operator, because the Schwarz inequality shows that

$$\left| \int_\Omega G(x, x') f(x')d\tau \right|^2 \leq M\|f\|^2,$$

and a further integration over Ω, with respect to x, shows that

$$\|G_0 f\|^2 \leq M\|f\|^2 \times \text{volume}(\Omega).$$

Furthermore, the Green's function satisfies the equation $G(\mathbf{x}, \mathbf{x}') = G(\mathbf{x}, \mathbf{x}')$; hence, G_0 is a symmetric operator. Since it is also bounded and defined on a domain dense in L^2, its closure, which will be called G, is self-adjoint.

The resolvent of A_0 can be expressed in terms of G: Let g be any function continuous in $\tilde{\Omega}$, and let λ be any nonreal number. Then the equation

$$Gf - \frac{4\pi}{\lambda} f = -\frac{1}{\lambda} Gg \qquad (11.7\text{-}5)$$

has a unique solution f in L^2, because $4\pi/\lambda$ is nonreal, hence is in the resolvent set of G. By Exercise 2 below, Gf and Gg are continuous, hence f is continuous; by (11.7-2), then,

$$-\nabla^2 \frac{4\pi}{\lambda} f = 4\pi \left(f + \frac{1}{\lambda} g \right). \qquad (11.7\text{-}6)$$

This equation shows that $\nabla^2 f$ is a continuous function. Furthermore, since Gf and Gg vanish on the boundary, so does f by (11.7-5), and we conclude that f is in the domain of A_0. Hence (11.7-6) can be written as $A_0 f - \lambda f = g$, that is, as

$$f = (A_0 - \lambda)^{-1} g.$$

Lastly, since G and its resolvent $(G - (4\pi/\lambda))^{-1}$ are both bounded, (11.7-5) shows that $\| f \| \leq$ constant $\|g\|$, and it is concluded that any nonreal λ is in the resolvent set of A_0; hence A_0 is essentially self-adjoint.

G is a positive integral operator of the Fredholm type, and it is known (see Courant and Hilbert) that G has a pure point spectrum and that its eigenvalues $\mu_i (i = 1, 2, \ldots)$ are positive and accumulate only at 0. If $Gf_i = \mu_i f_i$, then clearly $A_0 f_i = (4\pi/\mu_i) f_i$. Furthermore, if λ is real, but $4\pi/\lambda$ is not equal to any μ_i, then (11.7-5) again has a solution, and the above argument shows that this λ is in the resolvent set of A_0.

Conclusion. A_0 has a pure point spectrum; its eigenvalues λ_i are positive, and $\lambda_i \to +\infty$, as $i \to \infty$.

EXERCISES

1. Using the maximum principle for harmonic functions, show that

$$0 \leq G(\mathbf{x}, \mathbf{x}') < \frac{1}{|\mathbf{x} - \mathbf{x}'|} \quad \text{for } \mathbf{x}, \mathbf{x}' \in \Omega.$$

Conclude that if Ω is a bounded region, there is a constant M such that

$$\int G(\mathbf{x}, \mathbf{x}')^2 \, d^3\mathbf{x}' \leq M \quad \text{for all } \mathbf{x} \text{ in } \Omega.$$

2. Let $\{\varphi_k\}$ be a Cauchy sequence of test functions which converge in $L^2 = L^2(\Omega)$ to a distribution f in L^2. Show that the functions $(G\varphi_k)(\mathbf{x})$ converge uniformly, as $k \to \infty$, hence that Gf is a continuous function.

CHAPTER 12

Compact, Hilbert–Schmidt, and Trace-Class Operators

Canonical representation of a compact operator; the spectrum; the Hilbert–Schmidt norm; the trace; the spectra of operators with compact resolvent; applications to differential and integral operators.

Prerequisite: Chapters 1–9

This chapter deals with several classes of bounded operators: the compact, Hilbert–Schmidt, trace-class and degenerate operators; they are related by the inclusions

bounded $\supset$ compact $\supset$ Hilbert–Schmidt $\supset$ trace-class $\supset$ degenerate.

The operators in each of the classes have some properties in common with finite-dimensional operators or matrices. In a finite-dimensional space the classes all coincide.

12.1 Some Properties of Matrices

According to one form of the polar decomposition formula, any $n \times n$ matrix A can be written as $= UR$, where U is unitary and R is positive semi-definite. Diagonalization of R by a further unitary matrix then shows that an arbitrary matrix A can be written as

$$A = U_1 D U_2 \qquad (12.1\text{-}1)$$

where U_1 and U_2 are unitary and D is diagonal with nonnegative diagonal elements. This is the finite-dimensional case of the standard form of a compact operator given below.

If λ is an eigenvalue of A and $\mathbf{v}$ is a vector such that for some positive integer m

$$(A - \lambda)^m \mathbf{v} = 0, \qquad (A - \lambda)^{m-1}\mathbf{v} \neq 0, \qquad (12.1\text{-}2)$$

then **v** is a *generalized eigenvector* of order m. A generalized eigenvector of order 1 is an ordinary eigenvector. The dimension of the nullspace $\mathfrak{N}(A - \lambda)$ is the *geometric multiplicity* of λ; it is the maximum number of linearly independent ordinary eigenvectors corresponding to the eigenvalue λ. The space spanned by the ordinary and generalized eigenvectors of all orders corresponding to λ is called the *(algebraic) eigenspace*; its dimension is called the *algebraic multiplicity* of λ; it is equal to the multiplicity of λ as a root of the characteristic equation $\det(A - \lambda) = 0$ and is equal to the number of occurrences of λ on the main diagonal of the Jordan normal form of A. The highest order of eigenvector for a given λ is the *index* of λ. These definitions apply to any bounded operator in a Hilbert or Banach space, λ being an isolated point of the point spectrum. If A is a normal matrix ($AA^* = A^*A$), the geometric and algebraic multiplicities are the same for each eigenvalue, the index is $= 1$, and all eigenvectors are ordinary ones.

One of the standard matrix norms (see Gantmacher 1959), often denoted by $\|\cdot\|_2$, is given by

$$\|A\|_2 = \left(\sum_{j,k=1}^{n} |A_{jk}|^2 \right)^{1/2} = (\mathrm{tr}(A^*A))^{1/2}; \qquad (12.1\text{-}3)$$

it is the finite-dimensional case of the Hilbert–Schmidt norm given below.

The sum of the eigenvalues of A, each counted by its algebraic multiplicity, is the *trace* of A, denoted by tr A; it is equal to the sum of the diagonal elements

$$\mathrm{tr}\, A = \sum_{j=1}^{n} A_{jj};$$

also tr $AB = $ tr BA, and, more generally, $\mathrm{tr}(A_1 A_2 \cdots A_k) = \mathrm{tr}(A_2 \cdots A_k A_1)$ (cyclic permutation). The trace of A is equal to $(-1)^{n-1}$ times the coefficient of λ^{n-1} in the characteristic equation. If P is nonsingular, $P^{-1}AP$ has the same eigenvalues as A, hence the same trace; that can also be seen by noting that tr $P^{-1}AP = $ tr $APP^{-1} = $ tr A.

The *rank* $r(A)$ of an $n \times n$ matrix A is the maximum number of linearly independent columns (or rows) of A. It is also the dimension of the range of A; in this form, the definition applies also to any degenerate operator.

12.2 Compact Operators

Almost everything in this chapter is valid for operators in a Banach space, or from one Banach space to another (see Kato 1966), but we shall discuss only bounded operators in a separable Hilbert space $\mathfrak{H}$. It will be evident from the definition below that only a bounded operator can be compact.

An operator A is *compact* or *completely continuous* if for every bounded sequence $\{\varphi_j\}$ of elements in $\mathfrak{H}$, the sequence $\{A\varphi_j\}$ contains a convergent subsequence. *Note*: If A were not bounded, we could find a bounded sequence $\{\varphi_i\}$ such that $\|A\varphi_i\| \to \infty$, hence A would not be compact. As will be seen in a moment, a compact operator can also be characterized as one that converts a weakly convergent sequence into a strongly convergent one.

The operator A^*A is positive semidefinite, and we define $R = (A^*A)^{1/2}$, as in Section 9.10. Since $\|Rx\| = \|Ax\|$ for all x in $\mathfrak{H}$, R is compact if A is, because if $\{A\varphi_i\}$ converges, then

$$\|R\varphi_j - R\varphi_k\| = \|A\varphi_j - A\varphi_k\|,$$

hence $\{R\varphi_j\}$ also converges.

Lemma 1. *If A is compact, R has a pure point spectrum with eigenvalues ≥ 0, either finite in number or accumulating only at the origin; the eigenvalues that are >0 have finite multiplicity (and of course index $= 1$, because R is self-adjoint).*

PROOF. If the continuous spectrum in the sense of Hilbert were not empty, it would contain a number $\lambda_0 \neq 0$ (in fact >0) since that spectrum is a perfect set (by Exercise 2 in Section 11.5). Then there would be an infinite orthonormal sequence $\{u_k\}$ such that $\|Au_k - \lambda_0 u_k\| \to 0$ as $k \to \infty$, but that would contradict the compactness of A, for the sequence $\{u_k\}$ is bounded and if it had a subsequence $\{v_k\}$ such that $\{Av_k\}$ converged, say to w, then $\lambda_0 v_k$ would $\to w$, but $(\lambda_0 v_k, x) \to 0$ for every x, since $\{v_k\}$ is orthonormal, which would imply $w = 0$, while $\|\lambda_0 v_k\| = |\lambda_0|$ for all k, which would imply $w \neq 0$. Therefore A has a pure point spectrum. Similar arguments show that the eigenvalues cannot accumulate at any $\lambda \neq 0$ and that no eigenvalue can have infinite multiplicity.

Now let $\{\varphi_j\}$ be a maximal orthonormal set of eigenvectors of R corresponding to the positive eigenvalues $\{\alpha_j\}$; that is, the $\{\varphi_j\}$ are a complete orthonormal set in $\mathfrak{N}(R)^\perp$. For each j, define $\psi_j = (1/\alpha_j)A\varphi_j$. Then,

$$(\psi_j, \psi_k) = \frac{1}{\alpha_j \alpha_k}(A\varphi_j, A\varphi_k) = \frac{1}{\alpha_j \alpha_k}(\varphi_j, R^2\varphi_k) = \frac{1}{\alpha_j \alpha_k}(\varphi_j, \alpha_k^2\varphi_k) = \delta_{jk}.$$

Hence the set $\{\psi_j\}$ is also orthonormal. If x is a linear combination of the φ_j, then

$$Ax = \sum (\varphi_j, x)\alpha_j\psi_j. \tag{12.2-1}$$

This equation is valid also if $x \in \mathfrak{N}(R) = \mathfrak{N}(A)$, for then all the (φ_j, x) are $=0$, hence it is valid for all x.

Theorem. *A is compact if and only if there are orthonormal sequences $\{\varphi_j\}$ and $\{\psi_j\}$ of the same (finite or infinite) length and a corresponding sequence $\{\alpha_j\}$ of positive numbers, which converges to zero (unless it is a finite sequence), such that A is given by (12.2-1). The adjoint of A is given by*

$$A^*y = \sum (\psi_j, y)\alpha_j\varphi_j. \tag{12.2-2}$$

PROOF. If A is compact, the existence of the sequences $\{\varphi_j\}$, $\{\psi_j\}$, and $\{\alpha_j\}$ was established above. We prove next that conversely if such sequences are given, the operator A determined by (12.2-1) is compact. To that end, let $\{x_r\}_1^\infty$ be a bounded sequence in $\mathfrak{H}$. We have to show that $\{Ax_r\}$ has a convergent subsequence. Now, $\{x_r\}$, being bounded, has a weakly convergent subsequence, by Exercise 4 in Section 1.9, which we call simply $\{x_r\}$ after renaming. We shall show that the weak convergence of $\{x_r\}$ implies the strong convergence of $\{Ax_r\}$. Since the x_r are bounded, there is an M such that

$$\sum_{(j)} |(\varphi_j, x_r)|^2 \leq M^2 \quad \text{for all } r.$$

Let $\varepsilon > 0$ be given. Choose J so that $\alpha_j < \varepsilon$ for all $j > J$; then

$$\left\| \sum_{j>J} (\varphi_j, x_r) \alpha_j \psi_j \right\| \leq M\varepsilon \quad \text{for all } r. \tag{12.2-3}$$

The x_r converge weakly. Hence with ε and J fixed, there is an R such that

$$|(\varphi_j, x_r - x_s)| \leq \frac{\varepsilon}{\sqrt{J}} (1 \leq j \leq J) \quad \text{for all } r, s > R.$$

It follows that

$$\left\| \sum_{j=1}^{J} (\varphi_j, x_r - x_s) \alpha_j \psi_j \right\| \leq \varepsilon \max(\alpha_j) \quad \text{for } r, s > R. \tag{12.2-4}$$

We conclude from (12.2-3 and 4) that $\{Ax_r\}$ is a Cauchy sequence, as required. Lastly, to establish (12.2-2), we need only observe that with A^* so defined $(y, Ax) = (A^*y, x)$ for all x and y in $\mathfrak{H}$.

Note. Even if the sequences $\{\varphi_j\}$ and $\{\psi_j\}$ are infinite, they are not necessarily complete. Furthermore, one of them may be complete and the other not.

If A is a nonsingular $n \times n$ matrix, then (12.2-1) agrees with (12.1-1), provided the φ_j are taken as the rows of U_2 and the ψ_j as the columns of U_1. (If A is singular, some of those rows and columns have to be ignored; they do not contribute to (12.1-1) because of zeros in the diagonal matrix D.)

We state without proof that the spectrum of a compact operator T is a countable set that accumulates only, if at all, at 0. Each nonzero λ in $\sigma(T)$ is an eigenvalue of finite multiplicity. If λ is an eigenvalue of T, then $\bar{\lambda}$ is an eigenvalue of T^*. (See Kato 1966, Section III.6.7).

EXERCISE

1.   Let $\{\varphi_j\}_{-\infty}^{\infty}$ be a complete orthonormal set and $\{\alpha_j\}_{-\infty}^{\infty}$ a corresponding set of positive numbers that $\to 0$ as $j \to \pm\infty$. Determine the point and continuous spectra of the operator A given by

$$Ax = \sum_{-\infty}^{\infty} (\varphi_j, x) \alpha_j \varphi_{j+1}.$$

12.3 Hilbert–Schmidt and Trace-Class Operators

If A is any bounded operator and $\{\chi_k\}$ is any complete orthonormal sequence in $\mathfrak{H}$, it is natural to think of the quantities $(\chi_k, A\chi_l)$ $(k, l = 1, 2, \ldots)$ as matrix elements of A. Hence, we should like to define the norm $\|A\|_2$ in analogy with (12.1-3) by

$$\left[\sum_{(k\,l)} |(\chi_k, A\chi_l)|^2 \right]^{1/2} \tag{12.3-1}$$

and the trace by

$$\sum_{(k)} (\chi_k, A\chi_k). \tag{12.3-2}$$

Therefore we wish to know for what operators A these sums are convergent and are independent of the choice of the sequence $\{\chi_k\}$.

First, an operator A is said to be a *Hilbert–Schmidt operator* if it is compact and if the positive numbers α_j that appear in (12.2-1) are such that

$$\|A\|_2 \stackrel{\text{def}}{=} \left(\sum \alpha_j^2\right)^{1/2} < \infty. \tag{12.3-3}$$

We investigate the sum in (12.3-1) when A satisfies this condition. Let $\{\varphi_j\}$ be extended to a complete orthonormal sequence by including eigenvectors in the nullspace $\mathfrak{N}(R)$ and correspondingly extending $\{\alpha_j\}$ by including zeros. Then

$$
\begin{aligned}
\sum_{(k\,l)} |(\chi_k, A\chi_l)|^2 &= \sum_{(l)} \|A\chi_l\|^2 \\
&= \sum_{(k\,l)} |(\varphi_k, A\chi_l)|^2 = \sum_{(k,l)} |(A^*\varphi_k, \chi_l)|^2 \\
&= \sum_{(k)} \|A^*\varphi_k\|^2 = \sum_{(k)} \left\| \sum_{(j)} (\psi_j, \varphi_k)\alpha_j\varphi_j \right\|^2 \\
&= \sum_{(k)}\sum_{(j)} \alpha_j^2 |(\psi_j, \varphi_k)|^2 = \sum_{(j)} \alpha_j^2.
\end{aligned}
\tag{12.3-4}
$$

(Note that we have rearranged series only when they are absolutely convergent.) Therefore, if A is a Hilbert–Schmidt operator, the sum in (12.3-1) is convergent and is independent of the choice of the sequence. Furthermore,

$$\|A\|_2^2 = \sum_{(l)} \|A\chi_l\|^2. \tag{12.3-5}$$

The converse is given by the following lemma, whose proof is left to Exercise 1 below.

Lemma.   *If A is any bounded operator such that $\sum \|A\chi_l\|^2$ converges for some complete orthonormal sequence $\{\chi_l\}$, then (1) the sum converges to the same value if $\{\chi_l\}$ is replaced by any other complete orthonormal sequence, and (2) A^*A has the required kind of spectrum for A to be compact (see Lemma in the preceding section), and $\sum \alpha_j^2 < \infty$ so that A is a Hilbert–Schmidt operator, hence (12.3-4 and 5) hold.*

If the more stringent condition

$$\sum \alpha_j < \infty \tag{12.3-6}$$

is also satisfied, A is of the *trace class*. We show that if A satisfies this condition, the sum (12.3-2) is independent of the choice of the sequence $\{\chi_k\}$, provided the sequence is orthonormal and complete. First,

$$\sum_{(k)} (\chi_k, A\chi_k) = \sum_{(k)} \left[\sum_{(j)} (\varphi_j, \chi_k)\alpha_j(\chi_k, \psi_j) \right]. \tag{12.3-7}$$

If condition (12.3-6) is satisfied, this double series converges absolutely, because each α_j is ≥ 0 and

$$\sum_{(k)} |(\varphi_j, \chi_k)(\chi_k, \psi_j)| \leq \left[\sum_{(k)} |(\varphi_j, \chi_k)|^2 \sum_{(k)} |(\chi_k, \psi_j)|^2\right]^{1/2} = \|\varphi_j\| \|\psi_j\| = 1,$$

hence the sum of the absolute values of the terms in (12.3-7) is $\leq \sum \alpha_j < \infty$. Therefore we can sum (12.3-7) on k first, with the result

$$\sum_{(k)} (\chi_k, A\chi_k) = \sum_{(j)} \alpha_j(\varphi_j, \psi_j) \overset{\text{def}}{=} \text{tr } A. \tag{12.3-8}$$

Operators of the trace class play a role in quantum statistics; see Chapter 14.

We note that the second sum in (12.3-8) may converge, even absolutely, for a compact operator A that is not of the trace class.

Lastly, if there are only finitely many, say r, terms in the sum (12.2-1), that is, if A^*A has only finitely many nonzero eigenvalues, each of finite multiplicity, then A is a *degenerate operator of rank* r; then, $r = \dim \mathfrak{R}(A) = \dim \mathfrak{R}(A^*)$.

EXERCISES

1.   Prove the above lemma. *Hint* for part (1): Rewrite the sum as

$$\sum_{(k,\ l)} |(\omega_k, A\chi_l)|^2,$$

where $\{\omega_k\}$ is any other complete orthonormal sequence. *Hint* for part (2): Choose the sequence $\{\omega_k\}$ suitably, using eigenvectors and approximate eigenvectors (if any) of A^*A.

2.   Show that if A is compact and B is bounded, then AB and BA are compact.

3.   Show that if A is a Hilbert–Schmidt operator and U is unitary, then A^*, $(A^*A)^{1/2}$, and U^*AU are Hilbert–Schmidt operators and

$$\|A\|_2 = \|A^*\|_2 = \|(A^*A)^{1/2}\|_2 = \|U^*AU\|_2.$$

4.   Show that if A is a Hilbert–Schmidt operator then $\|A\| \leq \|A\|_2$; also that A^*A is of the trace class, and $\|A\|_2 = \text{tr } A^*A$.

5.   Show that if A is a Hilbert–Schmidt operator and B is bounded, then AB and BA are Hilbert–Schmidt operators and

$$\|BA\|_2 \leq \|B\| \|A\|_2$$

$$\|AB\|_2 \leq \|B\| \|A\|_2$$

6.   Show that if A and B are Hilbert–Schmidt operators, then $A + B$ is, too, and

$$\|A + B\|_2 \leq \|A\|_2 + \|B\|_2.$$

7.   Show that if A and B are Hilbert–Schmidt operators, then AB and BA are in the trace class, and

$$\text{tr } AB = \text{tr } BA$$

$$|\text{tr } AB| \leq \|A\|_2 \|B\|_2$$

8.   Show that if A is of the trace class, then A^* is also, and

$$\operatorname{tr} A^* = \overline{\operatorname{tr} A}.$$

9.   Show that if A is in the trace class and B is bounded, then AB and BA are in the trace class and $\operatorname{tr} AB = \operatorname{tr} BA$. *Hint*: Write $A = UR$ by polar decomposition. Then the operators $C = UR^{1/2}$ and $D = R^{1/2}$ are Hilbert–Schmidt operators and $A = CD$.

10.   Show that if A is degenerate and B is bounded, then A^*, AB, and BA are degenerate and

$$r(A^*) = r(A); \qquad r(AB) \le \min\{r(A), r(B)\}.$$

Some of the above may be summarized in the statements:

1. $\operatorname{tr}(A_1 A_2 \cdots A_k) = \operatorname{tr}(A_2 \cdots A_k A_1)$ whenever $A_1, \ldots, A_k$ are bounded and at least one of them is of the trace class or at least two of them are Hilbert–Schmidt operators.
2.   The Hilbert–Schmidt norm fulfills all the requirements of a norm.
3.   The Hilbert–Schmidt operators constitute a Hilbert space with the inner product defined as

$$(A, B) = \operatorname{tr} A^* B = \sum (A\chi_i, B\chi_i).$$

For the last of these, the completeness of the space has to be proved. For that, either see Kato, Section V.2.4, or regard it as an additional, slightly more difficult exercise.

12.4 Hilbert–Schmidt Integral Operators

If $K(x, y)$ is continuous in the unit square in the x, y plane, then the operator K defined in $L^2(0, 1)$ by

$$(Kf)(x) = \int_0^1 K(x, y) f(y) dy$$

is the prototype of an operator of the Hilbert–Schmidt class. Operators of this kind appear as the resolvents of regular Sturm–Liouville operators, where $K(x, y)$ is a Green's function—see Section 10.6.

A considerable generalization is this: Let $\mathfrak{H}$ denote the Hilbert space $L^2(\mathbb{R}^n)$ and let $K = K(\mathbf{x}, \mathbf{y})$ be a distribution in $L^2(\mathbb{R}^{2n})$. To define the analogue of the above integral, we let $\kappa_m(\mathbf{x}, \mathbf{y})$ $(m = 1, 2, \ldots)$ be functions in $C_0^\infty(\mathbb{R}^{2n})$ that converge to $K(\mathbf{x}, \mathbf{y})$ in L^2 as $m \to \infty$. Then, for any g in $\mathfrak{H}$ we let $\psi_l(\mathbf{y})$ $(l = 1, 2, \ldots)$ be functions in $C_0^\infty(\mathbb{R}^n)$ that converge to $g(\mathbf{y})$ in L^2 as $l \to \infty$.

EXERCISES

1.   Show that the functions

$$\int \kappa_m(\mathbf{x}, \mathbf{y}) \psi_l(\mathbf{y}) d^n\mathbf{y} \qquad m, l = 1, 2, \ldots \tag{12.4-1}$$

form a Cauchy sequence in $L^2(\mathbb{R}^n) = \mathfrak{H}$. We call $f(\mathbf{x})$ the limit of the sequence, and we write

$$f(\mathbf{x}) = (Kg)(\mathbf{x}) = \int K(\mathbf{x}, \mathbf{y})g(\mathbf{y})d^n\mathbf{y}. \qquad (12.4\text{-}2)$$

2. Show that

$$\|f\| = \lim_{\substack{m \to \infty \\ l \to \infty}} \left\| \int \kappa_m(\mathbf{x}, \mathbf{y})\psi_l(\mathbf{y})d^n\mathbf{y} \right\| \le \|K\|_0 \|g\|, \qquad (12.4\text{-}3)$$

where $\|K\|_0$ denotes the norm of $K(\mathbf{x}, \mathbf{y})$ in $L^2(\mathbb{R}^{2n})$. This shows that K, as defined by (12.4-2) is a bounded operator, and its bound $\|K\|$ cannot exceed $\|K\|_0$.

Now let $\{\chi_k\}_1^\infty$ be a complete orthonormal sequence in $L^2(\mathbb{R}^n) = \mathfrak{H}$. For simplicity, we assume that the $\chi_k(\mathbf{x})$ are smooth real functions. Then the functions

$$\chi_k(\mathbf{x})\chi_l(\mathbf{y}) \qquad k, l = 1, 2, \ldots$$

form an orthonormal sequence in $L^2(\mathbb{R}^{2n})$. That sequence is complete, but we use only the Bessel inequality

$$\sum_{(k\,l)} |(\chi_k, K\chi_l)|^2 \le \|K\|_0^2 \qquad (12.4\text{-}4)$$

from which it follows from the lemma in the preceding section that K is a Hilbert–Schmidt operator.

12.5 Operators with Compact Resolvent

Let T be a closed operator whose resolvent $R_\lambda = (T - \lambda)^{-1}$ is compact for some λ_0 in the resolvent set $\rho(T)$. The resolvent equation in the form

$$R_\lambda = R_{\lambda_0}(I + (\lambda - \lambda_0)R_\lambda), \qquad (12.5\text{-}1)$$

then shows that R_λ is compact for any other λ in $\rho(T)$, because the second factor on the right side of the above equation is bounded. Such an operator T is called an *operator with a compact resolvent*.

We know that the spectrum $\sigma(R_{\lambda_0})$ consists of a bounded countable set with no accumulation point except at the origin, and that any $\mu \ne 0$ in $\sigma(R_{\lambda_0})$ is an eigenvalue of finite multiplicity. That leads immediately to knowledge of the spectrum of T. First, suppose that $\mu \ne 0$ is in the resolvent set $\rho(R_{\lambda_0})$. Then, for any y in $\mathfrak{H}$ the equation

$$R_{\lambda_0}x - \mu x = y$$

has a solution x. In particular, if z is an arbitrary element of $\mathfrak{H}$ and we set $y = -\mu R_{\lambda_0}z$, the equation

$$R_{\lambda_0}x - \mu x = -\mu R_{\lambda_0}z$$

has a solution x. This equation shows that x is in the domain of $T - \lambda_0$ (which is the range of R_{λ_0}), hence the equation

$$x - \mu(T - \lambda_0)x = -\mu z,$$

that is

$$Tx - \left(\lambda_0 + \frac{1}{\mu}\right)x = z,$$

has a solution x for every z; in other words, $\lambda_0 + (1/\mu)$ is in the resolvent set $\rho(T)$.

On the other hand, if $\mu \neq 0$ is an eigenvalue of R_{λ_0} and x is a corresponding eigenvector, that is, if

$$R_{\lambda_0}x - \mu x = 0,$$

then $x - \mu(T - \lambda_0)x = 0$, or

$$Tx - \left(\lambda_0 + \frac{1}{\mu}\right)x = 0,$$

which shows that the same x is an eigenvector of T corresponding to eigenvalue $\lambda_0 + (1/\mu)$. Hence, any $\lambda \neq \lambda_0$ is either in $P\sigma(T)$ or in $\rho(T)$, and we already know that λ_0 is in $\rho(T)$. So T has a pure point spectrum.

Let μ_k be one of the nonzero eigenvalues of R_{λ_0}, so that $\lambda_k = \lambda_0 + 1/\mu_k$ is an eigenvalue of T. The associated spectral projector for R_{λ_0} is

$$P_k = \frac{1}{2\pi i} \oint^{(\mu_k -)} (R_{\lambda_0} - \mu)^{-1}\, d\mu; \tag{12.5-2}$$

it has a finite dimensional range, because μ_k is an eigenvalue of finite multiplicity. The path of integration is a small enough contour about μ_k so that it does not encircle any other eigenvalue and does not encircle the origin. A direct calculation shows that under the transformation $\lambda = \lambda_0 + (1/\mu)$, (12.5-2) goes into

$$P_k = \frac{1}{2\pi i} \oint^{(\lambda_k -)} (T - \lambda)^{-1}\, d\lambda. \tag{12.5-3}$$

That is, P_k is also the projector for T corresponding to eigenvalue λ, and we see that λ_k has the same multiplicity as μ_k. We conclude:

Theorem 1. *If T is an operator with compact resolvent, the spectrum $\sigma(T)$ consists just of eigenvalues λ_k of finite multiplicity accumulating only at ∞ in the λ plane.*

As noted above, the resolvent of a regular Sturm–Liousville operator T is an integral operator of Hilbert–Schmidt type, hence the theorem applies, and T has a pure point spectrum, as already found in Section 10.6. A further example of a differential operator with a compact resolvent is the Laplacian in a bounded region Ω of $\mathbb{R}^3$ discussed in Section 11.7. An essentially self-adjoint version of the Laplacian in Ω is the operator A_0 defined by equation (11.7-1). Its inverse A_0^{-1}, which is the resolvent of A_0 at $\lambda = 0$, is the integral operator in (11.7-3), whose kernel is the Green's function $G(\mathbf{x}, \mathbf{x}')$ for the region Ω. Because of the integrable nature of the singularity at $\mathbf{x} = \mathbf{x}'$, as indicated in (11.7-4), the integral

$$\iint |G(\mathbf{x}, \mathbf{x}')|^2 \, d^3\mathbf{x} \, d^3\mathbf{x}'$$

is finite; hence A_0^{-1} is a Hilbert–Schmidt operator, hence compact, and we conclude from Theorem 1 above that A_0 has a pure point spectrum consisting only of eigenvalues of finite multiplicity accumulating only at infinity, in agreement with the statement made in Section 11.7 on the basis of the Fredholm theory.

If an operator T with compact resolvent is also self-adjoint then its eigenvectors form a complete orthonormal set. If T is not self-adjoint, a complete set of vectors is obtained if (1) the generalized eigenvectors are included, and (2) certain further conditions are satisfied. We state below two theorems on completeness, which have been used in the theory of hydrodynamic stability to establish the completeness of the normal modes of small perturbation of a steady flow; the main operators of that theory are not self-adjoint.

Let λ_k $(k = 1, 2, \ldots)$ be the eigenvalues of T. For each k, the range $\mathfrak{C}_k$ of the projector P_k given by (12.5-3) is a finite-dimensional space invariant under T; x is in $\mathfrak{C}_k$ if and only if Tx is in $\mathfrak{C}_k$. Therefore, T restricted to $\mathfrak{C}_k$ can be represented by an $n_k \times n_k$ matrix T_k, where $n_k = \dim \mathfrak{C}_k$. The eigenvectors and generalized eigenvectors of T_k (see Section 12.1) correspond to a set of vectors $x_{k\,s}$ $(s = 1, \ldots, n_k)$ in $\mathfrak{H}$ that span $\mathfrak{C}_k$. Under the further conditions stated in the theorems below, the eigenspaces $\mathfrak{C}_k$ $(k = 1, 2, \ldots)$ span $\mathfrak{H}$. Then, if v is any vector in $\mathfrak{H}$, the component of v in $\mathfrak{C}_k$ is of the form

$$\sum_{s=1}^{n_k} c_{k\,s} x_{k\,s}, \tag{12.5-4}$$

hence

$$v = \sum_{k=1}^{\infty} \left(\sum_{s=1}^{n_k} c_{k\,s} x_{k\,s} \right). \tag{12.5-5}$$

Although this sum cannot in general be arbitrarily rearranged, as it can when T is self-adjoint, we can say that the vectors $\{x_{k\,s}\}$ form a complete set in the sense that their finite linear combinations are dense in $\mathfrak{H}$.

It is generally believed that in the hydrodynamic stability problem only eigenvectors of order 1 (ordinary eigenvectors) occur, but that has not been proved. As in the finite-dimensional case, an eigenvector of order m corresponding to an eigenvalue λ_k is a vector x in $\mathfrak{H}$ such that

$$(T - \lambda_k)^m x = 0, \qquad (T - \lambda_k)^{m-1} x \neq 0.$$

For $m > 1$ it can also be characterized as the solution x of the equation

$$(T - \lambda_k)x = y,$$

where y is some eigenvector of order $m - 1$. Since T is a differential operator in the hydrodynamic problem, finding x entails solving an inhomogeneous differential equation, once y is known. Computer codes have been devised for that purpose, but apparently no generalized eigenfunctions have been found so far.

The following theorem of Naïmark has been used by DiPrima and Habetler (1969) to establish the completeness of the solutions of the Orr-Sommerfeld problem, which is a non-self-adjoint eigenvalue problem of a fourth order ordinary differential equation for two-dimensional normal mode disturbances of plane laminar flow.

Theorem 2 (Naïmark). *Suppose T is an operator with compact resolvent, and suppose there is a sequence of concentric circles $|\lambda| = r_l \, (l = 1, 2, \ldots)$ in the resolvent set of T (i.e., not passing through any eigenvalues) such that*

$$\sup\{\|R_\lambda\| : |\lambda| = r_l\} \to 0 \quad \text{as } l \to \infty. \tag{12.5-6}$$

Then the eigenvectors and generalized eigenvectors $\{x_{k\,s}\}$ of T form a complete set in $\mathfrak{H}$.

Remark. Since $\|R_\lambda\| \geq [\text{dist}(\lambda, \sigma(T))]^{-1}$, it is clear that the hypothesis (12.5-6) can be satisfied only if the eigenvalues λ_k become more and more widely separated as $k \to \infty$.

The idea of the proof is this: The operator

$$P^{(l)} = \frac{1}{2\pi i} \oint_{|\lambda| = r_l} (-R_\lambda) d\lambda$$

is the sum of the projectors P_k for all eigenvalue λ_k that lie inside the circle $|\lambda| = r_l$. Therefore $P^{(l)}v$ is a partial sum of (12.5-5), and it becomes the whole sum in the limit as $l \to \infty$. From the definition of the resolvent, we have $(T - \lambda)R_\lambda = I$, hence

$$-R_\lambda = \frac{1}{\lambda} I - \frac{1}{\lambda} TR_\lambda.$$

If v is in the domain $\mathfrak{D}(T)$, we have

$$TR_\lambda v = R_\lambda Tv,$$

hence

$$P^{(l)}v = v - \frac{1}{2\pi i} \oint_{|\lambda| = r_l} \frac{1}{\lambda} R_\lambda \, d\lambda Tv.$$

Owing to the hypothesis (12.5-6), the second term in the last equation goes to zero in the limit $l \to \infty$. Therefore, any v in $\mathfrak{D}(T)$ can be expanded as (12.5-5); but $\mathfrak{D}(T)$ is dense in $\mathfrak{H}$, and the completeness follows.

A similar theorem of Carleman is somewhat more powerful in that it doesn't require the eigenvalues λ_k to be more and more separated as $k \to \infty$. It merely requires that they be concentrated in certain directions from the origin in the complex plane, as $k \to \infty$, and that T have a Hilbert–Schmidt resolvent. It has been used by Sattinger (1970) to prove the completeness of normal mode disturbances of general three-dimensional steady flow. A general form of the theorem is given in Dunford and Schwartz (vol. II, 1963, p. 1042); we shall give a simplified form in which the eigenvalues are required to lie near the real axis; in that respect T resembles a self-adjoint operator, whose eigenvalues are real.

Theorem 3 (Carleman). *If T is an operator with Hilbert–Schmidt resolvent and if along each ray $\lambda = re^{i\theta}$ (θ fixed) except the real axis ($\theta = 0$ or π), $\|R_\lambda\| = O(|\lambda|^{-1})$ for large λ, then the eigenvectors and generalized eigenvectors of T form a complete set in $\mathfrak{H}$.*

In the hydrodynamic application, the eigenvalues λ_k lie in a parabolic region

$$\text{Re } \lambda \geq \text{const.} + \text{const.}(\text{Im } \lambda)^2.$$

The hypothesis is stronger than necessary; it is only necessary that $\|R_\lambda\|$ have the asymptotic behavior stated above on each of five rays from the origin such that the angle between adjacent rays is in each case $< \pi/2$ (see Dunford and Schwartz).

CHAPTER 13

Probability; Measures

Univariate and multivariate probability distributions; cumulative
probabilities; densities; canonical decomposition of a nondecreasing
function; discrete, atomic, singular, continuous, and absolutely
continuous probability distributions; nondecreasing functions of
several variables; mean: expectation; moments; standard deviation;
characteristic function; correlation coefficient and matrix; measures;
set functions; the extension and Riesz representation theorems for
measures; sampling; sample mean; sample variance; marginal and
conditional probabilities; normal distribution; central limit theorem;
the Monte Carlo method; probability and measure in a Hilbert space.

Prerequisites: Chapters 1 to 3; Stieltjes integral

The concept of cumulative probability provides the basis for discussion of
probabilities in finite-dimensional spaces. Most probabilities in physics are
either discrete or absolutely continuous or a mixture of the two, but the
notion of singular continuous distributions is needed to complete the
conceptual framework. The main applications are to quantum mechanics,
discussed in the next chapter, statistical mechanics, error analysis, and
Monte Carlo. The outstanding phenomenon is the trend toward the universal
so-called normal distributions on averaging, described in the central limit
theorem. Cumulative, marginal, and conditional probabilities are used in
Monte Carlo simulation by computer of natural random phenomena too
complicated for analysis. The ideas of probability and measure as set functions are needed for discussion of probability in infinite-dimensional and
abstract spaces, such as appear in statistical mechanics and in the theory of
stochastic processes.

13.1 Univariate or One-Dimensional Probability Distributions: Cumulative Probability; Density

The results of many independent repetitions of an elementary observation or experiment (for example, on an atomic system) are described in terms of a probability distribution. The result of a given repetition is a set of values of certain quantities, say n in number, representing angles of deflection, voltages, energies, counter readings, and the like; hence, the result may be represented as a point in an n-dimensional space. It is characteristic of most elementary processes that no matter how carefully the conditions are controlled the result varies significantly and randomly from one repetition to another. The resulting distribution of points in the n-dimensional space is described by a cumulative probability in that space, defined below. In the present discussion, the probability concept is regarded as intuitive; for example, the statement that a neutron has probability 0.316 of passing through a certain foil without collision is taken to indicate that (1) neutrons did so in 31.6% of a large number of tries, and (2) it is believed that in future tries neutrons will behave similarly, the corresponding percentage tending to some value near 31.6% as the number of tries increases indefinitely.

In probability theory, the cases $n = 1$ and $n > 1$ are usually called *univariate* and *multivariate*, respectively. We start with the univariate case. If ξ is a quantity determined in an experiment, then the function F defined by the equation

$$F(x) = \mathbf{P}\{\xi \leq x\}, \tag{13.1-1}$$

where, for any real x, $\mathbf{P}\{\xi \leq x\}$ denotes the probability that the measured value of ξ is $\leq x$, is called the *cumulative probability* of the *random variable* ξ. Other probabilities can be expressed in terms of F; in particular, the probability that ξ lies in an interval $(x_1, x_2]$ is given by

$$\mathbf{P}\{x_1 < \xi \leq x_2\} = F(x_2) - F(x_1); \tag{13.1-2}$$

also, if F has jumps, it describes probabilities concentrated at points; namely

$$\mathbf{P}\{\xi = x_0\} = F(x_0 + 0) - F(x_0 - 0). \tag{13.1-3}$$

The terms on the right side of this equation denote the limiting values approached by $F(x)$ as $x \to x_0$ from above and from below, respectively.

It follows from the definition (13.1-1) that $F(x)$ is nondecreasing and continuous on the right and ranges from 0 to 1; namely,

$$F(b) \geq F(a) \quad \text{for } b \geq a \tag{13.1-4}$$

$$F(a + \varepsilon) \to F(a) \quad \text{as } \varepsilon \downarrow 0 \tag{13.1-5}$$

$$F(-\infty) = 0, \qquad F(+\infty) = 1 \tag{13.1-6}$$

Conversely, any function having these properties describes a probability distribution. For example, if F is known, a set of random numbers having the corresponding distribution can be generated with any desired accuracy by means of a computer.

The requirement of right continuity (13.1-5) is a course arbitrary, and we often write equations, like (13.1-3), in such a manner that they are independent of that requirement. (The right member of (13.1-3) does not depend on the convention that $F(x_0 + 0) = F(x_0)$.)

EXAMPLE 1

If ξ has only finitely many possible values $x_1, \ldots, x_N$, assumed written in increasing order, and if $\mathbf{P}\{\xi = x_i\} = p_i$, then $F(x)$ is the step function

$$F(x) = \begin{cases} 0, & \text{for } -\infty < x < x_1, \\ \sum_{k=1}^{i-1} p_k & \text{for } x_{i-1} \le x < x_i, i = 2, \ldots, N, \\ 1, & \text{for } x_N \le x < \infty. \end{cases} \qquad (13.1\text{-}7)$$

This type of function is appropriate for considerations of coin and die tossing and of finite games generally. It is also appropriate for describing the distribution of photon energies emitted when at atom jumps from some particular excited state to lower energy levels. (See Figure 13-1).

EXAMPLE 2

A neutron or photon enters matter at uniform density and travels a distance ξ before colliding (Figure 13-2); the cumulative probability of the random variable ξ is

$$F(x) = \begin{cases} 0, & \text{for } x < 0, \\ 1 - e^{-x/\lambda}, & \text{for } x > 0, \end{cases}$$

where λ is the mean free path. See Figure 13-3. Here, F has a derivative $f(x) = F'(x)$, which is continuous except for a jump at $x = 0$ and is called the *probability density*, because, for any x_0,

$$f(x_0) = \lim_{\Delta x \to 0} \frac{1}{\Delta x} \mathbf{P}\{x_0 < \xi < x_0 + \Delta x\}; \qquad (13.1\text{-}8)$$

such a probability distribution is called *continuous* or more properly *absolutely continuous* (definition below).

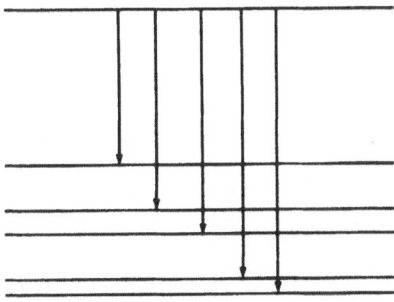

Figure 13.1 Energy level diagram.

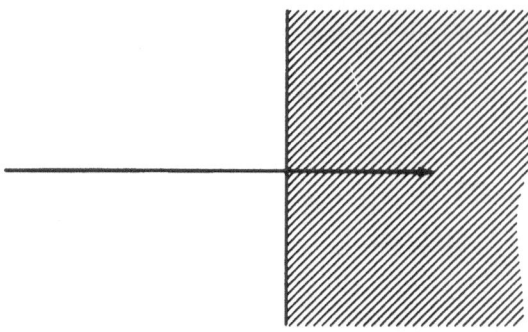

Figure 13.2 Neutron path for Example 2.

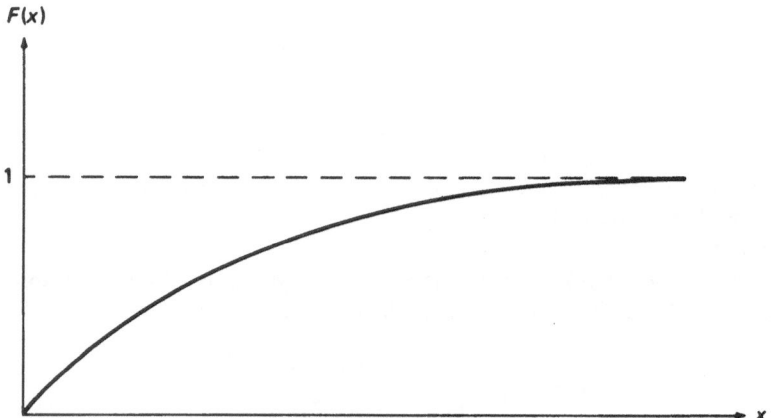

Figure 13.3 The cumulative probability for Example 2.

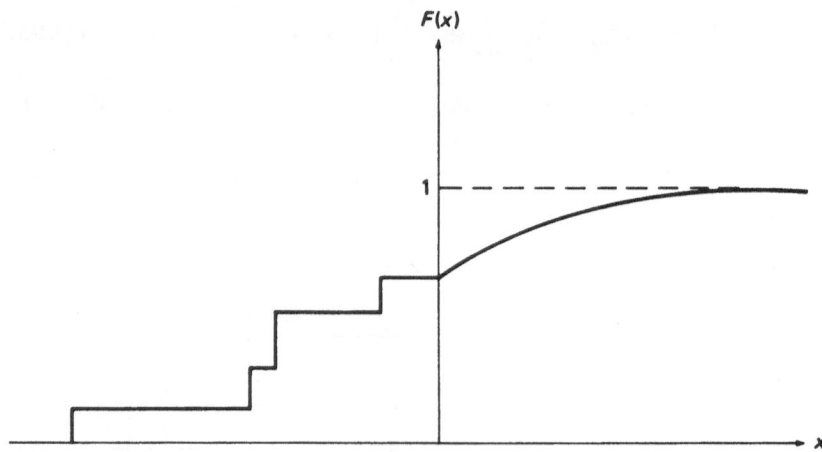

Figure 13.4 The cumulative probability for Example 3.

EXAMPLE 3

Consider an atom in its ground state, immersed in a radiation field that has a continuous spectrum. After absorbing a quantum, the atom can be in any one of various excited states or in the continuum above the ionization limit, which can be taken as the zero of energy. The probability distribution of the energy E after absorbtion of a single quantum is partly discrete and partly continuous, as in Figure 13-4; the cumulative probability $F(x) = \mathbf{P}\{E \leq x\}$ is a step function for $x < 0$ and is continuous for $x \geq 0$.

EXAMPLE 4

A neutron or photon passes through an infinite succession of parallel thin absorbing foils, uniformly spaced, as in Figure 13-5; if ξ is the distance from the first foil to the one at which absorption occurs, and if α is the probability of absorption at each foil $(0 < \alpha < 1)$, then the cumulative probability of ξ is

$$F(x) = 1 - \alpha^n \quad \text{for } (n - 1)d < x \leq nd, \qquad n = 1, 2, \ldots,$$

where d is the separation of successive foils; this is a step function with infinitely many steps and is shown in Figure 13-6.

EXAMPLE 5

In the preceding example, let $\varphi = \varphi(t) = \varphi_0 \sin \omega t$ be an alternating voltage present in a circuit while the particle is moving down the line of foils; let τ be the time required to go from one foil to the next, and let $\theta = \omega\tau$. Then the voltage φ at the instant of absorption of the particle has the value 0 with the probability $1 - \alpha$, the value $\varphi_0 \sin \theta$ with probability $\alpha - \alpha^2, \ldots$, the value $\varphi_0 \sin n\theta$ with probability $\alpha^n(1 - \alpha)$, etc. The cumulative probability $F(x) = \mathbf{P}\{\varphi \leq x\}$ is then

$$F(x) = \sum_{n=0}^{\infty} \alpha^n(1 - \alpha) \qquad (\varphi_0 \sin n\theta \leq x);$$

the sum is over all n such that $\varphi_0 \sin n\theta \leq x$. If θ is an irrational multiple of π, $F(x)$ has infinitely many jumps, densely spaced in the interval $[-\varphi_0, \varphi_0]$.

EXAMPLE 6: The Cantor function

Suppose that a digital computer has an attachment (using radioactive decay, thermal noise or the like) which generates endlessly on demand a succession of independent

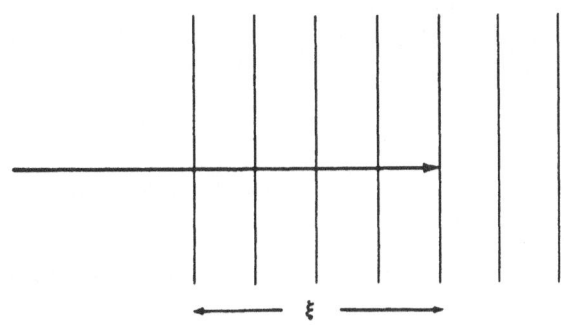

Figure 13.5 Neutron path for Example 4.

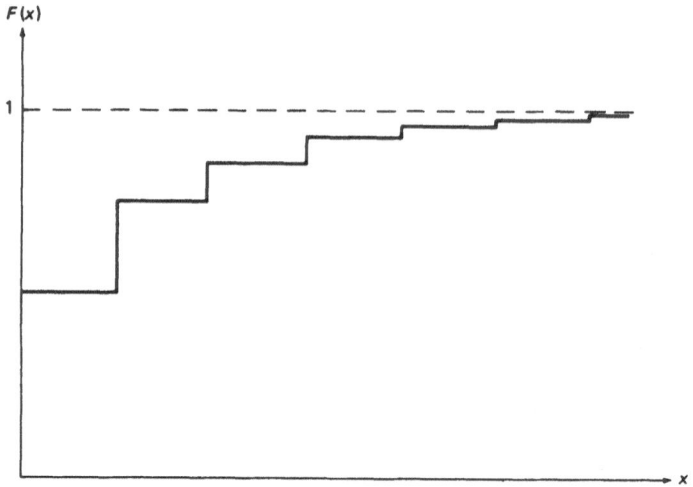

Figure 13.6 Cumulative probability for Example 4.

random numbers $x_1, x_2, \ldots$ uniformly distributed in the interval $[0, 1]$; these numbers are taken to be the values of a random variable ξ. Each x is supposed to be expressed as an infinite binary expansion

$$x = .abcd \ldots, \qquad (13.1\text{-}9)$$

where each digit $a, b, c, d, \ldots$ is either 0 or 1. Suppose that the computer contains a subroutine that transforms each such x into a corresponding number

$$y = .aabbcc \ldots, \qquad (13.1\text{-}10)$$

by duplicating each digit; the numbers y are taken to be the values of another random variable η. The cumulative probability function F for the values of η is easily found. Any number y of the form (13.1-10) is necessarily either less than $0.01 = \frac{1}{4}$ or greater than (or equal to) $0.11 = \frac{3}{4}$, according to whether x is less than or greater than (or equal to) $0.1 = \frac{1}{2}$. Hence, $F(y)$ has the constant value $\frac{1}{2}$ for $\frac{1}{4} \le y \le \frac{3}{4}$. Similarly, $F(y)$ has the value $\frac{1}{4}$ for $\frac{1}{16} \le y \le \frac{3}{16}$, and the value $\frac{3}{4}$ for $\frac{13}{16} \le y \le \frac{15}{16}$, etc. If y has the form (13.1-10), where the digits are equal in pairs out to infinity in the binary expansion, then, for that y, $F(y)) = .abc \ldots$, i.e.,

$$P\{\eta < y = .aabbcc \ldots\} = .abc \ldots. \qquad (13.1\text{-}11)$$

If y does not have that form, then it lies in one of the intervals of constancy of F described above.

The sum of the lengths of all the intervals of constancy of F is

$$\tfrac{1}{2} + 2\tfrac{1}{8} + 4\tfrac{1}{32} + \cdots + \frac{1}{2^n} + \cdots = 1,$$

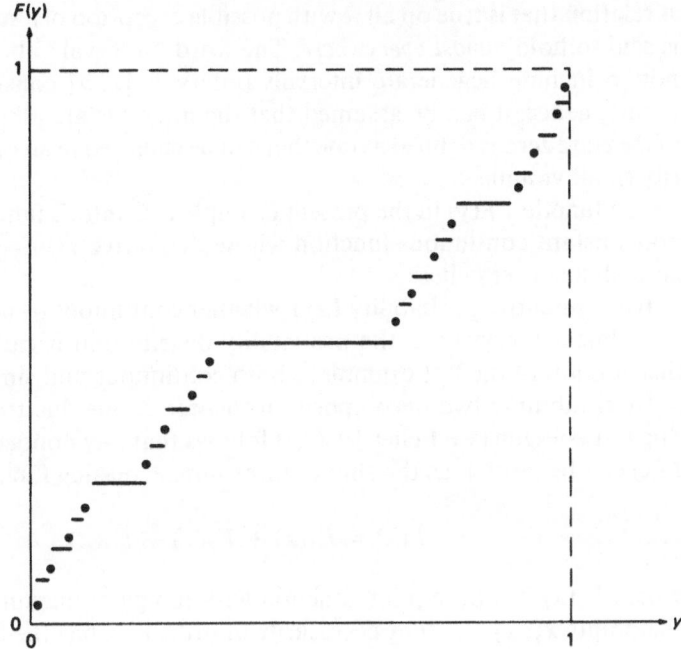

Figure 13.7 The Cantor function.

hence these intervals just fill up the interval [0, 1], and what is left over has measure zero. On the other hand, F is continuous, because (1) if y_0 has the form (13.1-10), then the statement $y \to y_0$ clearly implies $x \to x_0$, according to (13.1-11), and (2) any other y_0 is in an interval of constancy, hence *a fortiori* of continuity of F. $F(y)$ is sketched in Figure 13-7.

Digression on Sets of Measure Zero

A set S on $\mathbb{R}$ has *measure zero* if it can be enclosed in a collection of intervals the sum of whose lengths is arbitrarily small. For example, let S consist of the rational numbers in (0, 1). They can be written in a sequence $\alpha_1, \alpha_2, \alpha_3, \ldots,$ for example in the order $\frac{1}{2}, \frac{1}{3}, \frac{2}{3}, \frac{1}{4}, \frac{3}{4}, \frac{1}{5}, \frac{2}{5}, \frac{3}{5}$, etc. Given any $\varepsilon > 0$, α_1 can be enclosed in an interval of length $\varepsilon/2$, α_2 in an interval of length $\varepsilon/4, \ldots, \alpha_n$ in an interval of length $\varepsilon/2^n$, etc. These intervals cover S, and the sum of their lengths is equal to ε. Hence the rational numbers are a set of measure zero. In Example 6 above, if the first $1 + 2 + 4 + \cdots + 2^{n-1}$ intervals of constancy of $F(y)$ (taken in the order described above) are removed, the remainder of the interval [0, 1] consists of intervals having total length $= 1/2^n$, which can be made arbitrarily small, by taking n large enough. That is, the set of values of y at which F is not constant has measure zero. Stated differently, the derivative $F'(y)$ exists and is $=0$ everywhere except on a set of measure zero

A relation that is true on all $\mathbb{R}$ with possible exception of a set of measure zero is said to hold *almost everywhere*. The word "interval" above is understood not to include degenerate intervals (intervals $[a, a]$ consisting of a single point); hence, it can be assumed that the intervals are all open. In $\mathbb{R}^n$, a set of measure zero is defined as one that can be enclosed in an open set of arbitrarily small volume.

The function $F(y)$ in the present example is Cantor's famous example of a nonconstant continuous function whose derivative $f(y) = F'(y)$ is equal to zero almost everywhere.

If a cumulative probability $F(x)$, whether continuous or not, has derivative $=0$ almost everywhere, the probability distribution is called *singular*. The distribution of the last example is both continuous and singular.

By combining two decomposition theorems, one due to Jordan and one due to Lebesgue (see Feller 1966), it follows that any nondecreasing function $F(x)$ can be written as the sum of three nondecreasing functions

$$F(x) = F_1(x) + F_2(x) + F_3(x),$$

where $F_1(x)$ is a pure jump function with jumps of magnitude $p_1, p_2, \ldots$ at the points $x_1, x_2, \ldots$ (not necessarily in order) i.e. has the form

$$F_1(x) = \sum_{x_i \leq x} p_i \quad (\text{each } p_i > 0),$$

where, furthermore, $F_2(x)$ is continuous and is the integral (in the Lebesgue sense) of its derivative $f(x) \geq 0$;

$$F_2(x) = \int^x f(y)dy, \qquad f(x) = F'_2(x),$$

and where $F_3(x)$ is singular and continuous. If $F(x)$ is a pure jump function, as in Examples (1), (4), and (5) above, the probability distribution is called *atomic*; in Examples (1) and (4), it is also called *discrete* (the jumps are at isolated points). $F_2(x)$ belongs to the class of functions called *absolutely continuous*; see Section 13.10.

13.2 Means and Expectations

Suppose that $F(x)$ is the cumulative probability of a random variable ξ and that it is desired to find the average value of ξ obtained from a long sequence of measurements. The case in which ξ is a bounded random variable is considered first; that is, it is assumed that $F(x) = 0$ for $x < a$ and $F(x) = 1$ for $x \geq b$, so that all measured values of ξ are in the interval $[a, b]$. To approximate the average value, the interval $[a, b]$ is partitioned into N subintervals by subdivisions at $x_i (i = 0$ to $N)$ such that

$$a = x_0 < x_1 < \cdots < x_N = b.$$

Denote by x_i' an arbitrary point in the subinterval $[x_{i-1}, x_i]$ $(i = 1, 2, \ldots, N)$. The fraction of the measurements in which ξ lies between x_{i-1} and x_i is $F(x_i) - F(x_{i-1})$; hence the average value is approximately

$$\sum_{i=1}^{N} x_i'[F(x_i) - F(x_{i-1})].$$

In the limit, as the partition is refined indefinitely $(N \to \infty)$, this approaches the Stieltjes integral

$$\int_a^b x \, dF(x). \tag{13.2-1}$$

The theory of the Stieltjes integral of a continuous function is almost identical with the corresponding theory of the Riemann integral. See Natanson 1955, Chapter 8. Stieltjes integrals are often useful in physics; in Example 3, above, it permits one to include a summation over the discrete states and an integration over the continuous states in a single formula. The following notation for the Riemann–Stieltjes sums is often used: The interval $x_{i-1} < x \le x_i$ is denoted by Δ_i, and $F(x_i) - F(x_{i-1})$ is denoted by $F(\Delta_i)$; then the above sum is written as $\sum x_i' F(\Delta_i)$, and the corresponding Riemann–Stieltjes sum for a continuous function $\varphi(x)$ is $\sum \varphi(x_i') F(\Delta_i)$. A notation of this type is especially convenient in the multivariate case to be discussed in the next section.

The integral (13.2-1) is called the *mean*, or the *expected value*, or the *expectation* of ξ and is noted by $E(\xi)$ or μ. In the general case of an unbounded random variable, the expectation is

$$\mu = E(\xi) = \int_{-\infty}^{\infty} x \, dF(x) = \lim_{\substack{a \to -\infty \\ b \to +\infty}} \int_a^b x \, dF(x), \tag{13.2-2}$$

provided that the limit exists. If the limit does not exist, the average of the measurements cannot be expected to settle down to any fixed limiting value as the number of repetitions of the experiment is increased indefinitely.

Now let φ be a continuous function. Each measured value x of ξ corresponds to a number $\varphi(x)$, and these numbers are the values of a random variable denoted by $\varphi(\xi)$; the expected value of $\varphi(\xi)$ is

$$E(\varphi(\xi)) = \int_{-\infty}^{\infty} \varphi(x) dF(x), \tag{13.2-3}$$

again provided the integral exists. Several important cases are:

Case 1. $\varphi(x)$ is a power of x; the quantities $E(\xi^k)$ $k = 1, 2, \ldots$, insofar as they exist, are called the *moments* of the distribution; the first one is $E(\xi) = \mu$.

An important combination of the first two moments is the *variance* $\sigma^2 = \mathbf{E}((\xi - \mu)^2)$; since $\mathbf{E}(1) = 1$ and $\mathbf{E}(\xi) = \mu$, σ^2 can also be written as $\sigma^2 = \mathbf{E}(\xi^2) - 2\mu\mathbf{E}(\xi) + \mu^2\mathbf{E}(1) = \mathbf{E}(\xi^2) - \mu^2$; σ itself is called the *standard deviation* of ξ. If ξ is a bounded random variable, the moments of all orders exist and are finite. An important classical problem, the *moment problem*, is to compute F when all the moments are known. (See Feller, Vol. 2).

Case 2. φ is a bounded continuous function; then $\mathbf{E}(\varphi(\xi))$ always exists. The most important instance is $\varphi(x) = e^{-i\lambda x}$, λ real; the complex-valued function

$$\chi(\lambda) = \int_{-\infty}^{\infty} e^{-i\lambda x} \, dF(x) \tag{13.2-4}$$

of the real parameter λ is called the *characteristic function* of the probability distribution or of the random variable ξ; it will play an important role in the proof of the central limit theorem in Section 13.6.

Case 3. φ is a test function, $\varphi \in C_0^\infty$; then, a distribution f (in the sense of Schwartz) can be defined as the linear functional

$$\langle f, \varphi \rangle = \mathbf{E}(\varphi(\xi)) = \int_{-\infty}^{\infty} \varphi(x) dF(x), \quad \text{for all } \varphi \in C_0^\infty. \tag{13.2-5}$$

Integration by parts (see Natanson 1955 for integration by parts in Stieltjes integrals) gives

$$\langle f, \varphi \rangle = \varphi(x)F(x) \Big|_{-\infty}^{\infty} - \int_{-\infty}^{\infty} F(x)\varphi'(x)dx,$$

but the integrated part is zero, hence

$$\langle f, \varphi \rangle = -\langle F, \varphi' \rangle,$$

where F denotes the distribution determined by the function $F(x)$ through the equation

$$\langle F, \psi \rangle = \int_{-\infty}^{\infty} F(x)\psi(x)dx, \quad \forall \psi \text{ in } C_0^\infty.$$

That is, $f = F'$. The distribution f is called the *probability measure* of the random variable ξ (f is an ordinary function if and only if $F(x)$ is absolutely continuous, and in that case $f(x)$ is the probability density $F'(x)$ of ξ).

As a distribution, the function $F(x)$ is tempered, because it is bounded, hence $f = F'$ is also a tempered distribution, and the characteristic function $\chi(\lambda)$ is $= \sqrt{2\pi}$ times the Fourier transform of f.

13.3 Bivariate and Multivariate Distributions; Nondecreasing Functions of Several Variables

If each repetition of an experiment yields two numbers x, y, namely the values of the random variables ξ and η, then the distribution of the resulting points in the x, y plane is called *bivariate*. Bivariate distributions will now be discussed; the multivariate case will then be an obvious generalization and is discussed briefly at the end of this section.

The *joint cumulative probability* $F(\cdot, \cdot)$, is given by

$$F(x, y) = \mathbf{P}\{\xi \leq x \text{ and } \eta \leq y\}; \qquad (13.3\text{-}1)$$

it is the fraction of the experiments for which the resulting points in the x, y plane lie in the quadrant to the left of and below the point x, y. See Figure 13-8. Other probabilities can be expressed in terms of $F(\cdot, \cdot)$. If $x_1 < x_2$, then $F(x_2, y) - F(x_1, y)$ is the probability that $x_1 < \xi \leq x_2$ while η has any value $\leq y$. Therefore, if also $y_1 < y_2$, then

$$\mathbf{P}\{x_1 < \xi \leq x_2 \text{ and } y_1 < \eta \leq y_2\} =$$
$$F(x_2, y_2) - F(x_1, y_2) - F(x_2, y_1) + F(x_1, y_1). \quad (13.3\text{-}2)$$

If $\square$ denotes the indicated rectangular region of the plane, namely

$$\square = \{(x, y): x_1 < x \leq x_2 \text{ and } y_1 < y \leq y_2\}$$

then (13.3-2) is paraphrased by writing

$$\mathbf{P}\{(\eta, \xi) \in \square\} = F(\square), \qquad (13.3\text{-}3)$$

where $F(\square)$ stands for the combination in the right member of (13.3-2).

Clearly, $F(\infty, \infty) = 1$, while $F(x, -\infty) = 0$ for any x and $F(-\infty, y) = 0$ for any y. The requirement of the univariate case that $F(\cdot)$ be nondecreasing

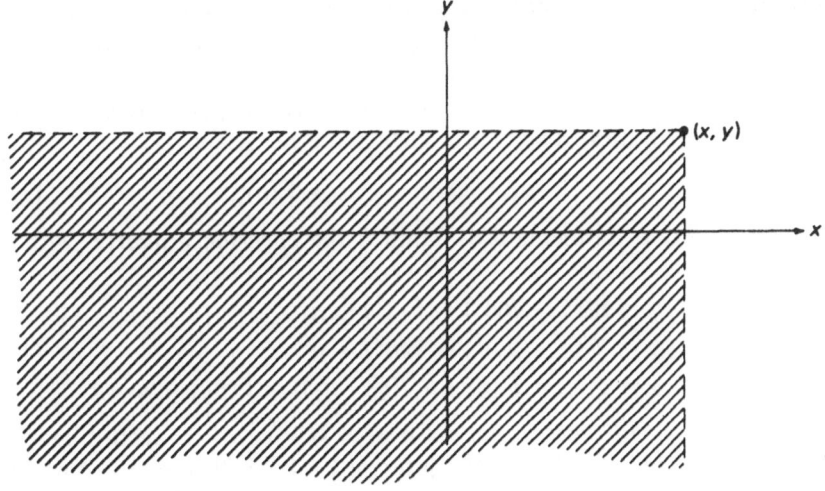

Figure 13.8 The region of the plane for equation (13.3-1).

is here replaced by the requirement that $F(\square) \geq 0$ for any $\square$ with $x_1 \leq x_2$ and $y_1 \leq y_2$. This requirement also includes the limiting case $F(x_2, y_2) - F(x_1, y_2) \geq 0$, obtained by letting $y_1 \to -\infty$ and the limiting case $F(x_2, y_2) - F(x_2, y_1) \geq 0$ obtained by letting $x_1 \to -\infty$. Under these circumstances, $F(x, y)$ is called a *nondecreasing function* of the two variables x and y.

To find the expected value of a continuous function $\varphi(\xi, \eta)$, assume first that ξ and η are bounded, so that the points (x, y) that can result from an experiment lie in a rectangle $a < x < b$, $c < y < d$. This rectangle is partitioned into many small rectangles $\square_{jk}$ by horizontal and vertical lines $x = x_0, x_1, x_2, \ldots, x_N$ and $y = y_0, y_1, y_2, \ldots, y_M$; if x'_j, y'_k is a point in $\square_{jk}$, then the average of $\varphi(\xi, \eta)$ is approximated by the double Riemann–Stieltjes sum

$$\sum_{jk} \varphi(x'_j, y'_k) F(\square_{jk}), \tag{13.3-4}$$

which converges to the double Stieltjes integral $\int_a^b \int_c^d \varphi(x, y) d^2 F(x, y)$ as the partitioning is refined ($N, M \to \infty$). If, in the unbounded case, this integral has a limit as $b, d \to \infty$ and $a, c \to -\infty$, then the expectation of $\varphi(\xi, \eta)$ is defined as

$$\mathbf{E}(\varphi(\xi, \eta)) = \int_{-\infty}^{\infty} \int_{-\infty}^{\infty} \varphi(x, y) d^2 F(x, y). \tag{13.3-5}$$

If the second derivative $f(x, y) = \partial^2 F / \partial x \partial y$ exists, and is piecewise continuous, it is called the *density* of the distribution, which is then called *absolutely continuous*; in this case,

$$\mathbf{E}(\varphi(\xi, \eta)) = \int_{-\infty}^{\infty} \int_{-\infty}^{\infty} \varphi(x, y) f(x, y) dx \, dy \tag{13.3-6}$$

At the other extreme, the distribution may be discrete with respect to both variables, i.e., the probability may be concentrated on isolated points in the x, y plane. Between these extremes there are so many possibilities that a complete classification will not be given.

The numbers $\mathbf{E}(\xi^k \eta^l)$, insofar as the integrals exist, are the *moments* of the bivariate distribution. The first moments are the means

$$\mu_1 = \mathbf{E}(\xi), \qquad \mu_2 = \mathbf{E}(\eta); \tag{13.3-7}$$

from the second moments are obtained the *covariance matrix*, with components

$$\begin{aligned} \rho_{11} &= \mathbf{E}((\xi - \mu_1)^2) = \mathbf{E}(\xi^2) - \mu_1^2 \\ \rho_{22} &= \mathbf{E}((\eta - \mu_2)^2) = \mathbf{E}(\eta^2) - \mu_2^2 \\ \rho_{12} &= \rho_{21} = \mathbf{E}((\xi - \mu_1)(\eta - \mu_2)) = \mathbf{E}(\xi\eta) - \mu_1\mu_2. \end{aligned} \tag{13.3-8}$$

The *coefficient of correlation* is $\rho = \rho_{12}/\sqrt{\rho_{11}\rho_{22}}$; according to the Schwartz inequality ρ lies in the interval $[-1, 1]$. If $\rho = \pm 1$, ξ and η are completely correlated, and the points x, y that can result from an experiment all lie on a straight line through the point μ_1, μ_2—that is also true if $\rho_{11} = 0$

or $\rho_{2\,2} = 0$, in which case ρ is undefined. If ξ and η are *independent* random variables, which means that $F(x, y)$ has the form $F_1(x)F_2(y)$, then $\rho = 0$ and the variables are uncorrelated (large ξ does not imply large η or small η). However, ρ can be $=0$ when ξ and η are not independent; an example is given by the distribution with the density

$$f(x, y) = \begin{cases} (4\pi)^{-1} & \text{for } x^2 + y^2 < 1, \\ 0 & \text{for } x^2 + y^2 > 1, \end{cases}$$

here $\rho_{1\,2}$ $(=\rho_{2\,1})$ vanishes, owing to the symmetry of f, but f cannot be written as $f_1(x)f_2(y)$.

From the cumulative probability function $F(x, y)$ we define the *probability measure* on $\mathbb{R}^2$ as the distribution f given by

$$\langle f, \varphi \rangle = \iint \varphi(x, y)d^2F(x, y) \qquad \forall \varphi \in C_0^\infty(\mathbb{R}^2); \qquad (13.3\text{-}9)$$

f is the distribution derivative $\partial^2 F/\partial x \partial y$ and is an ordinary function if and only if F is absolutely continuous. The two-dimensional *characteristic function* is

$$\chi(\lambda) = \iint e^{-i\lambda \cdot x} d^2F(x, y); \qquad (13.3\text{-}10)$$

it is $=2\pi$ times the Fourier transform of f.

In the remainder of this section we make a few remarks about the multi-dimensional case, which is a straightforward generalization of the bivariate case. The *(joint) cumulative probability* of random variables $\xi_1, \dots, \xi_n$ is the function

$$F(x_1, \dots, x_n) = \mathbf{P}\{\xi_1 \le x_1, \dots, \xi_n \le x_n\}. \qquad (13.3\text{-}11)$$

We let $\square$ denote the rectangular box

$$\square = \{\mathbf{x} : a_1 < x_1 \le b_1, \cdots, a_n < x_n \le b_n\} \qquad (13.3\text{-}12)$$

and we introduce the notation

$$F(\square) = \mathbf{P}\{\xi \in \square\}, \qquad (13.3\text{-}13)$$

where ξ denotes the vector-valued random variable whose components are $\xi_1, \dots, \xi_n$. The explicit formula for $F(\square)$ is the generalization of (13.3-2). We define a *vertex* $\mathbf{v}$ of $\square$ as a point $\mathbf{x}$ where each x_j is either $=a_j$ or $=b_j$, and we call $N_a(\mathbf{v})$ the number of a's among the x_j in the vertex $\mathbf{v}$. Then

$$F(\square) = \sum_{(\mathbf{v})} (-1)^{N_a(\mathbf{v})} F(\mathbf{v}), \qquad (13.3\text{-}14)$$

summed over all 2^n vertices of $\square$. From the probability interpretation (13.3-11) of F, it is clear that

$$F(\mathbf{x}) \text{ is } nondecreasing: \quad F(\square) \ge 0 \quad \text{for every } \square \qquad (13.3\text{-}15)$$

$$F(\mathbf{x}) \text{ is } normalized: \quad F(\infty, \dots, \infty) = 1 \quad \text{and} \quad F(x_1, \dots, x_n) = 0 \quad \text{if any}$$
$$x_j = -\infty. \qquad (13.3\text{-}16)$$

If $\varphi(\mathbf{x})$ is a continuous function, the Stieltjes integral

$$\int_\square \varphi(\mathbf{x}) d^n F(\mathbf{x}) \tag{13.3-17}$$

is the limit of a Riemann–Stieltjes sum obtained by partitioning $\square$ into a large number of small rectangular boxes $\square_\mathbf{j}$ by hyperplanes $x_j = x_{j\,p}$ ($p = 0, 1, \ldots, N$), where

$$a_j = x_{j\,0} < x_{j\,1} < \cdots < x_{j\,N} = b_j$$

and then setting the integral (13.3-17) equal to

$$\lim_{(J)} \sum \varphi(\mathbf{x}'_\mathbf{j}) F(\square_\mathbf{j}),$$

where, for each $\mathbf{j}$, $\mathbf{x}'_\mathbf{j}$ is a point in $\square_\mathbf{j}$, and where "lim" means the limit as the partition of $\square$ is refined indefinitely. If $\varphi(\mathbf{x})$ is bounded in all $\mathbb{R}^n$, then, owing to the properties (13.3-15, 16) of F, the integral (13.3-17) has a limit, denoted by $\int_{\mathbb{R}^n}$, as the b_j go independently to $+\infty$ and the a_j to $-\infty$.

The *probability measure* of the ξ_j is the distribution f defined by

$$\langle f, \varphi \rangle = \int_{\mathbb{R}^n} \varphi(\mathbf{x}) d^n F(\mathbf{x}) \qquad \forall \varphi \in C_0^\infty(\mathbb{R}^n), \tag{13.3-18}$$

and the characteristic function is defined by

$$\chi(\lambda) = \int_{\mathbb{R}^n} e^{-i\lambda \cdot \mathbf{x}} \, d^n F(\mathbf{x}). \tag{13.3-19}$$

EXERCISES

1.   Find the cumulative probability $F(x, y)$ of random variables ξ, η whose values are distributed uniformly around the unit circle $\xi^2 + \eta^2 = 1$.

2.   Show that the characteristic function of the distribution in the preceding example is

$$\chi(\lambda) = J_0(|\lambda|).$$

13.4 The Normal Distributions

The most prominent feature of elementary probability theory is the central limit theorem in the next section, which says that under the influence of averaging all probability distributions tend toward the normal or Gaussian distributions, which are discussed here.

We start with the one-dimensional case. The *normal* or *Gaussian* distribution on $\mathbb{R}$ is the distribution with cumulative probability

$$\Phi(x) = \frac{1}{\sqrt{2\pi}} \int_{-\infty}^x e^{-t^2/2} \, dt \tag{13.4-1}$$

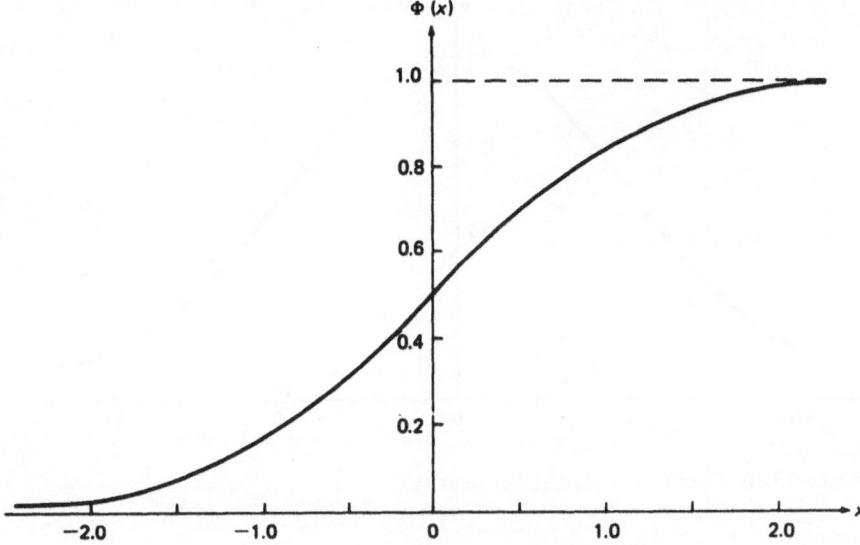

Figure 13.9 The normal probability function.

and density

$$\phi(x) = \frac{1}{\sqrt{2\pi}} e^{-x^2/2} \tag{13.4-2}$$

as shown in Figures 13-9 and 13-10.

EXERCISE

1. Show that $\Phi(\infty) = 1$ and that the normal distribution has zero mean and unit variance, and, more generally, moments of all orders given, for any positive integer k, by

$$E(\xi^{2k-1}) = 0, \qquad E(\xi^{2k}) = 1 \cdot 3 \cdot 5 \cdots (2k - 1), \tag{13.4-3}$$

where ξ is a random variable having the normal distribution. Show that the characteristic function is given by

$$\chi(\lambda) = \int_{-\infty}^{\infty} e^{-i\lambda x} d\Phi(x) = e^{-\lambda^2/2}.$$

Extensive tables of $\Phi(x)$ and related functions are found in Abramowitz and Stegun 1964. It is found that

$$\Phi(-0.67449) = \tfrac{1}{4}, \qquad \Phi(+0.67449) = \tfrac{3}{4}, \tag{13.4-4}$$

from which it follows that if ξ is normally distributed, then

$$P(-0.67449 \le \xi \le 0.67449) = \tfrac{1}{2}. \tag{13.4-5}$$

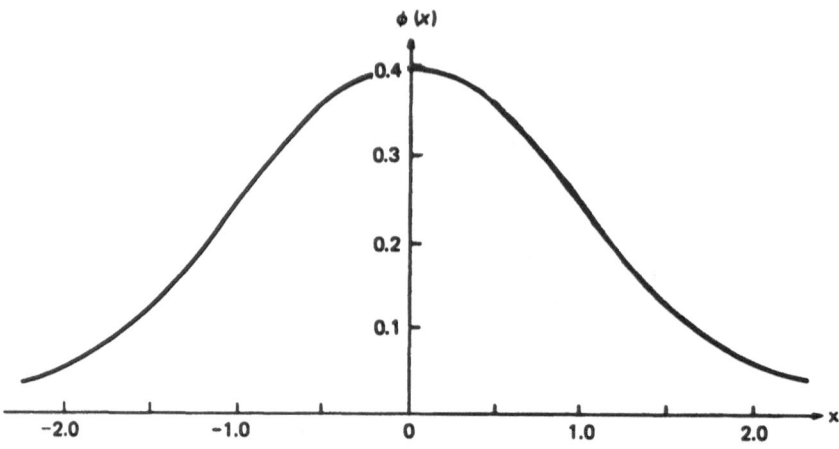

Figure 13.10 The normal probability density.

More generally, ξ is said to be *normally distributed with mean μ and variance σ^2* if

$$P\{\xi \leq x\} = F(x) = \Phi\left(\frac{x - \mu}{\sigma}\right), \qquad (13.4\text{-}6)$$

for then $E(\xi) = \mu$, $E((\xi - \mu)^2) = \sigma^2$.

A *bivariate normal distribution* for random variables, ξ, η, with given means μ_1, μ_2 and a given covariance matrix $\rho = (\rho_{kl})$—see Section 13.3, has a density of the form

$$f(x, y) = e^{-(a_{1\,1}x^2 + 2a_{1\,2}xy + a_{2\,2}y^2 + b_1 x + b_2 y + c)}, \qquad (13.4\text{-}7)$$

where

$$a_{1\,1}a_{2\,2} - a_{1\,2}^2 > 0, \qquad (13.4\text{-}8)$$

and where c is so chosen that $\int\int_{-\infty}^{\infty} f(x, y)dx\, dy = 1$. To find the relations between the μ's, ρ's, a's and b's, note that, because of (13.4-8), the exponent can be expressed as minus a sum of squares, which is written as $-\frac{1}{2}(u^2 + v^2)$, by a linear transformation

$$u = c_{1\,1}x + c_{1\,2}y + u_0,$$
$$v = c_{2\,1}x + c_{2\,2}y + v_0. \qquad (13.4\text{-}9)$$

In vector-matrix notation, these equations and their solutions are written as

$$\mathbf{u} = C\mathbf{x} + \mathbf{u}_0, \qquad \mathbf{x} = D\mathbf{u} + \mathbf{x}_0, \qquad (13.4\text{-}10)$$

where $CD = I$ and $D\mathbf{u}_0 = -\mathbf{x}_0$. If new random variables α and β are obtained from ξ and η by the transformation (13.4-9), that is, if

$$\alpha = c_{1\,1}\xi + c_{1\,2}\eta + u_0$$
$$\beta = c_{2\,1}\xi + c_{2\,2}\eta + v_0,$$

then the probability density $g(u, v)$ for (α, β) is given by $g(u, v) = f(x, y) \times |\partial(x, y)/\partial(u, v)|$, but the Jacobian is a constant, hence $g(u, v)$ is proportional to $\exp\{-\frac{1}{2}u^2 - \frac{1}{2}v^2\}$. Hence, upon normalizing, we find

$$g(u, v) = \frac{1}{2\pi} e^{-(u^2 + v^2)/2} = f(x, y) \left| \frac{\partial(x, y)}{\partial(u, v)} \right|,$$

from which it is easily seen that the means μ_1 and μ_2 of ξ and η are the components of the vector x_0, and the covariance matrix is $\rho = DD^T$ (T denotes transpose).

13.5 The Central Limit Theorem

Let a be the average of a large number n of independent measurements of a random variable ξ; it may be thought of as the value of another random variable α that results from a compound experiment consisting of n repetitions of the original experiment together with a calculation of the average; α has the same expected value as ξ and a smaller variance (by a factor $\sqrt{n}$). It will be shown that furthermore, if n is large, α has very nearly a normal distribution. If ξ has cumulative probability F, its characteristic function is denoted by

$$\chi_\xi(\lambda) = \int_{-\alpha}^{\infty} e^{-i\lambda x} \, dF(x).$$

Since there is a one-to-one correspondence between tempered distributions and their Fourier transforms, the function $\chi_\xi(\lambda)$, λ real, completely determines the measure $f = F'$.

Now let F_1 and F_2 be the cumulative probabilities of two independent random variables ξ and η. Clearly, $\xi + \eta$ may also be regarded as a random variable. It will be shown that the characteristic function of $\xi + \eta$ is the product of the characteristic functions of ξ and η. Let F_3 be the cumulative probability of the random variable $\zeta = \xi + \eta$. Then $\mathbf{E}(\varphi(\xi + \eta))$ is given by

$$\mathbf{E}(\varphi(\xi + \eta)) = \int_{-\infty}^{\infty} \int_{-\infty}^{\infty} \varphi(x + y) dF_1(x) dF_2(y); \qquad (13.5\text{-}1)$$

but also,

$$\mathbf{E}(\varphi(\xi + \eta)) = \mathbf{E}(\varphi(\zeta)) = \int_{-\infty}^{\infty} \varphi(w) dF_3(w). \qquad (13.5\text{-}2)$$

If $\varphi(x) = e^{-i\lambda x}$, it is seen that

$$\chi_{\xi + \eta}(\lambda) = \chi_\xi(\lambda)\chi_\eta(\lambda), \qquad (13.5\text{-}3)$$

as was to be proved.

Note that the absolute value of a characteristic function can never be ≥ 1, because

$$|\chi(\lambda)| \leq \int_{-\infty}^{\infty} |e^{-i\lambda x}| \, dF(x) = F(\infty) - F(-\infty) = 1$$

(F is nondecreasing).

Now, let ξ be a random variable and F its cumulative probability. By adding a suitable constant, if necessary, to each measured value of ξ, one can arrange that the resulting values have mean zero:

$$\mu = \mathbf{E}(\xi) = \int_{-\infty}^{\infty} x \, dF(x) = 0; \tag{13.5-4}$$

then, by multiplying each value by another constant, one can arrange that the resulting values have unit variance:

$$\sigma^2 = \mathbf{E}(\xi^2) = \int_{-\infty}^{\infty} x^2 \, dF(x) = 1, \tag{13.5-5}$$

and we assume that these changes have been made. It has been tacitly assumed that these integrals converge; the further assumption is now made that

$$\rho \overset{\text{def}}{=} \int_{-\infty}^{\infty} |x|^3 \, dF(x) < \infty \tag{13.5-6}$$

Let $\xi_1, \ldots, \xi_n$ be independent random variables all having the cumulative probability F, for example resulting from n independent repetitions of the measurement of ξ, so that $\eta \overset{\text{def}}{=} \xi_1 + \cdots + \xi_n$ is a random variable resulting from a compound measurement, which in turn may be repeated indefinitely often (always with the same n) to give a distribution of values of η. The variance of η is $= n$, because

$$\mathbf{E}(\eta^2) = \sum_{j=1}^{n} \sum_{k=1}^{n} \mathbf{E}(\xi_j \xi_k),$$

and

$$\mathbf{E}(\xi_j \xi_k) = \begin{cases} \displaystyle\iint xy \, dF(x) dF(y) = 0 & \text{if } j \neq k, \\[2mm] \displaystyle\int x^2 \, dF(x) = 1 & \text{if } j = k. \end{cases}$$

Consequently, the random variable

$$\zeta_n = \frac{\xi_1 + \cdots + \xi_n}{\sqrt{n}} \tag{13.5-7}$$

has zero mean and unit variance.

It is a curious and fundamental fact of probability theory that, as $n \to \infty$, the distribution of ζ_n tends to a universal probability distribution, namely the normal distribution, discussed in the preceding section, which is completely independent of the original distribution of ξ, so long as the conditions (13.5-4, 5, 6) are satisfied. This result is the famous central limit theorem. We state and prove a simple version of the theorem and then state the Berry-Essen version without proof.

Central Limit Theorem. *Let ξ and ζ_n be as above. Then the probability distribution of ζ_n converges (as a distribution) to the normal distribution, as $n \to \infty$. That is, if F_n is the cumulative probability function of ζ_n, then*

$$\int_{-\infty}^{\infty} \varphi(x)dF_n(x) \to \int_{-\infty}^{\infty} \varphi(x)d\Phi(x), \quad \text{for every test function } \varphi \qquad (13.5\text{-}8)$$

as $n \to \infty$, where Φ is given by (13.4-1).

Comments. (1) The right member, above, can be written as

$$\int \varphi(x)(2\pi)^{-1/2} \exp(-x^2/2)dx,$$

because the normal distribution has the density $(2\pi)^{-1/2} \exp(-x^2/2)$. (2) According to the Berry–Esseen theorem, below, $F_n(x) \to \Phi(x)$ pointwise uniformly.

PROOF OF THE THEOREM. If $\mathbf{P}\{\xi \leq x\}$ is $F(x)$, then $\mathbf{P}\{a\xi \leq y\}$ is $F(y/a)$, from which it follows that if $\chi(\lambda)$ is the characteristic function of ξ, then $\chi(a\lambda)$ is the characteristic function of $a\xi$. According to (13.5-3), the characteristic function of

$$\zeta_n = n^{-1/2}(\xi_1 + \cdots + \xi_n)$$

is

$$\chi_{\zeta_n}(\lambda) = \left[\chi_\xi\left(\frac{\lambda}{\sqrt{n}}\right)\right]^n. \qquad (13.5\text{-}9)$$

For given λ, it follows from (13.5-4, 5, 6) that

$$\chi_\xi(\lambda) = \int_{-\infty}^{\infty} e^{-i\lambda x} \, dF(x) = \int_{-\infty}^{\infty} [1 - i\lambda x - \tfrac{1}{2}\lambda^2 x^2 + O(\lambda^3 x^3)]dF(x)$$

$$= 1 - \tfrac{1}{2}\lambda^2 + O(\lambda^3).$$

Therefore,

$$\chi_\xi\left(\frac{\lambda}{\sqrt{n}}\right) = 1 - \frac{\lambda^2}{2n} + O\left(\frac{\lambda^3}{n^{3/2}}\right),$$

which can also be written as

$$\chi_\xi\left(\frac{\lambda}{\sqrt{n}}\right) = \left(1 - \frac{\lambda^2}{2n}\right)\left[1 + O\left(\frac{\lambda^3}{n^{3/2}}\right)\right]; \qquad (13.5\text{-}10)$$

hence,

$$\chi_{\zeta_n}(\lambda) = \left(1 - \frac{\lambda^2}{2n}\right)^n\left[1 + O\left(\frac{\lambda^3}{\sqrt{n}}\right)\right]. \qquad (13.5\text{-}11)$$

[The binomial theorem has been applied to the nth power of the square bracket in (13.5-10).] Finally, then

$$\chi_{\zeta_n}(\lambda) \rightarrow e^{-\lambda^2/2}, \quad \text{as } n \rightarrow \infty, \tag{13.5-12}$$

and the convergence is uniform with respect to λ in any finite interval. It will now be shown that if φ is any function in the class $\mathscr{S}$ of test functions for tempered distributions, then

$$\int_{-\infty}^{\infty} \varphi(\lambda)[\chi_{\zeta_n}(\lambda) - e^{-\lambda^2/2}]d\lambda \rightarrow 0, \quad \text{as } n \rightarrow \infty; \tag{13.5-13}$$

that is, if f_n and f denote the distribution of ζ_n and the normal distribution, then $\langle \hat{\varphi}, \hat{f}_n \rangle \rightarrow \langle \varphi, \hat{f} \rangle$, for any φ in $\mathscr{S}$, where the circumflex denotes the Fourier transform. Recall that $\varphi \rightarrow \hat{\varphi}$ is a one-to-one mapping of $\mathscr{S}$ onto itself and that the characteristic function of the normal distribution is $\exp(-\lambda^2/2)$. Since, for tempered distributions, $\langle \hat{\varphi}, \hat{f} \rangle$ is always equal to $\langle \hat{\varphi}, f \rangle$, it follows that $\langle \varphi, f_n \rangle \rightarrow \langle \varphi, f \rangle$, for every test function $\hat{\varphi}$; this is precisely the required result (13.5-8). To prove (13.5-13), split the range of integration into two parts: $|\lambda| < a$ and $|\lambda| > a$. Since $|\chi(\lambda)| \leq 1$ for any characteristic function, the contribution to (13.5-13) from $|\lambda| > a$ can be made arbitrarily small (independently of n) by choosing a large enough, because φ is in $\mathscr{S}$; then, because of the uniform convergence of (13.5-12) for $|\lambda| < a$, the contribution of $|x| < a$ can be made arbitrarily small by choosing n large enough. This completes the proof.

The theorem as stated says nothing about the manner or rapidity of convergence of $F_n(x)$ to $\Phi(x)$. For theorems on that aspect of the subject, see Feller, Vol. 2. Generally, rapid convergence as $n \rightarrow \infty$ requires the existence of higher moments of the distribution of ξ than the third, which is all that has been assumed here; it also requires a certain minimal degree of smoothness of $F(x)$, which is usually expressed in terms of the behavior of the characteristic function $\chi(\lambda)$ for large λ. To get uniformity with respect to x for these higher rates of convergence, a greater degree of smoothness of $F(x)$ must be postulated. A famous and remarkable theorem discovered independently by Berry in 1941 and Esseen in 1942—see Feller 1966—gives uniform convergence at the rate $O(n^{-1/2})$, with no assumptions other than (13.5-4, 5, 6). (The variance σ^2 appears explicitly instead of being set $=1$.)

Theorem. (Berry–Esseen). *Under the assumptions of this section,*

$$\left| F_n(x) - \Phi\left(\frac{x}{\sigma}\right) \right| \leq C \frac{\rho}{\sigma^3 \sqrt{n}} \quad \text{(all } x, \text{ all } n), \tag{13.5-14}$$

where C is a universal constant.

In his original paper, Esseen gave the limits

$$0.410 \leq C \leq 7.59 \tag{13.5-15}$$

which meant that with $C = 7.59$, (13.5-14) was sure to hold in all cases, while with $C < 0.410$ it was known to fail in at least one case. The result has been improved subsequently; see Feller 1966. More recent work has given the limits

$$0.410 \leq C \leq 0.800 \tag{13.5-16}$$

according to Professor Esseen (private communication 1971).

Now ρ cannot be smaller than σ^3 (see the Exercise below), but in some applications ρ and σ^3 can be expected to be comparable. Then from (13.5-16) we see that F_n differs from the normal distribution function by not more than about 0.05 for $n \approx 256$ and about 0.01 for $n \approx 6400$.

In many versions of the central limit theorem (including the Berry–Esseen version), the variables $\xi_1, \xi_2, \ldots, \xi_n, \ldots$ are not required to have identical distributions, but merely distributions that don't differ from each other too violently, in a sense specified in the theorems. The result is again that the distribution of $(\xi_1 + \cdots + \xi_n)/\sqrt{n}$ tends to the normal distribution, as $n \to \infty$.

EXERCISE

1.   By use of the Schwarz inequality show that $\int |x| dF(x) \le \sigma$. By a further use of the inequality, show that

$$\sigma^4 \le \int |x| dF(x) \int |x|^3 \, dF(x)$$

and conclude that $\sigma^3 \le \rho$. Generalize this result.

13.6 Sampling

Sometimes, a probability distribution can be calculated from physical laws; more often, inferences are drawn concerning the physical laws from an observed probability distribution. For example, an observed nonzero correlation between two random variables (i.e., a correlation coefficient $\neq 0$) is taken to indicate a causal relation between the variables. For another example, suppose that two different conjectures predict different values x_1 and x_2 of a quantity ξ, which has random variations that are due to experimental or observational errors. If a large number of observed values of ξ indicate that the expected value $E(\xi)$ agrees with one of the numbers x_1, x_2, but not with the other, that is taken as support for the corresponding conjecture.

A finite (but large) number n of observed values of a random variable ξ, or of a set of correlated variables $\xi, \eta, \ldots$ coming from n independent repetitions of an experiment or of an observed phenomenon, is called a *sample* of the distribution of ξ, or of $\xi, \eta, \ldots$. The theory of sampling is concerned with getting information from a sample, when the underlying probability distribution is unknown.

Suppose one wishes to find the expectation $E(\xi) = \mu$ of some single random variable ξ, and suppose that n independent measurements of ξ give the values $x_1, \ldots, x_n$. Then one takes the *sample average* (or *sample mean*)

$$a = \frac{x_1 + \cdots + x_n}{n} \tag{13.6-1}$$

as an approximation to μ, and one wishes to attach a "probable error" to this approximation. Clearly, a is a measurement of the random variable

$$\alpha = \frac{\xi_1 + \cdots + \xi_n}{n} \tag{13.6-2}$$

where $\xi_1, \ldots, \xi_n$ are independent and all have the same distribution as ξ. One may think of n independent repetitions of the experiment as constituting a compound experiment yielding a value of α, and one may ask how the observed values of α are distributed if the compound experiment is repeated many times. The expectation $E(\alpha)$ is simply $(1/n)[E(\xi_1) + \cdots + E(\xi_n)] = \mu$, but the variance $E((\alpha - \mu)^2)$ is readily seen to be smaller, by a factor $1/n$, than the variance σ^2 of ξ itself, because

$$E((\alpha - \mu)^2) = E\left(\left(\frac{1}{n}\sum \xi_i - \mu\right)^2\right) = \frac{1}{n^2} E((\sum[\xi_i - \mu])^2)$$

$$= \frac{1}{n^2} E(\sum[\xi_i - \mu] \sum [\xi_j - \mu]) = \frac{1}{n^2} \sum E((\xi_i - \mu)^2) = \frac{1}{n} \sigma^2;$$

$$\tag{13.6-3}$$

the next to the last equality depends on the vanishing of the correlations $E((\xi_i - \mu)(\xi_j - \mu))$ for $i \neq j$, which results from the independence of the ξ_i's.

The standard deviation $\sigma/\sqrt{n}$ of α could be taken as a measure of the probable error of the sample mean a if σ were known. However, since the underlying probability distribution is unknown, σ also has to be estimated; for this purpose the *sample variance*

$$V = \frac{1}{n} \sum_{j=1}^{n} (x_j - a)^2 \tag{13.6-4}$$

is introduced; intuition suggests that V should be rather close to σ^2. Because of (13.6-1), V can also be written as $(1/n) \sum x_j^2 - a^2$; hence, V is a measured value of the random variable

$$\beta = \frac{1}{n} \sum \xi_i^2 - \alpha^2, \tag{13.6-5}$$

and a short calculation shows that

$$E(\beta) = \frac{n-1}{n} \sigma^2. \tag{13.6-6}$$

Therefore, it seems reasonable to suppose that $\sigma^2 \approx (n/(n-1)V)$ and to take $\sqrt{V/(n-1)}$ as $\approx$ the standard deviation of α.

The Central Limit Theorem then shows that if n is large the cumulative probability of α is given approximately by

$$F(a) \approx \Phi\left(\frac{a - \mu}{\sqrt{V/(n-1)}}\right),$$

where $\mu = \mathbf{E}(\alpha) = \mathbf{E}(\xi)$. Hence if we define the probable error as

$$\delta a = 0.6745\sqrt{\frac{V}{n-1}}, \tag{13.6-7}$$

formula (13.4-4) shows that

$$\mathbf{P}\{|a - \mu| < \delta a\} = 0.5. \tag{13.6-8}$$

The interpretation of this result is as follows: suppose we make not just one set of n measurements but many sets of n measurements each. Then a and δa will have different numerical values after the various sets, while μ has always the same unknown value, and the statement $|a - \mu| < \delta a$ will be correct for about half of those sets and false for the other half.

For a more conservative error estimate, we can put

$$\delta' a = 1.6449\sqrt{\frac{V}{n-1}}, \tag{13.6-9}$$

and then

$$\mathbf{P}\{|a - \mu| < \delta' a\} = 0.9. \tag{13.6-10}$$

Although these error formulas are the ones customarily used, it should be mentioned that similar formulas can be obtained from the much more elementary Chebyshev inequality (see Exercise 1 below). Namely, if the numerical constants in (13.6-7 and 9) are increased by a factor ≈ 2, then (13.6-8 and 10). with $=$ replaced by $\geq$, follow from Chebyshev's inequality, if σ is still assumed to be $= \sqrt{V/(n-1)}$.

EXERCISES

1. Let $F(x)$ be the cumulative probability of a random variable ξ with mean μ and variance $\sigma^2 < \infty$. By considering the integral

$$\int_{|x-\mu|>t} (x - \mu)^2 dF(x),$$

derive Chebyshev's inequality

$$\mathbf{P}\{|x - \mu| > t\} \leq \sigma^2/t^2.$$

2. Verify (13.6-6).

In a sample of a distribution of two random variables ξ, η, the measured values are the pairs (x_i, y_i), $i = 1, 2, \ldots, n$; the *sample covariance matrix* is the 2×2 matrix whose elements are

$$V_{1\,1} = \frac{1}{n}\sum(x_i - a)^2, \qquad V_{2\,2} = \frac{1}{n}\sum(y_i - b)^2,$$

$$V_{1\,2} = V_{2\,1} = \frac{1}{n}\sum(x_i - a)(y_i - b), \tag{13.6-11}$$

where a and b are the sample means of ξ and η, respectively. The *sample correlation coefficient* is $W = V_{1\,2}/\sqrt{V_{1\,1}V_{2\,2}}$. The Cauchy inequality shows that $-1 \le W \le 1$ just as the Schwarz inequality shows that $-1 \le \rho \le 1$.

EXERCISE

3. Find random variables (associated with the compound experiment consisting of n determinations of ξ and η) of which $V_{1\,1}$, $V_{1\,2}$, and $V_{2\,2}$ are the expected values. Find approximate probable errors and devise a way of deciding roughly whether a given nonzero sample correlation coefficient W is significant.

13.7 Marginal and Conditional Probabilities

Suppose that each repetition of an experiment yields values x and y of the random variables ξ and η, and that the joint cumulative probability of ξ and η is

$$F(x, y) = \mathbf{P}\{\xi \le x \text{ and } \eta \le y\},$$

as in (13.3-1). The distribution of the values of x so obtained, ignoring the values of y completely, has the cumulative probability

$$F(x, \infty) = \textit{marginal cumulative probability of } \xi; \qquad (13.7\text{-}1)$$

Similarly,

$$F(\infty, y) = \textit{marginal cumulative probability of } \eta. \qquad (13.7\text{-}2)$$

If, at the other extreme, all the experiments are discarded except those in which η is found to lie in some small interval $y \le \eta < y + \Delta y$, then the resulting distribution of the values x of ξ is called the *conditional distribution* of ξ, the condition being that η lie in the interval mentioned.

To discuss conditional probabilities further, the simplifying assumption will be made that $F(x, y)$ has a continuous density

$$f(x, y) = \frac{\partial^2}{\partial x \partial y} F(x, y).$$

We shall return to the general case later.

The marginal distribution of ξ has the density

$$f_1(x) = \int_{-\infty}^{\infty} f(x, y)dy, \qquad (13.7\text{-}3)$$

which is the x-derivative of $F(x, \infty)$, because generally

$$F(x, y) = \int_{-\infty}^{x} \int_{-\infty}^{y} f(x', y')dx'\, dy',$$

owing to the boundary condition $F(-\infty, y) = F(x, -\infty) = 0$. Similarly the marginal distribution of η has the density

$$f_2(y) = \int_{-\infty}^{\infty} f(x, y)dx. \tag{13.7-4}$$

To find the conditional probabilities, we note that the fraction of the experiments in which $x \le \xi \le x + \Delta x$ and $y \le \eta \le y + \Delta y$ is approximately $f(x, y)\Delta x \Delta y$, while the total fraction in which $y \le \eta \le y + \Delta y$, regardless of the value of ξ, is $\int_{-\infty}^{\infty} f(x, y)dx\Delta y$. If all the experiments for which $\eta \notin [y, y + \Delta y]$ are discarded, then the distribution of ξ in the remaining experiments has a density given, in the limit $\Delta x, \Delta y \to 0$, by

$$\frac{1}{\Delta x} \frac{f(x, y)\Delta x \Delta y}{\int_{-\infty}^{\infty} f(x, y)dx\Delta y} = \frac{f(x, y)}{\int_{-\infty}^{\infty} f(x, y)dx} \stackrel{\text{def}}{=} f(x|y), \tag{13.7-5}$$

provided the denominator is $\ne 0$. $f(x|y)$ is called the *conditional probability density* of ξ, given that $\eta = y$. If the denominator vanishes, so does the numerator, and $f(x|y)$ is undetermined. However, the conditional probability often has a physical meaning, and it may be that it can be found from a modified experiment. For example, in an experiment to determine the distribution of scattered particles for a given primary energy E_0, it may be that the spectrum of the primary particles contains no particles at energies $\approx E_0$; then a different source of the primary particles is called for.

Physically, all probabilities are conditional, because any outcome of any experiment may depend on any of the conditions under which it is performed. Only an appeal to physics can predict whether, for example, the phase of the moon could influence a nuclear physics experiment. [For those who think this example extreme, it ought to be pointed out that the phase of the moon has a small effect on the earth's magentic field, through the influence of atmospheric tides on ionospheric currents.]

Equation (13.7-5) can be written as

$$f(x, y) = f_2(y)f(x|y), \tag{13.7-6}$$

where f_2 is the marginal density of η given by (13.7-4). Decompositions of this sort are important in Monte Carlo calculations. In the trivariate case of random variables ξ, η, ζ, we can write, with a slight change of notation.

$$f(x, y, z) = f(z)g(y|z)h(x|y, z).$$

Here $g(y|z)$ is the density of the values of η (at $\eta = y$), given that $\zeta = z$ but ignoring the values of ξ. The Monte Carlo work uses the corresponding cumulative probabilities:

$$F(z) = \int_{-\infty}^{z} f(z')dz' = \mathbf{P}\{\zeta \le z\} \qquad (\xi \text{ and } \eta \text{ ignored}) \tag{13.7-7}$$

$$G(y|z) = \int_{-\infty}^{y} f(y'|z)dy' = \mathbf{P}\{\eta \le y|\zeta = z\} \qquad (\xi \text{ ignored}) \tag{13.7-8}$$

$$H(x|y, z) = \int_{-\infty}^{x} h(x'|y, z)dx' = \mathbf{P}\{\xi \le x|\eta = y, \zeta = z\}. \tag{13.7-9}$$

It is clear that the three functions F, G, H completely determine the joint distribution of ξ, η, ζ. The description is not completely general, however, because the existence of the functions G and H depends on a certain smoothness of the distribution with respect to η and ζ (in the present derivation, we have even assumed the existence of the density $f(x, y, z)$), but it suffices for most Monte Carlo work. See Feller 1966 for a more general treatment.

As noted above, in connection with the possible vanishing of numerator and denominator in (13.7-5), there is the following lack of uniqueness of the functions G and H. If z_0 is in an interval of constancy of $F(z)$, then the distribution contains no triples ξ, η, ζ with $\zeta = z_0$, hence $G(y|z_0)$ and $H(x|y, z_0)$ are undetermined but irrelevant. Similarly, if for some z_1, y_0 is in an interval of constancy of $G(y|z_1)$, then $H(x|y_0, z_1)$ is undetermined but irrelevant.

The use of these functions in Monte Carlo calculations is explained in the next section.

13.8 Simulation: The Monte Carlo Method

The Monte Carlo method was originally developed for the study of a neutron chain reaction in a multiplicative system and is most easily described with that application in mind. The system consists of several fixed regions of space containing materials having known properties. The individual experiment (to continue the terminology of the preceding sections) consists in injecting a single neutron into the system at position $\mathbf{x}_0$ with velocity $\mathbf{v}_0$ at time t_0 and allowing the resulting branching neutron chain to develop until a certain "census" time $T > t_0$. The random variables ξ, η, ... are then the positions and velocities $\mathbf{x}_j$, $\mathbf{v}_j$ ($j = 1, \ldots, v$) of the neutrons present at time T. The number v of neutrons is also one of the random variables.

The experiment as described is to be performed repeatedly and independently a large number n of times. Wanted are the expected values of various quantities $\varphi(\xi, \eta, \ldots)$ such as the total kinetic energy of the neutrons at time T, certain moments of their spatial distributions, and so on.

However, the experiment is not performed in a laboratory, but is simulated in a computer, using random numbers. Here we are in an intermediate case between the two considered in the preceding sections. In one of those cases, the cumulative probability $F(x, y, \ldots)$ of the random variables ξ, η, ... was known and we merely wished to calculate certain expected values; in the other, $F(x, y, \ldots)$ was completely unknown, and we wished to learn something about it from a large sample of measured values.

Here the probability laws of the elementary processes (neutron–nucleus interaction) are fully known, but they become compounded in such a complicated way in the branching chain that it is almost impossible to write down, much less use, a formula for the cumulative probability $F(x, y, \ldots)$ of the result. The elementary processes are however accurately enough known for a precise simulation of the branching chains by computer, which is not only cheaper and safer, but more flexible and easier to measure than a chain reaction in a laboratory.

A Monte Carlo computer program contains a subroutine called a "random number generator." Each time the subroutine is actuated (by a CALL statement or the like) it produces a number r in the interval $0 < r < 1$. The successive numbers $r_1, r_2, \ldots$ produced in this way behave for practical purposes like independent values of a random variable ρ uniformly distributed in $(0, 1)$, i.e. with a cumulative probability $F(r) = 0$ for $r < 0$, $= r$ for $0 \leq r < 1$, and $= 1$ for $1 \leq r$. Strictly speaking, the numbers generated are neither random nor independent, since each one is computed somehow from its predecessors, but they pass all the standard statistical tests for randomness, uniformity, and independence with considerably more accuracy than needed for the Monte Carlo work. One of the earliest such subroutines used the simple formula

$$r_{k+1} \equiv 7^{13} r_k (\text{mod } 1), \tag{13.8-1}$$

where the r's are eleven-decimal-digit fractions and where the product is formed with 22 decimal accuracy but then truncated modulo 1. This scheme produces around 10^{10} different numbers before repeating, if we start with, say, $r_0 = 10^{-11}$ (obviously, r_0 should not be $= 0$ or have zero as its least significant digit).

Much work has gone into the study of random number generators, probably more than is justified since even the simplest methods, like (13.8-1), have been found to be fully satisfactory in practice.

In the simulation procedure each branching chain (i.e., each repetition of the "experiment") is constructed step-by-step using the random numbers and the known probability laws for the elementary processes.

The first step, after the initial neutron has been injected at position x_0 with velocity v_0, is to find the location x_1 of the first collision. The first free path length, i.e., the distance $|x_1 - x_0|$ to be travelled before the first collision, is just the random variable denoted by ξ in Example 2 of Section 13.1, with cumulative probability $F(x)$ shown in Figure 13-3. It is easily seen that if we draw a random number r from the generator and set the distance $= -\lambda \log r$, where λ is the mean free path, the correct probability distribution for the distance is obtained. Since the motion is in the direction of the unit vector $v_0/|v_0|$, we set

$$x_1 = x_0 + \left(\frac{-\lambda \log r}{|v_0|} \right) v_0. \tag{13.8-2}$$

The time of the first collision is

$$t_1 = t_0 + \left(\frac{-\lambda \log r}{|v_0|} \right). \tag{13.8-3}$$

This example illustrates the general principle that if $F(x)$ is the cumulative probability of a random variable ξ and if F^{-1} denotes the function inverse to F so that

$$r = F(x) \quad \Leftrightarrow \quad x = F^{-1}(r),$$

and if ρ has the uniform distribution in (0, 1), then the random variable

$$\xi = F^{-1}(\rho) \qquad (13.8\text{-}4)$$

has the distribution determined by F, because

$$\mathbf{P}\{\xi \leq x\} = \mathbf{P}\{\rho \leq r\} = r = F(x)$$

as required. Therefore, the rule for sampling a univariate distribution is to put a random number r into the inverse of the cumulative probability function.

Multivariate distributions can be sampled in a similar way using the decomposition into marginal, mixed, and conditional probabilities discussed in Section 13.7. For example, in the trivariate case, if the functions F, G, H given in (13.7-7, 8, 9) describe the distribution of variables ξ, η, ζ, and if F^{-1}, G^{-1}, H^{-1} denote the inverses of those functions with respect to the first argument, i.e., if

$$F(z) = r \quad \Leftrightarrow \quad z = F^{-1}(r),$$

$$G(y|z) = r \quad \Leftrightarrow \quad y = G^{-1}(r|z),$$

$$H(x|y, z) = r \quad \Leftrightarrow \quad x = H^{-1}(r|y, z),$$

then the sample values x, y, z of ξ, η, ζ are given by

$$z = F^{-1}(r_1)$$

$$y = G^{-1}(r_2|z)$$

$$x = H^{-1}(r_3|y, z),$$

where r_1, r_2, r_3 are independent random numbers given by the generator.

The second step in the simulation of a chain is to decide how many neutrons emerge from the collision at x_1. The number of emergent neutrons is a random variable whose cumulative probability is a step function, assumed known from laboratory measurements, and its value is determined by putting a random number into the inverse of that step function. The directions and energies of the emergent neutrons are then obtained by sampling still further elementary distributions, also known from laboratory measurements, and so on. This is continued until all branches of the chain have reached the census time T.

After a sufficiently large number of independent chains have been simulated in this way, say 1000 to 10,000, the desired statistical properties of the chain reaction are obtained by averaging, and the probable errors of the averages are computed from the formulas in Section 13.6.

The art of practical Monte Carlo calculation is based on many techniques that have been developed over the years for simplifying the sampling procedures and for reducing the variance—see Spanier and Gelbard 1969. Although the method is inherently limited in accuracy to around 1 % error, if often gives useful answers to problems of statistical physics that are completely out of the question by analytic methods because of the complicated nature of the physical systems.

13.9 Measures

Although the description in terms of the cumulative probability is the most appropriate one for the finite-dimensional cases, probability distributions can also be described in the framework of the general theory of distributions or of classical measure theory. Such descriptions can be generalized to the infinite-dimensional cases, to abstract sample spaces, and to the modern theory of stochastic processes.

Probability distributions in $\mathbb{R}^n$ belong to a class of distributions on $\mathbb{R}^n$ called *measures*, which we now discuss, mostly without proofs.

The classes $C_0^\infty = \mathscr{D}$ and $\mathscr{S}$ of test functions are often narrower than necessary for defining a particular distribution $\langle f, \cdot \rangle$, and the convergence modes $\varphi_j \overset{\mathscr{D}}{\to} \psi$ and $\varphi_j \overset{\mathscr{S}}{\to} \psi$ of test functions are often stronger than necessary for the linear functional to be continuous. For instance, the distribution $f(x) = \delta(x - x_0)$ can be considered as the functional defined by $\langle f, \varphi \rangle = \varphi(x_0)$ for all continuous φ; in this case pointwise convergence of φ_j to ψ suffices to guarantee that $\langle f, \varphi_j \rangle \to \langle f, \psi \rangle$.

In the following discussion, the convergence mode of test functions in $\mathbb{R}^n$ will be weakened so as to single out a class of distributions called measures, and it will turn out that any distribution in that class can be put in the form (13.3-18) where $F(\mathbf{x})$ is an arbitrary function of locally bounded variation. Then the domain of the functional will be extended to include all continuous functions $\varphi(\mathbf{x})$ of bounded support.

Definition 1. $\varphi_j \overset{C_0}{\to} \psi$ means that (1) the functions $\{\varphi_j(\mathbf{x})\}$ have their supports in some common bounded region of $\mathbb{R}^n$, and (2) $\varphi_j(\mathbf{x}) \to \psi(\mathbf{x})$ uniformly in $\mathbb{R}^n$.

Note. This is the same as the definition of $\varphi_j \overset{\mathscr{D}}{\to} \psi$, except that the derivatives of $\varphi_j(\mathbf{x})$ are not required to converge. Below, this definition will be applied generally to test functions in the space C_0, which are continuous but not necessarily differentiable.

Definition 2. A *measure* on $\mathbb{R}^n$ is a distribution f on $\mathbb{R}^n$ such that the functional $\langle f, \cdot \rangle$ is continuous with respect to the convergence mode just defined, that is, such that $\varphi_j \overset{C_0}{\to} \psi$ implies $\langle f, \varphi_j \rangle \to \langle f, \psi \rangle$.

In particular a probability measure f as defined by (13.3-18) is a measure, because if a sequence $\varphi_j(\mathbf{x})$ converges uniformly to $\psi(\mathbf{x})$, then

$$|\langle f, \varphi_j \rangle - \langle f, \psi \rangle| \leq \sup_{(\mathbf{x})} |\varphi_j(\mathbf{x}) - \psi(\mathbf{x})| \int_{\mathbb{R}^n} d^n F(\mathbf{x}), \qquad (13.9\text{-}1)$$

which $\to 0$, as $j \to \infty$, because $\int_{\mathbb{R}^n} d^n F(\mathbf{x}) = 1$. Therefore, f is a measure.

A distribution f on $\mathbb{R}^n$ such that $f \geq 0$ on all $\mathbb{R}^n$ is called *positive*.

Theorem.   *A positive distribution f on $\mathbb{R}^n$ is a measure. (Hence $\delta(x)$ is a measure; $\delta'(x)$, $\delta''(x)$, etc., are not measures.)*

PROOF.   Suppose $\{\varphi_k\}$ and ψ are in $^{real}C_0^\infty$ and are such that $\varphi_k \overset{C_0}{\to} \psi$. It must be proved that $\langle f, \varphi_k \rangle \to \langle f, \psi \rangle$. Then the same will hold for sequences in $^{cpx}C_0^\infty$. Let χ be a nonnegative function in C_0^∞ such that $\chi(x) = 1$ in a region containing the supports of all the φ_k and ψ. Given any $\varepsilon > 0$; for k large enough

$$-\varepsilon\chi(x) < \varphi_k(x) - \psi(x) < \varepsilon\chi(x);$$

hence the functions $\varepsilon\chi \pm (\varphi_k - \psi)$ are both nonnegative, hence

$$\varepsilon\langle f, \chi \rangle \pm [\langle f, \varphi_k \rangle - \langle f, \psi \rangle] \geq 0,$$

because f is a positive distribution; therefore $\langle f, \varphi_k \rangle \to \langle f, \psi \rangle$.

Furthermore, a distribution f defined by the Stieltjes integral (13.3-18) is a measure even if F does not satisfy the requirements (13.3-15 and 16) of a cumulative probability, but is merely any (generally complex) function of locally bounded variation. According to the Appendix to this chapter, F can be written as $F_1 - F_2 + iF_3 - iF_4$, where each $F_k(x)$ is nondecreasing. Then $\langle f, \varphi_j \rangle$ and $\langle f, \psi \rangle$ are decomposed into four terms each, and (13.9-1) holds for each term, after replacing $\int_{\mathbb{R}^n}$ by $\int_\square$, where $\square$ is a box that contains the supports of all the φ_j; hence, f is a measure.

According to the Riesz representation theorem, which will now be stated without proof, every measure is of the form (13.3-18) for a suitably chosen function $F(x)$. Hence, in one dimension, every measure f on $\mathbb{R}$ is the derivative F' (in the sense of distribution theory) of a function $F(x)$ of locally bounded variation. In n dimensions, $f = \partial^n F / \partial x_1 \cdots \partial x_n$.

Theorem (Riesz). *If f is a measure on $\mathbb{R}^n$, that is, if the functional $\langle f, \cdot \rangle$ is continuous with respect to the convergence mode $\overset{C_0}{\to}$, then there is a function $\sigma(x)$ of locally bounded variation such that*

$$\langle f, \varphi \rangle = \int_{\mathbb{R}^n} \varphi(x) d^n \sigma(x) \quad \text{for all } \varphi \text{ in } C_0(\mathbb{R}^n). \tag{13.9-2}$$

Furthermore, if σ is normalized by requiring

$$\sigma(0) = 0, \qquad \sigma(x) = \lim_{\substack{\varepsilon \to 0 \\ \varepsilon > 0}} \sigma(x + \varepsilon)$$

(or in lots of other possible ways), then σ is unique, for given f. ($\varepsilon > 0$ means that each $\varepsilon_j > 0$, $j = 1, \ldots, n$.)

Riesz's original version of the theorem, given in 1909 (see Riesz and Nagy 1953, Section 50), was for the case of a finite interval $[a, b]$ in one dimension. The form of the theorem given above is discussed for example in Laurent

Schwartz 1950, Section I.1 (see also Riesz–Nagy Section 59). The corresponding theorem for a measure on an abstract compact Hausdorf space is given in Dunford and Schwartz 1966, IV. 6.3.

It will now be shown that if f is a measure, the domain of definition of the functional $\langle f, \cdot \rangle$ can be extended from C_0^∞ to the class C_0 of all continuous functions of bounded support on $\mathbb{R}^n$, and the functional remains continuous with respect to the convergence mode $\overset{C_0}{\to}$ defined above.

It is first shown that f is locally bounded. It is recalled that a linear functional on a Banach or Hilbert space is continuous (i.e. with respect to convergence in the norm of the space) if and only if it is bounded. Here we have slightly less:

Lemma.   *If a linear functional $\langle f, \cdot \rangle$ on $\mathbb{R}^n$ is continuous in the sense of Definition 2, i.e., if f is a measure, and if Ω is a bounded region in $\mathbb{R}^n$, then there is a constant $K = K(\Omega)$ such that*

$$|\langle f, \varphi \rangle| \leq K(\Omega) \sup_{(x)} |\varphi(x)|$$

for all test functions φ with support in Ω.

The proof is by contradiction and is left as an exercise for the reader.

Extension Theorem.   *If f is a measure, the domain of definition of the functional $\langle f, \cdot \rangle$ can be enlarged to include all functions φ in the class C_0: the resulting extended functional, say $\langle f_1, \cdot \rangle$, remains continuous, in the sense that if $g_n(x)$ and $g(x)$ are in C_0 and $g_n(x) \overset{C_0}{\to} g(x)$ then $\langle f_1, g_n \rangle \to \langle f_1, g \rangle$. The extension f_1 of f is unique.*

PROOF, IN PART.   Given any $g = g(x)$ in C_0, $\langle f_1, g \rangle$ must be defined. Let $\{\varphi_n(x)\}$ be a sequence of C^∞ functions with support in a bounded region Ω (which contains the support of g) and such that $\varphi_n(x) \to g(x)$ uniformly, for all x, constructed for example by mollifiers applied to g—see Section 2.6. Then, according to the Lemma,

$$|\langle f, \varphi_k \rangle - \langle f, \varphi_l \rangle| \leq K(\Omega) \sup_{(x)} |\varphi_k(x) - \varphi_l(x)|,$$

so that $\{\langle f, \varphi_n \rangle\}$ is a Cauchy sequence of numbers. Define

$$\langle f_1, g \rangle = \lim_{n \to \infty} \langle f, \varphi_n \rangle.$$

EXERCISE

1. Complete the proof, by showing that
    1. If $\{\psi_n(x)\}$ is any other such sequence also converging uniformly to $g(x)$, then the sequence $\{\langle f, \psi_n \rangle\}$ has the same limit as $\{\langle f, \varphi_n \rangle\}$,
    2. The functional $\langle f_1, \cdot \rangle$ thus defined is linear,
    3. If $g(x)$ is in C_0^∞, then $\langle f_1, g \rangle = \langle f, g \rangle$,

4. If $g_n(x)$ and $g(x)$ are in C_0, and $g_n \overset{C_0}{\to} g$, then $\langle f_1, g_n \rangle \to \langle f_1, g \rangle$.
5. If $\langle f_2, \cdot \rangle$ is any other functional on C_0 having the stated properties, then $f_1 = f_2$.

The subscript on f_1 is usually omitted, unless it is necessary to distinguish between the functional $\langle f, \cdot \rangle$ and its extension.

For a given class of distributions there is a natural space of test functions. For the class of Schwartzian distributions, it is $C_0^\infty = \mathscr{D}$; for tempered distributions it is $\mathscr{S}$; for measures, it is C_0; and for distributions in L^2 (Chapter 5), it is L^2 itself. In each of these examples we have a space X of test functions with a convergence mode $\overset{X}{\to}$, in which C_0^∞ is densely embedded; that is; (1) for functions in C_0^∞, $\varphi_j \overset{\mathscr{D}}{\to} \varphi$ implies $\varphi_j \overset{X}{\to} \varphi$, and (2) for any function χ in X there are functions ψ_j in C_0^∞ such that $\psi_j \overset{X}{\to} \chi$. A continuous linear functional on X determines and is completely determined by its restriction to C_0^∞, which is a distribution in the sense of Chapter 2, and we call X the *natural space of test functions* for such distributions.

A measure f is called *bounded* if there is a constant K such that $|\langle f, \varphi \rangle| \le K \sup|\varphi(x)|$ for all φ in C_0. [According to the Lemma, every measure is *locally* bounded.]

Corollary to the Extension Theorem. *If f is a* bounded *measure, then the domain of definition of the functional $\langle f, \cdot \rangle$ can be extended to all $C = \{bounded\ continuous\ functions\}$, with preservation of the bound of $\langle f, \cdot \rangle$.*

PROOF IN ONE DIMENSION. Given $\psi \in C$; $\langle f, \psi \rangle$ must be defined. Let $\chi_n(x)$ be the continuous function $= 1$ for $|x| \le n$, $= 0$ for $|x| \ge n + 1$, and linear for $-n - 1 < x < -n$ and for $n < x < n + 1$. Then $\chi_n(x)\psi(x)$ is in C_0 for each n. The distribution f and the function ψ can be decomposed as

$$f = f_1 - f_2 + if_3 - if_4, \qquad \psi = \psi_1 - \psi_2 + i\psi_3 - i\psi_4,$$

where the f_i and the ψ_i are ≥ 0. For each i, j, $\langle f_i, \chi_n\psi_j \rangle$ is a bounded nondecreasing function of n. Consequently,

$$\lim \langle f, \chi_n\psi \rangle \text{ exists, } \overset{\text{def}}{=} \langle f, \psi \rangle.$$

Now

$$|\langle f, \chi_n\psi \rangle| \le K \sup|\chi_n(x)\psi(x)| \le K \sup|\psi(x)|;$$

therefore,

$$|\langle f, \psi \rangle| \le K \sup|\psi(x)|,$$

as claimed.

When $\langle f, \varphi \rangle$ is regarded as the value of the integral $\int_{-\infty}^{\infty} f(x)\varphi(x)dx$ in a generalized sense, then the extension of the domain of definition of the functional $\langle f, \cdot \rangle$ amounts to an extension of the concept of an integral of the form

$$\int_{-\infty}^{\infty} f(x)g(x)dx.$$

If f is an arbitrary Schwartzian distribution, then the integral is defined only for $g \in C_0^\infty$; if f is a tempered distribution, then the integral is defined for all $g \in \mathscr{S}$; if f is a measure, then the integral is defined for all $g \in C_0$; if f is a bounded measure, then the integral is defined for all bounded continuous functions g; if f is in L^2, then the integral is defined for all g in L^2. Similar remarks apply to distributions on $\mathbb{R}^n$.

13.10 Measures as Set Functions

The mathematical theory of probability is usually treated as a branch of measure theory. Although that is unnecessary for most applications, including the ones considered in this book (where the sample space is finite-dimensional), a brief introduction to the classical measure-theoretic approach is given in this section, because the reader will encounter it in the general literature, and because it is the only available approach to certain infinite-dimensional problems, for example where each event is a path in space-time and the totality of all paths is an infinite-dimensional space.

It is recalled that a nondecreasing function $F(x)$ on $\mathbb{R}$ determines a (positive) measure on $\mathbb{R}$. Classically, this measure is defined as the assignment of a nonnegative number $\mu(S)$ to each set S in a certain class of point sets on $\mathbb{R}$; $\mu(S)$ is the *measure* of S. The measure of an interval is defined as follows:

$$\mu(\Delta) = F(b - 0) - F(a + 0), \quad \text{If } \Delta = (a, b),$$

$$\mu(\Delta) = F(b + 0) - F(a - 0), \quad \text{if } \Delta = [a, b],$$

$$\mu(\Delta) = F(b + 0) - F(a + 0), \quad \text{if } \Delta = (a, b],$$ (13.10-1)

$$\mu(\Delta) = F(b - 0) - F(a - 0), \quad \text{if } \Delta = [a, b),$$

where $F(x \pm 0)$ denotes the limit of $F(x \pm \varepsilon)$, as $\varepsilon \downarrow 0$.

Note 1.   It is really only necessary to define $\mu(\Delta)$ for open intervals, because the measures of complements of measurable sets are defined below.

Note 2.   The measure of a degenerate interval $[a, a]$, containing only one point, is well-defined and may be > 0.) If the total variation of F is $= 1$, so that F may be regarded as the cumulative probability of a random variable ξ, then $\mu(\Delta)$ is the probability of ξ lying in Δ. We wish to extend this notion so that $\mu(S)$ is the probability of ξ lying in S, for more general sets S. If Δ_1 and Δ_2 are any two disjoint intervals, then clearly $\mu(\Delta_1 \cup \Delta_2)$ must be defined as $\mu(\Delta_1) + \mu(\Delta_2)$. Since the union of any two intervals can always be written as the union of two disjoint intervals, $\mu(\Delta_1 \cup \Delta_2)$ is then defined for all Δ_1 and Δ_2 and is $\leq \mu(\Delta_1) + \mu(\Delta_2)$; hence $\mu(S)$ is defined if S is any finite union of intervals. If S is a countable union $\bigcup_{j=1}^{\infty} \Delta_j$, then $\mu(\bigcup_{j=1}^{n} \Delta_j)$ is a nondecreasing function of n, and one defines

$$\mu(S) = \lim_{n \to \infty} \mu\left(\bigcup_{j=1}^{n} \Delta_j\right).$$ (13.10-2)

Note.   If the total variation of F on $\mathbb{R}$ is infinite and S is an unbounded set, then $\mu(S)$ may be $+\infty$.) Now suppose that a set $S = \bigcup_{j=1}^{\infty} S_j$ is contained in an interval Δ, and call S' the complement of S in Δ, i.e. the set of all points that are in Δ but not in S. The probability interpretation shows that $\mu(S')$ must be $= \mu(\Delta) - \mu(S)$, even when S' is not itself a countable union of intervals. For example, S may be dense in Δ (so that S' contains no non-degenerate intervals), while S' is still an uncountable set.

The procedure just described repeated indefinitely, defines $\mu(S)$ for all sets S in the *Borel class* (or *field*) $\mathfrak{B}$ of sets on $\mathbb{R}$, which is the smallest class of sets such that

1.   Every interval is in $\mathfrak{B}$,
2.   A countable union of sets in $\mathfrak{B}$ is in $\mathfrak{B}$,
3.   The complement of a set in $\mathfrak{B}$ is in $\mathfrak{B}$.

The measure μ is a function of sets defined for all S in $\mathfrak{B}$, and such that $\mu(S) \geq 0$ for all S in $\mathfrak{B}$

$$\mu\left(\bigcup_{j=1}^{\infty} S_j\right) \leq \sum_{j=1}^{\infty} \mu(S_j)$$

$$\mu\left(\bigcup_{j=1}^{\infty} S_j\right) = \sum_{j=1}^{\infty} \mu(S_j), \quad \text{if the } S_j \text{ are pairwise disjoint.}$$

(13.10-3)

A set S has *measure zero* if, for every $\varepsilon > 0$ S can be enclosed in a Borel set S' of measure $\leq \varepsilon$ or equivalently in an open set S' of measure $\leq \varepsilon$, as stated in Section 13.1. If S is any set which differs from a Borel set S_0 by a set of measure zero (i.e., if S can be obtained from S_0 by deleting the points of S_1 and adding the points of S_2, where S_1 and S_2 both have measure zero) then $\mu(S)$ is defined as $\mu(S_0)$. This last step is known as *completing the measure*, and the sets S obtainable in this way are called *measurable* or *μ-measurable* sets.

Note that the class of Borel sets is independent of the choice of the function F, while the class of sets having measure zero is not. If F is constant in an interval Δ, then $\mu(\Delta) = 0$, even if the length Δ is > 0. On the other hand, if F has a jump at $x = \xi$, and if $\{\xi\}$ denotes the set consisting of the single point ξ, then $\mu(\{\xi\}) > 0$. If $F(x) \equiv x$ (in this case the measure of any interval is its length), the measure μ is called *Lebesgue measure*, and is often denoted by m. It is known that there are sets S that are not Lebesgue measurable; if such an S lies in an interval of constancy of a function F that determines a measure μ, then S is μ-measurable, and $\mu(S) = 0$. Hence, the class of measurable sets depends on μ (i.e., on F), although the class of Borel sets does not.

If F is a real function of locally bounded variation, not necessarily decreasing, then μ is a real, not necessarily positive, measure on $\mathbb{R}$. If F is a complex function of locally bounded variation, then μ is a complex measure on $\mathbb{R}$.

The n-dimensional case will now be described for $n = 2$. Let $F = F(x, y)$ be a nondecreasing function in the plane $\mathbb{R}^2$ (see Section 13.3). The measure μ determined by F is first defined for open sets. If Ω is any open set in the plane, then $\mathscr{A}(\Omega)$ denotes the set of *admissible* functions for Ω, namely real functions $\varphi(x, y)$ such that

1. $\varphi(x, y)$ is continuous in all $\mathbb{R}^2$,
2. $\varphi(x, y) = 0$, for $(x, y) \notin \Omega$,
3. $\varphi(x, y) \leq 1$ in all $\mathbb{R}^2$.

Then $\mu(\Omega)$ is defined as follows:

$$\mu(\Omega) = \sup_{\varphi \in \mathscr{A}(\Omega)} \iint_{\mathbb{R}^2} \varphi(x, y) d^2 F(x, y). \tag{13.10-4}$$

[The double Stieltjes integral of a continuous junction φ with respect to a nondecreasing function F was discussed in Section 13.3]. The function φ can be arbitrarily "close," in some sense, to the characteristic function of Ω, i.e., the function $= 1$ in Ω and $= 0$ elsewhere. This suggests that the above should be written as

$$\mu(\Omega) = \iint_{\Omega} d^2 F(x, y), \tag{13.10-5}$$

and it is easily verified that that is correct if Ω is a region (rectangle, disk, ellipse, triangle, annulus, etc.) for which one knows how to define the integral; otherwise, (13.10-4) may be regarded as the definition of (13.10-5). If $F(x, y) = x + y$, then μ is *Lebesgue measure* on $\mathbb{R}^2$, and $\mu(\Omega)$ is the area of Ω and can be written as $\iint_\Omega dx\, dy$. If F is the cumulative probability of random variables ξ, η, then $\mu(\Omega)$ is the probability of the point (ξ, η) lying in Ω.

From here on, the discussion is the same as for $\mathbb{R}$, except that "open set" replaces "interval" throughout. The Borel sets in $\mathbb{R}^2$ are those that can be obtained from the open sets by the operations of complementation and countable union (hence also countable intersection). The set function $\mu(S)$ defined as described above for the Borel sets has again the properties (13.10-3) and is called a *positive measure* on $\mathbb{R}^2$.

In the abstract theory, one starts with an abstract *sample space* $\mathfrak{S}$ (a set of otherwise undefined points x, which are called *events* or *simple events* and are thought of as the possible outcomes of a trial or experiment) and a so-called *σ-algebra* $\mathfrak{A}$ of subsets of $\mathfrak{S}$; this is a collection of subsets such that the complement of any S in $\mathfrak{A}$ is in $\mathfrak{A}$, and such that if $\{S_n\}$ is any countable collection of the sets in $\mathfrak{A}$, then the union and the intersection of the sets $\{S_n\}$ are in $\mathfrak{A}$. There is then given a nonnegative set function $\mathbf{P}$ on $\mathfrak{A}$, for which $\mathbf{P}(\mathfrak{S}) = 1$, and which is *countably additive*; this means that if $\{S_n\}$ is a countable collection of disjoint sets in $\mathfrak{A}$, then $\mathbf{P}(\bigcup S_n) = \sum \mathbf{P}(S_n)$. (It follows that if "disjoint" is dropped, then " $=$ " is replaced by " $\leq$ ".) The triple $\{\mathfrak{S}, \mathfrak{A}, \mathbf{P}\}$ is called a *probability space*. See Feller 1966.

$\mathbf{P}(S)$ is regarded as the probability that the outcome of a trial is one of the points in the set S. If $\mathfrak{S}$ is a topological space, and $\mathfrak{A}$ is the smallest σ-algebra that contains the open sets of $\mathfrak{S}$, then the sets in the collection $\mathfrak{A}$ are called the *Borel sets* of $\mathfrak{S}$.

The terminology varies somewhat in the literature. A σ-algebra is sometimes called a "Borel algebra" or a "σ-field" or a "Borel field."

If $\varphi(x)$ is a real-valued function defined on $\mathfrak{S}$, we should like to define its expectation as the Stieltjes integral

$$\mathbf{E}(\varphi(x)) = \int_{\mathfrak{S}} \varphi(x) d\mathbf{P}(x). \tag{13.10-6}$$

In order to define such an integral in an abstract probability space, we restrict φ to the following class of functions:

Definition. $\varphi(x)$ is called a *random variable* on the probability space $\{\mathfrak{S}, \mathfrak{A}, \mathbf{P}\}$ if for every real t the set of all x for which $\varphi(x) \le t$ is in the σ-algebra $\mathfrak{A}$.

Note. x denotes an undefined point in an abstract space $\mathfrak{S}$, but if $\mathfrak{S}$ is covered by any *reasonable* coordinate system (finite or infinite), then each coordinate of x is a random variable, in agreement with the usage in the preceding sections.

In the present context, because $\mathbf{P}$ is a positive measure, it is not necessary to use the full theory of the Stieltjes integral in an abstract measure space to interpret (13.10-6). Instead, we proceed as follows (see Feller, 1966, Section IV.4): First, $\varphi(x)$ is a *simple* function if it assumes just the values $\varphi^{(1)}, \varphi^{(2)}, \ldots$ on the sets $S_1, S_2, \ldots$ in the algebra. Then we define

$$\mathbf{E}(\varphi(x)) = \sum_{j=1}^{\infty} \varphi^{(j)} \mathbf{P}(S_j) \tag{13.10-7}$$

provided that the sum converges absolutely; otherwise the expectation is undefined. Any random variable $\varphi(x)$ is the uniform limit of a sequence $\{\varphi_k(x)\}$ of simple functions, and we set

$$\mathbf{E}(\varphi(x)) = \lim_{k \to \infty} \mathbf{E}(\varphi_k(x)). \tag{13.10-8}$$

The interpretation is this: It can be proved that either the quantities $\mathbf{E}(\varphi_k(x))$ all exist for sufficiently large k and their limit exists or they are all undefined. Furthermore, if the limit exists, it is the same for all sequences $\{\varphi_k\}$ converging to φ.

This completes the description of the abstract conceptual framework of the mathematical theory of probability. For further developments, see Feller, Vol. 2, 1966.

We return now to the finite-dimensional case and consider briefly the question when two different measures determine the same measurable sets. This equation leads to the notion of absolutely continuous functions and measures and to the Radon–Nikodym Theorem. We shall discuss mainly the one-dimensional case. As noted above, the Borel sets on $\mathbb{R}$ (or on $\mathbb{R}^n$) are independent of the choice of the function $F(x)$ (or $F(x)$) used for defining a measure μ, but the sets of measure zero, which are used in completing the measure, depend on F. We are therefore concerned with the question when the sets of μ measure zero are the same as the sets of Lebesgue measure zero, and with the further question when, given two measures μ_1 and μ_2, obtained from functions F_1 and F_2, the sets of μ_1 measure zero are the same as those of μ_2 measure zero.

Theorem 1.   *Every set on $\mathbb{R}$ of Lebesgue measure zero has μ measure zero if and only if the function $F(x)$ can be written as a Lebesgue integral*

$$F(x) = \int^x f(x')dx'; \qquad (13.10\text{-}9)$$

then, the derivative $F'(x)$ exists almost everywhere and is equal almost everywhere to $f(x)$.

Feller 1966 defines $F(x)$ to be absolutely continuous if it can be expressed in the above form, and he calls the above theorem, slightly generalized, the Radon–Nikodym theorem. Following that point of view, we called a probability distribution **P** absolutely continuous (Section 13.1) if its cumulative probability can be expressed in the form (13.10-9); the theorem shows that for such a distribution, if S is a set of Lebesgue measure zero, then the probability $\mathbf{P}\{x \in S\}$ is $=0$. The statement of the above theorem applies more generally, however, to any real or complex function $F(x)$ of locally bounded variation. Another and more frequent definition of absolute continuity is this:

Definition 1.   A function $F(x)$ is called *absolutely continuous* on $\mathbb{R}$ if for every $\varepsilon > 0$ there is a $\delta > 0$ such that whenever a union of finitely many disjoint intervals (a_k, b_k) has Lebesgue measure less than δ, i.e.,

$$\sum (b_k - a_k) < \delta, \qquad (13.10\text{-}10)$$

then

$$\sum |F(b_k) - F(a_k)| < \varepsilon. \qquad (13.10\text{-}11)$$

Absolute continuity in an interval (α, β) on $\mathbb{R}$ is similarly defined.

The two concepts of absolute continuity are equivalent for a finite interval but not for all $\mathbb{R}$; for example the function $F(x) = x^2$ can be written in the form (13.10-9), but is not absolutely continuous as here defined, because for any $\delta > 0$ we can find an interval $(a, a + \delta)$ such that $(a + \delta)^2 - a^2$ is arbitrarily large by simply taking a large enough. However, under the additional restriction that the total variation of F on $\mathbb{R}$ be finite, (13.10-9) implies absolute continuity. Conversely, if $F(x)$ is absolutely continuous on $\mathbb{R}$ according to the above definition, then (1) it is easily proved that $F(x)$ has bounded variation on $\mathbb{R}$, and (2) it follows from the Radon–Nikodym Theorem below that $F(x)$ can be represented in the form (13.10-9). Lastly, a theorem of Banach and Zarecki (see Natanson 1955) says that if $F(x)$ has bounded variation on $\mathbb{R}$ and if $\mu(S) = 0$ for every set S of Lebesgue measure zero, then $F(x)$ is absolutely continuous, as defined above. (If $F(x)$ does not have bounded variation, the measure μ is of course not defined.)

If $F(x)$ is the Cantor function described in Example 6 of Section 13.1 and sketched in Figure 13-7, it was shown that for any $\delta > 0$ we can find a finite collection of open intervals on $\mathbb{R}$ of total length less than δ but such that the total increase of $F(x)$ takes place in those intervals; then the sum in (13.10-11) is $= 1$. Hence $F(x)$ is not absolutely continuous. It was also pointed out that $F'(x)$ is $= 0$ almost everywhere, so that the right member of (13.10-9), when interpreted as a Lebesgue integral, which is the only way it can be interpreted, is a constant, hence is not equal to $F(x)$.

In the Radon–Nikodym Theorem, Legesgue measure is replaced by an arbitrary positive measure μ_1 on $\mathbb{R}$.

Definition 2. A measure μ_2 on $\mathbb{R}$ is called *absolutely continuous with respect to* a positive measure μ_1 if for every $\varepsilon > 0$ there is a $\delta > 0$ such that whenever a union of finitely many disjoint intervals Δ_k has μ_1 measure less than δ, i.e.,

$$\sum \mu_1(\Delta_k) = \sum F_1(\Delta_k) < \delta,$$

then

$$\sum |\mu_2(\Delta_k)| = \sum |F_2(\Delta_k)| < \varepsilon.$$

(Note that $\sum \mu_1(\Delta_k)$ could have been written as $\sum |\mu_1(\Delta_k)|$, because μ_1 is a positive measure.)

Theorem (Radon–Nikodym).   *If μ_2 is absolutely continuous with respect to μ_1, then F_2 can be written in the form*

$$F_2(x) = \int^x f(x')dF_1(x').$$

An application to the representation of a physical system in which a given observable is diagonal is found in Section 14.8.

The multidimensional version is the same except that Δ_k represents a rectangular box (also called an *interval* in $\mathbb{R}^n$), $F(\Delta_k)$ is as defined in Section 13.3, and the equation in the Radom–Nikodym Theorem is interpreted as

$$F_2(\mathbf{x}) = \int^{x_1} \cdots \int^{x_n} f(\mathbf{x}')d^n F_1(\mathbf{x}').$$

For an abstract version of the theorem, see Halmos 1950 or Dunford and Schwartz 1966.

The answer to the question with which we began this discussion is that a sufficient condition for two measures μ_1 and μ_2 to determine the same sets of measure zero is that they be positive measures and each be absolutely continuous with respect to the other.

1.   Let $f(x)$ and $F(y)$ be absolutely continuous functions. Show that $F(f(x))$ is absolutely continuous if $f(x)$ is monotonic or if $F(y)$ is Lipschitz continuous.

2.   Show that if

$$f(x) = \left\{ \begin{array}{ll} \left(x \sin \dfrac{1}{x}\right)^3 & \text{for } x \neq 0, \\[2mm] 0 & \text{for } x = 0, \end{array} \right\}$$

$$F(y) = y^{1/3} \quad \text{(real root)},$$

then $f(x)$ and $F(y)$ are absolutely continuous in $(-1, 1)$, but $F(f(x))$ is not. Why does this not contradict Exercise 1? Note that the integrals

$$\int_{x_0}^{1} \frac{d}{dx} F(f(x)) dx, \qquad \int_{-1}^{-x_0} \text{same,}$$

diverge (fail to converge) as Riemann integrals at $x_0 = 0$; calling them Lebesgue integrals does not help.

13.11 Probability in Hilbert Space; Cylinder Sets; Gaussian Measures

We present here some of the simplest ideas concerning measures in topological vector spaces, in order to illustrate the ideas given in the preceding section and to show some of the things that can happen in the infinite-dimensional case. For further details, see Gel'fand and Vilenkin 1961 (Volume 4 of Generalized Functions).

In a finite-dimensional space $\mathbb{R}^n$, Lebesgue measure m has the property that it is the same for congruent sets: If S is measurable and S' is obtained from S by translation and rotation in $\mathbb{R}^n$, then $m(S') = m(S)$. If Ω is an open set, $m(\Omega)$ is its volume; hence, if Ω is also bounded and nonempty, then

$$0 < m(\Omega) < \infty. \tag{13.11-1}$$

There is no measure having these properties in the infinite-dimensional case. Let $\{\varphi_k\}_1^\infty$ be a complete orthonormal set in a separable real Hilbert space $\mathfrak{H}$. For any x in $\mathfrak{H}$ we write

$$x = \sum_1^\infty x_k \varphi_k,$$

and we call the x_k the coordinates of x. Consider the bounded open sets

$$\Omega_1 = \left\{ x : 0 < x_k < \frac{1}{k} \ (k = 1, 2, \ldots) \right\}$$

$$\Omega_{1/2} = \left\{ x : 0 < x_k < \frac{1}{2k} \ (k = 1, 2, \ldots) \right\}. \tag{13.11-2}$$

By translation of $\Omega_{1/2}$ in various directions we can generate infinitely many disjoint copies of it in Ω_1. Hence, if $m(\Omega_{1/2}) > 0$, then $M(\Omega_1) = \infty$, while if $m(\Omega_1) < \infty$, then $m(\Omega_{1/2}) = 0$, so that (13.11-1) is violated. Hence there is no concept of volume in $\mathfrak{H}$ and no concept of probability density. There are however probability distributions, including continuous ones, based on the so-called cylinder sets, which, according to Gel'fand and Vilenkin, were introduced by Kolmogorov in 1936.

If $\mathfrak{M}$ is any finite-dimensional subspace of $\mathfrak{H}$ and S is any Borel set in $\mathfrak{M}$, then the set

$$Z = S + \mathfrak{M}^{\perp}, \qquad (13.11\text{-}3)$$

i.e. the set of all points $x + y$ in $\mathfrak{H}$, where $x \in S$ and $y \in \mathfrak{M}^{\perp}$, is called a *cylinder set*; S is its *base* and $\mathfrak{M}^{\perp}$ its *generator*. (If $\mathfrak{M}$ were two-dimensional and $\mathfrak{M}^{\perp}$ one-dimensional, Z would be an ordinary three-dimensional cylinder, but of course in the above definition $\mathfrak{M}^{\perp}$ is necessarily infinite-dimensional.)

To say that S is a Borel set in $\mathfrak{M}$ refers to the usual topology of $\mathfrak{M}$, which is isomorphic to $\mathbb{R}^m$ for some m. The Borel sets are generated by the operations of complementation (with respect to $\mathfrak{M}$) and countable union, starting with the open sets of $\mathfrak{M}$.

The base S and the generator $\mathfrak{M}^{\perp}$ of a cylinder set Z are not uniquely determined by Z. For example, we can always replace the subspace $\mathfrak{M}$ by a larger one $\mathfrak{M}'$ (i.e., one of larger dimension) that contains $\mathfrak{M}$ and then replace the set S by the set S' in $\mathfrak{M}'$ given by $S + (\mathfrak{M}' \ominus \mathfrak{M})$, where $\mathfrak{M}' \ominus \mathfrak{M}$ denotes the orthogonal complement of $\mathfrak{M}$ in $\mathfrak{M}'$. Then clearly $S' + \mathfrak{M}'^{\perp}$ is the same as $S + \mathfrak{M}^{\perp}$. It follows that any two cylinder sets Z_1 and Z_2 (or any finite number of them) can be described as having bases in a common finite-dimensional subspace $\mathfrak{M}$ and a common generator $\mathfrak{M}^{\perp}$; namely, if $\mathfrak{M}_1$ and $\mathfrak{M}_2$ are the subspaces for Z_1 and Z_2, we take $\mathfrak{M}$ as the subspace spanned by $\mathfrak{M}_1$ and $\mathfrak{M}_2$.

In this way we see that the cylinder sets form an *algebra* $\mathfrak{A}_0$ of sets, that is, a collection of sets having the following properties:

1. The union and intersection of any two cylinder sets are cylinder sets.
2. The complement (in $\mathfrak{H}$) of a cylinder set is a cylinder set.

The algebra has the further property that if a countable collection $\{Z_i\}_1^{\infty}$ of cylinder sets all have their bases in a common finite-dimensional subspace, then their union $\bigcup_{i=1}^{\infty} Z_i$ and their intersection $\bigcap_{i=1}^{\infty} Z_i$ are cylinder sets.

$\mathfrak{A}_0$ is not a σ-algebra because a countable union $\bigcup_{i=1}^{\infty} Z_i$ is not in general a cylinder set unless the Z_i all have their bases in a common finite-dimensional subspace, but we define $\mathfrak{A}$ as the smallest σ-algebra containing $\mathfrak{A}_0$. $\mathfrak{A}$ is obtained by the operations of complementation and countable union starting with the cylinder sets. The sets Ω_1 and $\Omega_{1/2}$ given by (13.11-2) are in $\mathfrak{A}$.

Now suppose that $\mathbf{P}$ is a countable additive set function defined on the σ-algebra $\mathfrak{A}$ and satisfying the axioms

$$0 \le \mathbf{P}(X) \le 1 = \mathbf{P}(\mathfrak{H}) \qquad (13.11\text{-}4)$$

as in the preceding section. Then the triple $\{\mathfrak{H}, \mathfrak{A}, \mathbf{P}\}$ is a probability space.

If Z is a cylinder set, $\mathbf{P}(Z)$ can be interpreted as a marginal probability. Namely, if $\mathfrak{M}$ is the subspace that contains the base S of Z, we let $\{\varphi_k\}_1^\infty$ be an orthonormal basis in $\mathfrak{H}$ such that $\{\varphi_1, \ldots \varphi_m\}$ is a basis in $\mathfrak{M}$ and $\{\varphi_{m+1}, \ldots\}$ is a basis in $\mathfrak{M}^\perp$. Then the coordinates $\{x_k\}_1^\infty$ of a point x in $\mathfrak{H}$ relative to the basis $\{\varphi_k\}_1^\infty$ can be thought of as the random variables that describe the outcome of an experiment, and $\mathbf{P}(Z)$ is the probability that the point $(x_1, \ldots, x_m)$ lies in S_1 while the values of $x_{m+1}, \ldots$ are ignored completely. In that sense $\mathbf{P}(Z)$ is a marginal probability. Specification of $\mathbf{P}(Z)$ for all Z in $\mathfrak{A}_0$ amounts to specifying all possible finite-dimensional marginal probabilities. It is assumed that the specification is consistent with the probability interpretation; namely, $\mathbf{P}(Z)$ satisfies (13.11-4), is finitely additive, and is also countably additive in the sense that for disjoint Z_i in $\mathfrak{A}_0$

$$\mathbf{P}\left(\bigcup_1^\infty Z_i\right) = \sum_1^\infty \mathbf{P}(Z_i), \qquad (13.11\text{-}5)$$

whenever the union $\bigcup_1^\infty Z_i$ is also in $\mathfrak{A}_0$. We then call $\mathbf{P}$ a *probability measure* on $\mathfrak{A}_0$. The following is a fundamental theorem in the theory of probability (see Feller 1966, Section IV.5):

Extension Theorem. *If $\mathfrak{A}_0$ is an algebra of sets and $\mathbf{P}$ is a probability measure on $\mathfrak{A}_0$, then $\mathbf{P}$ has a unique extension to a probability measure on the σ-algebra $\mathfrak{A}$ generated by $\mathfrak{A}_0$.*

If $\mathbf{P}$ is restricted to a subalgebra of $\mathfrak{A}_0$ consisting of all Z with base S in some fixed finite-dimensional $\mathfrak{M}$, we can define

$$\mathbf{P}_{\mathfrak{M}}(S) = \mathbf{P}(Z);$$

then clearly $\mathbf{P}_{\mathfrak{M}}$ is a probability measure in $\mathfrak{M}$. But $\mathfrak{M}$ is finite-dimensional, hence $\mathbf{P}_{\mathfrak{M}}$ can be described by the methods of the preceding sections, for example by a cumulative probability $F(x_1, \ldots, x_m)$ or, in case it is absolutely continuous, by a density $f(x_1, \ldots, x_m)$. To show that a given set function $\mathbf{P}$ is a probability measure in $\mathfrak{H}$, after $\mathbf{P}_{\mathfrak{M}}$ has been shown to be a probability measure in each $\mathfrak{M}$, it remains only to show that the countable additivity (13.11-5) holds when the Z_i don't all have bases in a common $\mathfrak{M}$, even though their union is a cylinder set. An example of that is given in Exercise 4 below.

Gel'fand and Vilenkin (1966, Section IV.2) give various circumstances under which $\mathbf{P}$ is countably additive when $\mathbf{P}_{\mathfrak{M}}$ is a probability measure in each $\mathfrak{M}$.

In case $\mathbf{P}$ is defined on $\mathfrak{A}_0$ but is only assumed to be finitely additive (hence may not be, strictly speaking, a probability), it is called a *cylinder-set measure*.

The main examples are the so-called Gaussian measures in $\mathfrak{H}$; they correspond to the normal distributions in a finite-dimensional space. In Section 13.4 a bivariate normal distribution was defined having prescribed means μ_1 and μ_2 of the random variables ξ and η and a prescribed covariance matrix ρ. It was shown that by a linear transformation from ξ, η to α, β a normal distribution of α, β was obtained with zero means and with $\rho =$ the unit matrix.

The probability density of α, β was then $\exp\{-\frac{1}{2}(\alpha^2 + \beta^2)\}$. The generalization of that case to $\mathfrak{H}$ will be described first. It turns out not to be countably additive, while certain other Gaussian measures are.

Let Z be a cylinder set whose base S lies in an m-dimensional subspace $\mathfrak{M}$ of $\mathfrak{H}$ and let $x_1, \ldots, x_m$ be Cartesian coordinates in $\mathfrak{M}$. A cylinder-set measure is defined by setting

$$\mathbf{P}(Z) = (2\pi)^{-m/2} \int_S e^{-(1/2)(x_1^2 + \cdots + x_m^2)} dx_1 \cdots dx_m. \qquad (13.11\text{-}6)$$

Hence $\mathbf{P}_{\mathfrak{M}}$ has density $(2\pi)^{-m/2} \exp\{-\frac{1}{2}(x_1^2 + \cdots + x_m^2)\}$ in $\mathfrak{M}$. (Recall that $\mathfrak{H}$ is a *real* Hilbert space, hence the coordinates x_j are real.) Exercise 2 below shows that $\mathbf{P}(Z)$ is independent of the choice of $\mathfrak{M}$ and S for given Z, and Exercise 3 shows that $\mathbf{P}$ is not countably additive.

EXERCISES

1. (To show that marginal probabilities can give a lot of information): Let $F(x, y)$ be the cumulative probability of random variables ξ, η, and consider the transformations

$$\xi' = \xi \cos \theta + \eta \sin \theta$$

$$\eta' = -\xi \sin \theta + \eta \cos \theta \qquad (\theta \text{ real}).$$

Suppose that for every such transformation the marginal distribution of ξ', ignoring η', is known. Show that $F(x, y)$ is then determined.

2. Show that if $\mathfrak{M}$ and S are replaced by $\mathfrak{M}'$ and S' for given Z as described above, the value of $\mathbf{P}(Z)$ given by (13.11-6) is unaltered.

3. Let x_j ($j = 1, \ldots$) be coordinates with respect to a complete orthonormal set $\{\varphi_j\}_1^\infty$ in a real Hilbert space $\mathfrak{H}$. Consider the cylinder sets

$$Z_1: \ |x_1| < a \qquad (x_2, x_3, \ldots \text{ arbitrary}),$$

$$\vdots$$

$$Z_k: \ \sum_{j=1}^k x_j^2 < a^2 \qquad (x_{k+1}, \ldots \text{ arbitrary}),$$

$$\vdots$$

where a is a positive constant, and show that (13.11-6) gives

$$\mathbf{P}(Z_k) = \frac{\Gamma(k/2, a^2/2)}{\Gamma(k/2)},$$

where $\Gamma(x, y)$ is the incomplete gamma function given by

$$\Gamma(x, y) = \int_0^y e^{-t} t^{x-1} dt$$

(so that $\Gamma(x) = \Gamma(x, \infty)$). From the obvious inequality $\Gamma(x, y) < y^{x-1}$ for x and y positive and Stirling's asymptotic formula for $\Gamma(x)$ show that for fixed $a > 0$ $\mathbf{P}(Z_k) \to 0$ as $k \to \infty$. The intersection $\bigcap_{k=1}^\infty Z_k$ is the ball B_a of radius a in $\mathfrak{H}$ and is not a cylinder set, but if the function $\mathbf{P}$ could be extended to the σ-algebra $\mathfrak{A}$ (which contains B_a), we should conclude that $\mathbf{P}(B_a) = 0$ for any a. Lastly, since $\mathfrak{H}$ is the countable union $\bigcup_{a=1}^\infty B_a$, we see that the Gaussian measure $\mathbf{P}$ is not countably additive.

4.   Let x_j $(j = 1, \ldots)$ be coordinates as above, and consider the following cylinder sets (in each case, the unspecified coordinates are arbitrary):

$$Z_1: \quad x_1 < 1$$

$$Z_2: \quad x_1 \geq 1, \qquad x_2 < 1$$

$$\vdots$$

$$Z_k: \quad x_j \geq 1 \quad (j = 1, \ldots, k-1), \qquad x_k < 1$$

$$\vdots$$

show that the Z_k are disjoint, their bases do not lie in any common finite-dimensional subspace, and their union is a cylinder set.

Roughly speaking, the reason the Gaussian measure described above fails to behave like a probability is that it has unit variance in every one of the infinitely many directions in $\mathfrak{H}$; that tends to push the probability out to large distances, to such an extent in fact that the probability of finding x in any finite ball is zero (see Exercise 3 above). That way of speaking is of course only heuristic, because there is no such thing as a probability density in $\mathfrak{H}$, hence it is meaningless to talk about "where" the probability is located, but it suggests that perhaps we can obtain a more reasonable Gaussian measure if we choose a fixed orthonormal basis $\{\psi_l\}_1^\infty$ in $\mathfrak{H}$ and require that the variance in direction ψ_l go to zero at a sufficient rate as $l \to \infty$. The resulting measure will of course be highly anisotropic.

To that end, let B be a positive definite compact operator, and call $A = B^{-1}$. (The orthonormal vectors ψ_l referred to above will be the eigenvectors of B and the variance in direction ψ_l will be the corresponding eigenvalue.) If Z is any cylinder set, with base S in an m-dimensional subspace $\mathfrak{M}$, we let $\{\varphi_1, \ldots \varphi_m\}$ be an orthonormal basis in $\mathfrak{M}$ and we define a Hermitian matrix $A(\mathfrak{M})$ by writing

$$A(\mathfrak{M})_{jk} = (\varphi_j, A\varphi_k) \qquad (j, k = 1, \ldots, m).$$

We then set

$$P(Z) = P_{\mathfrak{M}}(S) = \frac{\sqrt{\det A(\mathfrak{M})}}{(2\pi)^{m/2}} \int_S \exp\left\{ -\frac{1}{2} \sum_{jk} x_j A(\mathfrak{M})_{jk} x_k \right\} dx_1 \ldots dx_m.$$

We state without proof that this determines a cylinder set measure and that it is countably additive if B is an operator of the trace class; then $\{\mathfrak{H}, \mathfrak{A}, \mathbf{P}\}$ is a probability space.

For more details, see Gel'fand and Vilenkin 1961.

Appendix to Chapter 13—Functions of Bounded Variation

The total variation of a real function $f(x)$ of one variable in an interval $[a, b]$ is roughly the total amount of vertical motion required to trace a graph of $f(x)$ from $x = a$ to $x = b$, counting upward and downward motions both positively. If $f(x)$ has a continuous derivative, then the total variation is $\int_a^b |f'(x)|\, dx$. However, for the concept to be meaningful, it is not necessary for $f(x)$ to be differentiable or even continuous. The total variation of a step function is the sum of the magnitudes of its steps.

A *partition* P_N of $[a, b]$ is a set of points x_j such that

$$a = x_0 < x_1 < \cdots < x_N = b.$$

The *total variation* of f in $[a, b]$ is defined as

$$V_a^b = V_a^b(f) = \sup\left\{ \sum_{j=1}^{N} |f(x_j) - f(x_{j-1})| : \text{all } P_N, \text{ all } N \right\}. \quad (13.\text{A-}1)$$

This definition clearly agrees with the special cases described above. The function $f(x)$ has *bounded variation* in $[a, b]$ if $V_a^b < \infty$; it has *bounded variation on* $\mathbb{R}$ if V_a^b is finite and remains bounded as $a, b \to -\infty, +\infty$; it has *locally bounded variation* if V_a^b is finite for any finite interval $[a, b]$.

EXAMPLE 1
The function $f(x) = 1$ for x rational and $= 0$ for x irrational has unbounded variation in any interval.

EXAMPLE 2
The continuous function

$$f(x) = \begin{cases} 0 & \text{for } x = 0, \\ x \sin \dfrac{1}{x} & \text{for } x \neq 0, \end{cases} \quad (13.\text{A-}2)$$

has unbounded variation in any interval containing the origin.

EXAMPLE 3
A monotonic function (nondecreasing or nonincreasing) has bounded variation in any interval, and in fact $V_a^b = |f(b) - f(a)|$.

A partial converse of the third example is that any function $f(x)$ of locally bounded variation can be written as the difference of two nondecreasing (or two nonincreasing) functions:

$$f(x) = f_1(x) - f_2(x). \quad (13.\text{A-}3)$$

To show that, we first separate V_a^b into two parts, the *positive* and *negative* variations $\overset{+}{V}_a^b$ and $\bar{V}_a^b$ by introducing the notation

$$[y]^+ = \max\{0, y\},$$
$$[y]^- = \max\{0, -y\}, \quad (13.\text{A-}4)$$

for any real y, and by defining

$$\overset{+}{V}{}_a^b = \sup\{\textstyle\sum[f(x_j) - f(x_{j-1})]^+ : \text{all } P_N, \text{all } N\},$$

$$\overset{-}{V}{}_a^b = \sup\{\textstyle\sum[f(x_j) - f(x_{j-1})]^- : \text{all } P_N, \text{all } N\}. \qquad (13.\text{A-}5)$$

Clearly

$$\overset{+}{V}{}_a^b + \overset{-}{V}{}_a^b = V_a^b$$

$$\overset{+}{V}{}_a^b - \overset{-}{V}{}_a^b = f(b) - f(a).$$

Finally, f_1 and f_2 in (13.A-3) can be taken as

$$f_1(x) = \overset{+}{V}{}_a^x + \text{const.},$$
$$\qquad\qquad\qquad\qquad (13.\text{A-}6)$$
$$f_2(x) = \overset{-}{V}{}_a^x + \text{const.},$$

where the two constants are so chosen that their difference is equal to $f(a)$.

The representation (13.A-3) is highly nonunique, because f_1 and f_2 can be replaced by $f_1 + g$ and $f_2 + g$, where g is any other nondecreasing function. The functions given by (13.A-6), which are called the *rising* and *falling parts* of $f(x)$, have the special property that when one increases, the other remains constant, so that the increase of $f_1(x) + f_2(x)$ in an interval $[a, b]$ is $= V_a^b$, not merely $\geq V_a^b$.

If $f(x)$ is complex valued, the definition (13.A-1) remains valid, but the separation into rising and falling parts must be done separately for Re $f(x)$ and Im $f(x)$; then we have

$$f(x) = f_1(x) - f_2(x) + if_3(x) - if_4(x),$$

where each f_k is nondecreasing.

Now consider a real function $f(x, y)$ of two variables. The *total variation* of f in a rectangle $a \leq x \leq c, b \leq y \leq d$ is

$$V_{ab}^{cd} = \sup\left\{\textstyle\sum_{jk} |f(\square_{jk})| : \text{all } P_N, \text{all } N\right\}, \qquad (13.\text{A-}7)$$

where P_N denotes a partition of the given rectangle into smaller rectangles $\square_{jk}$ by vertical and horizontal lines with coordinates x_j and y_k such that

$$a = x_0 < x_1 < \cdots < x_N = c$$
$$b = y_0 < y_1 < \cdots < y_N = d;$$

the notation $f(\square)$ was explained in Section 13.3.

The positive and negative variations $\overset{+}{V}{}^{c\,d}_{a\,b}$ and $\bar{V}{}^{c\,d}_{a\,b}$ are obtained by replacing $|\cdot|$ in (13.A-7) by $[\cdot]^+$ and $[\cdot]^-$, as in the one-variable case. When the quantities $f(\square_{j\,k})$ are added, for a collection of small rectangles that fill out a larger one, there is mutual cancellation of all terms except at the corners of the large rectangle; therefore

$$\overset{+}{V}{}^{c\,d}_{a\,b} - \bar{V}{}^{c\,d}_{a\,b} = f(c, d) - f(c, b) - f(a, d) + f(a, b).$$

If we replace c and d by x and y and regard a and b as constants, we can write

$$f(x, y) = (\overset{+}{V}{}^{x\,y}_{a\,b} + f(x, b) + f(a, y) + \text{const.}) - (\bar{V}{}^{x\,y}_{a\,b} + \text{const.})$$
$$= f_1(x, y) - f_2(x, y), \qquad (13.\text{A-8})$$

where f_1 and f_2 are nondecreasing functions of x, y as defined in Section 13.3.

Note. The terms $f(x, b)$ and $f(a, y)$ in this equation, which depend on only one variable each, can be included either with f_1 or with f_2, because for them $f(\square)$ is always zero.

The n-dimensional case described at the end of Section 13.3 follows the same pattern. We use the notation of that section and define

$$V^b_a = \sup\left\{ \sum_j |f(\square_j)|: \text{all } P_N, \text{all } N \right\}. \qquad (13.\text{A-9})$$

Then a complex-valued function $f(\mathbf{x})$ of locally bounded variation can be written as

$$f = f_1 - f_2 + if_3 - if_4, \qquad (13.\text{A-10})$$

where each $f_k(\mathbf{x})$ is nondecreasing.

CHAPTER 14

Probability and Operators in Quantum Mechanics

States of a system; observables; measurement; probabilistic axioms
for a quantum mechanical system; spectral projectors; orthogonality
of the projectors as a consequence of the probabilistic axioms;
cumulative probability for an observable; probability basis for
representation of an observable as a self-adjoint operator;
expectation; variance; uncertainty; quantum mechanical ensemble;
the density matrix—a positive definite operator of unit trace;
expectation function; expression of unbounded observables in terms
of bounded ones; commutation relations; Banach algebra; C^*- and
B^*-algebras; representation of B^*- by C^*-algebras; positive definite
element; positive functional; observable with a simple spectrum;
generating vector; spectral representations of a Hilbert space;
complete set of commuting observables; unitary "observables."

Prerequisite: Chapters 1–9, 12, 13

This chapter contains a discussion of the role played by probability in the
foundations of quantum mechanics and an introduction to the role it plays in
quantum statistical mechanics.

14.1 States of a System: Observables

In the quantum-mechanical description of a physical system (e.g., an atom),
each possible state is supposed to correspond to a nonzero element (vector) φ
in a Hilbert space $\mathfrak{H}$, or, more precisely, to the ray $\{a\varphi : a \in \mathbb{C}\}$ consisting of
all multiples of φ. Conversely, each ray in $\mathfrak{H}$ corresponds to a possible state
of the system. It is often convenient to assume φ normalized ($\|\varphi\| = 1$); φ
is then still not unique, because any vector of the form $e^{i\alpha}\varphi$ (α real) can equally
well be taken as representative of the ray in question, i.e. of the state of the
system.

Let $\mathscr{A}$ denote a real observable quantity in the classical sense, e.g., the energy of the system. Classically, $\mathscr{A}$ has a definite value λ in each state φ of the system. In quantum theory, on the other hand, it is supposed that if the system is put repeatedly into a state φ, and $\mathscr{A}$ is measured each time, then generally different values $\lambda, \lambda', \lambda'', \ldots$ are obtained, governed, however, by a probability law. Let $F(\cdot)$ be the cumulative probability for those values (see Chapter 13), so that, for any real λ_0,

$$F(\lambda_0) = \mathbf{P}\{\lambda \leq \lambda_0\}$$

where $\mathbf{P}$ denotes probability; the function F depends, of course, on the initial state φ.

Furthermore, a measurement of the observable $\mathscr{A}$ generally disturbs the system, so that it is in different states $\psi, \psi', \psi'', \ldots$ after the respective measurements; these final states are distributed (in $\mathfrak{H}$) according to a probability law (also depending on the initial state φ), and the ψ's are correlated, in a manner to be explained below, with the λ's. It has been assumed that the system is put back again into the initial state φ, after each measurement of $\mathscr{A}$, before the next measurement is made.

It will follow from certain simple axioms of quantum mechanics, described below, that a self-adjoint operator A in $\mathfrak{H}$ is associated with each observal $\mathscr{A}$; the cumulative probability $F(\cdot)$ of the values of $\mathscr{A}$ for the given state φ is then given in terms of A and φ by the equation (14.3-7), below.

14.2 Probabilities—A Finite Model

The axioms will be stated first for a model in which there are only finitely many possible final states (normalized vectors) $\psi_1, \psi_2, \ldots, \psi_n$, called *eigenstates* of the observable $\mathscr{A}$, that can result from a measurement. In each state ψ_i, $\mathscr{A}$ has a unique real value λ_i, which is supposed to be a precise property of the system when it is in the state ψ_i. The numbers $\lambda_1, \ldots, \lambda_n$ will be provisionally called *eigenvalues*, although so far no matrix or operator has been associated with $\mathscr{A}$. We tentatively assume that the λ_i are all different; that assumption can be avoided by slight modification of the formulation, as will be seen. It will follow from the axioms below that the eigenstates span $\mathfrak{H}$, which is therefore n-dimensional in this model.

The main axiom has two parts: (a) if $\mathscr{A}$ is measured when the system is in a linear superposition of the eigenstates, i.e. in a state of the form

$$\varphi = \sum_{j=1}^{n} c_j \psi_j, \qquad (14.2\text{-}1)$$

then the probability of obtaining the measured value λ_k is proportional to the square of the norm of the kth term of the sum, that is,

$$\mathbf{P}\{\lambda = \lambda_k\} = \text{const.} \, |c_k|^2, \qquad (14.2\text{-}2)$$

and (b) after a measurement yielding the value λ_k the system is in the state ψ_k.

In case $\varphi = c_k \psi_k$ (the other c_j being zero), the measured value is certain to be λ_k (and the system is undisturbed by the measurement), hence the proportionality constant in (14.2-2) is $= \|\varphi\|^{-1}$ (recall that $\|\psi_k\| = 1$). Hence, in general

$$\mathbf{P}\{\lambda = \lambda_k\} = \frac{|c_k|^2}{\|\varphi\|^2}. \qquad (14.2\text{-}3)$$

The axioms imply that the ψ_j are orthogonal, as follows: Since the sum of the probabilities (14.2-3) is $= 1$,

$$\sum_{j=1}^{n} |c_j|^2 = \|\varphi\|^2 = (\varphi, \varphi) = \sum_{j\,k=1}^{n} \overline{c_j} c_k (\psi_j, \psi_k). \qquad (14.2\text{-}4)$$

This is an identity in the c_i and therefore

$$(\psi_j, \psi_k) = \delta_{j\,k}, \qquad (14.2\text{-}5)$$

as claimed. It follows that $c_j = (\psi_j, \varphi)$ for each j; hence axiom (a) takes the form

$$\mathbf{P}\{\lambda = \lambda_k\} = \frac{|(\psi_k, \varphi)|^2}{\|\varphi\|^2}. \qquad (14.2\text{-}6)$$

(If the λ_i are not all different, the right member must be summed for all k such that $\lambda_k = \lambda$). We now assume that (14.2-6) is valid for all φ in the Hilbert space $\mathfrak{H}$ (which is so far unspecified); since the probabilities sum to 1,

$$\sum_{j=1}^{n} |(\psi_j, \varphi)|^2 = \|\varphi\|^2, \quad \text{for any } \varphi \text{ in } \mathfrak{H}, \qquad (14.2\text{-}7)$$

which shows, according to Theorem 1 of Section 1.6, that the vectors $\psi_1, \ldots, \psi_n$ form a complete orthonormal set in $\mathfrak{H}$. In this model, $\mathfrak{H}$ is the n-dimensional complex vector space V^n, and every φ has the form (14.2-1).

The axiom will now be put into a form, for the n-dimensional case, that can be easily generalized to the case of infinitely many (perhaps continuously distributed) possible measured values λ of $\mathscr{A}$. Henceforth, the initial state will be assumed represented by a normalized vector; i.e., $\|\varphi\| = 1$.

Let Δ denote a (finite or infinite) interval on $\mathbb{R}$, and call

$$\mathfrak{M}(\Delta) = \text{span of } \{\psi_j : \lambda_j \in \Delta\} \qquad (14.2\text{-}8)$$

i.e., $\mathfrak{M}(\Delta)$ is the subspace of the n-dimensional space consisting of linear combinations of those eigenstates for which the corresponding eigenvalues lie in Δ. Let $P(\Delta)$ denote the orthogonal projector of $\mathfrak{H}$ onto $\mathfrak{M}(\Delta)$; specifically, if φ is given by (14.2-1), then

$$P(\Delta)\varphi = \sum_{\lambda_j \in \Delta} c_j \psi_j. \qquad (14.2\text{-}9)$$

It follows from (14.2-6) since the ψ_j are orthonormal, and φ is normalized, that the probability distribution is given by

$$\mathbf{P}\{\lambda \in \Delta\} = \|P(\Delta)\varphi\|^2, \tag{14.2-10}$$

which is the desired form of the probability axiom. Furthermore, if the measured value lies in Δ, then the state of the system afterward is a vector in the range of the projector $P(\Delta)$, i.e., in $\mathfrak{M}(\Delta)$.

Now let Δ and Δ' be any two disjoint intervals ($\Delta \cap \Delta'$ is empty); from the orthogonality of the ψ_j it follows that

$$\mathfrak{M}(\Delta) \perp \mathfrak{M}(\Delta') \tag{14.2-11}$$

or

$$P(\Delta)P(\Delta') = P(\Delta')P(\Delta) = 0. \tag{14.2-12}$$

Finally, the completeness relation (14.2-7) is equivalent to the requirement that

$$P(\Delta) = I \quad \text{if } \Delta = (-\infty, \infty). \tag{14.2-13}$$

so that the $\mathfrak{M}(\mathbb{R}) = \mathfrak{H}$. The last four equations (14.2-10, 11, 12, 13) are taken over directly to the infinite-dimensional case.

14.3 Probabilities—The General Case ($\mathfrak{H}$ Infinite-Dimensional)

To motivate the axioms, we consider the following pair of physical operations, supposed repeated infinitely many times:

1. The system is put into the state φ,
2. The observable $\mathscr{A}$ is measured.

Call $\lambda^{(i)}$ the measured value of $\mathscr{A}$ (a real number) obtained in the ith repetition, and $\psi^{(i)}$ the resulting final state of the system. For any interval Δ on $\mathbb{R}$, we call

$$\mathfrak{M}(\Delta) = \text{subspace of } \mathfrak{H} \text{ spanned by } \{\psi^{(i)}: \text{all } i \text{ such that } \lambda^{(i)} \in \Delta\}. \tag{14.3-1}$$

As in the finite model, we assume that λ_i is uniquely determined by the final state $\psi^{(i)}$, so that no two different λ's can correspond to the same ψ. Hence, if Δ_1 and Δ_2 are disjoint intervals, then $\mathfrak{M}(\Delta_1)$ and $\mathfrak{M}(\Delta_2)$ are disjoint manifolds, by which is meant that they have only the zero vector in common. In further analogy with the finite model, we assume that every ψ in $\mathfrak{H}$ is a possible result of the measurement, so that these manifolds together span $\mathfrak{H}$. That makes it possible to define projectors $P(\Delta)$ as follows: Let $\Delta_1, \Delta_2,$ and Δ_3 be disjoint intervals whose union is $\mathbb{R}$, for example $(-\infty, a]$, $(a, b]$, and (b, ∞). Then $\mathfrak{M}(\Delta_1), \mathfrak{M}(\Delta_2),$ and $\mathfrak{M}(\Delta_3)$ are disjoint manifolds that span $\mathfrak{H}$. If an arbitrary x in $\mathfrak{H}$ is decomposed as $x_1 + x_2 + x_3$, where $x_i \in \mathfrak{M}(\Delta_i)$, we set $x_i = P(\Delta_i)x$ ($i = 1, 2, 3$) and thus define the projector $P(\Delta)$ for an arbitrary interval Δ. It will soon be seen that the $P(\Delta)$ are *orthogonal* projectors.

The above considerations lead to these axioms concerning the distribution of the values (λ) of an observable $\mathscr{A}$:

1. For each interval Δ on $\mathbb{R}$ there is a projector $P(\Delta)$ (depending on $\mathscr{A}$) such that if the initial state is φ, with $\|\varphi\| = 1$, then

$$\mathbf{P}\{\lambda \in \Delta\} = \|P(\Delta)\varphi\|^2. \tag{14.3-2}$$

2. If a measurement of $\mathscr{A}$ yields a value in Δ, the resulting final state of the system lies in the manifold

$$\mathfrak{M}(\Delta) = \mathfrak{R}(P(\Delta)).$$

It follows from these axioms that if Δ_1 and Δ_2 are disjoint intervals, then

$$P(\Delta_1)P(\Delta_2) = 0 = P(\Delta_2)P(\Delta_1),$$

while if $\Delta_1 \subset \Delta_2$, then

$$P(\Delta_1)P(\Delta_2) = P(\Delta_1) = P(\Delta_2)P(\Delta_1).$$

Hence, it suffices to consider the projectors

$$E_{\lambda_0} \overset{\text{def}}{=} P((-\infty, \lambda_0])(-\infty < \lambda_0 < \infty), \tag{14.3-3}$$

since then, for example,

$$P((a, b]) = E_b - E_a.$$

We now show that the projectors are orthogonal, i.e. are self-adjoint operators, so that for example if Δ_1 and Δ_2 are disjoint intervals, then $\mathfrak{M}(\Delta_1)$ and $\mathfrak{M}(\Delta_2)$ are orthogonal manifolds (see Section 9.2). Let φ and ψ denote arbitrary elements in the manifolds that correspond to the intervals $(-\infty, \lambda_0]$ and (λ_0, ∞), respectively, i.e.,

$$\begin{aligned} &\varphi \in \mathfrak{M}((-\infty, \lambda_0]), \\ &\psi \in \mathfrak{M}((\lambda_0, \infty)). \end{aligned} \tag{14.3-4}$$

If the system is initially in the state $\alpha\varphi + \beta\psi$, where α and β are arbitrary numbers, then

$$\begin{aligned} \mathbf{P}\,(\lambda \le \lambda_0) &= \frac{\|\alpha\varphi\|^2}{\|\alpha\varphi + \beta\psi\|^2}, \\ \mathbf{P}\,(\lambda > \lambda_0) &= \frac{\|\beta\psi\|^2}{\|\alpha\varphi + \beta\psi\|^2} \end{aligned} \tag{14.3-5}$$

hence

$$\|\alpha\varphi\|^2 + \|\beta\psi\|^2 = \|\alpha\varphi + \beta\psi\|^2 = (\alpha\varphi + \beta\psi, \alpha\varphi + \beta\psi).$$

On expanding and cancelling certain terms, we find

$$2\,\text{Re}\,\bar{\alpha}\beta\,(\varphi, \psi) = 0.$$

Since α and β were arbitrary, $(\varphi, \psi) = 0$; therefore, the two manifolds (14.3-4) are orthogonal; hence E_{λ_0} given by (14.3-3) is an orthogonal projector (see the Exercise below). It is seen that the family of projectors $E_\lambda(-\infty < \lambda < \infty)$ has all the properties of a resolution of the identity (see Section 9.6); in particular, the relations

$$E_\lambda \to 0 \text{ (strongly)}, \quad \text{as } \lambda \to -\infty$$

$$E_\lambda \to I \text{ (strongly)}, \quad \text{as } \lambda \to +\infty$$

result from the assumption that $\mathscr{A}$ can always be measured (or observed) when the system is in any arbitrary state φ in $\mathfrak{H}$ and that the result is always a real number λ between $-\infty$ and $+\infty$.

The equation

$$A = \int_{-\infty}^{\infty} \lambda \, dE_\lambda \tag{14.3-6}$$

then defines a self-adjoint operator A, which is regarded as the mathematical representation of the observable $\mathscr{A}$. The possible measured values of $\mathscr{A}$ constitute the spectrum of A.

From (14.3-2), it is seen that if the system is in a state φ, with $\|\varphi\| = 1$, then the cumulative probability of λ is given by

$$F(\lambda_0) \stackrel{\text{def}}{=} \mathbf{P}\{\lambda \leq \lambda_0\} = \|E_{\lambda_0}\varphi\|^2 = (\varphi, E_{\lambda_0}\varphi). \tag{14.3-7}$$

This equation gives the main physical interpretation of the family of projectors E_λ.

EXERCISE

1.   Show that if the ranges of projectors P and $I - P$ are orthogonal, then P is self-adjoint.

14.4 Expectations: The Domain of A

In this section, we show that the domain of the self-adjoint operator A that represents an observable $\mathscr{A}$ has the following interpretation: A given φ in $\mathfrak{H}$ is in the domain of A if and only if the probability distribution of the values of $\mathscr{A}$ has a finite variance when the system is in the state φ.

The cumulative probability for the distribution of the measured values of $\mathscr{A}$, when the system is in the state φ, is the function $F(\cdot)$ given by (14.3-7). Therefore, expectations are obtained as follows: The expectation or mean of $\mathscr{A}$ itself, denoted by $\mathbf{E}(\mathscr{A}; \varphi)$ is

$$\mathbf{E}(\mathscr{A}; \varphi) = \int_{-\infty}^{\infty} \lambda \, d(\varphi, E_\lambda \varphi) \stackrel{\text{def}}{=} \lambda_0, \tag{14.4-1}$$

the variance is

$$E((\mathscr{A} - \lambda_0)^2; \varphi) = \int_{-\infty}^{\infty} (\lambda - \lambda_0)^2 \, d(\varphi, E_\lambda \varphi), \qquad (14.4\text{-}2)$$

and, if $f(\cdot)$ is any continuous function, then

$$E(f(\mathscr{A}); \varphi) = \int_{-\infty}^{\infty} f(\lambda) d(\varphi, E_\lambda \varphi), \qquad (14.4\text{-}3)$$

in each case on the assumption that the integral is convergent at $\lambda = \pm\infty$.

According to Section 9.10, if A is a self-adjoint operator with E_λ as its resolution of the identity, and if $f(\lambda)$ is any continuous function defined for real values of λ, then an operator $f(A)$ is defined as follows: first,

$$\mathfrak{D}(f(A)) = \left\{ \varphi \colon \int_{-\infty}^{\infty} |f(\lambda)|^2 \, d(\varphi, E_\lambda \varphi) < \infty \right\} \qquad (14.4\text{-}4)$$

(this domain is dense in $\mathfrak{H}$); then for any φ in $\mathfrak{D}(f(A))$,

$$f(A)\varphi = \int_{-\infty}^{\infty} f(\lambda) dE_\lambda \varphi. \qquad (14.4\text{-}5)$$

Therefore, the integral in (14.4-3) converges in particular if $\varphi \in \mathfrak{D}(f(A))$. For such φ, the integral in (14.4-5) converges strongly, i.e., it is the strong limit in $\mathfrak{H}$ of $\int_{-n}^{n} f(\lambda) dE_\lambda \varphi$, as $n \to \infty$. Hence, it follows from equations (14.4-1, 2, 3) that

$$E(\mathscr{A}; \varphi) = (\varphi, A\varphi), \quad \text{for } \varphi \in \mathfrak{D}(A) \qquad (14.4\text{-}6)$$

$$E((\mathscr{A} - \lambda_0)^2; \varphi) = (\varphi, (A - \lambda_0)^2 \varphi), \quad \text{for } \varphi \in \mathfrak{D}(A^2), \qquad (14.4\text{-}7)$$

$$E(f(\mathscr{A}); \varphi) = (\varphi, f(A)\varphi), \quad \text{for } \varphi \in \mathfrak{D}(f(A)). \qquad (14.4\text{-}8)$$

[Since $\mathscr{A}$ can be any observable, the last two equations are merely the special cases of (14.4-6) for the observables $(\mathscr{A} - \lambda_0)^2$ and $f(\mathscr{A})$, respectively.] These equations are sometimes taken as the basis of an axiomatic treatment of quantum theory, but the treatment in the preceding section is somewhat preferable, because in it the *existence* of a self-adjoint operator A corresponding to an observable $\mathscr{A}$ is shown to follow from the probabilistic interpretation of quantum mechanics. Furthermore, the probability distribution given by (14.3-7) for the measured values of $\mathscr{A}$ in a state φ is well defined for any φ, not merely for those φ in $\mathfrak{D}(A)$. According to (9.8-5), φ is in $\mathfrak{D}(A)$ if and only if the second moment of that distribution is finite, i.e., the variance (14.4-2) is finite, as stated at the beginning of this section; in this case, the variance is equal to $\|(A - \lambda_0)\varphi\|^2/\|\varphi\|^2$, where λ_0 is the mean of A, given by (14.4-1). The *uncertainty of $\mathscr{A}$* (or A) *in the state φ* is the square root of the variance (called the "standard deviation" in statistics), hence is equal to $\|(A - \lambda_0)\varphi\|$ if φ is normalized and $\in \mathfrak{D}(A)$, and is infinite if $\varphi \notin \mathfrak{D}(A)$.

EXERCISE

1. In the commutation relation

$$pq - qp = -i\hbar$$

for a coordinate q and corresponding momentum p, p and q are unbounded operators, and the above equation is meaningless as a relation among operators. What is meant, of course, is that for every ψ that is in the domain of pq and also in the domain of qp, i.e., $\psi \in \mathfrak{D}(pq) \cap \mathfrak{D}(qp)$, the equation

$$(pq - qp)\psi = -i\hbar\psi \tag{14.4-9}$$

holds; furthermore, to give this requirement adequate strength, it is assumed that $\mathfrak{D}(pq) \cap \mathfrak{D}(qp)$ is dense in $\mathfrak{H}$. Show that if Δp and Δq are the uncertainties of p and q in the state ψ, as defined above, then

$$\Delta p \Delta q \geq \frac{\hbar}{2}. \tag{14.4-10}$$

At the other extreme, if ψ is not even in $\mathfrak{D}(p) \cap \mathfrak{D}(q)$, then at least one of Δp, Δq is infinite, and neither is zero, so (14.4-10) still holds. What can you say about the case where ψ is in $\mathfrak{D}(p) \cap \mathfrak{D}(q)$ but not in $\mathfrak{D}(pq) \cap \mathfrak{D}(qp)$?

See Section 14.6 for a further discussion of the commutation relations.

The definition (14.4-4, 5) of $f(A)$ is valid also for complex valued $f(\cdot)$. In particular, if $f(\lambda) = e^{it\lambda}$, then the characteristic function of the distribution is obtained, namely

$$\chi(t) = \mathbf{E}(e^{it\mathscr{A}}; \varphi) = (\varphi, e^{itA}\varphi);$$

e^{itA} is a unitary operator defined in all $\mathfrak{H}$. [In this case, the condition on the integral in (14.4-4) is satisfied for all φ—in fact, the value of the integral is $\|\varphi\|^2$.] If A is the Hamiltonian operator H, then e^{itH} is the operator that gives the evolution of the physical system in time: $\varphi(t) = e^{-i\hbar t H}\varphi(0)$.

14.5 The Density Matrix

Up to now, two distinct kinds of variability have been encountered: (1) the classical kind, due only to inadequate knowledge of the state of the system (the complete motion of a tossed coin could in principle be predicted, if the initial data were accurate enough, but in practice the outcome appears random) and (2) the quantum mechanical kind, where there is generally an inherent uncertainty of the value of an observable $\mathscr{A}$ even when the system is in a precisely fixed state φ. For purposes of quantum statistical mechanics, it is necessary to combine the two kinds of variability; this is done by considering large ensembles of identical, noninteracting physical systems. Just as in the classical case, each system of the ensemble is supposed to be in a precise state, but different systems are in different states, and when we choose one of the systems at random, we get a random result, with a probability that depends on the make-up of the ensemble as well as on the quantum mechanical uncertainties associated with the individual systems.

Consider first the finite model of Section 14.2, in which each state vector φ can be written as $\sum_{j=1}^{n} c_j \psi_j$, where $\{\psi_1, \ldots, \psi_n\}$ is an orthonormal system. Then we write $c_j = (x_j + iy_j)$, and each state corresponds to a point in $\mathbb{R}^{2n}$ with coordinates $(x_1, \ldots, x_n, y_1, \ldots, y_n) = \mathbf{x}$. If $\|\varphi\| = 1$, then $\sum (x_j^2 + y_j^2) = 1$, and the point lies on the unit sphere in $\mathbb{R}^{2n}$. To describe an ensemble of N systems ($N \gg 1$), one can assign to each system a point on the unit sphere and then specify the density $\rho = \rho(\mathbf{x})$ of points there, i.e., the number of points in each unit of $(2n - 1)$-dimensional volume of "area" on the sphere.

The method just described fails in the infinite-dimensional case, because of the impossibility of defining volume in a reasonable way, as discussed in Section 13.11. The way out of the difficulty is to deal directly with probabilities and thus avoid attempting to describe the distribution of states in $\mathfrak{H}$. First, a discrete model is considered. Suppose that $\{\varphi_k\}$ is some finite or countably infinite collection of normalized states (not assumed orthogonal or even linearly independent), and suppose that each of the systems in the ensemble is in one of those states. Let the fraction of the total number $N(\gg 1)$ of the systems that are in the state φ_k be p_k, so that if a system is chosen at random from the ensemble, p_k is the probability of finding it in the state φ_k. If A is the self-adjoint operator corresponding to some bounded observable, then the expectation of A, when the system is known to be in a state φ, is $(\varphi, A\varphi)$, according to the preceding section; hence the expectation of A for the whole ensemble is

$$E(A) = \sum_k p_k(\varphi_k, A\varphi_k). \tag{14.5-1}$$

According to Section 12.3, $(\varphi_k, A\varphi_k)$ is equal to $\mathrm{tr}(P_{\varphi_k} A)$, where P_{φ_k} is the orthogonal projector of $\mathfrak{H}$ onto the one-dimensional supspace that contains φ_k, and where "tr" stands for "trace." Therefore, $E(A) = \mathrm{tr}(DA)$, where D is the operator

$$D = \sum_k p_k P_{\varphi_k}. \tag{14.5-2}$$

Each P_{φ_k} is self-adjoint, positive definite, and of the trace class, with trace $= 1$, according to Section 12.3. Since each probability p_k is ≥ 0, and since $\sum p_k = 1$, it is seen that D is also positive definite and of trace $= 1$.

If the initial collection $\{\varphi_k\}$ of states is dense on the unit sphere in $\mathfrak{H}$, then any ensemble of systems can be approximated by a discrete model, as above; therefore, the main axiom of statistical quantum mechanics, due to von Neumann, is obtained by simply dropping the requirement of discreteness.

Axiom. With any ensemble of identical noninteracting quantum mechanical systems there is associated a positive definite operator D, of trace $= 1$, called the *density matrix* of the ensemble. The expectation of any bounded self-adjoint operator (observable) A is given, for the ensemble, by

$$E(A) = \mathrm{tr}(DA). \tag{14.5-3}$$

The following was proved by von Neumann: Suppose that $\mathbf{E}(\cdot)$ is any real-valued function, defined for all bounded self-adjoint operators in a Hilbert space, which has the following properties (they are obviously necessary properties if $\mathbf{E}(\cdot)$ is to be interpreted as an expectation), namely

$$\mathbf{E}(aA + bB) = a\mathbf{E}(A) + b\mathbf{E}(B) \quad (a, b \text{ real}),$$

$$\mathbf{E}(A) \geq 0 \quad \text{if } A \text{ is positive definite};$$

then, there is a positive definite operator D of the trace class such that $\mathbf{E}(A) = \text{tr}(DA)$, for all A. If also $\mathbf{E}(I) = 1$, where I is the identity operator, then $\text{tr } D = 1$. This shows that the description in terms of a density matrix is completely general.

Other statistical quantities can be expressed in terms of $\mathbf{E}(\cdot)$. For example, the variance of A is $\mathbf{E}(A^2) - (\mathbf{E}(A))^2$. If $f_x(t)$ is the function equal to 1 for $t \leq x$ and equal to zero for $t > x$, then $F(x)$, defined as $\mathbf{E}(f_x(A))$, is the cumulative probability of A for the given ensemble.

The restriction to bounded observables may seem at first glance to be a defect of this theory, but, for the physical interpretation, it is really no restriction at all. If A is an unbounded self-adjoint operator, and if $f(t)$ is a bounded increasing function, say $f(t) = \tanh t$, then the operator $B = f(A)$ is bounded, and a measurement of B gives exactly the same information about the system as a measurement of A; if b is the measured value of B, then the corresponding measured value of A is $\tanh^{-1} b$. (See also Exercise 4 in the next section).

EXERCISE

1. Suppose the system is an atom which has precisely two energy states, with state vectors ψ_1 and ψ_2. The operator D can be represented by a 2×2 matrix with matrix elements $D_{jk} = (\psi_j, D\psi_k)$. Compute this matrix for the following cases and verify that it is positive definite and has trace $= 1$: (1) Half the systems (atoms) in the ensemble are in the state ψ_1 and the rest are in ψ_2. (2) All the systems of the ensemble are in the same state, whose state vector is $\alpha\psi_1 + \beta\psi_2$, where α and β are certain constants such that $|\alpha|^2 + |\beta|^2 = 1$. Show that in this case the eigenvalues of D_{jk} are 0 and 1, and generalize this result to systems (atoms) having any finite number of energy states. (3) The systems are in different states such that if the state vector is written as $\psi_1 \sin \theta e^{i\varphi_1} + \psi_2 \cos \theta e^{i\varphi_2}$, then all values of the angles θ, φ_1, and φ_2 are equally likely in the ensemble.

It is notable that the first and third of the ensembles in this exercise have identical statistical properties, even though entirely different state vectors were used in their construction. This is one reason why many workers in statistical quantum mechanics (or quantum statistical mechanics) prefer the purely algebraic formulation of quantum mechanics, described briefly in the next section, in which Hilbert spaces and state vectors do not appear at all.

The density matrix can be interpreted as referring not to an ensemble but to a single system, concerning whose actual state we are in ignorance; in the first example in the exercise, we believe that the system is equally likely to be in state 1 or in state 2, but we don't know which—note that that is quite different from knowing with certainty that the system is in state $(1/\sqrt{2}) \times (\psi_1 + \psi_2)$.

14.6 Algebras of Bounded Operators; Canonical Commutation Relations

It was noted in the preceding section that the properties of quantum-mechanical systems can be expressed in terms of bounded observables alone, and that from some points of view a description is more basic if it does not refer to state vectors in a Hilbert space.

The set $\mathfrak{A}$ of all bounded (not necessarily self-adjoint) operators in a Hilbert space is an algebra, called a *C*-algebra*. More generally, $\mathfrak{A}$ need not contain *all* the bounded operators, but is required to be closed under the operations of multiplication (AB) and the formation of linear combinations ($aA + bB$; a, b in $\mathbb{C}$), it is required to contain the adjoint A^* of each A in $\mathfrak{A}$, and it is required to be a complete space. The last means that if $\|A_n - A_m\| \to 0$, as $n, m \to \infty$, then there is an A in $\mathfrak{A}$ such that $\|A_n - A\| \to 0$, as $n \to \infty$. This algebra has the following properties:

1. It is a complete normed linear space, hence a Banach space—see Chapter 15. The norm of an operator A is its bound $\|A\|$.
2. An associative multiplication is defined in it: if A, B, and C are in $\mathfrak{A}$, then AB, etc., are in $\mathfrak{A}$, and $(AB)C = A(BC)$; $\|AB\| \le \|A\|\,\|B\|$; also $(aA)(bB) = (ab)(AB)$.

The multiplication is distributive:

3. $A(B + C) = AB + AC$, and $(A + B)C = AC + BC$.
4. An *involution* $A \to A^*$ is defined in it such that $(A^*)^* = A$, $(A + B)^* = A^* + B^*$, $(AB)^* = B^*A^*$, $(aA)^* = \bar{a}A^*$.
5. $\|A^*A\| = \|A\|^2$ (from which it follows that $\|A^*\| = \|A\|$).
6. It contains an identity I such that $AI = IA = A$.

A (complex) *B*-algebra* is defined as an *abstract* algebra having properties 1 through 5. For the present brief discussion it will also be assumed to have an identity (property 6).

Comments. Corresponding real algebras are sometimes considered. For example, an algebra of commuting self-adjoint bounded operators is one such, if the scalars a, b are restricted to the real field $\mathbb{R}$. In the noncommutative case, non-self-adjoint elements are unavoidable, for suppose that $A^* = A$ and $B^* = B$; then $(AB)^* = AB$ if and only if $AB = BA$. The familiar equation $pq - qp = \hbar/i$ shows that nonreal scalars are then also unavoidable. The distinction that a B^*-algebra is abstract while a C^*-algebra is an algebra of bounded operators in a Hilbert space is made by some authors (e.g. Rickart), but not by all. A basic theorem says that any B^*-algebra can be realized as a C^*-algebra; that is, it is isometrically isomorphic to some algebra of bounded linear operators in some Hilbert space (not unique). These algebras are special cases of more general *Banach* algebras, in which some or all of properties 4., 5., and 6. may be missing.

EXERCISES

1. For bounded operators, verify that $\|AB\| \leq \|A\|\|B\|$ and that $\|A^*A\| = \|A\|^2$.

2. In a B^*-algebra with identity, show that $I^* = I, \|I\| = 1$; if an element A has an inverse A^{-1} (an element B is called the *inverse* of A and denoted by A^{-1} if $AB = I$ *and* $BA = I$), then $(A^{-1})^* = (A^*)^{-1}$.

3. Suppose that $F(\cdot)$ is a real linear functional defined for the self-adjoint elements (elements such that $A^* = A$) of a B^*-algebra $\mathfrak{A}$. [For instance, $F(\cdot)$ might be the expectation functional $\mathbf{E}(\cdot)$ discussed in the preceding section.] For any A in $\mathfrak{A}$, let $B = (1/2)(A + A^*)$, $C = (1/2i)(A - A^*)$, define $F_1(A) = F(B) + iF(C)$ and show that $F_1(\cdot)$ is a linear extension (no longer real) to all of $\mathfrak{A}$. [In particular, show that $F_1(aA) = aF_1(A)$, for any complex number a.]

A few basic facts about B^*-algebras are now stated without proof.

In any Banach algebra $\mathfrak{A}$ with identity, the *spectrum* of an element A, $\sigma(A)$, is the set of all complex numbers λ such that the element $\lambda I - A$ has no inverse in $\mathfrak{A}$. [This agrees with the definition in Chapter 7, if $\mathfrak{A}$ is the algebra of all bounded operators in a Hilbert space.] General Banach algebras have the following defect, from our point of view: if $\mathfrak{A}$ is a subalgebra of a Banach algebra $\mathfrak{B}$, then the spectra of a given element A with respect to the two algebras may not be the same, because $\lambda I - A$ may have an inverse in $\mathfrak{B}$ but not in $\mathfrak{A}$. Therefore, one must write $\sigma_{\mathfrak{A}}(A)$ and $\sigma_{\mathfrak{B}}(A)$. However, in the special case of B^*-algebras with identity, and if $\mathfrak{A} \subset \mathfrak{B}$, in the sense that $\mathfrak{A}$ is a subset of $\mathfrak{B}$ and is a B^*-algebra in its own right, with the same identity, the same norm, the same multiplication, and the same conjugation as in $\mathfrak{B}$, then the spectra $\sigma_{\mathfrak{A}}(A)$ and $\sigma_{\mathfrak{B}}(A)$ are identical, for any A in $\mathfrak{A}$. The importance of this for quantum mechanics is evident, because the spectrum of A gives the possible measured values of A, which ought not to depend on whether A is thought of as belonging to $\mathfrak{A}$ or to $\mathfrak{B}$.

If A is a self-adjoint element (one such that $A^* = A$), then the spectrum of A lies on the real axis in the λ plane.

A linear function F on $\mathfrak{A}$ is called *bounded* if there is a number $\|F\|$ such that $|F(A)| \leq \|F\|\|A\|$, for all A. An element A is called *positive definite* if it can be written in the form B^*B, for some B in $\mathfrak{A}$. A linear functional F on $\mathfrak{A}$ is called *positive* if, for every positive definite element A, $F(A)$ is real and ≥ 0, i.e., $F(B^*B) \geq 0$ for every B. It can be proved that F is positive if and only if $\|F\| = F(I)$. It is recalled that the expectation functional $\mathbf{E}(\cdot)$ for observables in a quantum-mechanical statistical ensemble is a positive linear functional such that $\mathbf{E}(I) = 1$.

In the abstract algebraic formulation of quantum statistical mechanics, a *dynamical system* is described by a B^*-algebra $\mathfrak{A}$. The self-adjoint elements of $\mathfrak{A}$ represent the observables of the system, and they are subject to various (generally nonlinear) relations, which describe the make-up of the system;

they describe, for example, commutation relations, the dependence of the Hamiltonian on the various coordinates and momenta, and so on; these relations, in turn, determine the algebraic structure of $\mathfrak{A}$. An *ensemble* of such systems, all identical and non-interacting, is then described by a positive definite functional $\mathbf{E}(\cdot)$ on $\mathfrak{A}$, such that $\mathbf{E}(I) = 1$. The possible measured values of an observable A are the points of the spectrum of A, and the expected value of A in the given ensemble is $\mathbf{E}(A)$.

EXERCISE

4. Let q and p be a coordinate and the corresponding momentum, as in the exercise in Section 14.4. Since these are self-adjoint, the unitary operators $U(\alpha) = e^{ip\alpha}$ and $V(\beta) = e^{iq\beta}$ are well defined (see Section 9.10) and bounded, hence are in the C^*-algebra of all bounded operators in $\mathfrak{H}$. Show that the commutation relation (14.4-9) can be expressed in terms of the bounded operators as

$$U(\alpha)V(\beta)U(-\alpha)V(-\beta) = e^{i\hbar\alpha\beta}. \tag{14.6-1}$$

For this purpose, assume that there is a set S of vectors ψ, dense in $\mathfrak{H}$, for which all expressions such as $pqU(\alpha)\psi$, etc. are well defined (but see Exercise 6 below). Show first that for such ψ

$$\frac{d}{d\alpha}[U(\alpha)q - qU(\alpha)]\psi = ip[U(\alpha)q - qU(\alpha)]\psi + \hbar U(\alpha)\psi,$$

hence that the vector $\psi(\alpha) \overset{\text{def}}{=} [U(\alpha)q - qU(\alpha)]\psi$ satisfies the same differential equation and the same initial conditions as the vector $\hbar\alpha U(\alpha)\psi$, hence that

$$[U(\alpha)q - qU(\alpha)]\psi = \hbar\alpha U(\alpha)\psi.$$

Then show that both sides of (14.6-1), when applied to ψ, satisfy the same differential equation with respect to the variable β, and the same initial condition.

The classical commutation relations for p and q were first put into the form (14.6-1) by Hermann Weyl. It was then proved by von Neumann (1931), along lines suggested by M. H. Stone, that any operators p and q that satisfy the commutation relations in Weyl's form are equivalent respectively to the operator $(\hbar/i)(d/dx)$ and the operator of multiplication by x, applied to functions of x; specifically, von Neumann proved that if p and q are operators in a separable Hilbert space $\mathfrak{H}$ and if the resulting $U(\alpha)$ and $V(\beta)$ satisfy (14.6-1), then $\mathfrak{H}$ can be written as a finite or countable direct sum of Hilbert spaces $\mathfrak{H}_n$, $n = 1, 2, \ldots$, each of which is invariant under p and q and each of which can be mapped onto $L^2(\mathbb{R})$ by a unitary transformation W such that $WpW^{-1} = (\hbar/i)(d/dx)$ and $WqW^{-1} = x$ (operator of multiplication by x). In this sense, $(\hbar/i)(d/dx)$ and x are the only possible representations of such operators p and q, up to a unitary transformation.

EXERCISES

5. Find explicit expressions for $U(\alpha)$ and $V(\beta)$, when $\mathfrak{H} = L^2(\mathbb{R})$, $p = (\hbar/i)(d/dx)$, $q = x$ (the domains being suitably chosen), and verify (14.6-1) directly. In this case, the set S can be taken as the Schwartz class $\mathscr{S} = \mathscr{S}(\mathbb{R})$ of test functions for tempered

distributions; any ψ in $\mathscr{S}$ is in the domain of any finite product of operators chosen from the set

$$\{p^m, q^n, U(\alpha), V(\beta): \text{all positive integers } m, n, \text{ all real } \alpha \text{ and } \beta\}.$$

6.   Show that the assumption in Exercise 4 of the existence of the set S of vectors ψ cannot be entirely done away with, by considering the following example (from B. Misra, private communication) in the Hilbert space $\mathfrak{H} = L^2(0, 1)$: p and q are the self-adjoint operators defined by the equations

$$\mathfrak{D}(p) = \{f \in L^2 : f' \in L^2, f(0) = f(1)\}; \quad \text{for } f \in \mathfrak{D}(p), \quad pf = -i\hbar f'$$

(where the derivative f' is understood in the distribution sense), and

$$\mathfrak{D}(q) = \mathfrak{H}, \quad (qf)(x) = xf(x).$$

Show that $(pq - qp)\psi = i\hbar\psi$ for all ψ in a set dense in $\mathfrak{H}$, while (14.6-1) fails to hold, except for certain values of β. Note that if ψ is in $\mathfrak{D}(p)$, then $q\psi$ and $V(\beta)\psi$ are not generally in $\mathfrak{D}(p)$. Note also that p and q cannot possibly be unitarily equivalent to the corresponding operators on $L^2(\mathbb{R})$ because the operator q is bounded on $L^2(0, 1)$ but unbounded on $L^2(\mathbb{R})$.

A consequence of the Stone–von Neumann theorem is that if self-adjoint operators p and q satisfy (14.6-1), then each has a purely continuous spectrum consisting of all real numbers.

In Putnam's book (1967), there are theorems on conditions under which the equation $pq - qp = -i\hbar$ on a dense set in $\mathfrak{H}$ implies the Weyl condition (14.6-1), and corresponding theorems for sets of canonical variables p_j, q_j $j = 1, 2, \ldots$. When there are infinitely many such pairs $\{p_j, q_j\}$, the Stone–von Neumann theorem does not apply, and there are generally many different and inequivalent representations of the Weyl commutation relations. This fact plays a role in quantum field theory; see Jost (1965).

14.7 Self-Adjoint Operator with a Simple Spectrum

An $n \times n$ matrix M is said to have a simple spectrum if each of its n eigenvalues $\lambda_1, \ldots, \lambda_n$ is simple, i.e., the λ_i are all different. In this case, as is well known, the eigenvectors $v^{(1)}, \ldots, v^{(n)}$ span the entire n-dimensional complex vector space V^n or $\mathbb{C}^n$. [Proof: To prove that the $v^{(i)}$ are independent, suppose that

$$c_1 v^{(1)} + \cdots + c_n v^{(n)} = 0; \tag{14.7-1}$$

it must be proved that all c_i are zero. For any $k = 1, \ldots, n$ let $p_k(\cdot)$ be the Lagrange interpolation polynomial such that $p_k(\lambda_j) = 0$, for $j \neq k$, and $p_k(\lambda_k) \neq 0$. When the matrix $p_k(M)$ is applied to the vector (14.7-1), all terms on the left vanish except the kth term; hence, $c_k = 0$.]

The foregoing definition of simple spectrum does not extend to operators in $\mathfrak{H}$, because there are no eigenvectors in $\mathfrak{H}$ corresponding to the continuous spectrum. The definition will therefore be so rephrased that it can apply in $\mathfrak{H}$, as well. Let $\lambda_1, \ldots, \lambda_k$ be the distinct eigenvalues of M ($k = n$ if the spectrum is simple) and let $P_1, \ldots, P_k$ be the corresponding projectors onto the eigenspaces $\mathfrak{E}_1, \ldots, \mathfrak{E}_k$ ($P_j \mathfrak{E}_l = \delta_{jl} \mathfrak{E}_l$). The spectrum is simple if and only if there is some vector ξ in V^n such that the projections $P_j \xi$ ($j = 1, \ldots, k$) of ξ span V^n (for example, if the λ_j are all different, ξ can be taken as a linear combination $\sum c_j v^{(j)}$ with no $c_j = 0$).

If M is a Hermitian matrix (and hence represents a self-adjoint operator in V^n), the corresponding resolution of the identity is

$$E_t = \sum_{(i \text{ such that } \lambda_i \leq t)} P_i,$$

and the spectrum is simple if and only if the vectors $E_t \xi$, obtained by letting t vary on $\mathbb{R}$, span V^n.

Now let $A = \int_{-\infty}^{\infty} t \, dE_t$ be a self-adjoint operator in $\mathfrak{H}$. The spectrum of A is said to be *simple* if there is an element ξ in $\mathfrak{H}$ such that the closed linear span of the elements

$$\{E_t \xi : -\infty < t < \infty\} \tag{14.7-2}$$

is all of $\mathfrak{H}$, i.e., the finite linear combinations of these elements are dense in $\mathfrak{H}$; ξ is called a *generating vector* for A.

Similarly, the spectrum of a unitary operator

$$U = \int_0^{2\pi} e^{i\theta} \, dF_\theta$$

is *simple* if there is a *generating vector* ξ such that the elements $\{F_\theta \xi : 0 \leq \theta < 2\pi\}$ span $\mathfrak{H}$.

If A has a simple spectrum, then, in particular, for each λ in the point spectrum $P\sigma(A)$, the eigenspace $\mathfrak{E}_\lambda$ is one-dimensional, just as in the finite-dimensional case. To show that, let ξ be the generating vector, and let

$$P_\lambda = E_{\lambda+0} - E_{\lambda-0}$$

be the projector onto $\mathfrak{E}_\lambda$. Then, $P_\lambda E_t \xi$ is $= P_\lambda \xi$ if $t \geq \lambda$ and is $= 0$ if $t < \lambda$. Therefore, for any linear combination η of the elements $E_t \xi$, hence for any η in $\mathfrak{H}$, $P_\lambda \eta$ is a constant times $P_\lambda \xi$, so $\mathfrak{E}_\lambda$ is one-dimensional.

Roughly speaking, if A corresponds to an observable $\mathscr{A}$, then a measured value of $\mathscr{A}$ uniquely determines the state of the system (the state in which the system is left after the measurement), provided $\mathscr{A}$ has a simple spectrum; otherwise, the values of other observables $\mathscr{B}, \mathscr{C}$, etc. (which commute with $\mathscr{A}$) have to be given also, to specify a unique state of the system (see Section 14.9).

A Sturm–Liouville operator with one regular endpoint and one singular endpoint in the limit-point case, called A_α, was described in Sections 10.11 and 10.12. In the eigenfunction expansion (10.12-7) of a given function $g(x)$, only one function appears for any given value $s = \lambda$ of the eigenfunction parameter, namely $f_2(x, \lambda)$. That suggests that A_α has a simple spectrum. For given λ, $f_2(x, \lambda)$ satisfies the differential equation $-(pf')' + qf = \lambda f$ and the boundary condition (10.11-2), hence is an eigenfunction in a generalized sense, but not in the strict sense because it is not quadratically integrable in $(0, \infty)$ (unless λ is in the point spectrum), hence is not in the Hilbert space. To prove that the spectrum of A_α is simple, we must find a generating vector $\xi = \xi(x)$ in $\mathfrak{H} = L^2(0, \infty)$.

EXERCISES

1. Let $\eta(s)$ be a function of class $C^1(\mathbb{R})$ that is $\neq 0$ for all s and is in L_σ^2. Show that the function

$$\xi(x) = \int_{-\infty}^{\infty} f_2(x, s)\eta(s)d\rho(s)$$

is a generating vector for A_z. Hint: the functions $h(s)$ of class $C_0^1(\mathbb{R})$ are dense in L_σ^2, and the mapping (10.12-6 and 7) is an isomorphism of $L^2(0, \infty)$ and L_σ^2.

2. Formulate a definition of the multiplicity (a positive integer) of the spectrum of a self-adjoint operator whose spectrum is not necessarily simple.

3. Show that the operator $-id/dx$ of Section 10.1 has a simple spectrum, and exhibit a generating vector $\xi(x)$. The resolution of the identity is given in Section 10.1.

4. Show that the operator $-(d/dx)^2$ of Section 10.2 does not have a simple spectrum by showing that if $\xi(x)$ were a generating vector, then the formula

$$\varphi(x) = \int_{-\infty}^{\infty} f(t)d_t(E_t\xi)(x),$$

f arbitrary, could not give all φ in $L^2(\mathbb{R})$; in fact any $\varphi(x)$ given by the formula has a Fourier transform of the form $g(s)\hat{\xi}(s)$, where $g(s)$ is an *even* function.

14.8 Spectral Representation of $\mathfrak{H}$ for a Self-Adjoint Operator with a Simple Spectrum

It was shown in Chapter 1 that all infinite-dimensional separable Hilbert spaces are identical in structure. The Hilbert space of a quantum-mechanical system may be thought of as an abstract space, but it is often convenient to have a concrete realization of it, for example as an L^2 space of functions, just as it is often convenient to introduce Cartesian coordinates in a finite-dimensional space. The so-called spectral representation associated with a given self-adjoint operator A having a simple spectrum is one such realization. In quantum mechanics, it is called a *representation in which A is diagonal*. Some theorems on spectral representation are given in this section with only heuristic "proofs"—for details, the reader is referred to Akhiezer and Glazman 1963.

Let A be a self-adjoint operator in a Hilbert space (abstract or not) with simple spectrum and with a generating vector ξ. Let E_t be the resolution of the identity for A. Suppose that an element u of $\mathfrak{H}$ is given by an integral of the form

$$u = \int_{-\infty}^{\infty} f(t)dE_t\xi, \qquad (14.8\text{-}1)$$

where $f(t)$ is a complex-valued function of the real variable t. That is, it is supposed that the Riemann–Stieltjes sum

$$\sum_{(j)} f(t_j)E(\Delta_j)\xi \tag{14.8-2}$$

converges, in $\mathfrak{H}$, to the element u, as the partition of the t axis is refined. Since the spectrum of A is simple, any u in $\mathfrak{H}$ can be approximated by (14.8-2); hence, it is reasonable to suppose that any u can be represented in the form (14.8-1).

We now let

$$v = \int_{-\infty}^{\infty} g(t)dE_t\xi \tag{14.8-3}$$

be another such element, and we derive an expression for the inner product (u, v) in terms of the functions f and g. The inner product of the Riemann–Stieltjes sums corresponding to (14.8-1 and 3), when expanded, consists of terms containing $\overline{f(t_j)}g(t_k)$ multiplied by

$$(E(\Delta_j)\xi, E(\Delta_k)\xi);$$

owing to the properties of the projectors, this is equal to

$$\delta_{jk}(\xi, E(\Delta_k)\xi),$$

and if Δ_k is the interval (τ_k, τ_{k+1}), the above is equal to

$$\delta_{jk}[(\xi, E_{\tau_{k+1}}\xi) - (\xi, E_{\tau_k}\xi)].$$

We arrive thus at a sum

$$\sum_{(j)} \overline{f(t_j)}g(t_j)[(\xi, E_{\tau_{k+1}}\xi) - (\xi, E_{\tau_k}\xi)],$$

as an approximation to (u, v). This is also a Riemann–Stieltjes sum; hence, by refinement of the partition, we may suppose that

$$(u, v) = \int_{-\infty}^{\infty} \overline{f(t)}g(t)d(\xi, E_t\xi). \tag{14.8-4}$$

The function

$$\sigma(t) \overset{\text{def}}{=} (\xi, E_t\xi) \tag{14.8-5}$$

is real, because E_t is self-adjoint, and it is nondecreasing, because, if $t_2 > t_1$, then

$$\begin{aligned}
\sigma(t_2) - \sigma(t_1) &= (\xi, (E_{t_2} - E_{t_1})\xi) \\
&= (\xi, (E_{t_2} - E_{t_1})^2\xi) \\
&= ((E_{t_2} - E_{t_1})\xi, (E_{t_2} - E_{t_1})\xi),
\end{aligned}$$

which is nonnegative. In terms of $\sigma(\cdot)$,

$$(u, v) = \int_{-\infty}^{\infty} \overline{f(t)}g(t)d\sigma(t), \qquad (14.8\text{-}6)$$

$$\|u\|^2 = \int_{-\infty}^{\infty} |f(t)|^2 \, d\sigma(t). \qquad (14.8\text{-}7)$$

These are the desired expressions.

In Section 5.9, spaces of the type L_σ^2 were defined; inner product and norm were given by expressions identical to (14.8-6) and (14.8-7), when $f(\cdot)$ and $g(\cdot)$ are smooth functions. For smooth functions, the foregoing arguments can be easily made rigorous, hence the mapping $f(\cdot) \to u$ is an isometric isomorphism of a dense set in L_σ^2 onto a dense set in $\mathfrak{H}$. This mapping is a bounded linear transformation from the Hilbert space L_σ^2 to the Hilbert space $\mathfrak{H}$; by an obvious generalization of the Extension theorem of Section 7.1, it can be extended to all of each space.

Lastly, if the operator $A = \int_{-\infty}^{\infty} t \, dE_t$ is approximated by a Riemann–Stieltjes sum

$$\sum_{(j)} t_j E(\Delta_j), \qquad (14.8\text{-}8)$$

using the same partition of the t axis as for (14.8-2), and if the operator (14.8-8) is then applied to the vector (14.8-2), the resulting double sum reduces to a single sum (because of the properties of the projectors), namely, to

$$\sum_{(j)} t_j f(t_j) E(\Delta_j) \xi.$$

It is surmised that if u corresponds to $f(t)$, then Au corresponds to $tf(t)$. These considerations suggest the following theorem:

Theorem. *Let A be a self-adjoint operator with a simple spectrum in a Hilbert space $\mathfrak{H}$. Then, there is a nondecreasing function $\sigma(\cdot)$ and a one-to-one mapping $u \to f(\cdot)$ of $\mathfrak{H}$ onto the space L_σ^2 such that if $u \to f(\cdot)$ and $v \to g(\cdot)$, then $(u, v) = (f(\cdot), g(\cdot))$. Operators in $\mathfrak{H}$ correspond to operators in L_σ^2; In particular, A corresponds, in L_σ^2, to the operation of multiplying $f(t)$ by t; that is, if u in $\mathfrak{H}$ corresponds to the function $f(t)$, then Au corresponds to the function $tf(t)$.*

Note that A does not uniquely determine L_σ^2, because the generating vector ξ was not unique. If ξ_1 is another generating vector for A, and if $\sigma_1(t) = (\xi_1, E_t \xi_1)$ is the corresponding nondecreasing function, then there is a positive function $\rho(t)$ such that

$$\sigma_1(t) = \int_{-\infty}^{t} \rho(t')d\sigma(t');$$

hence the measure $d\sigma_1$, is absolutely continuous with respect to the measure $d\sigma$. If an element u in $\mathfrak{H}$ is represented by $f(t)$ in L_σ^2, then it is represented by

$\rho(t)^{-1/2}f(t)$ in $L_{\sigma_1}^2$. In quantum-mechanical terms, going from σ to σ_1 represents merely a change of normalization of the basis vectors. If A has a pure point spectrum, then σ can be so chosen that all the basis vectors are normalized to 1, but otherwise there is no unique choice of σ, because there is no general agreement on the most convenient normalization of the continuous state wave functions.

EXERCISE

1. With ξ and ξ_1 as above, show that if

$$\xi_1 = \int_{-\infty}^{\infty} a(t)dE_t\xi,$$

then

$$\rho(t) = |a(t)|^2.$$

14.9 Complete Set of Commuting Observables

Consider two self-adjoint operators

$$A = \int_{-\infty}^{\infty} t\,dE_t \quad \text{and} \quad B = \int_{-\infty}^{\infty} t\,dF_t; \qquad (14.9\text{-}1)$$

they are said to *commute* if

$$E_tF_s = F_sE_t, \quad \text{for all } s, t. \qquad (14.9\text{-}2)$$

[N.B. Since A and B are generally unbounded, one cannot say that $AB = BA$ unless the domains of AB and BA happen to be the same, whereas E_t and F_s are defined in all $\mathfrak{H}$; however, $ABu = BAu$ for all u (if any) such that both sides of the equation are meaningful.] Commuting operators A and B are said to have a *simple joint spectrum* or to form a *complete set of commuting observables* if there is an element ξ (a *generating vector*) in $\mathfrak{H}$ such that the closed linear span of the elements

$$\{E_sF_t\xi: -\infty < s, t < \infty\} \qquad (14.9\text{-}3)$$

is all of $\mathfrak{H}$.

The generalization to any finite number of self-adjoint or unitary operators is obvious.

If A and B form a complete set as defined above, a measure is defined in the s, t plane by setting

$$\sigma(s, t) = (\xi, E_sF_t\xi), \qquad (14.9\text{-}4)$$

where ξ is the generating vector; this is a nondecreasing function as defined in Chapter 13 on probability. Namely, if $\square$ denotes the rectangular region of the s, t plane defined as

$$\square = \{s, t: a \le s < b, c \le t < d\} \qquad (14.9\text{-}5)$$

and if $\sigma(\square)$ is defined as

$$\sigma(\square) = \sigma(b, d) - \sigma(a, d) - \sigma(b, c) + \sigma(a, c),$$

then $\sigma(\square)$ is ≥ 0 for all such $\square$. By means of double Stieltjes integrals (see Section 13.3), a space $L_\sigma^2(\mathbb{R}^2)$ is defined, in analogy with Section 5.9, as follows: In $\mathscr{S}(\mathbb{R}^2)$, an inner product

$$(\varphi, \psi) = \int_{-\infty}^{\infty} \int_{-\infty}^{\infty} \overline{\varphi(s, t)} \psi(s, t) d^2\sigma(s, t)$$

is defined. The resulting inner-product space is enlarged to a complete space $L_\sigma^2(\mathbb{R}^2)$ of distributions on $\mathbb{R}^2$ by the same method as for $L_\sigma^2 = L_\sigma^2(\mathbb{R})$ in Section 5.9.

The theorem of the preceding section is now restated for the case of a finite number of operators.

Theorem. *Let $A_1, \ldots, A_k$ be commuting self-adjoint operators in $\mathfrak{H}$ with a simple joint spectrum. Then, there is a nondecreasing function $\sigma(t_1, \ldots, t_k)$ and a one-to-one mapping $u \to f(t_1, \ldots, t_k)$ of $\mathfrak{H}$ onto the space $L_\sigma^2(\mathbb{R}^k)$ such that if $u \to f(\cdots)$ and $v \to g(\cdots)$, then $(u, v) = (f(\cdots), g(\cdots))$. Operators in $\mathfrak{H}$ correspond to operators in $L_\sigma^2(\mathbb{R}^k)$; in particular, for each $j = 1, \ldots, k$, A_j corresponds, in $L_\sigma^2(\mathbb{R}^k)$, to the operation of multiplying $f(t_1, \ldots, k_k)$ by t_j; that is, if u corresponds to the function $f(t_1, \ldots, t_k)$, then $A_j u$ corresponds to the function $t_j f(t_1, \ldots, t_k)$.*

In quantum-mechanical terms, the Hilbert space $L_\sigma^2(\mathbb{R}^k)$ provides a representation of a physical system in which the observables $A_1, \ldots, A_k$ are diagonal.

EXAMPLE 1

Let A be the operator $-(d/dx)^2$, as in Exercise 4 of Section 14.7, whose spectrum is not simple. A second operator B is defined by the equation

$$(Bf)(x) = f(-x)$$

for all f in L^2; it commutes with A. B has a pure point spectrum consisting of the two eigenvalues $\mu = \pm 1$, because $B^2 = I$; any even distribution in L^2 is an eigenfunction for $\mu = +1$, and any odd one for $\mu = -1$. The pair of equations $-f''(x) = \lambda f(x)$ and $f(-x) = \mu f(x)$ has only one solution, except for normalization, namely $\cos\sqrt{\lambda}x$ for $\mu = 1$, and $\sin\sqrt{\lambda}x$ for $\mu = -1$, so presumably A and B have a simple joint spectrum, i.e., A and B form a complete set of commuting operators. The resolution of the identity F_t for B is easily found to be

$$(F_t\varphi)(x) = \begin{cases} 0, & \text{for } t < -1, \\ \tfrac{1}{2}\varphi(x) - \tfrac{1}{2}\varphi(-x), & \text{for } -1 \leq t < 1, \\ \varphi(x), & \text{for } t \geq 1. \end{cases}$$

The problem of expressing an arbitrary u in $\mathfrak{H}$ in terms of elements of the form $E_s F_t \xi$ reduces to expressing $\varphi(x)$ as

$$\varphi(x) = \int_{-\infty}^{\infty} [g(s)e^{-ixs} + h(s)e^{ixs}]\hat{\xi}(s)ds$$

(details omitted); although $g(\cdot)$ and $h(\cdot)$ are required to be even functions, there is now enough freedom so that any $\varphi(x)$ can be represented in this way, if $\hat{\xi}(s)$ is, say, $=e^{-s}$ for $s \geq 0$ and $=0$ for $s < 0$. Hence the joint spectrum is simple.

CHAPTER 15

Problems of Evolution; Banach Spaces

Initial-value problem; initial data; boundary conditions and other auxiliary conditions; evolution; particle dynamics; heat flow; wave motion; state space; norm; Banach space; well-posed and ill-posed problems; generalized solutions; Lorentz invariance of well-posedness.

Prerequisite: Partial differential equations of physics; Chapters 1–8

The laws of classical physics are causal or deterministic, and that leads to the concept of a well-posed initial-value problem. Roughly speaking, a detailed knowledge of the state of a system at time $t = t_0$ enables one to predict the subsequent states for all $t > t_0$. This chapter and the next two are devoted to the study of such problems. Differential equations are usually involved, and one must decide what is physically acceptable as a solution of the equations, and what are the appropriate initial and auxiliary conditions. A physical principle that guides the proper formulation of the problems is that there should be exactly one solution for every initial state, and the solution should depend continuously on the initial state, in a sense to be explained. It will be seen that Banach spaces provide the appropriate abstract setting for these problems. Most of the discussion is for linear problems; non-linear ones, for which the theory is quite fragmentary, are discussed briefly in Chapter 17.

15.1 Initial-Value Problems in Mechanics

In many problems of theoretical physics, the time variable t plays a special role. The mathematical formulation includes *initial data*, which describe the state of the system at an initial instant $t = 0$, and the problem is to find the states at later time $t > 0$, that is, to find the *evolution* of the system from its initial state.

A finite-dimensional example in celestial mechanics is the problem of the dynamics of a system of N bodies, regarded as mass points and moving freely in space, subject only to their mutual gravitational action. An instantaneous state of the system is specified by giving the values of the three Cartesian coordinates and the three corresponding momentum components for each body. If these $6N$ quantities are known for $t = 0$, their values are then uniquely determined (barring collisions) for later times $t > 0$ by Newton's laws of gravitation and motion.

The foregoing problem is nonlinear, but there are many problems in elementary mechanics, involving rigid bodies, walls, springs, weights, and so on, in which the system has one or more states of equilibrium, and if the system is close to equilibrium the equations of motion can be linearized. For n degrees of freedom we let $\mathbf{y}$ be an n-component vector whose components indicate the departure from equilibrium. The resulting equation is of the form

$$\ddot{\mathbf{y}} = A\mathbf{y}, \tag{15.1-1}$$

where A is an $n \times n$ real matrix. An instantaneous configuration of the system corresponds to a point $\mathbf{y}$ in n-dimensional space, and if $\mathbf{y}$ and $\dot{\mathbf{y}}$ are given at time $t = 0$, then the subsequent motion of the point $\mathbf{y}$ in the n-dimensional space is completely determined by (15.1-1), in the linear approximation.

For a conservative (frictionless) system, the matrix A is symmetric, hence has real eigenvalues; if all eigenvalues are negative, the motion consists of sinusoidal oscillations about the equilibrium; if any eigenvalue is positive, the general solution has exponentially diverging terms in it, and the equilibrium configuration is unstable. Of course, for certain special choices of the initial data $\mathbf{y}$ and $\dot{\mathbf{y}}$ at $t = 0$, the exponentially diverging terms are absent, but then the slightest alteration of the initial data can make them reappear, and large departures from equilibrium eventually result; i.e., the equilibrium is unstable. It will be seen below, under the heading of well- and ill-posed problems, that a similar but more drastic kind of instability can appear in the infinite-dimensional cases.

15.2 Initial-Value Problem of Heat Flow

The finiteness of the number of degrees of freedom in the foregoing examples results from the idealization of the bodies as mass points and rigid bodies. Physics deals more generally with extended media. Then, an initial-value problem is based on a system of one or more partial-differential or integro-differential equations (abbreviated DE), together with initial conditions (IC) and boundary conditions or other auxiliary conditions (AC). The differential equations can be written with $\partial/\partial t$ as the operator appearing on the left sides and with operators either not containing t or containing t only parametrically on the right. Sometimes, a differential equation not containing t at all appears as an auxiliary condition; examples are the divergence conditions $\nabla \cdot \mathbf{E} = 0$ and $\nabla \cdot \mathbf{H} = 0$ of Maxwell's equations in empty space. When there are only partial differential equations and initial data (then it is usually necessary to give the data in all space, so as to avoid boundary conditions), the problem is often called a *Cauchy problem*.

A simple prototype is provided by one-dimensional heat flow. If $u = u(x, t)$ is the temperature at time t at position x along a thermally insulated rod, then the heat flux past point x is proportional to $-\partial u/\partial x$; the divergence of this flux causes a corresponding rate of decrease of the temperature (or increase, if the divergence is negative); hence, u satisfies the differential equation

$$\frac{\partial u}{\partial t} = \sigma \frac{\partial^2 u}{\partial x^2}, \qquad a \le x \le b, 0 \le t \tag{15.2-1}$$

where σ is a positive constant (it has been assumed that the thermal conductivity and the specific heat are constants), and where a and b are the x coordinates of the ends of the rod. The initial condition is

$$u(x, 0) = f(x) \quad \text{(a known function)}, \qquad a \le x \le b. \tag{15.2-2}$$

The problem as formulated so far has an infinity of solutions, hence boundary conditions are needed. The proper choice depends on the physical arrangement, but one possibility is

$$u(a, t) = u(b, t) = 0, \qquad 0 \le t, \tag{15.2-3}$$

which corresponds to maintaining the two ends of the rod at a fixed temperature, here taken as zero.

A solution of these equations in the classical sense is called a *strict solution* of the initial-value problem. A necessary condition for the existence of a strict solution is that the initial data be consistent with the differential equation and the boundary conditions, that is, that $f(x)$ be twice differentiable and vanish at $x = a$ and $x = b$. It is often desirable to have solutions in a more general sense.

The standard Fourier series method leads to more general solutions. It is:

$$u(x, t) = \sum_{n=1}^{\infty} b_n e^{-n^2 \sigma t} \sin nx, \tag{15.2-4}$$

where, for simplicity, the interval (a, b) has been taken as $(0, \pi)$, and where the b_n are the coefficients of the Fourier sine series for $f(x)$:

$$b_n = \frac{2}{\pi} \int_0^\pi f(x) \sin nx \, dx. \tag{15.2-5}$$

This permits, for example, certain discontinuous initial temperature distributions which are of physical interest. It raises the question whether (15.2-4) should be considered a solution whenever the integrals (15.2-5) exist, say in the Lebesgue sense, even though, for instance, $f(x)$ may have discontinuities densely distributed in $(0, \pi)$, and even though the series (15.2-4) may fail to converge for $t = 0$.

There are solutions corresponding to initial data that constitute a distribution rather than a function $f(x)$. To show this, we first consider the problem on the entire real line $\mathbb{R}$, so that the interval $[a, b]$ is replaced by $\mathbb{R}$ throughout, and there are no boundary conditions. For any fixed real y, the function

$$\psi(x, t; y) = \frac{1}{\sqrt{4\pi\sigma t}} e^{-(x-y)^2/4\sigma t} \tag{15.2-6}$$

satisfies the differential equation (15.2-1) for all x and all $t > 0$. According to equation (2.6-3),

$$\lim_{t \downarrow 0} \psi(x, t; y) = \delta(x - y) \tag{15.2-7}$$

in the sense of convergence of distributions; therefore, $\psi(x, t; y)$, which is called *the fundamental solution*, corresponds to initial data given by $f(x) = \delta(x - y)$. One may imagine that, at time $t = 0$, a unit amount of heat is suddenly put into the rod at $x = y$. Suppose now that $f = f(y)$ is any real tempered distribution on $\mathbb{R}$. For any $t > 0$, $\psi(x, t; y)$ as a function of y is in the Schwartz class $\mathscr{S}$ of test functions. As the reader can easily see, the function

$$u(x, y) = \langle f, \psi \rangle = \int_{-\infty}^{\infty} f(y)\psi(x, t; y)dy \tag{15.2-8}$$

satisfies the differential equation (15.2-1) for all x and all $t > 0$. Furthermore,

$$\lim_{t \downarrow 0} u(x, t) = f(x)$$

in the sense of convergence of distributions. Hence, we have a solution of the initial-value problem in a very general sense.

The corresponding solution for the finite interval $[a, b]$ with the boundary condition (15.2-3) can now be obtained by use of the well-known device of reflecting the solution in the lines $x = a$ and $x = b$ in the x, t plane. When $f(x)$ is given in $[a, b]$, we extend it to all $\mathbb{R}$ by requiring it to be an odd (generalized) function of $x - a$ and of $x - b$ (hence periodic with period $2(b - a)$); that is, we require that $f(x)$, $-f(2a - x)$, and $-f(2b - x)$ all be the same distribution. Then the solution given by (15.2-8) has these same reflection properties. Furthermore, owing to the special properties of the fundamental solution (15.2-6), it is easy to see that for $t > 0$ $u(x, t)$ is an ordinary function, even when $u(x, 0)$ is a distribution. (That is a special property of the heat flow problem, not of initial-value problems in general.) Hence, from the equation

$$u(x, t) = -u(2a - x, t) = -u(2b - x, t),$$

we see that $u = 0$ for $x = a$ and for $x = b$.

15.3 Well- and Ill-Posed Problem

With any reasonable choice of the class of admissible initial functions $f(x)$, the heat flow problem is *well posed* (*in the sense of Hadamard*), which means that (a) it has a solution for each $f(x)$ in the class, (b) the solution is unique for any given $f(x)$, and (c) the solution depends continuously on $f(x)$. The last means that if a perturbation δu of the solution is small at $t = 0$, it is also small for any given $t > 0$. For example, if we assume that $f(x)$ is piecewise continuous and has bounded variation in $[a, b]$, the Fourier series method can be used, and (15.2-4) shows that all Fourier coefficients of u, hence also of δu, decrease as t increases. Well-posedness will be defined in the Banach space context in the next chapter.

Suppose, however, that the sign is changed on the right of the differential equation (15.2-1); i.e.,

$$\frac{\partial u}{\partial t} = -\sigma \frac{\partial^2 u}{\partial x^2}, \qquad a \le x \le b, 0 \le t \ (\sigma > 0), \qquad (15.3\text{-}1)$$

the initial and boundary conditions being the same. Then the solution (15.2-4) is replaced by

$$u(x, t) = \sum_{n=1}^{\infty} b_n e^{+n^2 \sigma t} \sin nx,$$

whose terms increase exponentially with time. Unless the Fourier coefficients b_n of the initial function f decrease exceedingly rapidly as $n \to \infty$ (i.e. unless $f(x)$ is an exceedingly smooth function), the Fourier series diverges, for any positive t whatever, owing to the factor n^2 in the exponent, and there is no solution. Even when the solution exists, it does not depend continuously on the initial data. Given any positive number M (no matter how large) and any ε and δ (no matter how small), a perturbation

$$u_1(x, t) = \varepsilon e^{m^2 \sigma t} \sin mx$$

can be found which is no larger than ε at $t = 0$, but is larger than M at $t = \delta$, by simply taking m large enough. This initial value problem is therefore ill posed.

Ill-posedness is more drastic than mere instability. If the original heat-flow equation (15.2-1) is modified by adding a term ku on the right, where k is a constant, then there is a mode of perturbation, namely

$$u_1(x, t) = \text{const. } \exp\left[\left(k - \frac{\sigma \pi^2}{(b-a)^2}\right)t\right]\sin\frac{\pi(x-a)}{b-a}$$

which grows without limit as $t \to \infty$, if k is $> \sigma\pi^2/(b-a)^2$, hence the physical system is unstable in the usual sense. However, in a finite time, say $t \le t_1$, no perturbation can grow by a factor larger than

$$\exp\left[\left(k - \frac{\sigma\pi^2}{(b-a)^2}\right)t_1\right],$$

hence the solution depends continuously on the initial data. That is, if we take the above factor into account, we can decide how accurately $f(x)$ must be known to guarantee that the error be $<\varepsilon$ for $0 \le t \le t_1$, whereas in an ill-posed problem an error can be amplified by an arbitrarily large factor in any given time interval.

Ill-posed problems can be of physical interest. The heat-flow problem with sign reversed, as above, is equivalent to the problem of knowing a temperature distribution for $t = 0$ and wishing to know it for $t < 0$. There is no practical solution of such a problem unless more information is available about the thermal history of the system, usually in the form of inequalities.

EXERCISE

1. Find the Fourier series solution of (15.2-1) and (15.2-2) if (15.2-3) is replaced by

$$\frac{\partial u}{\partial x}(a, t) = \frac{\partial u}{\partial x}(b, t) = 0, \qquad 0 \le t, \tag{15.3-2}$$

and give the physical interpretation of these new boundary conditions. Show that if all four boundary conditions (15.2-3) and (15.3-2) are imposed simultaneously then there is no solution at all unless $f(x) \equiv 0$.

15.4 The Initial-Value Problem of Wave Motions

It is recalled that the one-dimensional wave equation

$$\frac{\partial^2 u}{\partial t^2} = c^2 \frac{\partial^2 u}{\partial x^2} \tag{15.4-1}$$

has solutions of the form

$$u(x, t) = f(x + ct) + g(x - ct). \tag{15.4-2}$$

We note in passing that the problem can be put into the canonical form mentioned at the beginning of Section 15.2 by introducing another function $v(x, t)$ and writing

$$\frac{\partial u}{\partial t} = c \frac{\partial v}{\partial x},$$

$$\frac{\partial v}{\partial t} = c \frac{\partial u}{\partial x}; \tag{15.4-3}$$

an instantaneous state of the system is described by giving the values of $u(x, 0)$ and $v(x, 0)$ as functions of x. (If only u appears one must give also $\partial u/\partial t$.)

If f and g are twice differentiable, then (15.4-2) gives a strict solution of (15.4-1). It is often desirable for physical reasons to admit more general "solutions" of this type, for example solutions having a saw-tooth wave form. Then f and g are permitted to have discontinuous derivatives; at a discontinuity of f' or g' the second derivatives appearing in (15.4-1) do not exist. Furthermore, by means of the famous example of Weierstrass, f and g can be chosen so that f'' and g'' do not exist at all, even if f' and g' are continuous, hence (15.4-1) would never be satisfied, for any x or any t, even though such a "solution" can represent the sort of wave motion that physicists call "white noise." Still worse, f and g themselves may be nondifferentiable or even everywhere discontinuous. Clearly, as soon as generalized solutions are considered, it becomes necessary to specify the class of functions or distributions that are physically admissible. The generalized solution (15.4-2) always satisfies the differential equation (15.4-1) in the distribution sense.

It is evident from (15.4-2) that if u and $\partial u/\partial t$ are both known at $t = 0$ for all x the functions f and g are fully determined by the initial data, and then the solution (15.4-2) is unique. It will be seen in Chapter 17 that in certain closely related nonlinear problems the generalized solutions (these called "weak" solutions) are not unique, until certain auxiliary conditions have been imposed, and, for a given differential equation and given initial data, different generalized solutions are obtained, depending on what auxiliary conditions are imposed.

15.5 The Function Space (State Space) of an Initial-Value Problem

For any fixed value of the variable t, in problems of the type described above, certain functions of the other variables (called *space variables*) describe an instantaneous state of the physical system. These functions will be denoted by a single symbol u, which is regarded as representing a point in an infinite-dimensional space $\mathfrak{B}$. As time goes on, the point $u = u(t)$ moves through $\mathfrak{B}$ in a way that corresponds to the evolution of the system.

Henceforth, until Chapter 17, only linear problems are discussed. The sum $u_1 + u_2$ or the difference $u_1 - u_2$ of two elements of $\mathfrak{B}$ is defined as the point in $\mathfrak{B}$ obtained by adding or subtracting the corresponding functions. If α is any number, αu is defined as the point obtained by multiplying all the functions that represent u by α. In this way, $\mathfrak{B}$ is made into a linear space. These operations may lead to functions that do not directly represent states of the physical system (e.g., containing negative densities), but it is convenient to regard such functions as representing generalized states of the system. Then, if a function in $\mathfrak{B}$ is expanded in a series of some kind, the individual terms and the partial sums of the series are also represented by points in $\mathfrak{B}$. For the same reason, it is convenient to admit complex-valued functions.

A concept of distance is introduced in the space $\mathfrak{B}$ in such a way that two points u_1 and u_2 are close together if and only if they represent nearly identical states of the physical system. For linear problems, this distance is taken as a function of the difference $u_1 - u_2$ ($= w$, say) and is called the *norm* of $u_1 - u_2$ or of w and is denoted by $\|u_1 - u_2\| = \|w\|$; it is a positive number if u_1 and u_2 are different states and is zero if $u_1 = u_2$. Familiar examples are the maximum norm and the root-mean-square or L^2 norm—see Section 15.7. The distance $\|\alpha w\|$ between states αu_1 and αu_2 is regarded as $|\alpha|$ times the distance between u_1 and u_2; i.e. $\|\alpha w\| = |\alpha| \|w\|$. The interpretation as distance suggests as a further requirement the triangle inequality $\|u_1 - u_3\| \le \|u_1 - u_2\| + \|u_2 - u_3\|$ or, more simply, $\|w_1 + w_2\| \le \|w_1\| + \|w_2\|$. It is perhaps not obvious whether this requirement is dictated by physical considerations (it is recalled that the triangle inequality does not hold in the geometry of relativity), but in fact it is satisfied by any choice of norm that is likely to be used in practice, and it is essential for certain important results in the theory of operators.

15.6 Completeness of the State Space; Banach Space

If one makes an infinite sequence of changes of the state of a system, and the magnitudes of the changes decrease sufficiently rapidly along the sequence, then it seems evident that there ought to be a possible state of the system which is the limit of the sequence of states. Stated mathematically, if $\|u_l - u_n\| \to 0$, as l and $n \to \infty$ independently, then the sequence $\{u_n\}$ of points in $\mathfrak{B}$ ought to have a limit in $\mathfrak{B}$. In this sense, $\mathfrak{B}$ is a continuum, like the real number system (every Cauchy sequence has a limit) rather than an imperfect continuum, like the rational numbers alone.

These ideas suggest taking $\mathfrak{B}$ as a *Banach space*, which was defined in Chapter 1 (Sections 1.2 and 1.3) as a *complete normed linear space*. A Banach space has all the attributes of a Hilbert space except those connected with the inner product, and as noted in Chapter 1, it may not be possible to define an inner product so as to satisfy the relation $(u, u)^{1/2} = \|u\| = \text{norm of } u$ in $\mathfrak{B}$.

Often there is a choice among various Banach spaces. Whether a given "solution" ought to be accepted as representing a possible evolution of the physical system has to be decided by physics, but it will be seen in the next chapter that, once a Banach space $\mathfrak{B}$ has been chosen, the appropriate generalizations of the strict solutions are the generalized solutions determined by a certain family of operators $E(t)$ in $\mathfrak{B}$.

15.7 Examples of Banach Spaces

Spaces of continuous functions with the so-called maximum norm are useful in problems of heat flow, diffusion, and transport. Examples are the following:

1. $C(a, b) = \{f : f(x) \text{ is continuous}, a \le x \le b\}$

$$\|f\| = \sup\{|f(x)| : a \le x \le b\},$$

   that is, $C(a, b)$ is the set of all functions of x that are defined and continuous for $a \le x \le b$; for any such function f, the norm is the supremum or least upper bound of the values of $|f(x)|$. (In this case, "max" could have been written in place of "sup".)

2. $C(\mathbb{R}^n) = \{f: f(\mathbf{x}) \text{ is continuous and bounded for all } \mathbf{x} \text{ in } \mathbb{R}^n\}$

$$\|f\| = \sup\{|f(\mathbf{x})|: \mathbf{x} \text{ in } \mathbb{R}^n\}$$

the vector $\mathbf{x}$ stands for a point $x_1, \ldots, x_n$ in the n-dimensional space $\mathbb{R}^n$.

3. $CP(\mathbb{R}, p) = \{f: f(x) \text{ is continuous for all } x \text{ and is periodic with period } p \; [f(x + p) \equiv f(x)]\}$

$$\|f\| = \sup\{|f(x)|: x \text{ in } \mathbb{R} \text{ (or } x \text{ in any period)}\}.$$

In each of these examples, two cases arise, depending on whether the functions and the scalars are real-valued or complex-valued. When necessary, these cases are distinguished by writing $C^{\text{real}}(a, b)$ or $C^{\text{complex}}(a, b)$ and the like.

All the axioms of a Banach space, with possible exception of completeness, are obviously true, for these spaces.

Assertion.   The above spaces are complete, hence are Banach spaces.

PROOF, FOR $C(\mathbb{R}^n)$.   Let $\{f_k\}$ be any Cauchy sequence in $C(\mathbb{R}^n)$. For any $\varepsilon > 0$, $\|f_k - f_l\|$ is $\leq \varepsilon$, for all sufficiently large k and l; that is

$$|f_k(\mathbf{x}) - f_l(\mathbf{x})| < \varepsilon, \quad \text{for all } \mathbf{x}, \tag{15.7-1}$$

for all sufficiently large k and l. Therefore, (1) $\{f_k(\mathbf{x})\}$ is a Cauchy sequence of numbers, for any $\mathbf{x}$, hence converges to a limit $f(\mathbf{x})$; (2) since the convergence is uniform, $f(\mathbf{x})$ is continuous; (3) letting $k \to \infty$ in (15.7-1) gives

$$|f(\mathbf{x}) - f_l(\mathbf{x})| < \varepsilon, \quad \text{for all } \mathbf{x},$$

for all sufficiently large l—this shows both that $f(\mathbf{x})$ is bounded and that $\|f - f_l\| \to 0$ as $l \to \infty$. Therefore, the sequence $\{f_k\}$ has the limit $f = f(\mathbf{x})$ in the space, and the space is complete.

A class of discrete coordinate spaces is typified by the Hilbert space l^2, discussed in Section 1.3, each point ξ of which is an infinite sequence $\{x_k\}$ of complex numbers such that the series $\sum_{k=1}^{\infty} |x_k|^2$ converges; namely,

$$l^2 = \left\{\xi = \{x_k\}: \sum_{k=1}^{\infty} |x_k|^2 < \infty\right\};$$

$$\|\xi\| = \|\{x_k\}\| = \left(\sum_{k=1}^{\infty} |x_k|^2\right)^{1/2}.$$

Similar spaces are

$$l^1 = \left\{\xi = \{x_k\}: \sum_{k=1}^{\infty} |x_k| < \infty\right\}$$

$$\|\xi\| = \sum_{k=1}^{\infty} |x_k|$$

and, more generally, for any $p \geq 1$,

$$l^p = \left\{ \xi = \{x_k\}: \sum_{k=1}^{\infty} |x_k|^p < \infty \right\}$$

$$\|\xi\| = \left(\sum_{k=1}^{\infty} |x_k|^p \right)^{1/p}.$$

For l^2, the proofs of the triangle inequality and of completeness were given in Chapter 1 on Hilbert spaces; the proofs for the l^p spaces are similar, but are omitted since these spaces will not be used.

These spaces can be generalized as follows: Let $\{\mathfrak{B}_k\}$ be a sequence of Banach spaces ($k = 1, 2, \ldots$); a space $l^2\{\mathfrak{B}_k\}$ is defined, each element ξ of which is a sequence $\{u_k\}$, where u_k is an element of $\mathfrak{B}_k$, such that the series $\sum_{k=1}^{\infty} \|u_k\|^2$ converges, where, in the kth term of the sum, $\|\cdot\|$ denotes the norm in the space $\mathfrak{B}_k$; that is

$$l^2\{\mathfrak{B}_k\} = \left\{ \xi = \{u_k\}: \sum_{k=1}^{\infty} \|u_k\|^2 < \infty \right\}$$

$$\|\xi\| = \left(\sum_{k=1}^{\infty} \|u_k\|^2 \right)^{1/2}.$$

The Fock spaces that appear in the theory of second quantization are of this type, with each $\mathfrak{B}_k$ a Hilbert space—see Chapter 1.

The spaces of distributions denoted by $L^p(\mathbb{R}^n)$, $L^p(\Omega)$, $L^p_\sigma(\mathbb{R})$ ($1 \leq p \leq \infty$) in Chapter 5 are all Banach spaces, the norm being given by

$$\|f\| = \left[\int_{\mathbb{R}^n \text{ or } \Omega} |f|^p \, d\tau \text{ or } \int |f|^p d\sigma(x) \right]^{1/p}.$$

If $\mathfrak{B}$ is any Banach space, then the set $B(\mathfrak{B})$ of bounded linear operators defined in all of $\mathfrak{B}$ is another Banach space (in fact, a Banach algebra, because multiplication is defined in it); the norm of an element A in $B(\mathfrak{B})$ is its bound $\|A\|$.

In each of the above spaces, a point represents a single function. In many problems, the description of the state of a physical system requires several functions (e.g., in fluid dynamics, the pressure $p(\mathbf{x})$, the density $\rho(\mathbf{x})$ and the three components of the velocity $\mathbf{v}(\mathbf{x})$). These functions can be written as the components $f_j(\mathbf{x})$ of a vector-valued function $\mathbf{f}(\mathbf{x})$ (which need not have the same number of components as $\mathbf{x}$). The above spaces can then all be generalized; for example,

$$C(\mathbb{R}^n \to \mathbb{R}^m) = \{f: \mathbf{f}(\mathbf{x}) \text{ is a bounded continuous mapping from } \mathbb{R}^n \text{ into } \mathbb{R}^m\}$$

$$\|\mathbf{f}\| = \sup\{|f_j(\mathbf{x})|: \mathbf{x} \text{ in } \mathbb{R}^n, j = 1, \ldots, m\};$$

if the functions are complex-valued, the space is called $C(\mathbb{R}^n \to \mathbb{C}^m)$. Similarly, in $L^p(\mathbb{R}^n \to \mathbb{R}^m)$ and $L^p(\mathbb{R}^n \to \mathbb{C}^m)$ each component $f_j(\mathbf{x})$ is a real or complex distribution in the real or complex space $L^p(\mathbb{R}^n)$, and

$$\|\mathbf{f}\| = \left[\sum_{j=1}^{m} \int_{\mathbb{R}^n} |f_j(\mathbf{x})|^p d\tau\right]^{1/p}.$$

If $\mathfrak{B}$ is any Banach space, another Banach space can be derived from it as follows: A one-parameter family $u(t)$ of elements of $\mathfrak{B}$ (a *curve* in $\mathfrak{B}$) is called *continuous* if $\|u(t + \delta) - u(t)\| \to 0$ as $\delta \to 0$ for all t. A Banach space $\mathfrak{B}_1(a, b)$ is then defined in this way:

$$\mathfrak{B}_1(a, b) = \{u(t) \text{ a continuous curve in } \mathfrak{B}: a \leq t \leq b\};$$

$$\text{for } u \text{ in } \mathfrak{B}_1(a, b), \qquad \|u\| = \max_{a \leq t \leq b} \|u(t)\|.$$

A space of this kind will play a role in Section 16.6 on inhomogeneous initial-value problems.

15.8 Inequivalence of Various Banach Spaces

It is recalled that all infinite-dimensional separable Hilbert spaces are isometrically isomorphic, i.e., are equivalent as abstract Hilbert spaces. The same is not true for Banach spaces. It is easily seen, for example, that the spaces L^1 and L^2 (say on $\mathbb{R}$) are not equivalent. Since L^2 is a Hilbert space, the parallelogram law (1.3-5) holds in it, whereas that law does not hold in L^1, according to Section 1.3. Since the parallelogram law involves only intrinsic properties (i.e. the properties of an abstract Banach space), there can be no isomorphism between L^1 and L^2. Furthermore, L^1 and L^p are inequivalent for any $p > 1$, because L^p is reflexive and L^1 is not (see Section 5.7), and reflexivity is also an intrinsic property. It can be proved that L^p and L^r are always inequivalent, for $p \neq r$.

The choice of norm in a function space is primarily a matter of physics; it places restrictions on the class of admissible functions. If $w(x)$ is to have a finite value of the maximum norm, it must be bounded; if it is to have a finite L^2 norm it must be quadratically integrable, and so on. The inequivalence of different norms has also an importance for convergence that it does not have in finite-dimensional spaces, where all the usual norms are topologically equivalent. For example, if $\mathbf{v}$ is a vector whose components are $v_1, \ldots, v_n$, then the norms

$$\|\mathbf{v}\|_{\max} = \max\{|v_1|, \ldots, |v_n|\}$$

and

$$\|\mathbf{v}\|_2 = (|v_1|^2 + \cdots + |v_n|^2)^{1/2}$$

both determine the same mode of convergence, namely, $\mathbf{v}_k \to \mathbf{w}$ if and only if each component of $\mathbf{v}_k$ converges to the corresponding component of $\mathbf{w}$. In function spaces, however, different norms are not generally equivalent, and a problem can be well posed with respect to one norm but not with respect to another. In problems of wave motion and the like, a root-mean-square type of norm is often appropriate, because the energy is the integral of the square of some field quantity (or a sum of such squares), hence such a norm is finite for physically admissible functions, and it remains bounded during the evolution of the system.

15.9 Linear Operators

Many of the concepts dealing with linear operators are the same in any Banach space as in a Hilbert space. The following terms have the same definitions as in Chapter 7: linear operator A; domain $\mathfrak{D}(A)$; range $\mathfrak{R}(A)$; extension; bound $\|A\|$; product; inverse; eigenvalue; eigenvector; point, continuous, and residual spectrum; resolvent set; resolvent; graph $\Gamma(A)$; closed operator; compact operator (but not Hilbert–Schmidt or trace-class operator).

The concepts symmetric, self-adjoint, unitary, and normal, do not apply. There is no general theory of spectral decomposition like that in Chapter 9.

The extension theorem holds (a bounded operator has a unique extension to the entire closure of its domain with unincreased bound); the proof in Section 7.1 is valid in any Banach space.

EXAMPLE 1
Let $\mathfrak{B}$ be the space $CP(\mathbb{R}, 2\pi)$ of continuous periodic functions on $\mathbb{R}$ with period 2π, in which $\|\cdot\|$ is taken as the maximum norm. Let T be an operator whose domain $\mathfrak{D}(T)$ is the set of functions in $\mathfrak{B}$ with absolutely convergent Fourier series and such that for any such function f, if

$$f(x) = \sum_{-\infty}^{\infty} {}_{(n)} C_n e^{inx},$$

then

$$(Tf)(x) = \sum_{-\infty}^{\infty} {}_{(n)} \frac{C_n}{1 + n^2} e^{inx}.$$

T is a bounded operator with domain dense in $\mathfrak{B}$, hence it has a unique bounded extension T' with $\mathfrak{D}(T') = \mathfrak{B}$. T' can in fact be expressed as an integral operator of the form

$$(T'f)(x) = \int_{-\infty}^{\infty} K(x - y) f(y) dy.$$

EXERCISE

1. Find an expression for the kernel $K(x - y)$ in the above example (it is not unique).

Whether a given operation, such as differentiation, leads to a bounded or unbounded operator depends on the choice of Banach space. Let $\mathfrak{B}$ be the space of infinitely differentiable functions on $\mathbb{R}$ such that the quantity

$$\|f\| \stackrel{\text{def}}{=} \sup_{(x)} \sup_{(q)} \left\{ \left| \left(\frac{d}{dx}\right)^q f(x) \right| : x \in \mathbb{R}, q = 0, 1, 2, \ldots \right\}$$

is finite. With minor modifications, the proof of the completeness of $C(\mathbb{R}^n)$ given in Section 15.7 shows that $\mathfrak{B}$ is a Banach space. In this space, the differentiation operator T given by

$$\mathfrak{D}(T) = \mathfrak{B}, \qquad (Tf)(x) = \frac{d}{dx} f(x)$$

is a bounded linear operator. It is noted in passing that this Banach space is probably not a very useful one, since, e.g., the function $\sin x$ is in it, while $\sin(1 + \varepsilon)x$ is not in it, for any $\varepsilon > 0$.

15.10   Linear Functionals; The Dual Space

As in a Hilbert space, a linear functional $l(u)$ on $\mathfrak{B}$ is *bounded* if the quantity

$$\|l\| \stackrel{\text{def}}{=} \sup_{u \neq 0} \frac{|l(u)|}{\|u\|}$$

is finite. The set of all bounded linear functionals on $\mathfrak{B}$ is another Banach space, in which the norm of l is its bound $\|l\|$; it is called the space *dual* to $\mathfrak{B}$ and is denoted by $\mathfrak{B}'$. According to Section 5.7, the dual of a space of type L^p is a corresponding space of type L^q, where $(1/p) + (1/q) = 1$. According to the Riesz–Frechet representation theorem of Section 1.8, a Hilbert space may be regarded as its own dual.

EXERCISE

1.   Let $\mathfrak{B}$ be the space $C(\mathbb{R})$ of bounded continuous real functions on $\mathbb{R}$ with the maximum norm. Show that the dual space $\mathfrak{B}'$ is isometric to the space of all real functions $\sigma(x)$ of bounded variation on $\mathbb{R}$ and such that $\sigma(0) = 0$ and $\sigma(x + 0) = \sigma(x)$, with the norm of a function $\sigma(x)$ in $\mathfrak{B}'$ being its total variation on $\mathbb{R}$. (*Hint*: Use the Riesz representation theorem of Section 13.9.)

15.11   Convergence of Vectors and Operators

The convergence modes defined for Hilbert spaces in Sections 1.9 and 9.9 are still applicable, except that, for weak convergence, the linear functionals $l(\cdot)$ in $\mathfrak{B}'$ replace the expressions $(v, \cdot)$ (which *are* the linear functionals if $\mathfrak{B}$ is a Hilbert space). A sequence $\{u_n\}$ converges *strongly* to w if $\|u_n - w\| \to 0$, and weakly if $l(u_n) \to l(w)$ for every l in $\mathfrak{B}'$. A sequence $\{A_n\}$ of bounded operators converges *uniformly* to B if $\|A_n - B\| \to 0$, *strongly* if $\|A_n v - Bv\| \to 0$ for every v in $\mathfrak{B}$, and *weakly* if $l(A_n v) \to l(Bv)$ for every v in $\mathfrak{B}$ and every l in $\mathfrak{B}'$. The weak convergence modes will not concern us much.

15.12 Inner Product; Hilbert Space

It is recalled that the matrix A of the problem in Section 15.1 is symmetric for a conservative system (i.e., $A_{jk} = A_{kj}$ for all j, k). The symmetry is best described in terms of the scalar product $\mathbf{x} \cdot \mathbf{y}$ of vectors; namely, A is symmetric if and only if $(A\mathbf{x}) \cdot \mathbf{y} = \mathbf{x} \cdot (A\mathbf{y})$ for all vectors $\mathbf{x}$ and $\mathbf{y}$, because, when written out, this equation says

$$\sum_{(j\,k)} A_{j\,k} x_k y_j = \sum_{(j\,k)} A_{k\,j} x_k y_j$$

It is sometimes possible (but not for all Banach spaces) to introduce similar notions in $\mathfrak{B}$ in terms of an *inner product* for two elements of $\mathfrak{B}$, written as (u, v), which is the analogue of the scalar product of vectors, and has similar properties. If an inner product is defined in $\mathfrak{B}$ and in such a way that $\|u\| = (u, u)^{1/2}$ (as for vectors), then $\mathfrak{B}$ is a Hilbert space (Chapter 1). The most important symmetric operators in a Hilbert space are the self-adjoint operators; they were discussed in Chapters 7–9.

15.13 Relativistic Problems

In the special theory of relativity of Einstein, there is no uniquely determined time variable t. However, relativistic theories are invariant *in form* under Lorentz transformations; hence if a problem is well posed in one Lorentz frame, then it is well posed in all Lorentz frames. The Dirac equation and Maxwell's equations will provide examples, in the next chapter.

   It is not always a trivial matter to decide which solution of the initial-value problem in one frame represents the same evolution of the physical system as a given solution in another frame.

15.14 Seminorms

As noted in Section 5.9, a *seminorm* is a function $\|\cdot\|$ that satisfies all the requirements of a norm except that it is only required to be semidefinite ($\|u\| \geq 0$ for all u) rather than definite ($\|u\| > 0$ except for $u = 0$). A space V that is complete with respect to a seminorm can be made into a Banach space as follows: u and u' in V are said to be *equivalent* (in symbols, $u \sim u'$) if $\|u - u'\| = 0$. The relation $\sim$ is reflexive ($u \sim u$), symmetric (if $u \sim u'$, then

$u' \sim u$), and transitive (if $u \sim u'$ and $u' \sim u''$, then $u \sim u''$, because of the triangle inequality); therefore, this relation separates V into disjoint equivalence classes. The equivalence class that contains u is denoted by $[u]$ and is given by

$$[u] = \{v: \|u - v\| = 0\}.$$

Furthermore, if $u \sim u'$, then $\|u\| = \|u'\|$ (also by the triangle inequality), hence $\|[u]\|$ can be unambiguously defined as equal to $\|u\|$. It is easy to show that, with this norm, the set of all equivalence classes is a Banach space $\mathfrak{B}$. In algebraic terminology, if V_0 is the subspace of V consisting of elements with norm $= 0$, then $\mathfrak{B}$ is the factor space V/V_0.

This process is used in the classical definition of L^p spaces, according to which a function $f(x)$ is in L^p if it is measurable and $|f(x)|^p$ is integrable. However, it is necessary to consider two functions $f(x)$ and $g(x)$ to be identical, as elements of L^p, if they differ only on a set of measure zero, for then

$$\|f - g\| = \left[\int |f(x) - g(x)|^p \, dx \right]^{1/p} = 0.$$

CHAPTER 16

Well-Posed Initial-Value Problems; Semigroups

Banach-space formulation; strict solution; Hadamard's concept of a well-posed problem; existence and uniqueness of the solution; continuous dependence on the initial state; generalized solution; semigroup; strongly continuous semigroup; infinitesimal generator; Hille–Yoŝida theorem; inhomogeneous problems; inhomogeneous boundary conditions; problems with explicit time dependence; applications to heat flow, wave motion, quantum mechanics, electromagnetism, and neutron transport.

Prerequisites: Chapter 15

This chapter gives the general theory of linear initial-value problems or problems of evolution in physics.

16.1 Banach-Space Formulation of an Initial-Value Problem

A linear initial-value problem of the kind considered in Chapter 15 can be formulated abstractly as the problem of finding a function $u(t)$, with values in a Banach space $\mathfrak{B}$, such that

$$\frac{d}{dt} u(t) = Au(t), \tag{16.1-1}$$

and

$$u(0) = u_0, \quad \text{given}, \tag{16.1-2}$$

where A is an operator involving the space variables, and where the derivative of $u(t)$ is the limit, as $\Delta t \to 0$, of $[u(t + \Delta t) - u(t)]/\Delta t$, in the sense of convergence in $\mathfrak{B}$. A *strict solution* of equation (16.1-1) is therefore defined as a function $u(t)$ $(u(t) \in \mathfrak{B}$ for all $t \geq 0)$ such that

$$u(t) \in \mathfrak{D}(A), \quad \text{for all } t \geq 0 \tag{16.1-3}$$

$$\lim_{\Delta t \to 0} \left\| \frac{u(t + \Delta t) - u(t)}{\Delta t} - Au(t) \right\| = 0, \quad \text{for all } t \geq 0. \tag{16.1-4}$$

The boundary conditions or other auxiliary conditions are taken care of by restricting the domain $\mathfrak{D}(A)$ to elements of $\mathfrak{B}$ that satisfy those conditions; they are assumed to be linear and homogeneous, so that the set $\mathfrak{S}$ of all u that satisfy them is a linear manifold; $\mathfrak{D}(A)$ is assumed to be contained in $\mathfrak{S}$. Then, equation (16.1-3) requires, among other things, that $u(t)$ satisfy the auxiliary conditions for all t. Equation (16.1-1) is a linear equation of evolution; it tells how the physical system evolves, once the initial state has been given.

The application of (16.1-1) is not restricted to first-order equations in t, since higher-order equations can be reduced to systems of first order ones by the introduction of additional dependent variables.

Much of the theory given in this chapter can be generalized to problems in which the operator A depends (sufficiently smoothly) on t; for example, A may be a differential operator whose coefficients depend on t. Such problems are considered briefly in Section 16.7; until that section, A is assumed independent of t.

The choice of the Banach space $\mathfrak{B}$, as well as of the domain of A, is an essential part of the formulation of the problem. It will be seen that a given problem can be well posed, in a sense to be explained in the next section, for one choice of $\mathfrak{B}$ but not for another. The choice is often dictated, at least in part, by physical considerations.

16.2 Well-Posed Problem; Generalized Solutions

The initial-value problem is called *well posed* (in the sense of Hadamard) if it has the following properties.

1. The strict solutions are uniquely determined by their initial elements;
2. The set $\mathfrak{U}$ of all initial elements of strict solutions is dense in the Banach space $\mathfrak{B}$;
3. For any finite interval $[0, t_0]$ there is a constant $K = K(t_0)$ such that every strict solution satisfies the inequality

$$\|u(t)\| \leq K\|u(0)\|, \quad \text{for } 0 \leq t \leq t_0. \tag{16.2-1}$$

In connection with the first requirement, note that if *any* strict solution is uniquely determined by its initial element, then all are, because of the linearity of the problem. Stated differently, if any solution is nonunique— then all are nonunique, for if two different solutions $u_1(t)$ and $u_2(t)$ are such that $u_1(0) = u_2(0)$, then the solution $u_3(t) = u_1(t) - u_2(t)$ can be added to any other solution $u(t)$, thereby destroying the uniqueness of $u(t)$.

The physical interpretation of requirements (2.) and (3.) above is as follows: First, even though a given state u_0 may not be the initial state of a strict solution (for example, owing to nondifferentiability of the initial data), there is required to be a state $\tilde{u}_0$ which approximates u_0 arbitrarily closely and which is the initial state of a strict solution. Second, it is required that a small error or perturbation of the initial state result in only a small change of the later states—that is, if $\|u_1(0) - u_2(0)\| < \varepsilon$, then $\|u_1(t) - u_2(t)\| < K\varepsilon$ for all t in $[0, t_0]$; i.e., the solution must depend continuously on the initial data.

Note.   Since a strict solution $u(t)$ is by definition in $\mathfrak{D}(A)$ for $t = 0$, $\mathfrak{U}$ is $\subset \mathfrak{D}(A)$; it will be seen in Section 16.7 that A can be so chosen, without altering the totality of generalized solutions, that any u_0 in $\mathfrak{D}(A)$ is the initial element of a strict solution, hence $\mathfrak{U}$ is $= \mathfrak{D}(A)$, but that is not assumed here.

For a well-posed problem, generalized solutions can be defined in a meaningful way, as follows: A strict solution $u(t)$ may be regarded as a function of t and of the initial element $u(0)$. For $u(0)$ fixed and t variable, $u(t)$ is a particular strict solution. For t fixed and $u(0)$ variable, the mapping $u(0) \to u(t)$ is a linear transformation in $\mathfrak{B}$, denoted by $E_0 = E_0(t)$, with domain $\mathfrak{D}(E_0)$ equal to the set $\mathfrak{U}$ of all initial elements of strict solutions. Then with t allowed to vary, $E_0(t)$ is a one-parameter family of operators, and, according to (16.2-1), the operators are bounded, and in fact are bounded uniformly with respect to t in $[0, t_0]$; namely, $\|E_0(t)\| \leq K$. According to the Extension Theorem, therefore, since $\mathfrak{U}$ is dense in $\mathfrak{B}$, $E_0(t)$ has, for each t, a unique bounded linear extension defined in all $\mathfrak{B}$; this we call $E(t)$; hence,

$$\mathfrak{D}(E(t)) = \mathfrak{B}, \tag{16.2-2}$$

$$\|E(t)\| = \|E_0(t)\| \leq K, \quad \text{for } 0 \leq t \leq t_0. \tag{16.2-3}$$

For any $u_0 \in \mathfrak{B}$ (even if u_0 is not in $\mathfrak{U}$), the function $u(t)$ given by

$$u(t) = E(t)u_0 \tag{16.2-4}$$

is called the *generalized solution* of the initial-value problem with initial element u_0. The properties of the generalized solutions of various physical problems are discussed below. $E(t)$ is called the *(generalized) solution operator* of the initial-value problem.

Whether a problem is well posed often depends critically on the auxiliary conditions, which influence the domain of A and hence that of $E_0(t)$. In the simple heat flow problem of Section 15.2, if there are too many boundary conditions, say $u(x, t) = 0$ and $(\partial/\partial x)u(x, t) = 0$ at both $x = a$ and $x = b$, then there is no strict solution at all, except for $u(x, 0) \equiv 0$; hence $\mathfrak{U}$ is not dense in $\mathfrak{B}$. On the other hand, if there are too few boundary conditions, then no solution is unique.

The requirement (2.) that $\mathfrak{U}$ be dense in $\mathfrak{B}$ can sometimes be relaxed. Then, the closure of $\mathfrak{U}$ is a proper subspace $\mathfrak{B}_0$ of $\mathfrak{B}$, and the problem may be well posed in $\mathfrak{B}_0$, though not in $\mathfrak{B}$. That is acceptable if $\mathfrak{B}_0$ is large enough to include all states of the physical system. In the heat flow problem, for example, if $\mathfrak{B}$ is the space of all continuous functions on $[a, b]$ with the maximum norm, then $\mathfrak{B}_0$ is the subspace

$$\mathfrak{B}_0 = \{u(x) \in \mathfrak{B} : u(a) = u(b) = 0\}. \qquad (16.2\text{-}5)$$

If the physical system is regarded as including the devices that maintain the temperature $u = 0$ at the ends of the rod, then indeed $\mathfrak{B}_0$ includes all states of the system.

We show that the heat flow problem is well posed in $\mathfrak{B}_0$ if we define the operator A as follows:

$$\mathfrak{D}(A) = \{u \in \mathfrak{B}_0 : u'' \in \mathfrak{B}_0\}$$
$$Au = \sigma u''. \qquad (16.2\text{-}6)$$

In this problem, $\mathfrak{U}$ is $= \mathfrak{D}(A)$, and if the function $u(x, 0) = f(x)$ is in $\mathfrak{D}(A)$, then, according to Section 15.2,

$$u(x, t) = \int_{-\infty}^{\infty} f(y)\psi(x, t; y)dy, \qquad (16.2\text{-}7)$$

where $f(y)$ has been extended to all $\mathbb{R}$ in such a way as to be an odd function of $y - a$ and of $y - b$, and where $\psi(x, t; y)$ is the fundamental solution given by (15.2-6). Since ψ is ≥ 0 and integrates to 1, we have

$$|u(x, t)| \leq \sup |f(y)|,$$

therefore

$$\|u(t)\| \leq \|u(0)\|,$$

from which we see that (1) the solution is unique, because if $u(t) = u_1(t) - u_2(t)$, where u_1 and u_2 are any solutions, then $u_1(0) = u_2(0)$ only if $u_1(t) = u_1(t)$ for all t, and (2) the solution depends continuously on $u(0)$. That is, the problem is well posed. The solution operator $E_0(t)$ is the integral operator in equation (16.2-7), and the generalized solutions are given by that equation with $f(y)$ allowed to be any element of the Banach space $\mathfrak{B}_0$.

The problem is also well posed in the Hilbert space $L^2(a, b)$ with A given by

$$\mathfrak{D}(A) = \{u \in L^2 : u'' \in L^2, u(a) = u(b) = 0\},$$
$$Au = \sigma u'', \qquad (16.2\text{-}8)$$

for if we multiply the partial differential equation (15.2-1) through by $\bar{u}(x, t)$ and integrate by parts, using the boundary condition, we find:

$$\frac{d}{dt} \int_a^b \frac{1}{2} |u|^2 \, dx = -\int_a^b \left|\frac{\partial u}{\partial x}\right|^2 \, dx \leq 0,$$

hence

$$\|u(t)\| \leq \|u(0)\|.$$

(Note that in this case it is not necessary to introduce a smaller Banach space $\mathfrak{B}_0$, because $\mathfrak{D}(A)$ is dense in L^2.)

The problem is well posed with respect to any of the common norms, such as L^p or norms based on $u(x)$ and one or more (or infinitely many) of the derivatives of $u(x)$, and so on; the solution is given by (16.2-7). The problem is well posed in all these norms because the heat equation is exceedingly "stable" in the forward direction (t increasing from 0); the solution decreases and becomes constantly smoother, as time goes on. In contrast, the equation is so "unstable" in the backward direction (t decreasing from 0) that the inverse problem is ill posed for every choice of norm.

The corresponding multidimensional problem is also well posed, for example in the Banach space $C(\mathbb{R}^n)$ with maximum norm or in $L^2(\mathbb{R}^n)$. The fundamental solution is

$$\psi(\mathbf{x}, t; \mathbf{y}) = (2\pi\sigma t)^{-n/2} e^{-|\mathbf{x}-\mathbf{y}|/4\sigma t}. \qquad (16.2\text{-}9)$$

16.3 Wave Motion

As stated in Section 15.4, the pure initial-value problem of wave motion in one dimension has always a solution, in the distribution sense, as follows (boundary conditions will be considered later): If φ and ψ are any distributions on $\mathbb{R}$, and if distributions f and g are defined by the equations $f = \frac{1}{2}\varphi + \frac{1}{2}\psi$, $g = \frac{1}{2}\varphi - \frac{1}{2}\psi$, then the distributions on $\mathbb{R}^2$ given by

$$\begin{aligned} u(x, t) &= f(x + ct) + g(x - ct) \\ v(x, t) &= f(x + ct) - g(x - ct) \end{aligned} \qquad (16.3\text{-}1)$$

satisfy the differential equations

$$\begin{aligned} \frac{\partial u}{\partial t} &= c\frac{\partial v}{\partial x} \\ \frac{\partial v}{\partial t} &= c\frac{\partial u}{\partial x} \end{aligned} \qquad (16.3\text{-}2)$$

in the distribution sense, and the initial conditions

$$\begin{aligned} u(x, 0) &= \varphi(x) \\ v(x, 0) &= \psi(x). \end{aligned} \qquad (16.3\text{-}3)$$

Note. For a general distribution $h(x, y)$ on $\mathbb{R}^2$, the expression $h(x, 0)$ would be meaningless. In the theory of initial-value problems, however, t is regarded as a parameter, and it is seen from (16.3-1) that, for any fixed value t_0 of the parameter, $u(x, t_0)$ and $v(x, t_0)$ are well defined distributions on $\mathbb{R}$. Hence, the initial conditions (16.3-3) make sense. In the present problem, one could regard x instead as the parameter, and then $u(x_0, t)$ and $v(x_0, t)$ would be well-defined distributions, for fixed x_0. More generally, any linear combination $t' = \alpha x + \beta t$ may be regarded as the parameter, provided that the lines $\alpha x + \beta t = $ constant are not characteristic lines—see next Chapter—i.e., provided that $\beta \neq \pm c\alpha$. In relativistic problems, different time variables t, t' are associated with different frames of reference.

To establish uniqueness of the solution, it must be proved that if $\varphi = \psi = 0$ for all x, then $u = v = 0$ for all x and all t. For this purpose, let ξ and η be new variables in the x, t plane, where $\xi = x + ct$ and $\eta = x - ct$. Then equations (16.3-2) become

$$\frac{\partial}{\partial \eta}(u + v) = 0, \qquad \frac{\partial}{\partial \xi}(u - v) = 0. \tag{16.3-4}$$

Therefore, $u + v$ is a distribution depending only on ξ, and $u - v$ is a distribution depending only on η (see Section 2.7); if these distributions are called $2f$ and $2g$, then equations (16.3-1) are obtained, showing that every solution of (16.3-2) has the form (16.3-1). In particular, if $\varphi = \psi = 0$, then $u = v = 0$. This argument is merely a simple special case of the method of characteristics, by which uniqueness of the solution of a general (nonlinear) hyperbolic system is proved (see next chapter); the values of $u + v$ and $u - v$ are propagated along the characteristic lines $\xi = \text{const.}$, $\eta = \text{const.}$ in the x, t plane.

To discuss the question of continuous dependence of the solution on the initial data, one must choose a norm $\|\cdot\|$ for the states that are represented, for given t, by the distributions $u(x, t)$ and $v(x, t)$—this entails also a decision as to what class of distributions can represent states of the system. That is, one must choose a Banach space; various choices will be discussed. In any case, it is convenient to introduce 2-component vectors

$$\mathbf{u} = \begin{pmatrix} u \\ v \end{pmatrix}, \qquad \boldsymbol{\varphi} = \begin{pmatrix} \varphi \\ \psi \end{pmatrix}, \text{ etc.};$$

then the differential equations (16.3-2) and the solution (16.3-1) can be written in condensed notation as

$$\frac{d}{dt}\mathbf{u}(t) = \tilde{A}\mathbf{u}(t) \quad \text{and} \quad \mathbf{u}(t) = \tilde{E}(t)\mathbf{u}(0),$$

where

$$\tilde{A} = \begin{pmatrix} 0 & 1 \\ 1 & 0 \end{pmatrix} c \frac{\partial}{\partial x} \tag{16.3-5}$$

$$\tilde{E}(t) = \frac{1}{2}\begin{pmatrix} 1 & 1 \\ 1 & 1 \end{pmatrix} T_{-ct} + \frac{1}{2}\begin{pmatrix} 1 & -1 \\ -1 & 1 \end{pmatrix} T_{ct}, \tag{16.3-6}$$

where, for any real a, T_a is the displacement operator defined by $(T_a f)(x) = f(x - a)$. In a particular Banach space, the operators A and $E(t)$ of the general theory are obtained from $\tilde{A}$ and $\tilde{E}(t)$ by suitable specification of their domains.

Consider first the choice

$$\mathcal{B} = \left\{ \mathbf{u} = \begin{pmatrix} u \\ v \end{pmatrix} : u(x) \text{ and } v(x) \text{ are bounded continuous functions} \right\}, \tag{16.3-7}$$

$$\|\mathbf{u}\| = \sup_{(x)}(\max\{|u(x)|, |v(x)|\}).$$

In this case, it is immediately seen from (16.3-1) that $\|\mathbf{u}(t)\| \leq \|\mathbf{u}(0)\|$; hence, the initial value problem is well posed in this Banach space. However, it will be seen below that the same is not true (i.e. in the maximum norm) in more dimensions; hence, this type of Banach space is not physically appropriate.

The problem is easily seen to be well-posed also in the Banach space

$$\mathscr{B} = \left\{ \mathbf{u} = \begin{pmatrix} u \\ v \end{pmatrix} : u, v \in L^2(\mathbb{R}) \right\},$$

(16.3-8)

$$\|\mathbf{u}\|^2 = \int_{-\infty}^{\infty} (|u(x)|^2 + |v(x)|^2) dx,$$

because a straightforward calculation starting with (16.3-1) shows that

$$\|\mathbf{u}(t)\|^2 \leq \int_{-\infty}^{\infty} (|\varphi|^2 + |\psi|^2) dx = \|\mathbf{u}(0)\|^2;$$

hence, the problem is well posed. The domain of the operator A, whose action is given by (16.3-5), may be taken as consisting of all $\mathbf{u}$ such that u and v are in C_0^∞, or as

$$\mathscr{D}(A) = \{ \mathbf{u} \in \mathscr{B} : \tilde{A}\mathbf{u} \in \mathscr{B} \},$$

or as anything between these extremes. In the former case, the strict solutions are C_0^∞ functions, for each t, and they satisfy the differential equation system everywhere in the x, t plane. In the latter case, they are continuous functions that are differentiable, in the classical sense, almost everywhere, and they satisfy the differential equation, in the classical sense, almost everywhere. The generalized solutions need not be differentiable, in the classical sense, for any x or any t; for each t, they are distributions in L^2. We have thus arrived, via Hadamard's concept of a well-posed problem, at the astonishing notion of a nowhere differentiable solution of a differential equation. The same is true for the previously considered Banach space given by (16.3-7). Since the generalized solutions often have physical significance, for example as white noise, one sometimes prefers an integral formulation of the physical laws, not dependent on differentiability, such as given in the Exercise below and for fluid dynamics more generally in the next chapter.

EXERCISE

1. Let $\mathbf{x}$ denote the vector with components x and y, where $y = ct$, and let $\mathscr{C}$ be the perimeter of a rectangle, regarded as a closed curve, in the x, y plane. Show that if $\mathbf{u}$ is a strict solution of the above problem, then

$$\oint_{\mathscr{C}} \mathbf{u} \cdot d\mathbf{x} = 0 \quad \text{and} \quad \oint_{\mathscr{C}} \mathbf{w} \cdot d\mathbf{x} = 0,$$

(16.3-9)

where

$$\mathbf{u} = \begin{pmatrix} u \\ v \end{pmatrix}, \quad \mathbf{w} = \begin{pmatrix} v \\ u \end{pmatrix}.$$

Show that for a strict solution the requirement that (16.3-9) holds for all rectangles $\mathscr{C}$ is equivalent to the differential equations. Show, using the Schwarz inequality for distributions in L^2, that equations (16.3-9) continue to hold for the generalized solutions.

If the equations are interpreted as describing the vibrations of the air in a pipe, then the quantity $\frac{1}{2}\|\mathbf{u}\|^2 = \frac{1}{2}\int(|u|^2 + |v|^2)dx$ is the total energy of the vibration, in suitable units. For a strict solution, u and $\partial u/\partial x$ are in L^2, for given t, hence $u(x, t) \to 0$, as $x \to \pm\infty$, according to Section 5.6. Therefore, from the differential equation and an integration by parts with respect to x, we see that

$$\frac{d}{dt}\|\mathbf{u}\|^2 = 0,$$

hence $\|\mathbf{u}(t)\|^2$ is $= \|\mathbf{u}(0)\|^2$, not merely $\le \|\mathbf{u}(0)\|^2$; i.e., the energy is conserved, for strict solutions, hence also for generalized solutions.

It is now shown that whether a problem is well posed may depend on the choice of norm. There are many ways of converting a second order equation to a first order system. In place of (16.3-2), one could write

$$\frac{\partial u}{\partial t} = w,$$

$$\frac{\partial w}{\partial t} = c^2\frac{\partial^2 u}{\partial x^2} \tag{16.3-10}$$

In the Banach space based on the norm

$$\|\mathbf{u}\| = \sup_{(x)} \max\{|u(x)|, |w(x)|\}, \tag{16.3-11}$$

the problem is now *ill* posed, as can be seen from the strict solution

$$u = \cos Kx \cos cKt$$

$$w = -cK \cos Kx \sin cKt,$$

for which

$$\frac{\|\mathbf{u}(t)\|}{\|\mathbf{u}(0)\|} = \max\{|\cos cKt|, cK|\sin cKt|\};$$

for any $t > 0$ this can be made arbitrarily large, by suitable choice of K; hence $E_0(t)$ is not a bounded operator. A knowledge of physics helps to exclude formulations of this kind, because u and w have different dimensions, and hence equation (16.3-11) does not check dimensionally.

In all cases in which the forward problem is well posed, the backward one is also, because the system (16.3-2) is invariant under the substitutions $-t \to t, u \to u, -v \to v$.

EXERCISES

2. Determine whether the initial-value problem is well posed or ill posed with respect to the following norms (u, v, and w are the quantities appearing in equations (16.3-2) and (16.3-10)). In each case, the domain $\mathfrak{D}(A)$ can be taken as either C_0^∞ or as the set of all $\mathbf{u}$ in $\mathfrak{B}$ such that $\tilde{A}\mathbf{u}$ is in $\mathfrak{B}$.

1. $\|\mathbf{u}\| = \left(\int_{-\infty}^{\infty}(|u(x)|^2 + |w(x)|^2)dx\right)^{1/2}$

2. $\|\mathbf{u}\| = \sup_{(x)}(|u(x)| + |v(x)|)$

3. $\|\mathbf{u}\| = \sup_{(x)}(|u(x)| + |w(x)|)$

Now suppose that the boundary condition $u = 0$ is imposed at $x = a$ and $x = b$;

$$u(a, t) = u(b, t) = 0 \qquad (t \geq 0), \qquad (16.3\text{-}12)$$

and that the initial values (16.3-2) are given only in the interval (a, b). If the waves are interpreted as sound waves in a pipe, and if u and v are quantities proportional to the fluid velocity and the pressure, respectively, then the conditions (16.3-12) represent closed ends of the pipe at $x = a$ and $x = b$. For strict solutions, the meaning of the conditions is clear, since $u(x, t)$ is a function, in fact a function of class C^1.

For a generalized solution, the conditions (16.3-12) are not directly meaningful, because u is only a distribution, but they can be reinterpreted by the reflection device of assuming $u(x, t)$ defined as a distribution on all $\mathbb{R}$ for each t and subject to the requirements

$$u(2a - x, t) = -u(x, t) \qquad \text{(all } x, \text{ all } t),$$

$$u(2b - x, t) = -u(x, t) \qquad \text{(all } x, \text{ all } t),$$

with similar requirements for $v(x, t)$. The details are left to the reader.

In more than one dimension, the problem of wave motion is ill posed in the maximum norm, owing to the rise in amplitude when a spherical wave converges to a point, which is shown as follows:

EXERCISE

3. Let r be the radial coordinate in 3-dimensional space V^3. Let $\psi(r)$ be a smooth (class C^2) function whose support lies in $[r_0, \infty)$, where $r_0 > 0$. Show that, for $t < r_0/c$, the function

$$u(r, t) = \psi(r + ct)\,\frac{r + ct}{r} \qquad (16.3\text{-}13)$$

satisfies the wave equation

$$\frac{\partial^2 u}{\partial t^2} = \frac{c^2}{r^2}\frac{\partial}{\partial r}\left(r^2\frac{\partial u}{\partial r}\right), \qquad (16.3\text{-}14)$$

hence represents an incoming wave. By letting $\psi(r)$ approximate a function with a discontinuity at r_0 (see Figure 16-1), show that $\sup_{(r)}|u(r, t)|$ can exceed $\sup_{(r)}|\psi(r)|$ by any factor, hence the problem is ill posed in the maximum norm. [One might think of redefining the norm as $\|f\| = \sup_{(r)}|f(r)|$, whereupon $\|u\| = \|\psi\|$ for all $t < r_0/c$, but this trick succeeds only for waves converging to the origin, not for ones that converge to other points in V^3.]

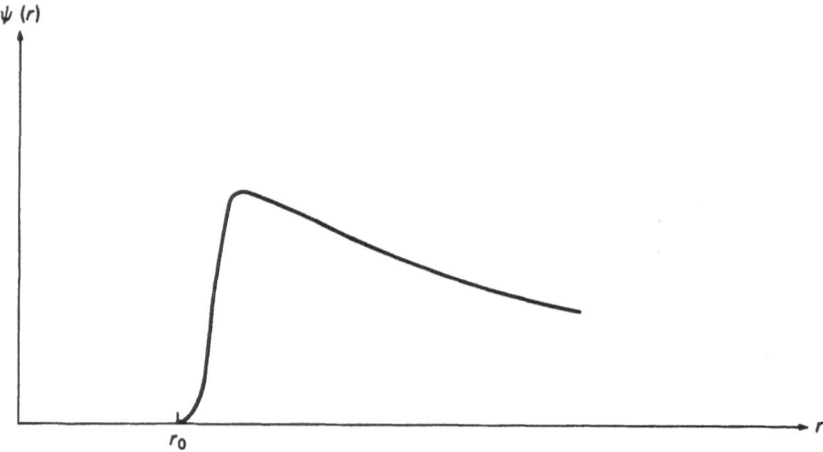

$\psi(r)$

r_0

r

Figure 16.1 The function $\psi(r)$.

There is no generalization to more dimensions of the method based on equations (16.3-1), but methods based on Fourier expansions can be used for the pure initial-value problem or the problem in a rectangular box in any number of dimensions, and the problem is well posed with respect to the L^2 norm that generalizes (16.3-8).

16.4 The Schrödinger Equation

Let $\psi = \psi(\mathbf{x}, t) = \psi(x_1, \ldots, x_n, t)$ be a complex-valued function (*wave function*). The Schrödinger equation is

$$i\hbar \frac{\partial \psi}{\partial t} = \left[\frac{-\hbar^2}{2m} \nabla^2 + V \right] \psi \tag{16.4-1}$$

where ∇^2 is the n-dimensional Laplacian, and $V = V(\mathbf{x})$ is a given real function of $x_1, \ldots, x_n$ (the *potential*); $\hbar$ and m are Planck's constant divided by 2π and the electron mass, respectively. The operator $H = (-\hbar^2/2m)\nabla^2 + V$ is called the quantum-mechanical *Hamiltonian* (*operator*) of the system. For a hydrogen-like atom, we have $n = 3$ and $V = -Ze^2/|\mathbf{x}|$; Ze is the charge of the nucleus and $-e$ is that of the electron. For many-electron atoms, n is three times the number of electrons, and V includes the interactions of the electrons with each other as well as with the nucleus. For an electron in a metal crystal, n is equal to 3, and $V(\mathbf{x})$ is a triply periodic function.

The initial-value problem can be illustrated by considering a free particle in one dimension. The problem is to find a complex-valued function $\psi(x, t)$

such that the following equations hold:

$$\text{DE:} \quad \frac{\partial \psi}{\partial t} = \frac{i\hbar}{2m} \frac{\partial^2 \psi}{\partial x^2}, \quad -\infty < x < \infty, t \geq 0, \tag{16.4-2}$$

$$\text{IC:} \quad \psi(x, 0) \text{ given,} \quad -\infty < x < \infty. \tag{16.4-3}$$

As a first choice, the Banach space $\mathfrak{B}$ will be taken as the space of continuous bounded functions of x with the maximum norm

$$\|\psi\| = \sup_{(x)} |\psi(x)|. \tag{16.4-4}$$

With this choice of norm, the initial-value problem is *not* well posed, as can be seen by considering a strict solution of the form

$$\psi(x, t) = \sqrt{\frac{t_0}{t_0 - t}} \exp \frac{-imx^2}{2\hbar(t_0 - t)}, \quad 0 \leq t < t_0, \tag{16.4-5}$$

where t_0 is a positive constant. [This solution is modelled after the fundamental solution (15.2-6) of the heat-flow equation.] Since $\|\psi(x, t)\| = [t_0/(t_0 - t)]^{1/2}$ while $\|\psi(x, 0)\| = 1$, it follows that

$$\|E_0(t)\| \geq \sqrt{\frac{t_0}{t_0 - t}}.$$

For any given $t > 0$, $E_0(t)$ is therefore unbounded, because the square root can be made arbitrarily large by taking t_0 sufficiently close to t, say $t_0 = t + \varepsilon$. [It is noted in passing that the solution (16.4-5) is quadratically integrable if the constant t_0 is given a small negative imaginary part.]

The well-known importance of quadratic integrability in quantum mechanics suggests that $\mathfrak{B}$ ought to be taken as the space $L^2(\mathbb{R})$ (see Chapter 5), for this problem, with the norm

$$\|\psi\| = \left\{ \int_{-\infty}^{\infty} |\psi(x)|^2 \, dx \right\}^{1/2}. \tag{16.4-6}$$

$\mathfrak{D}(A)$ can be taken as consisting of those distributions $\psi(x)$ in L^2 such that $\psi'(x)$ and $\psi''(x)$ are also in L^2. Then, if $\psi = \psi(x, t)$ is a strict solution of (16.4-2),

$$\frac{d}{dt} \|\psi\|^2 = \int_{-\infty}^{\infty} \left(\bar{\psi} \frac{\partial \psi}{\partial t} + \psi \frac{\partial \bar{\psi}}{\partial t} \right) dx$$

$$= \frac{i\hbar}{2m} \int_{-\infty}^{\infty} \left(\bar{\psi} \frac{\partial^2 \psi}{\partial x^2} - \psi \frac{\partial^2 \bar{\psi}}{\partial x^2} \right) dx \tag{16.4-7}$$

$$= \frac{i\hbar}{2m} \left(\bar{\psi} \frac{\partial \psi}{\partial x} - \psi \frac{\partial \bar{\psi}}{\partial x} \right) \Bigg|_{-\infty}^{\infty} = 0.$$

[It was proved in Section 5.6 that if ψ and ψ' are both in $L^2(\mathbb{R})$, then $\psi \to 0$ as $x \to \pm \infty$; if ψ'' is also in L^2, then ψ' also $\to 0$ as $x \to \pm \infty$.] Hence, the norm $\|\psi\|$ of strict solution is the same, for $t > 0$, as for $t = 0$; that is, $\|E_0(t)\| = 1$. Uniqueness of the strict solutions also follows, because $\|\psi_1 - \psi_2\| = 0$ if and only if $\psi_1 = \psi_2$, so that if $\psi_1 = \psi_2$ at $t = 0$, then $\psi_1 = \psi_2$ for all t.

EXERCISE

1. Perform a Fourier transformation of $\psi(x, t)$ with respect to x in (16.4-2), and find the explicit form of the strict solution of the transformed equation, corresponding to (16.4-5).

The argument based on (16.4-7) breaks down in higher dimensions, because $\psi(x)$ does not necessarily $\to 0$ as $|x| \to \infty$ even when ψ and all its first partial derivatives are in L^2. However the result holds; it will be shown in Section 16.8 that if H is any self-adjoint operator, then (1) the Hille–Yošida theorem can be applied to the operator $-(i/\hbar)H$, and (2) if $d\psi/dt = -(i/\hbar)H\psi$, then $(d/dt)\|\psi\|^2 = 0$, hence the initial-value problem of the Schrödinger equation is well posed in the L^2 norm. This result is very general: H can be the Hamiltonian for any number of particles subject to any internal and external forces, provided only that H is self-adjoint. The self-adjointness of various Schrödinger and Dirac Hamiltonians was discussed in Chapters 10 and 11.

For many-electron atoms, an important auxiliary condition comes from the Pauli exclusion principle. A complete discussion of it would require the introduction of the electron spins, but the effect is essentially as illustrated by the following: Let $\psi(x_1, x_2, t)$ be the wave function of a two-electron atom, where x_1 and x_2 are now 3-vectors. The Hilbert space is now $L^2(\mathbb{R}^6)$. Sometimes, the wave function is required to be symmetric with respect to interchange of the two electrons, i.e. $\psi(x_1, x_2, t) = \psi(x_2, x_1, t)$ and sometimes it is required to be antisymmetric, $\psi(x_1, x_2, t) = -\psi(x_2, x_1, t)$. To take these requirements into account, two subspaces L_+^2 and L_-^2 of $L^2(\mathbb{R}^6)$ are defined, consisting respectively of the symmetric and of the antisymmetric functions in $L^2(\mathbb{R}^6)$. The Hamiltonian H for a many-electron atom is always invariant under interchange of two electrons, so that if ψ is in L_+^2 (or in L_-^2), then $H\psi$ is also in L_+^2 (or in L_-^2). Therefore, if $\psi(x_1, x_2, t)$ is any solution of the Schrödinger equation, the symmetric function

$$\psi(x_1, x_2, t) + \psi(x_2, x_1, t)$$

and the antisymmetric function

$$\psi(x_1, x_2, t) - \psi(x_2, x_1, t)$$

are solutions of the same equation in the subspaces L_+^2 and L_-^2, respectively. By restricting consideration to one or the other of these subspaces, which are closed subspaces, hence Banach spaces in their own right, the auxiliary condition is taken care of automatically. With a suitable domain, the Hamiltonian is self-adjoint in each of the subspaces, hence again the initial-value problem is well-posed, because of the Hille–Yošida Theorem.

16.5 Maxwell's Equations in Empty Space

This section illustrates the use of the second definition of a self-adjoint operator, given in Section 8.6. We consider the differential equation system

$$
\text{D.E.} \qquad
\begin{aligned}
\frac{\partial \mathbf{E}}{\partial t} &= -c\nabla \times \mathbf{H} \\
\frac{\partial \mathbf{H}}{\partial t} &= c\nabla \times \mathbf{E}
\end{aligned}
\qquad (16.5\text{-}1)
$$

for the vector fields $\mathbf{E}(\mathbf{x}, t)$ and $\mathbf{H}(\mathbf{x}, t)$, subject to the auxiliary conditions

$$
\text{A.C.} \quad \nabla \cdot \mathbf{E} = 0 \quad \text{and} \quad \nabla \cdot \mathbf{H} = 0 \qquad (16.5\text{-}2)
$$

and the initial condition

$$
\text{I.C.} \quad \mathbf{E}(\mathbf{x}, 0) \quad \text{and} \quad \mathbf{H}(\mathbf{x}, 0) \text{ known.} \qquad (16.5\text{-}3)
$$

[One might consider a somewhat more general electromagnetic problem. However, if there are currents and charges, whose motions are influenced by the fields, or if ferromagnetic effects are present, then the equations are nonlinear.]

The appropriate Banach space for this problem is a subspace $\mathfrak{H}_0$ of the Hilbert space $\mathfrak{H}$, whose elements u are pairs of vector fields in $\mathbb{R}^3$ with the Euclidean norm:

$$
u = \begin{bmatrix} \mathbf{E}(\mathbf{x}) \\ \mathbf{H}(\mathbf{x}) \end{bmatrix}, \qquad (16.5\text{-}4)
$$

$$
\|u\| = \left(\int_{\mathbb{R}^3} (|\mathbf{E}|^2 + |\mathbf{H}|^2) d^3\mathbf{x} \right)^{1/2} \qquad (16.5\text{-}5)
$$

where

$$
|\mathbf{E}| = (|E_x|^2 + |E_y|^2 + |E_z|^2)^{1/2} \qquad (16.5\text{-}6)
$$

and similarly for $|\mathbf{H}|$. Then,

$$
\mathfrak{H} = \{u : \|u\| < \infty\}. \qquad (16.5\text{-}7)
$$

The inner product in $\mathfrak{H}$ is

$$
(u_1, u_2) = \int_{\mathbb{R}^3} (\bar{\mathbf{E}}_1 \cdot \mathbf{E}_2 + \bar{\mathbf{H}}_1 \cdot \mathbf{H}_2) d^3\mathbf{x}.
$$

$\mathfrak{H}$ is the cartesian product of six spaces of the form

$$
\left\{ E_x(\mathbf{x}) : \int_{\mathbb{R}^3} |E_x|^2 d^3\mathbf{x} < \infty \right\},
$$

i.e. of spaces of the type $L^2(\mathbb{R}^3)$; hence, one can write

$$\mathfrak{H} = (L^2(\mathbb{R}^3))^6,$$

it being understood that the square of the norm in the cartesian product is the sum of the squares of the norms in the six factor spaces.

The subspace $\mathfrak{H}_0$ is given by

$$\mathfrak{H}_0 = \{u \in \mathfrak{H}: \nabla \cdot \mathbf{E} = 0, \nabla \cdot \mathbf{H} = 0\}. \tag{16.5-8}$$

It is clear that this is a linear manifold in $\mathfrak{H}$, and it will be proved that it is a *closed* linear manifold, hence a subspace, hence a Hilbert space in its own right. The components of $\mathbf{E}$ and $\mathbf{H}$ are distributions in L^2, and the derivatives in (16.5-8) are distribution derivatives. Now $\nabla \cdot \mathbf{E} = 0$ if and only if $\int \bar{\varphi} \nabla \cdot \mathbf{E} \, d^3\mathbf{x} = 0$ for every test function $\bar{\varphi}$ in $C_0^\infty(\mathbb{R}^3)$, while $\nabla \cdot \mathbf{H} = 0$ if and only if $\int \bar{\psi} \nabla \cdot \mathbf{H} \, d^3\mathbf{x} = 0$ for every test function $\bar{\psi}$. By definition of the distribution derivative, this means that the element

$$u = \begin{bmatrix} \mathbf{E} \\ \mathbf{H} \end{bmatrix}$$

is orthogonal to every element of $\mathfrak{H}$ of the form

$$\begin{bmatrix} \nabla\varphi \\ \nabla\psi \end{bmatrix}.$$

That is, $\mathfrak{H}_0$ is the orthogonal complement of the set of all elements of this form. An orthogonal complement is always a closed linear manifold (see Chapter 1), hence $\mathfrak{H}_0$ is closed.

In order to put (16.5-1) into the form $\partial u/\partial t = Au$, an operator A is defined, first in $\mathfrak{H}$, as follows: a formal operator A is defined as

$$A_0 u = \begin{bmatrix} -c\nabla \times \mathbf{H} \\ c\nabla \times \mathbf{E} \end{bmatrix} = \begin{bmatrix} 0 & -c\nabla \times \\ c\nabla \times & 0 \end{bmatrix}\begin{bmatrix} \mathbf{E} \\ \mathbf{H} \end{bmatrix} \tag{16.5-9}$$

(the quantities appearing here are well defined as distributions); then

$$\mathfrak{D}(A) = \{u \in \mathfrak{H}: A_0 u \in \mathfrak{H}\}; \quad \text{for } u \in \mathfrak{D}(A), \quad Au = A_0 u. \tag{16.5-10}$$

Au is always in $\mathfrak{H}_0$, because the divergence of a curl is zero; hence, in particular, A transforms $\mathfrak{H}_0$ into itself. It is the restriction of A to $\mathfrak{H}_0$, which will also be called simply A, that is relevant for the present initial-value problem. The following will be proved.

Theorem. *A is a symmetric operator in $\mathfrak{H}_0$. The resolvent set $\rho(A)$ contains the upper and lower half planes; it follows that A is self-adjoint and hence, via the Hille–Yošida theorem, that the initial-value problem is well posed.*

The proofs of these statements are almost trivial, after a Fourier transformation is made. We define

$$\hat{E}(k, t) = \frac{1}{(2\pi)^{3/2}} \int_{\mathbb{R}^3} E(x, t)e^{-ik \cdot x} d^3x,$$

and similarly for H. The mapping

$$\begin{bmatrix} E \\ H \end{bmatrix} \to \begin{bmatrix} \hat{E} \\ \hat{H} \end{bmatrix}$$

is an isomorphism of $\mathfrak{H}$ onto a Hilbert space $\hat{\mathfrak{H}}$, whose definition is the same as that of $\mathfrak{H}$, except for the appearance of the circumflex on E and $H \cdot \mathfrak{H}_0$ is mapped onto the subspace

$$\hat{\mathfrak{H}}_0 = \{\hat{u} \in \hat{\mathfrak{H}} : k \cdot \hat{E} = 0, k \cdot \hat{H} = 0\}. \qquad (16.5-11)$$

[$\hat{\mathfrak{H}}_0$ is the orthogonal complement of the set of all vectors of the form

$$\begin{bmatrix} k\varphi \\ k\psi \end{bmatrix},$$

which gives another proof that $\mathfrak{H}_0$ and $\hat{\mathfrak{H}}_0$ are closed.]. The image of the operator A is given by

$$\hat{A} \begin{bmatrix} \hat{E} \\ \hat{H} \end{bmatrix} = \begin{bmatrix} -ck \times \hat{H} \\ ck \times \hat{E} \end{bmatrix}, \qquad (16.5-12)$$

and its domain consists of all vectors in $\hat{\mathfrak{H}}_0$ such that the right member of (16.5-12) is in $\hat{\mathfrak{H}}_0$.

It now follows from the vector identities

$$-u_1 \cdot (k \times v_2) = -(u_1 \times k) \cdot v_2 = (k \times u_1) \cdot v_2$$
$$u_2 \cdot (k \times v_1) = (u_2 \times k) \cdot v_1 = -(k \times u_2) \cdot v_1$$

that $\hat{A}$ is a symmetric operator, i.e., that

$$(u, \hat{A}v) = (\hat{A}u, v) \quad \text{for all } u, v \in \mathfrak{D}(\hat{A}).$$

To investigate the resolvent and the resolvent set, let

$$\hat{v} = \begin{bmatrix} \hat{v}_1 \\ \hat{v}_2 \end{bmatrix}$$

be any element of $\hat{\mathfrak{H}}_0$, where $\hat{v}_1$ and $\hat{v}_2$ are 3-component vector fields, and consider the equation

$$(\hat{A} - \lambda)\hat{u} = \hat{v}, \qquad (16.5-13)$$

where λ is any nonreal number. If this equation has a unique solution $\hat{u}$ for arbitrary $\hat{v}$, and if $\|\hat{u}\| \leq K\|\hat{v}\|$, for some $K = K(\lambda)$, then λ is in the resolvent set $\rho(\hat{A}) = \rho(A)$, and $\hat{v} = \hat{R}_\lambda \hat{u}$, where $\hat{R}_\lambda$ is the resolvent

$$\hat{R}_\lambda = (\hat{A} - \lambda)^{-1}.$$

Explicitly, (16.5-13) is (the circumflex is omitted henceforth)

$$-c\mathbf{k} \times \mathbf{u}_2 - \lambda\mathbf{u}_1 = \mathbf{v}_1$$
$$c\mathbf{k} \times \mathbf{u}_1 - \lambda\mathbf{u}_2 = \mathbf{v}_2$$

$$(16.5\text{-}14)$$

These linear equations have the unique solution

$$\mathbf{u}_1 = \frac{\lambda\mathbf{v}_1 - c\mathbf{k} \times \mathbf{v}_2}{c^2k^2 - \lambda^2},$$

$$\mathbf{u}_2 = \frac{\lambda\mathbf{v}_2 + c\mathbf{k} \times \mathbf{v}_1}{c^2k^2 - \lambda^2}.$$

The equations $\mathbf{k} \cdot \mathbf{v}_1 = \mathbf{k} \cdot \mathbf{v}_2 = 0$, which imply $\mathbf{k} \cdot \mathbf{u}_1 = \mathbf{k} \cdot \mathbf{u}_2 = 0$ via (16.5-14), have been used. The denominator is never zero, because Im $\lambda \neq 0$. The six components of u are thus linear combinations of the six components of v with coefficients that are bounded functions of $\mathbf{k}$ for any given nonreal λ. It is then elementary to show that there is a $K = K(\lambda)$ such that $\|u\| \leq K\|v\|$. Hence λ is in the resolvent set $\rho(A)$, as was to be proved.

16.6 Semigroups

It is recalled that if A is an operator in a finite dimensional space (i.e., A is a square matrix), then the solution of the initial value problem is $u(t) = e^{tA}u(0)$, where, for any matrix M, the exponential function e^M is defined by its power series. In the infinite dimensional case, e^{tA} cannot be defined so easily if A is an unbounded operator; it will now be shown that, nevertheless, the solution operator $E(t)$ has many of the properties of an exponential e^{tA}.

If $u(t)$ is any strict solution of the differential equation (16.1-1), then the function $\tilde{u}(t) \overset{\text{def}}{=} u(t + s)$, where s is a constant ≥ 0, is a strict solution with initial element $u(s)$; therefore, $\tilde{u}(t) = E(t)u(s)$; but $u(s) = E(s)u(0)$ and $u(t + s) = E(t + s)u(0)$; hence;

$$E(t + s)u(0) = E(t)E(s)u(0), \quad \text{for all } u(0) \text{ in } \mathfrak{U}.$$

Since $\mathfrak{U}$ is dense in $\mathfrak{B}$, and the operators are bounded, it follows that

$$E(t + s) = E(t)E(s) \qquad (t, s \geq 0). \qquad (16.6\text{-}1)$$

[This argument shows incidentally that $\mathfrak{U}$ is the set not merely of *initial* values of strict solution, but of *all* values assumed by strict solutions.]

A collection of objects $a, b, c, \ldots$ in which an associative binary law of composition $a \circ b$ etc., is defined is called a *semigroup*; it differs from a group only in that inverses are not assumed to exist. The collection of operators $\{E(t): t \geq 0\}$ is a one-parameter semigroup of linear operators. If the initial value problem is reversible in time, as for wave motion, then the collection $\{E(t): \text{all real } t\}$ is a one-parameter *group* of linear operators. For operators, the law of composition, written as a product in equation (16.6-1), is the ordinary law of composition of transformations, hence is automatically associative: $(T_1 T_2)T_3 = T_1(T_2 T_3)$.

The semigroup $\{E(t): t \geq 0\}$ of solution operators of a well-posed initial value problem has some special properties. First, it is commutative, for equation (16.6-1) shows that $E(t)E(s) = E(s)E(t)$.

Second, it has an identity element $E(0) = I$, because $E(0)u = u$ for all u.

Third, it is strongly continuous. Let $u(t)$ be any strict solution of the differential equation (16.1-1). For any $\varepsilon > 0$, according to (16.1-4),

$$\left\| \frac{u(t + \Delta t) - u(t)}{\Delta t} - Au(t) \right\| < \varepsilon,$$

if Δt is small enough. Therefore, by the triangle inequality,

$$\| u(t + \Delta t) - u(t) \| \leq (\| Au(t) \| + \varepsilon)\Delta t,$$

i.e., as $\Delta t \to 0$, $u(t + \Delta t) \to u(t)$ in the sense of convergence in $\mathfrak{B}$. Stated differently, for any u in $\mathfrak{U}$, $E(t + \Delta t)u \to E(t)u$. Since $\mathfrak{U}$ is dense in $\mathfrak{B}$ and the operators $E(t)$ are bounded, it is a simple exercise in the use of the triangle inequality to show that

$$E(t + \Delta t)u \to E(t)u, \quad \text{for any } u \text{ in } \mathfrak{B}, \quad \text{as } \Delta t \to 0 \qquad (t \geq 0). \quad (16.6\text{-}2)$$

[For $t = 0$, Δt is assumed to $\to 0$ through positive values, since $E(t)$ is not generally defined for $t < 0$.] The property (16.6-2) is called *strong continuity* of the family $E(t)$ of operators. Note that the semigroup does *not* generally have the property of continuity in norm, that is, $\| E(t + \Delta t) - E(t) \|$ generally does not $\to 0$, as $\Delta t \to 0$.

Fourth, $E(t)$ is uniformly bounded in any finite interval: given $[0, t_0]$, there is a constant K such that $\| E(t) \| \leq K$ for all t in $[0, t_0]$, because (16.2-1) shows that $E_0(t)$ is thus bounded, and, according to the extension theorem, the norm of $E(t)$ is the same as that of $E_0(t)$.

In general, different operators A_1 and A_2 can yield the same semigroup $E(t)$ (for example, A_2 may be an extension of A_1), in which case the corresponding initial-value problems are identical except for the purely terminological difference that some strict solutions of one are only generalized solutions of the other.

In the next section it is shown that any $E(t)$ having the above properties is the solution operator of some well-posed initial-value problem.

For the general theory of semigroups, see Hille and Phillips 1957.

16.7 The Infinitesimal Generator of a Semigroup

We now consider the converse of the result of the preceding section. Given any semigroup $E(t)$ that has the properties described there, we shall define an operator A', called the *infinitesimal generator* of $E(t)$ such that the initial-value problem

$$\frac{d}{dt} u(t) = A'u(t), \qquad u(0) \text{ given}, \qquad (16.7\text{-}1)$$

is well posed and has $E(t)$ as its solution operator. The operator A' is defined as the "derivative" of $E(t)$ at $t = 0$, in the following sense: The domain $\mathfrak{D}(A')$ is defined as the set of all u in $\mathfrak{B}$ such that $(1/\Delta t)[E(\Delta t) - I]u$ has a limit in $\mathfrak{B}$ as $\Delta t \to 0$, and then $A'u$ is defined to be that limit: i.e.,

$$\lim_{\Delta t \to 0} \frac{1}{\Delta t} [E(\Delta t) - I]u = A'u. \tag{16.7-2}$$

It is obvious that $\mathfrak{D}(A')$ is a linear subspace of $\mathfrak{B}$ and that A' is a linear operator.

Theorem 1. *If $E(t)$ has the properties described in the preceding section, i.e. is a bounded strongly continuous commutative semigroup with identity $E(0) = I$, then the initial-value problem (16.7-1), with A' given by (16.7-2), is well posed. The set $\mathfrak{U}'$ of possible initial elements $u(0)$ of strict solutions is $= \mathfrak{D}(A')$, and the solutions are of the form $u(t) = E(t)u(0)$. Lastly, A' is a closed operator.*

PROOF.   (a) Let u_0 be any element of the domain $\mathfrak{D}(A')$ defined above. Since $E(t)$ is bounded and commutes with $E(\Delta t)$, the quantity

$$E(t)(1/\Delta t)[E(\Delta t) - I]u_0 \quad \text{or} \quad (1/\Delta t)[E(\Delta t) - I]E(t)u_0$$

tends to a limit, namely $E(t)A'u_0$, as $\Delta t \to 0$. Therefore, $E(t)u_0$ is in $\mathfrak{D}(A')$, and $A'E(t)u_0 = E(t)A'u_0$; hence the function $u(t) = E(t)u_0$ is such that

$$\left\| \frac{u(t + \Delta t) - u(t)}{\Delta t} - A'u(t) \right\| \to 0, \quad \text{as } \Delta t \to 0,$$

i.e., is a strict solution of (16.7-1). (b) To show that this solution is unique, let $u(t)$ be *any* solution such that $u(0) = u_0$; we shall show that $u(t) = E(t)u_0$. For any $t > 0$, the function $g(s) = E(t - s)u(s)$, defined for s in $[0, t]$, will be shown to be constant. First,

$$\frac{d}{ds} g(s) \bigg|_{s=s_0} = \frac{d}{ds} E(t - s)u(s_0) \bigg|_{s=s_0} + \frac{d}{ds} E(t - s_0)u(s) \bigg|_{s=s_0}.$$

The first term on the right is equal to

$$-\frac{d}{dt} E(t - s)u(s_0) \bigg|_{s=s_0} = -A'E(t - s_0)u(s_0)$$

(because $E(t - s)u(s_0)$ satisfies (16.7-1) according to (a)), and the second term is

$$E(t - s_0) \frac{s}{ds} u(s) \bigg|_{s=s_0} = E(t - s_0)A'u(s_0)$$

(because $E(\cdot)$ is bounded and because $u(t)$ was assumed to be a solution of (16.7-1)). As noted above, $E(\cdot)$ and A' commute when applied to an element in $\mathfrak{D}(A')$; therefore $dg(s)/ds = 0$, or $g(s)$ is a constant; hence, $g(t) = g(0)$, i.e., $u(t) = E(t)u(0)$, as was to be shown. It will now be shown that the infinitesimal generator A' is a closed operator. If $w(s)$ is any continuous one-parameter family of elements in $\mathfrak{B}$, the integral $\int_a^b w(s)ds$ is defined (it is also an element of $\mathfrak{B}$) as the limit of Riemann sums

$$\sum_{(j)} w(s_j')(s_{j+1} - s_j),$$

where ... , s_j, s_{j+1}, ... is a partition of $[a, b]$, and $s_j' \in [s_j, s_{j+1}]$ for each j, just as for an ordinary continuous function. It is left as an exercise in the use of the triangle inequality to show that the limit is unique and that the integral has all the expected properties, such as

$$\frac{d}{db} \int_a^b w(s)ds = w(b) \qquad (16.7\text{-}3)$$

$$\left\| \int_a^b w(s)ds \right\| \leq \int_a^b \|w(s)\| ds. \qquad (16.7\text{-}4)$$

If u_0 is in $\mathfrak{D}(A')$ and $u(t) = E(t)u_0$ is the corresponding strict solution, then clearly

$$u(t) - u_0 = \int_0^t A'u(s)ds, \qquad (16.7\text{-}5)$$

i.e.

$$[E(t) - I]u_0 = \int_0^t A'E(s)u_0 \, ds \qquad (16.7\text{-}6)$$

[**Note.**   The function $A'u(s) = E(s)A'u_0$ is continuous.] Suppose that $\{v_n\}$ is a sequence of elements of $\mathfrak{D}(A')$ such that $v_n \to u$ and $A'v_n \to w$. It must be shown that u is in $\mathfrak{D}(A')$ and $w = A'u$. Now,

$$E(\delta)u - u = \lim_{n \to \infty} [E(\delta)v_n - v_n] = \lim_{n \to \infty} \int_0^\delta A'E(s)v_n \, ds$$

$$= \lim_{n \to \infty} \int_0^\delta E(s)A'v_n \, ds = \int_0^\delta E(s)\lim_{n \to \infty} A'v_n \, ds.$$

[The convergence is uniform with respect to s, because $E(s)$ is bounded uniformly in s.] Therefore, since $A'v_n \to w$,

$$\frac{E(\delta)u - u}{\delta} = \frac{1}{\delta} \int_0^\delta E(s)w \, ds,$$

which converges to $E(0)w = w$, as $\delta \to 0$, by continuity of the integrand. By the definitions of A' and $\mathfrak{D}(A')$ given at the beginning of this section, therefore, u is in $\mathfrak{D}(A')$, and $A'u = w$; i.e., A' is a closed operator.

The choice of A' for A maximizes the collection of strict solutions of the initial-value problem having $E(t)$ as solution operator. That is:

Theorem 2. *If $u(t)$ is any strict solution of a well-posed initial-value problem* (16.1-1), *then $u(0)$ is in $\mathfrak{D}(A')$, where A' is the infinitesimal generator of the solution operator $E(t)$, hence $u(t)$ is also a strict solution of the equation*

$$\frac{d}{dt} u(t) = A'u(t).$$

PROOF. (16.1-4), with A replaced by A', follows from (16.7-2).

16.8 The Hille–Yošida Theorem

In the special case in which A is a bounded operator, the solution operator $E(t)$ is given by

$$E(t) = e^{tA} = \sum_{k=0}^{\infty} \frac{1}{k!} (tA)^k; \qquad (16.8\text{-}1)$$

the sum converges in norm, for all t; that is,

$$\left\| E(t) - \sum_{k=0}^{n} \frac{1}{k!} (tA)^k \right\| \to 0 \quad \text{as } n \to \infty. \qquad (16.8\text{-}2)$$

If A is unbounded, as in most practical cases, then the series (16.8-1) may not converge at all, and if it does, it gives at best a restriction of $E(t)$ to a domain small enough so that the operators A^k are all defined ($k = 1, 2, \ldots$).

EXERCISE

1. Prove (16.8-2), for A bounded, and show that

$$\frac{d}{dt} e^{tA} = A e^{tA} = e^{tA} A.$$

Whether a given operator A is the infinitesimal generator of a semigroup at all is not easy to decide. The following theorem is the most generally useful one. According to the preceding Section, A can in any case be assumed to be closed.

Theorem (Hille–Yošida). *Let A be a closed linear operator with domain dense in $\mathfrak{B}$. If, for all $\lambda > \alpha$, $(A - \lambda)^{-1}$ exists, is bounded, and has its domain dense in $\mathfrak{B}$, and if*

$$(\lambda - \alpha)\|(A - \lambda)^{-1}\| \leq 1, \quad \text{for all } \lambda > \alpha, \qquad (16.8\text{-}3)$$

(i.e., if the resolvent $R_\lambda(A)$ exists for $\lambda > \alpha$ and is bounded by $1/(\lambda - \alpha)$) then A is the infinitesimal generator of a semigroup $E(t)$, which is strongly continuous, for $t \geq 0$, and is such that $E(0) = I$ and $\|E(t)\| \leq e^{\alpha t}$, for $t \geq 0$. See Hille and Phillips 1957 for the proof.

If the hypotheses of the theorem are satisfied, then, clearly, A determines a well-posed initial-value problem.

An application of the theorem to the neutron transport problem will be made in the next section. Two elementary applications are these: (1) If A is a bounded operator, then $\|(A - \lambda)^{-1}\| \leq (\lambda - \|A\|)^{-1}$, for all $\lambda > \|A\|$, hence the initial value problem is well posed, and $\|E(t)\| \leq \exp\{t\|A\|\}$, in agreement with (16.8-1). (2) Let H be self-adjoint and let $A = -iH$. Then $(A - \lambda)^{-1} = i(H - i\lambda)^{-1}$, and the hypotheses of the Hille–Yošida theorem follow from the well-known properties of self-adjoint operators—see Section 8.3. In this case the semigroup $E(t)$ is usually denoted by e^{-iHt}. Applications to Schrödinger operators were made in Section 16.4.

16.9 Neutron Transport in a Slab; An Application of the Hille–Yošida Theorem

This section gives a further example of the calculation and use of the resolvent. Consider neutron transport in a homogeneous slab of material occupying the region $-a \le x \le a$, all y, all z. Assume that scattering is elastic and isotropic, and that all neutrons have the same speed v. Let θ denote the angle between the neutron's velocity and the x axis, and let $\mu = \cos \theta$. Denote the neutron density (number density) in phase space at position x, direction θ, and time t, by $\Psi(x, \mu, t)$; the density is assumed to be independent of y and z and of the azimuthal angle φ around the x direction. The equation of evolution of this system is the so-called transport equation

$$\left(\frac{1}{v}\frac{\partial}{\partial t} + \mu \frac{\partial}{\partial x} + \sigma\right)\Psi(x, \mu, t) = \sigma \frac{c}{2}\int_{-1}^{1} \Psi(x, \mu', t)d\mu' \qquad (16.9\text{-}1)$$

(see Richtmyer and Morton 1967), where σ is the total nuclear cross section per unit volume ($1/\sigma$ is the mean free path), and c is the average number of neutrons that emerge from a collision ($c = 1$ for pure scattering, $c < 1$ for scattering plus absorption, $c > 1$ for a multiplying medium).

The term $\sigma\Psi(x, \mu, t)$ can be eliminated from (16.9-1) by writing

$$\Psi(x, \mu, t) = \psi(x, \mu, t)e^{-v\sigma t}.$$

In units of length and time such that $\sigma = 1$, $v = 1$, (in these units, $2a$ is the slab thickness in mean free paths), we then have

$$\frac{\partial}{\partial t}\psi(x, \mu, t) = -\mu\frac{\partial}{\partial x}\psi(x, \mu, t) + \frac{c}{2}\int_{-1}^{1}\psi(x, \mu', t)d\mu'. \qquad (16.9\text{-}2)$$

This equation is written as

$$\frac{\partial\psi}{\partial t} = A\psi, \qquad (16.9\text{-}3)$$

where A is the integrodifferential operator on the right of (16.9-2), with a suitable domain in a suitable Banach space.

In the original formulation of the problem, (16.9-2) was assumed to hold only for $-a \le x \le a$, and the domain of A was restricted to functions that satisfy the boundary condition

$$\psi(a, \mu, t) = 0 \quad \text{for } \mu < 0$$

$$\psi(-a, \mu, t) = 0 \quad \text{for } \mu > 0,$$

which specifies that no neutrons are incident on the slab from outside. In a more convenient formulation (K. O. Friedrichs, unpublished), which will be adopted here, it is assumed that (16.9-2) holds for all x, but the constant c is replaced by

$$c(x) = \begin{cases} c, & \text{for } -a \le x \le a, \\ 0, & \text{otherwise.} \end{cases} \qquad (16.9\text{-}4)$$

This is equivalent to assuming that the region outside the slab is filled with a purely absorbing medium having the same total cross section σ as the slab. For all problems in which the initial neutron distribution $\psi(x, \mu, 0)$ contains no neutrons moving toward the slab from outside, i.e., is zero for $x\mu < 0$ and $|x| > a$, this formulation has exactly the same solution in the slab as the original formulation, because any neutron that escapes from the slab is certain to be absorbed on its next collision and never return to the slab. In this formulation, no boundary condition is needed.

The operator is not symmetric in the appropriate Hilbert space, hence cannot be made self-adjoint by any choice of domain. The first term in (16.9-2) is antisymmetric and the second symmetric; the two terms do not commute, hence A cannot be made into a normal operator by any choice of domain. The spectrum and other properties of A were investigated by Lehner and Wing 1955, 1956 and by Lehner 1962. The point spectrum consists of finitely many positive eigenvalues, the continuous spectrum is the entire imaginary axis, and the rest of the plane is resolvent set.

In this section, A will be considered in a Banach space of continuous functions with the maximum norm, which is easier to work with than the Hilbert space. It will be shown that, when the domain of A is suitably chosen, the requirements of the Hille–Yošida theorem are satisfied; hence the initial-value problem of (16.9-3) is well posed.

Let $\mathfrak{B}$ be the Banach space whose elements are bounded continuous (generally complex-valued) functions $f(x, \mu)$, defined for all real x and all μ in $[-1, 1]$, with the norm

$$\|f\| = \sup\{|f(x, \mu)|: x \in \mathbb{R}, \mu \in [-1, 1]\}.$$

The domain $\mathfrak{D}(A)$ of A consists of all f in $\mathfrak{B}$ such that $\partial f/\partial x$ is also in $\mathfrak{B}$, and then

$$(Af)(x, \mu) = -\mu \frac{\partial f}{\partial x}(x, \mu) + \frac{c(x)}{2} \int_{-1}^{1} f(x, \mu')d\mu'. \qquad (16.9\text{-}5)$$

In order to use the Hille–Yošida theorem, we must show that A is a closed operator and that its resolvent $R_\lambda = (A - \lambda)^{-1}$ exists and is bounded by $(\lambda - \alpha)^{-1}$, for all real λ greater than some constant α.

The proof that A is a closed operator is based on uniform convergence of functions and is left as an exercise (the definition of a closed operator is given in Section 7.6).

To investigate the resolvent, let g be an arbitrary element of $\mathfrak{B}$, and λ a complex number. The problem is to solve the equation

$$Af - \lambda f = g \qquad (16.9\text{-}6)$$

for f, if possible, and find a constant $K = K(\lambda)$ such that $\|f\| \le K\|g\|$, for all g in $\mathfrak{B}$. It turns out to be sufficient to consider positive values of λ. To solve (16.9-6), call

$$\xi(x) = \int_{-1}^{1} f(x, \mu)d\mu; \qquad (16.9\text{-}7)$$

then (16.9-6) takes the form

$$\left(\lambda + \mu \frac{\partial}{\partial x}\right) f(x, \mu) = \frac{c(x)}{2} \xi(x) - g(x, \mu). \qquad (16.9\text{-}8)$$

The function $\xi(x)$ is of course unknown, but the first step in the solution of (16.9-6) is to solve (16.9-8) for f in terms of ξ and g. When the result is put into (16.9-7) a Fredholm integral equation for $\xi(x)$ is then obtained. By elementary methods, the solutions of (16.9-8) is seen to be

$$f(x, \mu) = \begin{cases} \dfrac{1}{\mu} \displaystyle\int_{-\infty}^{x} e^{\lambda(x'-x)/\mu}\left[\dfrac{c(x')}{2} \xi(x') - g(x', \mu)\right] dx', & \text{for } \mu > 0, \\[4mm] \dfrac{-1}{\mu} \displaystyle\int_{x}^{\infty} e^{\lambda(x'-x)/\mu}\left[\dfrac{c(x')}{2} \xi(x') - g(x', \mu)\right] dx', & \text{for } \mu < 0. \end{cases} \qquad (16.9\text{-}9)$$

It is easily seen that $f(x, \mu)$ is continuous in μ as well as in x and that $f(x, 0)$ can be obtained from either line of the formula by letting $\mu \to 0$. If the square bracket is replaced by its greatest possible value $\frac{1}{2}c\|\xi\| + \|g\|$ in the above integrals, it is seen that

$$\|f\| \le \frac{1}{\lambda}\left(\frac{c}{2} \|\xi\| + \|g\|\right), \qquad (16.9\text{-}10)$$

where $\|\xi\|$ denotes $\sup|\xi(x)|$.

The function $f(x, \mu)$ given above is now integrated with respect to μ from -1 to 1, to give $\xi(x)$ according to (16.9-7). The result is

$$\xi(x) = \frac{c}{2} \int_{-a}^{a} E(\lambda|x' - x|)\xi(x')dx' + G_\lambda(x), \qquad (16.9\text{-}11)$$

where

$$G_\lambda(x) = \left[-\int_{-1}^{0} \frac{d\mu}{\mu} \int_{x}^{\infty} dx' + \int_{0}^{1} \frac{d\mu}{\mu} \int_{-\infty}^{x} dx'\right] e^{\lambda(x'-x)/\mu} g(x', \mu), \qquad (16.9\text{-}12)$$

and where $E(\cdot)$ is the exponential integral function

$$E(z) = \int_{z}^{\infty} e^{-w} \frac{1}{w} dw$$

(see Abramowitz and Stegun 1964). If $g(x', \mu)$ is replaced in (16.9-12) by $\|g\|$, the integrations can be evaluated; they give $2/\lambda$; hence,

$$\|G_\lambda\| \le \frac{2}{\lambda} \|g\|. \qquad (16.9\text{-}13)$$

Equation (16.9-11) is a Fredholm integral equation with a symmetric Hilbert–Schmidt kernel. It has a unique solution ξ, for any G_λ, provided that $2/c$ is not an eigenvalue of the integral operator that appears there. It will now be shown that that is the case, in particular, if λ is large enough. The function

$E(|w|)$ has a logarithmic singularity at $w = 0$ and decreases exponentially to zero, as $w \to \pm\infty$. Therefore the integral $\int_{-\infty}^{\infty} E(|w|)dw$ is a finite constant C. E is also positive, hence

$$\int_{-a}^{a} E(\lambda|x' - x|)dx' < \frac{C}{\lambda}.$$

Therefore the norm of the first term on the right of (16.9-11) can be made less than $\frac{1}{2}\|\xi\|$ for all λ greater than some λ_0, and then the only solution of the homogeneous equation (with $G_\lambda = 0$) is $\xi = 0$; so $2/c$ is not an eigenvalue. For $\lambda > \lambda_0$, more generally,

$$\|\xi\| \leq \tfrac{1}{2}\|\xi\| + \|G_\lambda\|$$

or

$$\|\xi\| \leq 2\|G_\lambda\| \tag{16.9-14}$$

Combining the inequalities (16.9-10, 13, 14) gives

$$\|f\| \leq \frac{1}{\lambda}\left(\frac{2c}{\lambda} + 1\right)\|g\|, \quad \text{for all } \lambda > \lambda_0.$$

Now $f = R_\lambda g$, hence

$$\|R_\lambda\| \leq \frac{1}{\lambda}\left(\frac{2c}{\lambda} + 1\right),$$

from which it follows that there is a number α such that

$$\|R_\lambda\| \leq \frac{1}{\lambda - \alpha} \quad \text{for } \lambda > \alpha;$$

hence the Hille–Yošida theorem applies, and the initial-value problem of neutron transport in the slab is well posed.

16.10 Inhomogeneous Problems

Consider the initial-value problem

$$\frac{d}{dt} u(t) - Au(t) = g(t), \tag{16.10-1}$$

$$u(0) = u_0, \tag{16.10-2}$$

where A is a closed linear operator such that the corresponding homogeneous problem (with $g(t) \equiv 0$) is well posed, and where u_0 and $g(t)$ are given; $g(t)$ is a one-parameter family of elements of the Banach space $\mathfrak{B}$, or a *curve* in $\mathfrak{B}$.

As a first example, suppose that the one-dimensional heat flow problem of Section 15.2 is modified by the presence of a source of heat distributed along the rod. Then the term $g(t)$ in (16.10-1) represents a function of x and t that gives the density of the source.

As a second example, which appears at first to be of an entirely different kind, suppose that the heat flow equation itself is homogeneous, but that the boundary condition of zero temperature at $x = a$ and $x = b$ is replaced by the condition that the temperature at each end of the rod is a given function of t, say $h_a(t)$ and $h_b(t)$ at $x = a$ and $x = b$, respectively. By a standard device, this problem can be reduced to a problem of the type of the first example. Namely, let $w(x, t)$ be any smooth function such that $w(a, t) = h_a(t)$ and $w(b, t) = h_b(t)$, and call

$$g(x, t) = -\frac{\partial w}{\partial t} + Aw = -\frac{\partial w}{\partial t} + \sigma \frac{\partial^2 w}{\partial x^2}.$$

Then the function $u(x, t) - w(x, t)$ satisfies the homogeneous boundary condition and the inhomogeneous partial differential equation. This reduction is essential, to permit the boundary condition to be enforced by a restriction on the domain of A.

Often, the source term represents an interaction between the process under study and some other process occurring at the same time in the system, as in the problem of coupled sound and heat flow discussed by Richtmyer and Morton (1967), and in electromagnetic problems, where the current and charge densities are the source of the field and are in turn influenced by that field. Then, the several processes have to be taken together as constituting a larger (and often nonlinear) problem. Nevertheless, it is sometimes convenient to have computational algorithms, existence proofs, and the like, for a single process, its source being thought of as given.

The problem will be called *well posed* if it has a unique solution for all reasonable choices of u_0 and of $g(t)$ and if the solution depends continuously, in some sense, on those choices. To make this precise, we need a new Banach space $\mathfrak{B}_1$, with norm $\|\cdot\|_1$, whose elements are functions $w(t)$ having values in $\mathfrak{B}$ and defined on an interval $[0, t_0]$; we set

$\mathfrak{B}_1(0, t_0) = \{w(\cdot) \text{ any curve in } \mathfrak{B}: w(t) \text{ is continuous for } t \in [0, t_0]\};$

$$\text{for } w(\cdot) \in \mathfrak{B}_1, \quad \|w(\cdot)\|_1 = \max\{\|w(t)\| : t \text{ in } [0, t_0]\}.$$

It is evident that any solution is unique, because of the uniqueness of the solutions of the homogeneous problem. Namely, the difference of two solutions, for given u_0 and given $g(\cdot)$, is a solution of the homogeneous problem with zero as initial element, hence must be zero for all t. It will be shown below that strict solutions exist for sets of u_0 and of $g(\cdot)$ that are dense in $\mathfrak{B}$ and $\mathfrak{B}_1$, respectively—formula (16.10-3) below gives such a solution explicitly by means of the solution operator $E(t)$ of the homogeneous problem. What will remain, then, is to show the continuous dependence of the solution on u_1 and $g(\cdot)$.

If $\hat{u}_0$ and $\tilde{u}_0$ are two nearly equal initial elements (nearly identical initial states of the physical system), and if $\hat{g}(t)$ and $\tilde{g}(t)$ are two nearly coincident curves in $\mathfrak{B}$, and if $\hat{u}(t)$ and $\tilde{u}(t)$ are resulting strict solutions of the problem (16.10-1, 2), then the problem is called *well-posed* if $\|\tilde{u}(\cdot) - \hat{u}(\cdot)\|_1$ is less than a preassigned $\varepsilon > 0$ whenever $\|\tilde{u}_0 - \hat{u}_0\|$ and $\|\tilde{g}(\cdot) - \hat{g}(\cdot)\|_1$ are both less than some $\delta > 0$. The change of initial element from $\hat{u}_0$ to $\tilde{u}_0$ changes the solution by an amount $E(t)(\tilde{u}_0 - \hat{u}_0)$, which is known to be small, if $\|\tilde{u}_0 - \hat{u}_0\|$ is small, because the homogeneous problem is well posed. Therefore, it suffices to consider the case $\tilde{u}_0 - \hat{u}_0 = 0$ and to consider only changes in $g(\cdot)$. Calling $\tilde{g}(\cdot) - \hat{g}(\cdot)$ simply $g(\cdot)$ and $\tilde{u}(\cdot) - \hat{u}(\cdot)$ simply $u(\cdot)$, the problem is thus to show that $\|u(\cdot)\|_1$ is small if $\|g(\cdot)\|_1$ is small, where $u(\cdot)$ is the solution of (16.10-1) with $u(0) = 0$. The mapping $F_0: g(\cdot) \to u(\cdot)$ (for $u(0) = 0$), of $\mathfrak{B}_1$ into itself determined by these strict solutions is a linear transformation in $\mathfrak{B}_1$; what must be shown is that F_0 has a dense domain and is bounded. The entire discussion thus parallels that for the homogeneous case, and generalized solutions will be defined by the extension F of F_0 with domain all of $\mathfrak{B}_1$.

Purely formal manipulation suggests that the solution of (16.10-1) and (16.10-2) ought to be

$$u(t) = E(t)u_0 + \int_0^t E(t - s)g(s)ds, \qquad (16.10\text{-}3)$$

because then,

$$\frac{d}{dt} u(t) = \frac{d}{dt} E(t)u_0 + E(0)g(t) + \int_0^t \frac{d}{dt} E(t - s)g(s)ds$$

$$= AE(t)u_0 + g(t) + \int_0^t AE(t - s)g(s)ds$$

$$= A\left[E(t)u_0 + \int_0^t E(t - s)g(s)ds \right] + g(t)$$

$$= Au(t) + g(t), \qquad (16.10\text{-}4)$$

and obviously $u(0) = u_0$. The steps that have to be justified here are differentiation under the integral sign in the first line of (16.10-4) and pulling the operator A out of the integral in the third line. If u_0 is any element of the set $\mathfrak{U}$ of initial elements of strict solutions of the homogeneous problem, introduced in Section 16.2, and if $g(\cdot)$ is any element of a certain set $\mathfrak{G}$ in $\mathfrak{B}_1$, then the above steps can be justified, and consequently $u(t)$ is a strict solution of the problem (16.10-1, 2). The set $\mathfrak{G}$ is given by

$$\mathfrak{G} = \{g(\cdot) \in \mathfrak{B}_1 : g(t) \in \mathfrak{D}(A^2), 0 \le t \le t_0; g(t), Ag(t), \text{ and}$$

$$A^2 g(t) \text{ are continuous, } 0 \le t \le t_0\}. \quad (16.10\text{-}5)$$

["Continuous" means strongly continuous, i.e. continuous in the topology of $\mathfrak{B}$; e.g., $\|g(t + \delta) - g(t)\| \to 0$ as $\delta \to 0$.] It will be shown that this set $\mathfrak{G}$ is dense in $\mathfrak{B}_1$.

The foregoing statements will now be proved.

Assertion 1. (justification of the first line of (16.10-4). For fixed t_0,

$$\frac{\partial}{\partial t} \int_0^{t_0} E(t - s)g(s)ds = \int_0^{t_0} \frac{\partial}{\partial t} E(t - s)g(s)ds.$$

For $g(\cdot)$ in $\mathfrak{G}$, all the integrands in (16.10-3, 4) are continuous functions with values in $\mathfrak{B}$; just as in Section 16.7, the integrals are to be interpreted as the (strong) limits in $\mathfrak{B}$ of the corresponding Riemann sums. As for ordinary integrals, differentiation with respect to the parameter t under the integral sign above is permissible if the difference quotient converges uniformly in s to the derivative, that is, if

$$\left\| \frac{E(\delta) - I}{\delta} E(t - s)g(s) - AE(t - s)g(s) \right\| \to 0$$

uniformly in s $(0 \leq s \leq t)$, as $\delta \to 0$. Since $E(t - s)$ is bounded and commutes with A when applied to any element in $\mathfrak{D}(A)$, the requirement reduces to

$$\left\| \frac{E(\delta) - I}{\delta} g(s) - Ag(s) \right\| \to 0$$

uniformly in s $(0 \leq s \leq t)$, as $\delta \to 0$. According to (16.7-6),

$$\frac{E(\delta) - I}{\delta} g(s) - Ag(s) = \frac{1}{\delta} \int_0^\delta [AE(v)g(s) - Ag(s)]dv;$$

since $Ag(s)$ is also in $\mathfrak{D}(A)$ for any s, according to (16.10-5), equation (16.7-6) can be applied again, namely to the integrand above, and the result is

$$\frac{E(\delta) - I}{\delta} g(s) - Ag(s) = \frac{1}{\delta} \int_0^\delta \left[\int_0^v AE(w)Ag(s)dw \right] dv;$$

$$\therefore \quad \left\| \frac{E(\delta) - I}{\delta} g(s) - Ag(s) \right\| \leq \tfrac{1}{2}\delta \sup \| E(w)A^2 g(s) \|,$$

where the supremum is taken for $0 \leq w \leq t$, $0 \leq s \leq t$; the supremum is finite, because $A^2 g(s)$ is continuous and the operators $E(w)$ are uniformly bounded. This completes the justification of the first line of (16.10-4).

Assertion 2. (justification of the third line of (16.10-4)

$$\int_0^t Ah(s)ds = A \int_0^t h(s)ds,$$

where, for given t, $h(s)$ is the continuous function $E(t - s)g(s)$. It is only necessary to approximate the integrals by Riemann sums. Clearly, A times a Riemann sum for an integral $\int_0^t h(s)ds$ is a Riemann sum for $\int_0^t Ah(s)ds$; since A is a closed operator, passage to the limit as the sums approach the integrals gives $A \int_0^t h(s)ds = \int_0^t Ah(s)ds$.

It is concluded that, for any u_0 in $\mathfrak{U}$ and any $g(\cdot)$ in $\mathfrak{G}$, equation (16.10-3) yields a strict solution of the initial value problem.

Assertion 3. $\mathfrak{G}$ is dense in $\mathfrak{B}_1$. To show this, one needs to know that the domain $\mathfrak{D}(A^2)$ is dense in $\mathfrak{B}$, where A is the infinitesimal generator of a semigroup $E(t)$ that is strongly continuous for $t \geq 0$. It can be shown that $\mathfrak{D}(A^k)$

is dense in $\mathfrak{B}$ for any $k = 1, 2, \ldots$. For proof, see Hille and Phillips 1957, p. 308, or Richtmyer and Morton 1967, p. 52. Now, let $g(t)$ be any continuous curve in $\mathfrak{B}$, $0 \leq t \leq t_0$, so that $g(\cdot)$ is an element of the Banach space $\mathfrak{B}_1$ defined at the beginning of this section. Divide the interval $[0, t_0]$ into N subintervals of length $\delta = t_0/N$; approximate each $g(n\delta)$ by an element h_n in $\mathfrak{D}(A^2)$; then define $h(t)$ by linear interpolation among the h_n, i.e.

$$h(t) = h_n + \frac{t - n\delta}{\delta} (h_{n+1} - h_n) \quad \text{for } n\delta \leq t \leq (n + 1)\delta;$$

clearly, $h(t)$ is in $\mathfrak{D}(A^2)$ for all t, and $h(t)$, $Ah(t)$, and $A^2h(t)$ are continuous; that is, $h(\cdot)$ is in $\mathfrak{G}$. If δ is small enough and if $\|g(n\delta) - h_n\|$ is small enough, for each n, clearly the quantity

$$\|g(\cdot) - h(\cdot)\|_1 = \sup\{\|g(t) - h(t)\| : t \in [0, t_0]\}$$

can be made arbitrarily small; that is, an arbitrary $g(\cdot)$ in $\mathfrak{B}_1$ can be arbitrarily closely approximated, in the norm of $\mathfrak{B}_1$, by an h in $\mathfrak{G}$; i.e., $\mathfrak{G}$ is dense in $\mathfrak{B}_1$.

Finally, let F_0 be the linear transformation in $\mathfrak{B}_1$ given by

$$\mathfrak{D}(F_0) = \mathfrak{G}$$

$$F_0 : g(\cdot) \to u(\cdot), \quad \text{where } u(t) = \int_0^t E(t - s)g(s)ds.$$

It is evident that F_0 is bounded for $0 \leq t \leq t_0$ (hence, the inhomogeneous problem is well posed) and that its bounded extension F to all $\mathfrak{B}_1$ is given by the same integral. In analogy with Section 16.2, then, the function $u(t)$ given by (16.10-3) is called the *generalized solution* of the inhomogeneous problem (16.10-1, 2) for any u_0 in $\mathfrak{B}$ and any $g(\cdot)$ in $\mathfrak{B}_1$.

16.11 Problems in Which the Operator A Is Time-Dependent

In most practical problems, the dependence of A on t is of a simple kind, Normally, the following assumptions are valid:

1. The domain $\mathfrak{D}(A)$ is independent of t
2. For any $v \in \mathfrak{D}(A)$, $A(t)v$ is (strongly) continuous in t
3. For any fixed s, the initial-value problem $(d/dt)u(t) = A(s)u(t)$ $(t \geq 0)$, $u(0)$ given, is well posed; let the solution be called $u_s(t)$.
4. The constant $K = K(t_0)$ that appears in the inequality $\|u_s(t)\| \leq K\|u_s(0)\|$, $0 \leq t \leq t_0$, (see (16.2-1)) can be chosen independent of s, in any finite interval $0 \leq s \leq s_0$.

When all these assumptions are made, the foregoing theory can be generalized in an obvious way. The initial-value problem

$$\frac{d}{dt} u(t) = A(t)u(t), \quad t \geq s,$$

$$u(s) = u_0 \quad \text{(given)}$$

(16.11-1)

is then well posed. Its strict solutions determine a bounded densely defined operator $E_0(t, s)$, such that $u(t) = E_0(t, s)u(s)$. The extension $E(t, s)$ of this operator to all $\mathfrak{B}$ determines the generalized solutions.

The operators $E(t, s)$ do not form a semigroup, but they satisfy the identity

$$E(t_3, t_2)E(t_2, t_1) = E(t_3, t_1) \qquad (t_3 \geq t_2 \geq t_1). \qquad (16.11\text{-}2)$$

For any s, $A(s)$ may be assumed to be the infinitesimal generator of the semigroup determined by the initial value problem $(d/dt)u(t) = A(s)u(t)$ $(t \geq 0)$. Then,

$$A(t) = \lim_{\delta \to 0} \frac{E(t + \delta, t) - I}{\delta}$$

If the hypotheses of the Hille–Yošida theorem are satisfied uniformly in t, i.e., if the resolvent $R_\lambda(A(t))$ exists for $\lambda > \alpha$ and is bounded by $(\lambda - \alpha)^{-1}$, for all t in a finite interval $[0, t_0]$, then the initial-value problem (16.11-1) is well posed in $[0, t_0]$.

CHAPTER 17

Nonlinear Problems: Fluid Dynamics

Relation between linear and nonlinear problems of evolution; fluid dynamics as an example; system of conservation laws; quasilinear equations; weak solutions; jumps and jump conditions; shock; slip surface; contact discontinuity; Rankine-Hugoniot conditions; entropy condition; characteristics; hyperbolic equations; characteristic form; Riemann invariants; Cauchy–Kovalevski theorem; noncharacteristic initial surface or initial data; characteristic plane; the Riemann problem; spontaneous generation of shocks; Helmholtz and Taylor instabilities; piecewise analytic initial-value problem; Mach reflection; triple shock intersection; corner flow; power series calculation of detached shock; algebraic manipulation of power series by computer; significance arithmetic; analytic continuation by computer

Prerequisite: some knowledge of fluid dynamics

Nonlinear initial-value problems are mostly unexplored. The linear problems in the preceding chapters are generally all special cases of nonlinear ones or become nonlinear when interactions are taken into account. Acoustics becomes fluid dynamics when the amplitudes are finite; Maxwell's equations and the Dirac equation yield a nonlinear system when the coupling between them is included (see Gross 1966). The new phenomena that appear in nonlinear problems are many and varied. In this chapter, a few of these new phenomena are described, in connection with fluid dynamics. The main conclusion is that some sort of piecewise analytic formulation is needed; the details of such a formulation are likely to remain unclear until much more theoretical work has been done.

Nonlinear steady state problems also have many new features, such as bifurcation phenomena and solitary waves, but are omitted from the present discussion. Convection and turbulence are also omitted; they are of a different character from the initial-value problems because unpredictability of detail is an essential feature. Problems of dynamic meteorology are of an intermediate character. In them, random effects and inadequate knowledge of the initial data both contribute to long range unpredictability, but new data become available from observations as time goes on.

The nonlinear phenomena are so numerous and varied that one cannot expect any one subject, such as fluid dynamics, to exemplify them all; still, things appear that have wider applicability, especially the theory of characteristics, the development of jumps and other singularities in analytic solutions, and the Cauchy–Kovalevski theorem, all of which are important, for example, in general relativity; the Cauchy problem of the Einstein field equations is discussed in Volume 2.

It is rather likely that nonlinear effects will turn out to be important in other areas, for instance quantum field theory, in connection with the interactions of particles. It seems impossible to predict what kind of phenomena will be encountered.

17.1 Wave Propagation

The equation of acoustic waves,

$$\frac{\partial^2 u}{\partial t^2} = c^2 \nabla^2 u,$$

where u is the overpressure $p - p_0$ relative to the ambient pressure p_0, represents an idealization of physical reality in three respects:

1. It is assumed that the ambient state is uniform (homogeneous) and isotropic.
2. It is assumed that the acoustic disturbance is infinitesimal: $u \ll p_0$.
3. It is assumed that the mean free path of the gas molecules is infinitesimal with respect to the length scale of the disturbance, which is $(|\nabla p|/p)^{-1}$.

If we drop the first two assumptions but retain the third, we have fluid dynamics, which may be regarded as the nonlinear generalization of wave motion and is the subject of this chapter.

As indicated in Section 15.4, the one-dimensional wave equation has solutions of the form $u(x, t) = f(x + ct)$ or $f(x - ct)$. If f has bounded support, such a solution represents a wave packet moving with constant speed $\pm c$ and without change of size, shape, or amplitude. For more general linear equations with constant coefficients, such as the equations of elastic vibrations of a homogeneous medium or the Schrödinger equation for a free particle, a wave packet generally changes as it moves, owing to the phenomena of dispersion and damping (or growth). For linear equations with variable coefficients, the motion is still more complicated, but in the special case of hyperbolic systems (see Section 17.8 below), there is a kind of propagation without dispersion or damping associated with the so-called characteristics, which are the analogues of the trajectories $x = \text{const.} \pm ct$ in the one-dimensional case and of moving wave fronts in more dimensions.

The equations of fluid dynamics are nonlinear, but if we superpose a small disturbance on a given smooth solution the disturbance satisfies a linear hyperbolic system obtained by linearizing the equations of fluid dynamics about the given solution. Study of the characteristics plays a dominant role in the analysis of fluid dynamical problems.

In nonlinear problems there are new phenomena, such as shocks and slip surfaces, represented by the so-called weak solutions of the equations, for whose study it is necessary to put the equations into the appropriate conservation law form. There are generally various possible conservation law forms; they have the same smooth solutions but different weak solutions, and one must decide between them on physical grounds.

17.2 Fluid-Dynamical Conservation Laws

Consider the one-dimensional motion of an ideal nonviscous fluid, under conditions such that heat conduction can be neglected. The fluid may be thought of as moving in a long frictionless pipe of unit cross-sectional area. Let $\rho, u, p, \mathscr{E}$ be the fluid's density, velocity, pressure, and internal energy per unit mass, each as a function of x and t, where x is a Cartesian coordinate measured along the pipe, and t is the time. It is convenient to introduce as further dependent variables the momentum and (total) energy per unit volume, namely $m = m(x, t) = \rho u$ and $e = e(x, t) = \rho \mathscr{E} + \frac{1}{2}\rho u^2$. Let the trajectories of two of the fluid particles be $x = a = a(t)$ and $x = b = b(t)$. where $a(t) < b(t)$, and consider the part of the fluid that lies between $x = a(t)$ and $x = b(t)$. Its total mass, total momentum, and total energy are

$$M = \int_a^b \rho \, dx, \qquad P = \int_a^b m \, dx, \qquad E = \int_a^b e \, dx. \qquad (17.2\text{-}1)$$

According to the fundamental physical laws, $\dot{M}$ is zero, where the dot denotes d/dt, $\dot{P}$ is equal to the sum of the forces acting on this part of the fluid, and $\dot{E}$ is equal to the rate at which those forces do work. Therefore,

$$\dot{M} = 0$$
$$\dot{P} = -p(b, t) + p(a, t) \qquad (17.2\text{-}2)$$
$$\dot{E} = -p(b, t)u(b, t) + p(a, t)u(a, t).$$

Each of the foregoing equations is of the form

$$\frac{d}{dt} \int_{a(t)}^{b(t)} f(x, t)dx + g(x, t)\Big|_{x=a(t)}^{x=b(t)} = 0,$$

or, since $\dot{a} = u(a, t)$ and $\dot{b} = u(b, t)$, it is of the form

$$\int_a^b \frac{\partial f}{\partial t} \, dx + (uf + g)\Big|_{x=a}^{x=b} = 0,$$

or, finally, if differentiability of $uf + g$ is assumed, it is of the form

$$\int_a^b \left[\frac{\partial}{\partial t} f + \frac{\partial}{\partial x} (uf + g) \right] dx = 0.$$

This holds for every interval (a, b); hence, if f and g are of class C^1, it follows that the expression in the square bracket vanishes identically. This gives the partial differential equations of fluid dynamics, when the expressions for f and g are those that appear in (17.2-1) and (17.2-2). We call

$$U = \begin{pmatrix} \rho \\ m \\ e \end{pmatrix}, \quad F = \begin{pmatrix} m \\ (m^2/\rho) + p \\ (e + p)m/\rho \end{pmatrix}; \qquad (17.2\text{-}3)$$

then the equations can be written in condensed form as

$$\frac{\partial}{\partial t} U + \frac{\partial}{\partial x} F = 0.$$

At this point we have more unknown functions (ρ, m, e, and p) than equations, but according to the laws of thermodynamics, there is a functional relation between p, ρ, and $\mathscr{E}$, called the *equation of state* of the fluid. If it is written as $p = f(\mathscr{E}, \rho)$ (for a perfect gas it is $p = (\gamma - 1)\rho\mathscr{E}$, where γ is a constant), then, in terms of the variables appearing in (17.2-3),

$$p = f\left(\frac{1}{\rho} \left(e - \frac{1}{2} m^2/\rho \right), \rho \right). \qquad (17.2\text{-}4)$$

If the symbol p in (17.2-3) is understood as an abbreviation for this expression, then each component of F is a function of the components of U, and we write $F = F(U)$, hence

$$\frac{\partial}{\partial t} U + \frac{\partial}{\partial x} F(U) = 0. \qquad (17.2\text{-}5)$$

A system of equations of this general form in any number of dependent and any number of independent variables, is called *a system of conservation laws*; see Lax 1954, 1957. For fluid flow in *two* space variables, with Cartesian coordinates x and y, the momentum density has two components m and n, and the conservation law form of the equation is

$$\frac{\partial}{\partial t} U + \frac{\partial}{\partial x} F(U) + \frac{\partial}{\partial y} G(U) = 0. \qquad (17.2\text{-}6)$$

where

$$U = \begin{bmatrix} \rho \\ m \\ n \\ e \end{bmatrix}, \quad F(U) = \begin{bmatrix} m \\ (m^2/\rho) + p \\ mn/\rho \\ (e + p)m/\rho \end{bmatrix}, \quad G(U) = \begin{bmatrix} n \\ mn/\rho \\ (n^2/\rho) + p \\ (e + p)n/\rho \end{bmatrix}. \qquad (17.2\text{-}7)$$

Equations (17.2-5) can also be written as a system of quasi-linear equations in various ways, for example

$$\left(\frac{\partial}{\partial t} + u\frac{\partial}{\partial x}\right)\rho = -\rho\frac{\partial u}{\partial x}$$

$$\rho\left(\frac{\partial}{\partial t} + u\frac{\partial}{\partial x}\right)u = -\frac{\partial p}{\partial x} \qquad\qquad (17.2\text{-}8)$$

$$\rho\left(\frac{\partial}{\partial t} + u\frac{\partial}{\partial x}\right)\mathscr{E} = -p\frac{\partial u}{\partial x},$$

where again $p = f(\mathscr{E}, \rho)$. A system of equations is called *quasi-linear* if it is linear in the partial derivatives of the highest order (here, the first order) with coefficients that are functions of the undifferentiated quantities and the lower order derivatives (here, only of the undifferentiated quantities $u, p, \rho, \mathscr{E}$).

We now show how a different system of conservation laws can be derived. Let $T = T(x, t)$ and $S = S(x, t)$ be the absolute temperature and the specific entropy (entropy per unit mass) of the fluid. According to the laws of thermo-dynamics, S and T are also functions of $\mathscr{E}$ and ρ, and are so related that

$$d\mathscr{E} + p\,d\left(\frac{1}{\rho}\right) = T\,dS. \qquad\qquad (17.2\text{-}9)$$

The first and third equations (17.2-8) can be combined to give

$$\left(\frac{\partial}{\partial t} + u\frac{\partial}{\partial x}\right)\mathscr{E} + p\left(\frac{\partial}{\partial t} + u\frac{\partial}{\partial x}\right)\frac{1}{\rho} = 0.$$

or, using (17.2-9),

$$\left(\frac{\partial}{\partial t} + u\frac{\partial}{\partial x}\right)S = 0,$$

which shows that the entropy is constant along the particle trajectories. This equation can be combined once more with the first equation of (17.2-8) to give

$$\frac{\partial}{\partial t}(\rho S) + \frac{\partial}{\partial x}(\rho u S) = 0, \qquad\qquad (17.2\text{-}10)$$

which is a fourth equation in conservation law form. If ρS is called $s = s(x, t)$ (it is the entropy per unit volume), and if the relation between $s, \mathscr{E},$ and ρ is written as $s = f_1(\mathscr{E}, \rho)$, then

$$s = f_1\left(\frac{1}{\rho}\left(e - \frac{\frac{1}{2}m^2}{\rho}\right), \rho\right).$$

Therefore, if (17.2-3) is replaced by

$$U = \begin{bmatrix} \rho \\ m \\ s \end{bmatrix} \qquad F(U) = \begin{bmatrix} m \\ (m^2/\rho) + p \\ ms/\rho \end{bmatrix}, \qquad (17.2\text{-}11)$$

an alternative system of conservation laws is obtained.

As long as the flow is smooth, i.e., the functions p, u, ρ, etc., are of class C^1 in their dependence on x and t, the system (17.2-8) and the two systems of conservation laws, based on (17.2-3) and (17.2-11), are all equivalent. However, real flows are not always smooth, and ones that start out smooth can develop shocks and other singularities, as time goes on. Flows with singularities are described by the weak solutions of the differential equations, which will be discussed in the next section. For weak solutions, the three systems of differential equations are not equivalent, and the correct form must be determined by physical considerations. Only the conservation-law system based on (17.2-3) gives the correct weak solutions, as will be seen, because the conservation of mass, momentum, and energy are the primary physical laws; when shocks are present, the entropy is not conserved, but increases.

Each of the conservation-law systems can be put into quasi-linear form, for smooth solutions, by first defining a matrix $A = A(U)$ by the equation

$$A_{jk} = \frac{\partial}{\partial U_k} F_j(U), \qquad (17.2\text{-}12)$$

and then writing

$$\frac{\partial U}{\partial t} + A(U) \frac{\partial U}{\partial x} = 0. \qquad (17.2\text{-}13)$$

17.3 Weak Solutions

The case of one space variable is treated first. Integration of (17.2-2) with respect to t from t_1 to t_2 gives

$$M(t_2) - M(t_1) = 0$$

$$P(t_2) - P(t_1) = \int_{t_1}^{t_2} [-p(b, t) + p(a, t)]dt \qquad (17.3\text{-}1)$$

$$E(t_2) - E(t_1) = \int_{t_1}^{t_2} [-p(b, t)u(b, t) + p(a, t)u(a, t)]dt,$$

where, as in (17.2-2), a and b stand for $a(t)$ and $b(t)$. These equations are the *fundamental* expression of the physical laws of mass, momentum, and energy for the fluid, because they do not require differentiability of the functions. They relate the values of the mass, momentum, and energy of a part of the fluid under consideration at time t_2 to the values of those same quantities at time t_1. However, if the components of $U(x, t)$ and of $F(U(x, t))$ are regarded as distributions in the x, t plane, then the above equations are exactly equivalent to the conservation law equations (17.2-5), if the derivatives are taken in the distribution sense. If $W(x, t)$ is a vector-valued test function (with the same number of components as U), then, by definition of distribution derivative, (17.2-5) means that

$$\int_{-\infty}^{\infty} \int_{-\infty}^{\infty} \left[\frac{\partial W}{\partial t} \cdot U + \frac{\partial W}{\partial x} \cdot F(U) \right] dx \, dt = 0. \qquad (17.3\text{-}2)$$

Any function $U(x, t)$ that satisfies this equation for all such vector test functions $W(x, t)$ is called a *weak solution* of the system of conservation laws (17.2-5).

17.4 The Jump Conditions

A weak solution is generally piecewise smooth. The smooth parts satisfy the differential equations in any of the forms, but that does not generally suffice to determine the course of the motion from initial data, and the differential equations must be supplemented by jump conditions on the discontinuities.

Suppose that a weak solution $U(x, t)$ has a jump discontinuity across a curve $\mathscr{C} : x = x(t)$ in the x, t plane, while $U(x, t)$ is otherwise differentiable in some neighborhood $\mathscr{R}$ of $\mathscr{C}$; $x(t)$ is assumed differentiable. Let $W(x, t)$ be a test function with support in $\mathscr{N}$. Let $\mathscr{R}$ be the part of the support of $W(x, t)$ that lies on one side (say the left) of $\mathscr{C}$ (see Figure 17-1). Then by Gauss's theorem, since $W = 0$ on the boundary of $\mathscr{R}$ except along $\mathscr{C}$,

$$\iint_{\mathscr{R}} \left(\frac{\partial W}{\partial t} \cdot U + \frac{\partial W}{\partial x} \cdot F \right) dx \, dt + \iint_{\mathscr{R}} \left(\frac{\partial U}{\partial t} + \frac{\partial F}{\partial x} \right) \cdot W \, dx \, dt$$

$$= \iint_{\mathscr{R}} \left[\frac{\partial}{\partial t} (W \cdot U) + \frac{\partial}{\partial x} (W \cdot F) \right] dx \, dt$$

$$= \int_{\mathscr{C}} \left[-\frac{\dot{x}}{\sqrt{1 + \dot{x}^2}} W \cdot U + \frac{1}{\sqrt{1 + \dot{x}^2}} W \cdot F \right] ds,$$

where the left-hand limiting values of U and $F(U)$ on $\mathscr{C}$ are understood; $(1 + \dot{x}^2)^{-(1/2)}$ and $-\dot{x}(1 + \dot{x}^2)^{-(1/2)}$ are the x and t components of the unit normal vector to $\mathscr{C}$. The second integral in the first line of this equation is zero, because (17.2-5) holds (in the strict sense) in the interior of $\mathscr{R}$. Therefore, if we integrate similarly over the right-hand part of the support of W and add the results, and make use of (17.3-2), we find that

$$0 = \int_{\mathscr{C}} (\dot{x}[U] - [F]) \cdot \frac{W}{\sqrt{1 + \dot{x}^2}} \, ds$$

where [ ] denotes the difference of the two limiting values of a function on the two sides of $\mathscr{C}$, i.e., the jump of the function (say from left to right). We

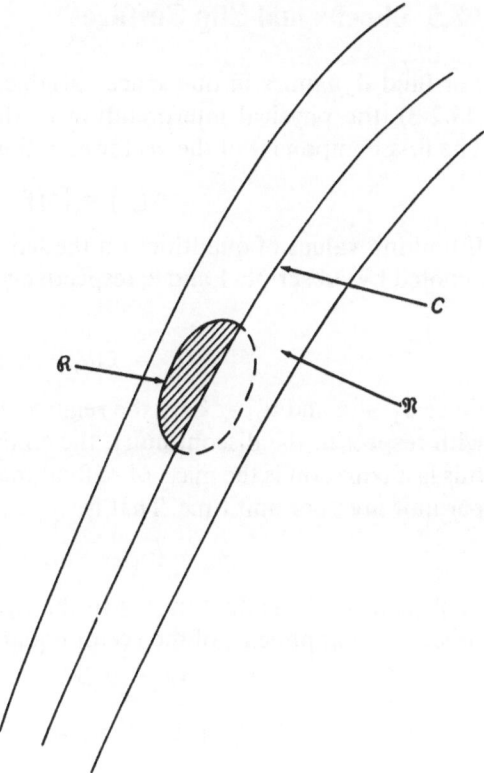

Figure 17.1 Diagram for the jump conditions.

get the *difference*, not the sum, because the direction of the normal unit vector
has to be reversed when Gauss's theorem is applied in the second part of the
support of **W**. Since **W** was arbitrary, the above equation implies the jump
condition:

$$\dot{x}[\mathbf{U}] - [\mathbf{F}(\mathbf{U})] = 0 \quad \text{on } \mathscr{C}. \tag{17.4-1}$$

The generalization to more than one space variable is straightforward.
$\mathbf{U}(x, y, t)$ is a weak solution of (17.2-6) if

$$\int\int\int_{-\infty}^{\infty} \left[\frac{\partial \mathbf{W}}{\partial t} \cdot \mathbf{U} + \frac{\partial \mathbf{W}}{\partial x} \cdot \mathbf{F}(\mathbf{U}) + \frac{\partial \mathbf{W}}{\partial y} \cdot \mathbf{G}(\mathbf{U}) \right] dx \, dy \, dt = 0 \tag{17.4-2}$$

for every vector test function $\mathbf{W}(x, y, t)$. To find the generalization of (17.4-1),
recall that -1 and $\dot{x}$ were the x and t components of a vector perpendicular
to $\mathscr{C}$ in the x, t plane. Hence, if $\mathbf{U}(x, y, t)$ is a weak solution of (17.2-6) and
is smooth except for a simple jump across a surface $\mathscr{S}$ in the 3-space with
coordinates x, y, t and if $(\lambda_x, \lambda_y, \lambda_t)$ is a vector perpendicular to $\mathscr{S}$, then $\mathbf{U}$
satisfies

$$\lambda_t[\mathbf{U}] + \lambda_x[\mathbf{F}(\mathbf{U})] + \lambda_y[\mathbf{G}(\mathbf{U})] = 0 \quad \text{on } \mathscr{S}, \tag{17.4-3}$$

instead of (17.4-1).

17.5 Shocks and Slip Surfaces

For fluid dynamics in one space variable, where $\mathbf{U}$ and $\mathbf{F}(\mathbf{U})$ are given by (17.2-3), the physical interpretation of the jump conditions is as follows: The first component of the vector equation (17.4-1) is

$$\dot{x}[\rho] = [m] = [\rho u]. \tag{17.5-1}$$

If limiting values of quantities on the left and right sides of the curve $\mathscr{C}$ are denoted by subscripts 1 and 2, respectively, then this equation can be written as

$$(u_1 - \dot{x})\rho_1 = (u_2 - \dot{x})\rho_2.$$

Since $u_1 - \dot{x}$ and $u_2 - \dot{x}$ are the relative velocities of the fluid on either side with respect to the discontinuity, the common value of the two members of this last equation is the mass M of fluid that passes through the discontinuity per unit area per unit time. That is,

$$M = (u_1 - \dot{x})\rho_1 = (u_2 - \dot{x})\rho_2; \tag{17.5-2}$$

M is positive if the fluid moves to the right through the discontinuity. The other two components of the vector equation (17.4-1) can then be written as

$$Mu_2 - Mu_1 = p_1 - p_2, \tag{17.5-3}$$

$$M(\mathscr{E}_2 + \tfrac{1}{2}u_2^2 - \mathscr{E}_1 - \tfrac{1}{2}u_1^2) = p_1 u_1 - p_2 u_2; \tag{17.5-4}$$

these equations say that the rate of increase of momentum due to the passage of the fluid through the discontinuity is equal to the difference of the pressure forces on the two sides and that the rate of increase of total energy (internal plus kinetic) is equal to the rate at which those forces do work.

For flows in two or three dimensions, the fluid velocity is a vector $\mathbf{u}$. There are, in consequence, a few minor changes to be made in equations (17.5-2, 3, 4); in particular, (17.5-3) is replaced by a vector equation. Suppose that $p, \rho, \mathbf{u}$ have simple jumps across a surface $\mathscr{S}$. Let P be a point of $\mathscr{S}$, and let U denote the speed of the surface, relative to the coordinate system, measured normally to the surface at P. That is, let $\boldsymbol{\lambda}$ be a unit vector normal to the surface at P, pointing into the region corresponding to the subscript 2; then, if a line is drawn through P in the direction of $\boldsymbol{\lambda}$, U is the speed of motion of the point of intersection of the surface with that line. Since $\boldsymbol{\lambda} \cdot \mathbf{u}$ is the component of the fluid velocity in this same direction, equation (17.5-2) is now replaced by

$$M = (\boldsymbol{\lambda} \cdot \mathbf{u}_1 - U)\rho_1 = (\boldsymbol{\lambda} \cdot \mathbf{u}_2 - U)\rho_2. \tag{17.5-5}$$

The rate of change of momentum is $M(\mathbf{u}_2 - \mathbf{u}_1)$ and the force is $\boldsymbol{\lambda}(p_1 - p_2)$; hence (17.5-3, 4) are replaced by

$$M(\mathbf{u}_2 - \mathbf{u}_1) = \boldsymbol{\lambda}(p_1 - p_2), \tag{17.5-6}$$

$$M(\mathscr{E}_2 + \tfrac{1}{2}u_2^2 - \mathscr{E}_1 - \tfrac{1}{2}u_1^2) = p_1 \boldsymbol{\lambda} \cdot \mathbf{u}_1 - p_2 \boldsymbol{\lambda} \cdot \mathbf{u}_2. \tag{17.5-7}$$

Equation (17.5-6) is equivalent to the two equations

$$M(\lambda \cdot \mathbf{u}_2 - \lambda \cdot \mathbf{u}_1) = p_1 - p_2 \qquad (17.5\text{-}8)$$

$$M(\lambda \times \mathbf{u}_2 - \lambda \times \mathbf{u}_1) = 0. \qquad (17.5\text{-}9)$$

From equation (17.5-9) it is seen that there are two possibilities: either the tangential velocity component $\lambda \times \mathbf{u}$ is continuous across $\mathscr{S}$, or $M = 0$. In the first case $\mathscr{S}$ represents the motion of a *shock* or *shock front*; in the second case, it is a *slip surface*, across which the pressure p and the normal velocity component are continuous ($\lambda \cdot \mathbf{u}_1 = \lambda \cdot \mathbf{u}_2 = U$), while the density ρ and the tangential velocity component $\lambda \times \mathbf{u}$ can have arbitrary jumps. In case M is $=0$ *and* $\lambda \times \mathbf{u}$ is continuous, the surface is a *contact discontinuity*, where only the density and the temperature are discontinuous, and there is no relative motion.

For the case of a shock, the above jump conditions can be written in terms of the specific volume $V = 1/\rho$ in the following form, among others:

$$\frac{1}{V_1}(\lambda \cdot \mathbf{u}_1 - U) = \frac{1}{V_2}(\lambda \cdot \mathbf{u}_2 - U) = \sqrt{\frac{p_2 - p_1}{V_1 - V_2}}, \qquad (17.5\text{-}10)$$

$$\mathscr{E}_2 - \mathscr{E}_1 = \tfrac{1}{2}(p_1 + p_2)(V_1 - V_2), \qquad (17.5\text{-}11)$$

$$\lambda \times (\mathbf{u}_1 - \mathbf{u}_2) = 0. \qquad (17.5\text{-}12)$$

In this form, they are known as the *Rankine–Hugoniot* conditions. The more general conditions (17.4-1) or (17.4-3) are often also called Rankine–Hugoniot conditions in a generalized sense. The positive square root in (17.5-10) corresponds to the case $M > 0$ (a chock moving with respect to the fluid toward the side of $\mathscr{S}$ denoted by subscript 1; the negative square root corresponds to $M < 0$.

17.6 Instability of Negative Shocks

If $M > 0$, the subscript 2 indicates fluid that has passed through (or has been passed over by) the shock front; one therefore expects that in this case $p_2 > p_1$ and $V_2 < V_1$, as is observed for real shocks. However, equations (17.5-10, 11, 12) remain valid if the subscripts 1 and 2 are interchanged (according to (17.5-5), that does not alter M); hence there are also solutions for which $p_2 < p_1$ and $V_2 > V_1$. Such solutions, called *negative shocks*, can be excluded by consideration of entropy, stability, or dissipative mechanisms, as will now be shown.

To investigate the entropy, the case of a γ-law gas is considered first; then

$$\mathscr{E} = \frac{pV}{\gamma - 1}, \qquad T \propto pV, \qquad S \propto \log(pV^\gamma). \qquad (17.6\text{-}1)$$

Generally, if $\mathscr{E} = \mathscr{E}(p, V)$, then equation (17.5-11), when written as

$$\mathscr{E}(p_2, V_2) - \mathscr{E}(p_1, V_1) = \frac{p_1 + p_2}{2}(V_1 - V_2),$$

establishes a relation between p_1, V_1, p_2, V_2; for fixed p_1, V_1, there results a one-parameter family of possible final states, each represented by a point p_2, V_2 in the p, V plane; these points lie on a so-called *Hugoniot curve*. For given p_1, V_1, call

$$\pi = \frac{p_2}{p_1}, \qquad \eta = \frac{V_1}{V_1} = \frac{\rho_2}{\rho_1};$$

π and η are the *pressure ratio* and the *compression ratio* of the shock. In the γ-law case, it is readily found that the Hugoniot curve is given by

$$\pi = \frac{\theta\eta - 1}{\theta - \eta} \qquad \left(\theta = \frac{\gamma + 1}{\gamma - 1}\right). \tag{17.6-2}$$

Since π and η are inherently positive, their ranges are $0 < \pi < \infty$, $1/\theta < \eta < \theta$. Positive and negative shocks correspond to $\eta > 1$ and to $\eta < 1$, respectively. An infinitely strong shock ($\pi \to \infty$) compresses the fluid by only the finite factor $\theta = (\gamma + 1)/(\gamma - 1)$ (the temperature $T_2 \to \infty$). The entropy, as a function of η on the Hugoniot curve is given by

$$S \propto \log p + \gamma \log V,$$

and the entropy change due to the shock is

$$\Delta S = S_2 - S_1 \propto \log \pi - \gamma \log \eta = \log \frac{\theta\eta - 1}{\theta - \eta} - \gamma \log \eta.$$

The derivative of ΔS with respect to η along the Hugoniot curve is

$$\frac{d}{d\eta} \Delta S = \frac{\theta}{\theta\eta - 1} + \frac{1}{\theta - \eta} - \frac{\gamma}{\eta} = \frac{\gamma\theta(\eta - 1)^2}{(\theta\eta - 1)(\theta - \eta)\eta}.$$

This quantity is positive on the entire curve ($1/\theta < \eta < \theta$), except at $\eta = 1$, where it is $=0$; hence, ΔS is >0 for $\eta > 1$ and <0 for $\eta < 0$. The shock can be present in a system in which all other processes are isentropic; hence, the case $\Delta S < 0$ can be ruled out and it is concluded on the basis of the second law of thermodynamics that negative shocks do not occur in nature. The conclusion that ΔS is an increasing function of η along the Hugoniot curve (except at $\eta = 1$) holds for a general equation of state, under mild assumptions—see Courant and Friedrichs 1948, Section 65.

The stability argument for excluding negative shocks is as follows: Consider two shocks, one following the other. For simplicity, suppose they are steady state plane parallel shocks, both moving in the same direction with respect to the fluid. It follows easily from the Rankine–Hugoniot equations that if the shocks are positive, then the one behind travels faster than the one in front, so that they coalesce into a single shock after a short time, whereas if they are negative shocks, they become more and more separated as time goes on. Owing to the lack of absolute precision in nature, a shock may equally well be thought of either as a single discontinuity or as a succession of many small discontinuities separated by infinitesimal distances. If the individual discontinuities are positive shocks, then they promptly coalesce so as to sharpen the total profile, whereas, if they are negative shocks, they move apart so as to make the profile a gradual transition.

Dissipative mechanisms such as heat conduction and viscosity are necessarily present, owing to the molecular nature of the fluid. Evidently, the transitions of pressure, density, and temperature from their initial to their final values cannot take place as mathematical discontinuities but must be spread out over a layer whose thickness is comparable with the mean free paths of the molecules of the fluid. Even if the transitions occur over several mean free paths, the temperature gradient is large enough to cause important heat flow from the hot side to the cold, and the shear is large enough to cause important viscous forces (recall that plane compression in one direction contains shear, as can be seen by considering axes at 45° to that direction). To get a qualitative estimate of the effect of these mechanisms, one can use the classical equations of heat flow and viscous forces, even though those equations are accurate only when the temperature and density change only slightly over a distance comparable with a mean free path. In the work of Becker 1922 (see also Richtmyer and Morton 1967, Section 12.10), the classical terms representing heat flow and viscous forces are included in the equations of fluid dynamics in one space variable x. Then a solution is sought in which $u, p\,\rho, \mathscr{E}$ depend on x and t only through the combination $w = x - Ut$, U being a constant, and in which $u, p, \rho, \mathscr{E}$ approach limiting values $u_1, p_1, \rho_1,$ $\mathscr{E}_1$ and $u_2, p_2, \rho_2, \mathscr{E}_2$, as $w \to -\infty$ and $w \to +\infty$, respectively, with a continuous but rather rapid transition from the one set to the other over a rather small w interval. It is found that the limiting values satisfy the jump conditions (17.5-2, 3, 4) exactly, with $\dot{x} = U$. (That was to be expected, because the jump conditions depend only on the overall conservation of mass, momentum, and energy, and the dissipative effects vanish in the limits $w \to \pm\infty$.) However, a solution of this kind exists only for positive shocks, i.e., if $u_1, u_2 > U$, then only for $p_2 > p_1$ (and hence $\rho_2 > \rho_1, \mathscr{E}_2 > \mathscr{E}_1, u_2 < u_1$). Therefore, there are no steady running solutions corresponding to negative shocks. Evidently, if a negative shock is started off, at $t = 0$, with an arbitrary profile, the profile then broadens out indefinitely and never settles down to a steady state.

17.7 Sound Waves and Characteristics in One Dimension

Sound waves are vibrations of small amplitude and small wavelength superimposed as a perturbation on a smooth fluid flow. If $U^0 = U^0(x, t)$ is a smooth solution of the system of conservation laws (17.2-5) and if $U^0 + \varepsilon U^1$

is another solution, where ε is a small quantity, and $\mathbf{U}^1$ represents the sound wave, then

$$\frac{\partial}{\partial t}(\mathbf{U}^0 + \varepsilon\mathbf{U}^1) + \frac{\partial}{\partial x}\mathbf{F}(\mathbf{U}^0 + \varepsilon\mathbf{U}^1) = 0.$$

Hence, to first order in ε,

$$\frac{\partial}{\partial t}\mathbf{U}^1 + \frac{\partial}{\partial x}(A(\mathbf{U}^0)\mathbf{U}^1) = 0,$$

where A is the matrix given by (17.2-12). The requirement that the wave lengths be small means that $\mathbf{U}^0$ should change by very little in a wavelength of $\mathbf{U}^1$, i.e., that $\partial\mathbf{U}^0/\partial x$ should be negligible compared to $\partial\mathbf{U}^1/\partial x$; therefore,

$$\frac{\partial\mathbf{U}^1}{\partial t} + A(\mathbf{U}^0)\frac{\partial\mathbf{U}^1}{\partial x} = 0. \tag{17.7-1}$$

For a given smooth flow $\mathbf{U}^0(x, t)$, the above is a linear equation for $\mathbf{U}^1(x, t)$; it is the same as (17.2-13) except that it has been linearized by replacing the unknown $\mathbf{U}(x, t)$ by the known function $\mathbf{U}^0(x, t)$ in the coefficient matrix A.

The elementary theory of sound waves suggests that the solution of (17.7-1) ought to represent waves propagating to the left and right through the fluid. If the coefficient matrix were constant and could be diagonalized by a transformation $A \to TAT^{-1} = D$, and if $T\mathbf{U}^1$ were called $\mathbf{V}^1$, then the equation (17.7-1) would be of the form

$$\frac{\partial\mathbf{V}^1}{\partial t} + D\frac{\partial\mathbf{V}^1}{\partial x} = 0. \tag{17.7-2}$$

In this form, the equations of the system would be mutually independent, the jth equation being

$$\frac{\partial V_j^1}{\partial t} + \lambda_j\frac{\partial V_j^1}{\partial x} = \left[\frac{\partial}{\partial t} + \lambda_j\frac{\partial}{\partial x}\right]V_j^1 = 0, \tag{17.7-3}$$

where λ_j is the jth eigenvalue of A. For λ_j real, the solution of this equation would represent a wave propagating with speed λ_j.

Since the coefficient matrix in (17.7-1) is not constant, the matrices T and D depend on $\mathbf{U}^0$; hence, in place of (17.7-2), we have

$$T\frac{\partial\mathbf{U}^1}{\partial t} + TA\frac{\partial\mathbf{U}^1}{\partial x} = 0,$$

that is,

$$T\frac{\partial\mathbf{U}^1}{\partial t} + DT\frac{\partial\mathbf{U}^1}{\partial x} = 0. \tag{17.7-4}$$

Hence, in place of (17.7-3), we have

$$\sum_{k=1}^{l} T_{jk}(\mathbf{U}^0)\left[\frac{\partial}{\partial t} + \lambda_j(\mathbf{U}^0)\frac{\partial}{\partial x}\right]U_k^1 = 0 \qquad (j = 1, \ldots, l) \tag{17.7-5}$$

where l is the number of equations in the system ($=3$ for fluid dynamics in one dimension). The solution now represents waves propagating in a medium with variable properties, and the equations of the system are mutually dependent because the matrix T_{jk} cannot be permuted with the differential operator in (17.7-5).

17.8 Hyperbolic Systems

It is recalled that a matrix A can be diagonalized by a similarity transformation $A \to TAT^{-1} = D$ if and only if it has a complete set of eigenvectors; then, the columns of T^{-1} are a complete set of right eigenvectors and the rows of T are a complete set of left eigenvectors. A system

$$\frac{\partial U}{\partial t} + A \frac{\partial U}{\partial x} = 0$$

of partial differential equations with constant coefficients is called *hyperbolic* if A has all real eigenvalues and a complete set of eigenvectors. A linear system with variable coefficients

$$\frac{\partial U}{\partial t} + A(x, t) \frac{\partial U}{\partial x} = 0$$

is called *hyperbolic in a region* $\mathscr{R}$ of the x, t plane if the matrix $A(x, t)$ has all real eigenvalues and a complete set of eigenvectors at each point of $\mathscr{R}$. For a nonlinear system, hyperbolicity depends not only on the equations but also on the solution. If $U(x, t)$ is a solution of

$$\frac{\partial U}{\partial t} + A(U) \frac{\partial U}{\partial x} = 0, \tag{17.8-1}$$

he system is *hyperbolic in $\mathscr{R}$ for the solution* $U(x, t)$ if the matrix $A(U(x, t))$ has the properties stated, in other words if the linearized system that results from linearizing (17.8-1) about the solution $U(x, t)$ is hyperbolic in $\mathscr{R}$.

Often, one imposes restrictions on the dependence of $A(U)$ on U or of $A(U(x, t))$ on x, t; for example, these functions should satisfy a Lipschitz condition; also, to avoid ill-conditioned matrices, one often requires that $\|T\| \|T^{-1}\|$ be bounded in $\mathscr{R}$, where $\|\cdot\|$ denotes a matrix norm, and where T is the matrix that diagonalizes A: $TAT^{-1} = D$.

If the matrix T is applied directly to the system (17.8-1) rather than to its linearized form, then,

$$\sum_{k=1}^{l} T_{jk}(U) \left(\frac{\partial}{\partial t} + \lambda_j(U) \frac{\partial}{\partial x} \right) U_k = 0 \qquad (j = 1, \ldots, l) \tag{17.8-2}$$

(compare with (17.7-5)). These equations are the *characteristic form* of the system. The system is hyperbolic if and only if it can be transformed into a system of real equations in characteristic form.

If the system is hyperbolic, then, for each $j = 1, \ldots, l$, there is a one-parameter family of curves $x(t)$ in the x, t plane, which are solutions of the equation $dx/dt = \lambda_j$, or

$$\frac{d}{dt} x(t) = \lambda_j(\mathbf{U}(x(t), t)). \qquad (17.8\text{-}3)$$

These curves are called the *characteristics* of the solution $\mathbf{U}(x, t)$; they are the rays (in the sense of geometrical acoustics) of sound waves superposed on the solution $\mathbf{U}(x, t)$. The essential feature of the characteristic form (17.8-2) is that, in the jth equation, all quantities are differentiated in the same direction in the x, t plane, at a given point x, t, namely in the direction of the jth characteristic through that point.

17.9 Fluid-Dynamical Equations in Characteristic Form

The equations of one-dimensional fluid dynamics can be put into characteristic form most easily by choosing the density ρ, the velocity u, and the specific entropy S as dependent variables. If the equation of state is written as

$$p = p(S, \rho),$$

and if $c = c(S, \rho)$ is defined by

$$c^2 = \frac{\partial}{\partial \rho} p(S, \rho), \qquad (17.9\text{-}1)$$

then the equations of Section 17.2 become

$$\left(\frac{\partial}{\partial t} + u \frac{\partial}{\partial x}\right)\rho + \rho \frac{\partial u}{\partial x} = 0$$

$$\left(\frac{\partial}{\partial t} + u \frac{\partial}{\partial x}\right)u + \frac{1}{\rho}\left(c^2 \frac{\partial \rho}{\partial x} + \frac{\partial p}{\partial S}\frac{\partial S}{\partial x}\right) = 0,$$

$$\left(\frac{\partial}{\partial t} + u \frac{\partial}{\partial x}\right)S = 0.$$

The matrix A is

$$A = \begin{bmatrix} u & \rho & 0 \\ \dfrac{c^2}{\rho} & u & \dfrac{1}{\rho}\dfrac{\partial}{\partial S}p(S, \rho) \\ 0 & 0 & u \end{bmatrix}; \qquad (17.9\text{-}2)$$

its eigenvalues $\lambda_1, \lambda_2, \lambda_3$ are the roots of the equation

$$\det(A - \lambda I) = [(u - \lambda)^2 - c^2](u - \lambda) = 0,$$

namely $\lambda = u \pm c, u$. The characteristics are the trajectories of forward and backward sound signals and of the fluid particles.

The equations in characteristic form are readily found to be

$$\left(\frac{\partial}{\partial t} + (u + c)\frac{\partial}{\partial x}\right)p + \rho c\left(\frac{\partial}{\partial t} + (u + c)\frac{\partial}{\partial x}\right)u = 0$$

$$\left(\frac{\partial}{\partial t} + (u - c)\frac{\partial}{\partial x}\right)p - \rho c\left(\frac{\partial}{\partial t} + (u - c)\frac{\partial}{\partial x}\right)u = 0 \qquad (17.9\text{-}3)$$

$$\left(\frac{\partial}{\partial t} + u\frac{\partial}{\partial x}\right)S = 0.$$

An important special case in that in which the entropy is constant at some initial time $t = 0$, i.e., in which $S(x, 0)$ is independent of x. Then the third equation of (17.9-3) shows that S is constant for all time (so long as there are no shocks); i.e., the flow is isentropic. Then, p and c are functions of ρ only and will be temporarily denoted by $p(\rho)$ and $c(\rho)$. If a new thermodynamic quantity $\sigma = \sigma(\rho)$ is defined by

$$\sigma = \int \frac{1}{\rho c(\rho)}\, dp(\rho),$$

then the first two of the equations, after dividing through by ρc, are

$$\left[\frac{\partial}{\partial t} + (u \pm c)\frac{\partial}{\partial x}\right](\sigma \pm u) = 0. \qquad (17.9\text{-}4)$$

The quantities $\sigma \pm u$, which are called *Riemann* invariants, are constant along the forward and backward characteristics. Various analytic and numerical methods for calculating one-dimensional isentropic flow have been based on these equations. An example is given in Section 17.14 below on the spontaneous generation of shocks.

In the above equation c and σ are functions of ρ, hence the dependent variables can be taken as $u(x, t)$ and $\rho(x, t)$. In some cases ρ can be eliminated from $c(\rho)$ and $\sigma(\rho)$; for a γ-law gas,

$$\sigma = \frac{2c}{\gamma - 1}, \qquad (17.9\text{-}5)$$

and then the dependent variables can be taken as $u(x, t)$ and $c(x, t)$.

17.10 Remarks on the Initial-Value Problem

For the nonlinear partial differential equations that arise in physical applications, the theory of the initial-value problems has two parts. The initial data and the solutions are generally piecewise smooth; hence, one part of the theory deals with the smooth parts of the solutions, and the other part with the jump discontinuities and other singularities. In a region of the x, t space where the solution is smooth, the evolution is governed by the differential equations. If the dependent quantities are all known at some time t_0 in the region, then we may expect them to be determined there at slightly later times $t_0 + \varepsilon$ from a local initial-value problem based on the differential equations. Such local initial-value problems are the subject of the Cauchy–Kovalevski theorem given in Section 17.12 below.

It is first recalled that a partial differential equation of any order higher than the first can always be reduced to a system of equations of lower order, by the introduction of new unknowns. For example, the equation

$$f\left(u, \frac{\partial u}{\partial t}, \frac{\partial^2 u}{\partial t^2}, \frac{\partial u}{\partial x}, \frac{\partial^2 u}{\partial x^2}, \frac{\partial^2 u}{\partial x\, \partial t}\right) = 0$$

can be rewritten as the system

$$f\left(u, v, \frac{\partial v}{\partial t}, \frac{\partial u}{\partial x}, \frac{\partial^2 u}{\partial x^2}, \frac{\partial v}{\partial x}\right) = 0,$$

$$\frac{\partial u}{\partial t} = v,$$

which is of the first order with respect to t.

This suggests that one ought to consider a system of the form

$$\frac{\partial u_i}{\partial t} = f_i \qquad (i = 1, \ldots, l), \tag{17.10-1}$$

where, for each i, f_i is a function of the unknowns $u_1, \ldots, u_l$ and of their derivatives with respect to the space variables $x, y, \ldots$. However, this system is too special in one respect: it assumes that the original equations, whatever they are, can be solved for all the first derivatives $\partial u_i/\partial t$ with respect to t. That is not always the case.

In order to get some insight into the physical interpretation of the condition of solvability with respect to the time derivatives, the linear case in two independent variables is considered. If the unknowns $u_i(x, t)$ $(i = 1, \ldots, l)$ are taken to be the components of a vector $\mathbf{U} = \mathbf{U}(x, t)$, and if the system is supposed reduced completely to a first order system with respect to both x and t, then the system can be written as

$$A\frac{\partial \mathbf{U}}{\partial t} + B\frac{\partial \mathbf{U}}{\partial x} + C\mathbf{U} = 0, \tag{17.10-2}$$

where A, B, and C are $l \times l$ matrices, whose elements are smooth functions of x and t. Note: This does not imply that the original equation was of the same order with respect to t as with respect to x. For example, the heat flow equation $\partial u/\partial t = \partial^2 u/\partial x^2$ can be written as (17.10-2) by introducing a new function $v = \partial u/\partial x$ and setting

$$A = \begin{pmatrix} 1 & 0 \\ 0 & 0 \end{pmatrix}, \qquad B = \begin{pmatrix} 0 & -1 \\ 1 & 0 \end{pmatrix}, \qquad C = \begin{pmatrix} 0 & 0 \\ 0 & -1 \end{pmatrix}. \tag{17.10-3}$$

Consider now the Cauchy problem or pure-initial-value problem consisting of (17.10-2) together with the initial condition that $U(x, 0)$ is given, for all x. If det A is $\neq 0$ in some region $\mathscr{R}$ of the x, t plane containing the x axis (or, more generally, containing a piece of the x axis), then (17.10-2) can be solved to give $\partial U/\partial t$ in terms of the given function $U(x, 0)$ on the x axis in $\mathscr{R}$. Differentiating equation (17.10-2) with respect to t then gives an equation for $\partial^2 U/\partial t^2$ in terms of the functions U and $\partial U/\partial t$ (now known) on the x axis, and so on. Hence, if $U(x, 0)$ is infinitely differentiable with respect to x, all derivatives $\partial^k U/\partial t^k$ are obtained, and they can be used to construct a power series.

$$\sum_{k=0}^{\infty} \frac{1}{k!} t^k \frac{\partial^k U}{\partial t^k}\bigg|_{x,0}. \qquad (17.10\text{-}4)$$

It can be proved (this is a special case of the Cauchy–Kowalevski theorem) that if $U(x, 0)$ is analytic and the matrices A, B, and C are analytic functions of x and t, then the series (17.10-4) converges for t in some interval $(-\varepsilon, \varepsilon)$, where ε can depend on x, and the resulting functions of x and t satisfy (17.10-2). Hence, a solution of the initial-value problem is obtained in some neighborhood of the x axis in $\mathscr{R}$.

The situation is quite different, if det $A = 0$ on the x axis in a region $\mathscr{R}$. In that case, let $V = V(x)$ be a left eigenvector of A corresponding to the eigenvalue zero. Then, if (17.10-2) is multiplied through by V^T on the left, it is seen that the initial function $U(x, 0)$ is required to satisfy the condition

$$\left(V^T B \frac{\partial}{\partial x} + V^T C\right) U(x, 0) = 0, \qquad (17.10\text{-}5)$$

for every such left eigenvector V; otherwise, the initial-value problem has no solution. Furthermore, if the initial function satisfies this condition, the power series method of solution described above generally breaks down, because the derivatives $\partial U/\partial t$, $\partial^2 U/\partial t^2$, etc. are not necessarily determined uniquely by (17.10-2) on the x axis; for instance, to any value of $\partial U/\partial t$ obtained from (17.10-2) there can be added an arbitrary multiple of a right eigenvector of A corresponding to the eigenvalue zero.

In summary, the initial-value problem has locally a unique solution if det $A \neq 0$ in $\mathscr{R}$, while if det $A = 0$ in $\mathscr{R}$ the solution generally does not exist and is generally nonunique when it does exist. This result will be interpreted and generalized in the next section.

The heat flow problem, as formulated above in terms of two functions u and v (that is admittedly not the best formulation), provides an example, for in that problem det $A = 0$, according to (17.10-3). There is no solution unless the initial values of u and v satisfy the condition $v = \partial u/\partial x$ on the x axis, and if they do the solution of the initial-value problem is not unique because an arbitrary function of t can be added to v.

17.11 Flow of Information Along the Characteristics in One Dimension

The conclusions of the preceding section will now be stated in terms of characteristics. As in Section 17.8, if there is a linear combination of the equations of the system (17.10-2) in which all quantities are differentiated in the same direction in the x, t plane, then the resulting equation (the linear combination) is said to be in *characteristic form*. That is the case if there is a vector $\mathbf{W} = \mathbf{W}(x, t)$ such that the vectors $\mathbf{W}^T A$ and $\mathbf{W}^T B$ are proportional, i.e., are such that

$$\lambda \mathbf{W}^T A = \mu \mathbf{W}^T B$$

for some numbers $\lambda = \lambda(x, t)$ and $\mu = \mu(x, t)$, not both zero. Then the linear combination in question is given by multiplying (17.10-2) through by $\mathbf{W}^T$ on the left. Assume first that one of the vectors $\mathbf{W}^T A$ and $\mathbf{W}^T B$ is not identically zero, hence is $\neq 0$ in some region $\mathcal{R}$ of the x, t plane. Then the linear combination can be written in $\mathcal{R}$ as

$$\mathbf{W}^T A \left(\mu \frac{\partial}{\partial t} + \lambda \frac{\partial}{\partial x} \right) \mathbf{U} + \mu \mathbf{W}^T C \mathbf{U} = 0$$

if $\mathbf{W}^T A \neq 0$, and as

$$\mathbf{W}^T B \left(\mu \frac{\partial}{\partial t} + \lambda \frac{\partial}{\partial x} \right) \mathbf{U} + \lambda \mathbf{W}^T C \mathbf{U} = 0$$

if $\mathbf{W}^T B \neq 0$. In either case, if a curve $\mathscr{C}: x = x(s), t = t(s)$ in $\mathcal{R}$ is determined by the equations

$$\dot{x} = \lambda(x, t), \qquad \dot{t} = \mu(x, t), \tag{17.11-1}$$

where the dot denotes indifferentiation with respect to the parameter s, then $\mathscr{C}$ is a *characteristic* or *characteristic curve*, and the linear combination takes the form

$$\mathbf{V}^T \frac{d}{ds} \mathbf{U} + \mathbf{X}^T \mathbf{U} = 0 \quad \text{on } \mathscr{C}. \tag{17.11-2}$$

This equation has the character of an ordinary differential equation. If $n - 1$ of the components of $\mathbf{U}$ are known on $\mathscr{C}$ and if the nth component is known at one point of $\mathscr{C}$, then that component can be obtained at all other points of $\mathscr{C}$ by integrating (17.11-2). This result is paraphrased by saying that *information about the solution flows along the characteristics*. Now consider the case in which both the vectors $\mathbf{W}^T A$ and $\mathbf{W}^T B$ are zero in a region $\mathcal{R}$, while $\mathbf{W}^T C$ is not. Then, by (17.10-2), $\mathbf{W}^T C \mathbf{U}$ is $=0$ in $\mathcal{R}$, hence again the components of $\mathbf{U}$ are dependent on $\mathscr{C}$. Lastly, if $\mathbf{W}^T A, \mathbf{W}^T B, \mathbf{W}^T C$ are all zero in a region $\mathcal{R}$, then the equations of the system (17.10-2) are not independent in $\mathcal{R}$; there are in effect fewer equations than unknowns, hence the local initial-value problem does not have a unique solution.

The characteristic curve $\mathscr{C}$ is parallel to the x axis if $\dot{t} = 0$ on $\mathscr{C}$, i.e., if $\mu = 0$ on $\mathscr{C}$, i.e., if the vector $\mathbf{W}^T A = 0$ on $\mathscr{C}$. Therefore, the preceding result concerning the initial-value problem of (17.10-2) and the initial condition $U(\mathbf{x}, 0)$ given can be stated as follows: This problem has a unique solution in some neighborhood of the x axis for an arbitrary initial function $U(\mathbf{x}, 0)$ if and only if the x axis is not a characteristic. If the x axis is a characteristic, then some of the information contained in the function $U(x, 0)$ simply flows along the x axis and puts a constraint on the initial function $U(x, 0)$, instead of flowing into the region $t > 0$ and thereby contributing to the determination of the solution for $t > 0$.

More generally, the initial data may be given along a curve $\mathscr{C}: x = x(s)$, $t = t(s)$; that is, $U(x(s)), t(s))$ is given as a function of s. Then, there is a unique analytic solution of the system (17.10-2) in some neighborhood of $\mathscr{C}$, for an arbitrary analytic initial function $U(x(s), t(s))$, if $\mathscr{C}$ is nowhere characteristic, i.e., nowhere tangent to a characteristic curve of the system (17.10-2). [It has been assumed that $\mathscr{C}$ is an analytic curve, i.e., that $x(s)$ and $t(s)$ analytic functions, so that power series expansions can be used.] This form of the result is appropriate, for example, in relativistic problems, where the time variable t is not physically unique. When special relativity is involved, $\mathscr{C}$ can be any space-like line in the x, t plane, or, more generally, a space-like hyperplane in space-time; when general relativity is involved, it can be any space-like hyper-*surface*. For a discussion of the Cauchy problem of the gravitational field equations, see Volume 2.

17.12 Characteristics in Several Dimensions; The Cauchy–Kovalevski Theorem

The case of three or more independent variables is similar. In place of (17.10-2) consider the system

$$A \frac{\partial \mathbf{U}}{\partial t} + B \frac{\partial \mathbf{U}}{\partial x} + C \frac{\partial \mathbf{U}}{\partial y} + D\mathbf{U} = 0, \qquad (17.12\text{-}1)$$

where the matrices A, B, C and D are smooth functions of x, y, and t. Let $\mathscr{S}$ be a smooth surface, given in terms of parameters α and β by

$$x = x(\alpha, \beta), \qquad y = y(\alpha, \beta), \qquad t = t(\alpha, \beta),$$

and consider an initial condition in which $\mathbf{U}$ is given on $\mathscr{S}$, i.e.,

$$U(x(\alpha, \beta), y(\alpha, \beta), t(\alpha, \beta)) = \text{given smooth function of } \alpha \text{ and } \beta. \quad (17.12\text{-}2)$$

In analogy with the two-variable case, the surface $\mathscr{S}$ is called *characteristic* if it is so oriented that the differential equations impose constraints on the initial function (17.12-2) on $\mathscr{S}$. Hence, we look for a linear combination of the equations of the system (17.12-1) in which all the unknowns (the components of $\mathbf{U}$) are differentiated in directions lying in a plane. If $\mathscr{S}$ is tangent to this plane at some point P, then the linear combination can be expressed, at

P, in terms of derivatives with respect to α and β, hence the resulting differential equation (the linear combination) imposes constraints on the initial function (17.12-2) at P. Under these circumstances the plane is a *characteristic plane* at P, and the surface $\mathscr{S}$ is *characteristic* at P. If $\mathscr{S}$ is characteristic at all its points, it is a *characteristic surface* of the equation system.

Suppose the linear combination in question is obtained by multiplying (17.12-1) through by a vector $\mathbf{W} = \mathbf{W}(x, y, t)$ on the left. In that linear combination, the unknown U_i is differentiated in a direction in the t, x, y space having direction cosines proportional to

$$(\mathbf{W}^T A)_i, (\mathbf{W}^T B)_i, (\mathbf{W}^T C)_i.$$

Therefore, if λ, μ, v are the direction cosines of the normal to $\mathscr{S}$ at P, the condition for $\mathscr{S}$ to be characteristic is that

$$\lambda \mathbf{W}^T A + \mu \mathbf{W}^T B + v \mathbf{W}^T C = 0. \tag{17.12-3}$$

Hence, the condition is that $\mathbf{W}$ be a left eigenvector of the matrix $\lambda A + \mu B + vC$ corresponding to the eigenvalue zero; the condition that zero be an eigenvalue is

$$\det(\lambda A + \mu B + vC) = 0. \tag{17.12-4}$$

The three unknowns λ, μ, v must satisfy also $\lambda^2 + \mu^2 + v^2 = 1$, hence two equations in all; therefore one can expect to have in general one or more one-parameter families of solutions. If these solutions are real, there are then corresponding one-parameter families of characteristic planes. For the equations of fluid dynamics, in two dimensions discussed in Exercise 3, below, there are two such families, one consisting of all planes tangent to the particle trajectory in the space t, x, y and one consisting of all planes tangent to the sonic cone. The fluid problem is of course nonlinear—see next paragraph. It is noted that no surface (plane) $t = $ constant can be characteristic, because if one of the characteristic planes coincided with a plane $t = $ constant, that would imply an infinite signal speed, whereas, in fluid dynamics, the fluid speed and the sound speed are both finite, in any choice of initial data, and the maximal signal speed is the sum of the two. This also follows from the fact that the matrix A of equation (17.12-1) is $=I$ in the fluid dynamical case.

Suppose now that the coefficient matrices A, B, C, and D in equation (17.12-1) depend on the components of $\mathbf{U}$ as well as on x, y, and t. (That is the case in fluid dynamics—see Section 17.2). Then the equations (17.12-1) are called *quasilinear*. The definitions and conclusions are the same as for the linear case, but the point of view is slightly different: for a given system of equations, a surface $\mathscr{S}$ may be characteristic or not, depending on the initial functions given on $\mathscr{S}$, i.e., on the components of the vector field (17.12-2) on $\mathscr{S}$, because of the dependence of A, B, and C on $\mathbf{U}$. One often speaks of the given initial functions being characteristic or noncharacteristic with respect to the given surface $\mathscr{S}$.

The Cauchy problem (the problem of determining $\mathbf{U}(x, y, t)$ from the

Cauchy data (17.12-2) and the differential equation (17.12-1) is called *analytic* if the surface $\mathscr{S}$ and all the functions involved are analytic. For $\mathscr{S}$ to be analytic, the functions $x(\alpha, \beta)$, $y(\alpha, \beta)$ and $t(\alpha, \beta)$ must be analytic, and the rank of the matrix

$$
\begin{pmatrix}
\dfrac{\partial x}{\partial \alpha} & \dfrac{\partial y}{\partial \alpha} & \dfrac{\partial t}{\partial \alpha} \\[2mm]
\dfrac{\partial x}{\partial \beta} & \dfrac{\partial y}{\partial \beta} & \dfrac{\partial t}{\partial \beta}
\end{pmatrix}
$$

must be $=2$ everywhere on $\mathscr{S}$. [To see that this latter condition is really necessary, note that the equations $x(\alpha, \beta) = \alpha^3$, $y(\alpha, \beta) = \beta$. $t(\alpha, \beta) = \alpha^2$ determine a surface which has a cusp on the y axis, where the rank of the above matrix is only $=1$.]

One version of the Cauchy–Kovalevski theorem is now stated, without proof, for the case of three independent variables x, y, t. The generalization to more independent variables will be evident.

Theorem. *It is assumed that analytic Cauchy data (17.12-2) are given on an analytic surface $\mathscr{S}$ and are noncharacteristic with respect to $\mathscr{S}$ in some (two-dimensional) neighborhood, on $\mathscr{S}$, of a point P. The matrices A, B, C, and D in (17.12-1) are assumed to be analytic functions of x, y, t, and the components of U. Then there is a three-dimensional neighborhood of P in which the Cauchy problem has a unique solution.*

In the most usual case, the surface $\mathscr{S}$ is the x, y plane ($t = 0$), and the conditions of the theorem are satisfied at all points of the plane. Then, if K is any compact region of the plane, there is an interval $(-\varepsilon, \varepsilon)$ such that the problem has a unique solution for all (x, y) in K and all t in $(-\varepsilon, \varepsilon)$.

Exercises

1.   Find the characteristics of the heat flow equation when it is written as the system (17.10-2.3). It was pointed out that when A is a singular matrix ($\det A = 0$), the power-series method generally breaks down, because the derivatives $\partial U/\partial t$, $\partial^2 U/\partial t^2$, etc., are not generally uniquely determined by the method described. Reconcile this statement with the fact that, for the heat flow equation, $u(x, t)$ is uniquely determined for $t > 0$ by $u(x, 0)$.

2. Discuss the characteristics of the Cauchy–Riemann system

$$
\frac{\partial u}{\partial t} = \frac{\partial v}{\partial x},
$$

$$
\frac{\partial v}{\partial t} = -\frac{\partial u}{\partial x}.
$$

3.   Consider a fluid flow in two dimensions, where the pressure p, the density ρ, and the velocity $\mathbf{u} = (u, v)$ are functions of x, y, and t. Starting with the equations of

Section 17.2, and taking the simple equation of state $p = (\gamma - 1)\rho\mathscr{E}$, show that the equations in characteristic form are

$$\frac{Dp}{Dt} - c^2 \frac{D\rho}{Dt} = 0, \tag{17.12-5}$$

$$\boldsymbol{\mu} \cdot \left(\frac{D\mathbf{u}}{Dt} + \frac{1}{\rho}\nabla p \right) = 0, \tag{17.12-6}$$

$$\rho c \left(\lambda \frac{D}{Dt} + c\nabla \right) \cdot \mathbf{u} + \left(\frac{D}{Dt} + c\lambda \cdot \nabla \right)p = 0. \tag{17.12-7}$$

All the vectors appearing in these equations have two components; in particular, $\nabla = (\partial/\partial x, \partial/\partial y)$, while λ and $\boldsymbol{\mu}$ are arbitrary unit vectors in the x, y plane. D/Dt denotes the operator

$$\frac{D}{Dt} = \frac{\partial}{\partial t} + \mathbf{u} \cdot \nabla = \frac{\partial}{\partial t} + u\frac{\partial}{\partial x} + v\frac{\partial}{\partial y},$$

which effects differentiation along the trajectories of the particles, and c is the adiabatic sound speed $\sqrt{\gamma p/\rho}$. This generalizes equations (17.9-3). In equation (17.12-6), the directions of differentiation are restricted to a plane tangent to the particle path and parallel to $\boldsymbol{\mu}$; in (17.12-7) they are restricted to a plane tangent to the sonic cone and such that the intersection of this plane with the x, y plane is perpendicular to λ.

For some problems, the word "analytic" can be replaced throughout by "smooth." That is true in particular for hyperbolic equations, including the equations of fluid dynamics, provided that "smooth" means once con-continuously differentiable, and certain reasonable restrictions are imposed. See Courant and Hilbert 1962 or Garabedian 1964. Discontinuities of the higher derivatives (and, under usual conditions, even of the first) can exist in the solution; they are in fact propagated along the characteristics. However, shocks and other major singularities are not covered by the Cauchy-Kovalevski theory. Furthermore, it appears that when a contact discontinuity or a slip surface is present, the surface and the flow on either side of it must generally be analytic, or at least piecewise analytic, not merely smooth, for a solution to exist, because of the Taylor and Helmholtz instabilities (see section 17.15).

17.13 The Riemann Problem and Its Generalizations

The simplest problem with nonsmooth initial data is the classical Riemann problem; this is a problem in one space variable x, in which the functions u, p, and ρ are initially constant except for discontinuities at a point $x = 0$, where they jump from values u_1, p_1, ρ_1 for $x < 0$ to u_2, p_2, ρ_2 for $x > 0$.

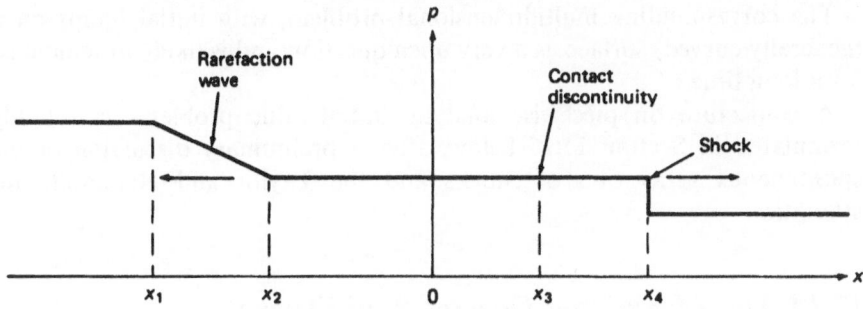

Figure 17.2 The pressure profile in the Riemann problem.

The shock tube provides an example. It is a long tube or pipe divided into two parts by a thin transverse diaphragm at $x = 0$. Air is pumped into one part (say the part $x < 0$) to a high pressure p_1, while the air in the other part remains at a lower pressure p_2. After the system comes to thermal and mechanical equilibrium, the temperature is constant, hence $p_1/\rho_1 = p_2/\rho_2$ by Boyle's law, and $u_1 = u_2 = 0$. Then, at time $t = 0$, the diaphragm is burst, or otherwise suddenly removed. A shock then moves through the air to the right, starting from $x = 0$, and a rare-faction wave to the left. The pressure profile at some time $t > 0$ is as shown in Figure 17.2. At $x = x_3$, where the fluids that were originally separated by the diaphragm are in contact, there is a contact discontinuity (see Section 17.5), where the pressure is continuous while the temperature and density are discontinuous. (The air just to the left of x_3 has expanded, while the air to the right has been compressed.) Each of the points x_1, x_2, x_3, x_4 moves away from the origin 0 with a constant speed. It is easily established that there is only one solution of this problem that satisfies the jump conditions and the entropy condition at the jumps and the differential equations between them. The solution is easily seen to be of the character just described.

Another example is provided by the collision of two initially cold interstellar gas clouds, whose surfaces are parallel planes. At the instant of collision, $p_1 = T_1 = p_2 = T_2 = 0$, while $u_1 > 0$ and $u_2 < 0$. In this case, there are two shocks, starting from the plane of collision, one moving into each cloud.

The general case, with arbitrary $u_1, u_2, p_1, p_2, \rho_1, \rho_2$ was solved by Riemann in 1860—see Courant and Friedrichs 1948.

Perhaps the next simplest problem is the same as the Riemann problem, except that the functions are only assumed to be analytic, on either side of $x = 0$ at $t = 0$, not necessarily constant. Much work has been done on this problem; existence and uniqueness have been proved for simplified versions of it (for example, when there is only one function and one differential equation), but the general case appears to be still an open question.

The corresponding multidimensional problem, with initial jumps on a (generally curved) surface, is a very open question and is likely to remain so for a long time.

A conjecture on piecewise analytic initial-value problems is roughly formulated in Section 17.16 below, after a preliminary discussion of the spontaneous generation of shocks and the Taylor and Helmholtz instabilities.

17.14 The Spontaneous Generation of Shocks

Suppose that a gas is initially at rest in a long tube closed at one end by a piston. Starting at $t = 0$, the piston is pushed into the tube with a continuous acceleration. We shall show that the resulting flow of the gas is smooth up to a certain time t^*, at which a shock wave forms at some point in the interior of the gas and starts moving through the gas away from the piston. The shock strength is zero at $t = t^*$, but positive and increasing for $t > t^*$. Thus, a flow that is initially smooth can develop a singularity as time goes on. This effect is observed in the atmospheres of certain pulsating stars: a shock is formed in each cycle, during the phase of outward acceleration, as the envelope is pushed outward by the expanding interior; the shock then moves outward through the remaining atmosphere and disappears in space, presumably heating the star's corona as it goes through.

We denote the x coordinate of the piston at time t by $\xi(t)$, and we assume that $\xi(t) = 0$ for $t < 0$. The gas occupies the region $x > \xi(t)$. We assume that the acceleration $\ddot{\xi}(t)$ is >0 for $t > 0$ and either is continuous for all t or has at most a jump at $t = 0$. For $t < 0$ we have $u = 0$, $c = c_0$ for all $x > 0$. The flow equations in characteristic form are given by (17.9-4 and 5). The characteristic curves $x_\pm(t)$ in the x, t plane are given by

$$\frac{dx_+}{dt} = u + c, \qquad \frac{dx_-}{dt} = u - c \qquad (17.14\text{-}1)$$

Along each backward characteristic $x_-(t)$ the quantity $u - \sigma = u - (2c/(\gamma - 1))$ is constant, and since $u = 0$, $c = c_0$, at $t = 0$ for all x, it follows that $u - (2c/(\gamma - 1))$ is constant in the entire flow until a shock forms:

$$u - \frac{2c}{\gamma - 1} = \frac{2c_0}{\gamma - 1}. \qquad (17.14\text{-}2)$$

Along each forward characteristic $x_+(t)$, $u + (2c/(\gamma - 1))$ is also constant, hence u and c are individually constant along it, hence each forward characteristic is a straight line in the x, t plane; its slope is $(u + c)^{-1}$. The backward characteristics are straight only up to their intersection with the forward characteristic that comes from the origin; see Figure 17-3.

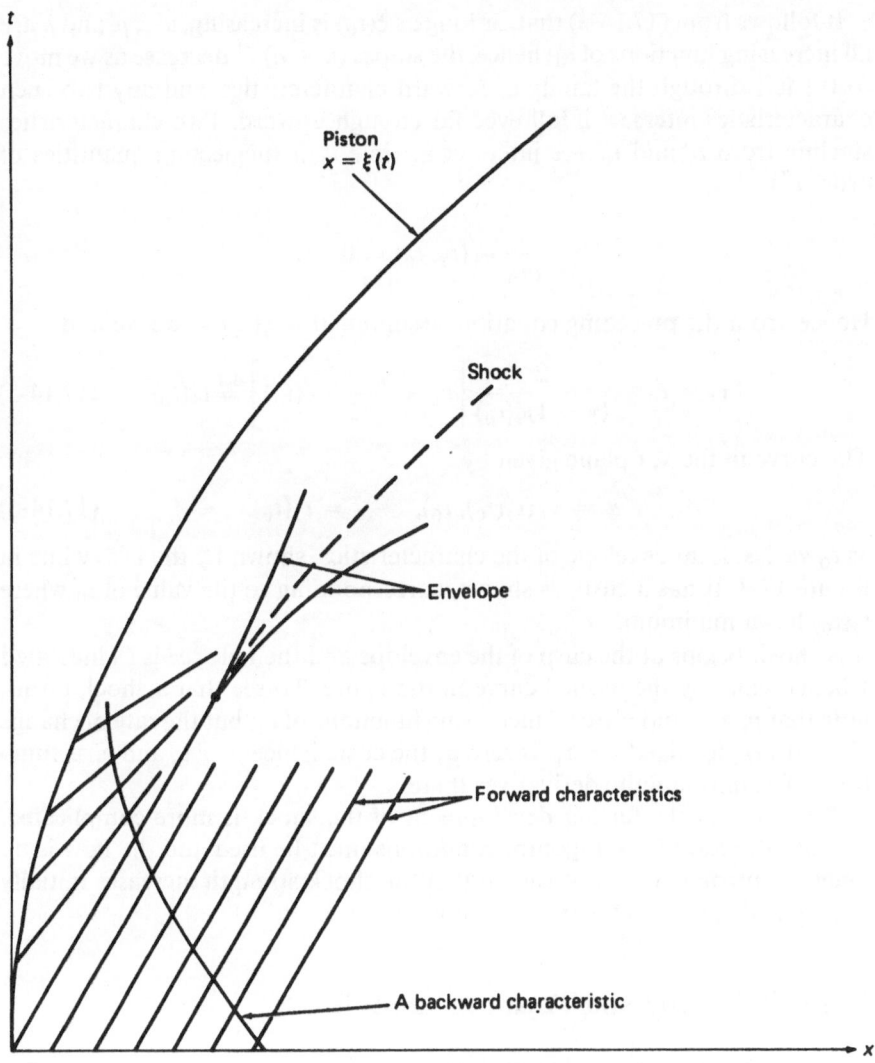

Figure 17.3 Spontaneous shock generation.

Along a forward characteristic that originates at the piston at a time t_0, the quantities u, c, p, and ρ are given by

$$u = \dot{\xi}(t_0), \qquad u - \frac{2(c - c_0)}{\gamma - 1} = 0, \qquad c^2 = \frac{\gamma p}{\rho}, \qquad p = K\rho^\gamma, \qquad (17.14\text{-}3)$$

and the equation of the characteristic itself is

$$x+(t) = \xi(t_0) + (u + c)(t - t_0) = \xi(t_0) + \left[c_0 + \frac{\gamma + 1}{2}\dot{\xi}(t_0)\right](t - t_0)$$
$$\overset{\text{def}}{=} x_+(t, t_0). \qquad (17.14\text{-}4)$$

It follows from (17.14-3) that, as long as $\dot{\xi}(t_0)$ is increasing, u, c, p, and ρ are all increasing functions of t_0; hence, the slopes $(u + c)^{-1}$ decrease as we move to the left through the family of forward characteristics, and any two such characteristics intersect if followed far enough upward. Two characteristics starting from t_0 and $t_0 + \varepsilon$ intersect at time t_1 if (neglecting quantities of order ε^2)

$$\frac{\partial}{\partial t_0} x_+(t_1, t_0) = 0.$$

Hence, from the preceding equation, assuming $0 < \ddot{\xi}(t_0) < \infty$, we find

$$t_1 = t_0 + \frac{2}{(\gamma - 1)\ddot{\xi}(t_0)} \left[c_0 + \frac{\gamma - 1}{2} \dot{\xi}(t_0) \right] \overset{\text{def}}{=} t_1(t_0). \qquad (17.14\text{-}5)$$

The curve in the x, t plane given by

$$x = x_+(t_1(t_0), t_0), \qquad t = t_1(t_0), \qquad (17.14\text{-}6)$$

as t_0 varies, is an envelope of the characteristics, shown by the heavy line in Figure 17-3. It has a cusp, as shown, corresponding to the value of t_0 where $t_1(t_0)$ has a minimum.

A shock begins at the cusp of the envelope and then proceeds as indicated schematically by the dashed curve in the figure. To see that a shock forms, note that p, ρ, u, and c are all increasing functions of t_0, but the rate of change of x with t_0, for fixed $t = t_1$, is zero at the cusp, hence p, ρ, u, and c, as functions of x, have infinite derivatives there.

The study of the further development of the shock is more complicated, because the Rankine–Hugoniot conditions must be used and the flow is no longer isentropic. It can be shown that the shock strength increases initially as $(t - t^*)^{1/2}$, where t^* is the minimum of $t_1(t_0)$.

17.15 Helmholtz and Taylor Instabilities

In order that an initial-value problem be well posed, in the sense of Hadamard, it is required not only that a unique solution exist for all initial states of the system defined by some "reasonable" class of initial functions, but also that the solution depend continuously, in some sense, on the initial data (see Chapter 16). Helmholtz instability provides an instance of discontinuous dependence on the initial data. A plane surface separates two regions (half spaces), in each of which a fluid is in uniform motion, with a different velocity in each, so that slippage (assumed frictionless) occurs on the plane. By introducing a small perturbation of the initial data, in the form of a small-amplitude sinusoidal corrugation of the surface and a corresponding slight modification of the flow near the surface, one can obtain a solution in which the amplitude of the perturbation increases with time; the increase is exponential as long as the amplitude is small compared to a wavelength, i.e., for the linearized problem—see below. This is the mechanism by which a breeze

produces waves on the surface of a pond; it is known as *Helmholtz instability*. (It has been assumed that the speed of slippage does not exceed a
certain fraction of the sound speed—at higher relative speeds, acoustic
effects damp the instabilities.) Moreover, the exponential growth rate of such
a perturbation increases without limit, as the wavelength λ of the corrugations
tends to zero. Consequently, given any $\varepsilon > 0$ and any $M > 0$, one can find a λ
small enough so that an initial "infinitesimal" perturbation of wavelength λ
increases by a factor $\geq M$ in a time interval $\leq \varepsilon$; that is, the solution of the
linearized problem does not depend continuously on the initial data.

If surface tension (more properly interface tension) exists between the two
fluids, then corrugations of wavelength less than some λ_0 are not amplified,
hence continuous dependence on the initial data is restored, even though
longer wavelengths are still unstable. Waves on the surface of a body of water
are stabilized both by surface tension and by gravity. On the other hand, when
a slip surface is generated within a given fluid, for example, by a moving
airfoil, or by the intersection of two shocks (see the discussion of the Mach
reflection problem in Section 17.17), there are generally no stabilizing effects.

The foregoing discussion is based on the linearized theory of the instability.
Although there is no quantitative theory of the nonlinear effects, one can
make a few qualitative and speculative remarks. Suppose that the unperturbed surface is the x, y plane, and that the perturbed one, $\mathscr{S}$, is given by
$z = z(x, y, t)$. [The linearized theory is based on the assumption that $\partial z/\partial x$
and $\partial z/\partial y$ are everywhere $\ll 1$.] For simplicity, effects of gravity, surface
tension, and compressibility will be omitted, and the flow will be assumed
irrotational both above and below $\mathscr{S}$, so that, at any t, the velocity is obtained
from a velocity potential φ as $\mathbf{u} = \nabla \varphi$, where $\nabla^2 \varphi = 0$ in each region.
The initial-value problem is then the following: The quantities

$$z(x, y, t) \quad \text{and} \quad \frac{\partial z}{\partial t}(x, y, t) \tag{17.15-1}$$

are given, at $t = 0$, as functions of x and y. The velocity potential $\varphi(x, y, z, t)$
is determined in each region, at any t, from the functions (17.15-1) by Laplace's equation $\nabla^2 \varphi = 0$ and by the requirements that (1) the normal
component of $\mathbf{u}$ be continuous across $\mathscr{S}$ and agree, on $\mathscr{S}$, with the normal
component of the velocity of $\mathscr{S}$ itself, and (2) that $\mathbf{u}$ tend to $(\pm\frac{1}{2}V, 0, 0)$ as
$z \to \pm \infty$, where V is the relative velocity or slip velocity. (*Note:* φ is not
continuous across $\mathscr{S}$). The motion is then governed by the requirement that
the pressure p be continuous across $\mathscr{S}$ and satisfy the equation

$$\rho_0 \left(\frac{\partial}{\partial t} + \mathbf{u} \cdot \nabla \right) \mathbf{u} + \nabla p = 0. \tag{17.15-2}$$

Consider a solution of the problem in which

$$z = z(x, t) = \sum_{-\infty}^{\infty} {}_{(k)} c_k(t) e^{ikx} \tag{17.15-3}$$

(Dependence on y could also be included, but has been omitted to simplify the writing.) In the linear approximation, each term of the sum can be treated by itself, and the time dependence is found (see Landau and Lifshitz 1959, Section 30) to be given by

$$c_k(t) = A_k e^{kVt/2} + B_k e^{-kVt/2}, \qquad (17.15\text{-}4)$$

as long as the amplitude is small compared with the wavelength, i.e., $|c_k| \ll 1/k$. The coefficient in one of the exponents, namely $\frac{1}{2}|k|V$, tends to infinity, as $k \to \infty$; hence, in order that a series of the form

$$\sum_{-\infty}^{\infty}{}_{(k)} d_k e^{ikx + (1/2)|k|Vt}$$

converge, for say $t = t_0 > 0$, it is necessary that

$$d_k = o(e^{-(1/2)|k|Vt_0}) \quad \text{as } k \to \infty,$$

which implies that the initial surface is analytic; if it is not analytic, no solution can be obtained by the Fourier series method.

The behavior at larger amplitudes (when the linear approximation is no longer valid) has been studied experimentally and by numerical calculation. For a surface initially of the form

$$z = a_k(t)\cos kx,$$

it is found that when the amplitude becomes comparable with the wavelength, i.e., when $a_k \approx 1/k$, the increase of $a_k(t)$ ceases to be exponential and becomes roughly linear, with $\dot{a}_k \approx \alpha V$, where α is a constant of the order of unity. The x dependence also changes.

Observations indicate that at still larger amplitudes ($a_k \gg 1/k$), the surface breaks up and is replaced by a turbulent layer whose thickness increases with time by eddy diffusion. A fine-grained initial irregularity of any sort causes an immediate local breakup of the surface.

On the other hand, if the initial surface is piecewise analytic, say with a zigzag cross section, as shown in Figure 17-4, the conjecture given in the next section, if correct, implies that there should be a piecewise analytic solution; the nature of this solution is unknown.

Taylor instability is similar. In this case, there is no slippage across $\mathscr{S}$. Instead, the two fluids have different densities, and there is an acceleration of the system in the direction from the lighter fluid to the heavier. An example

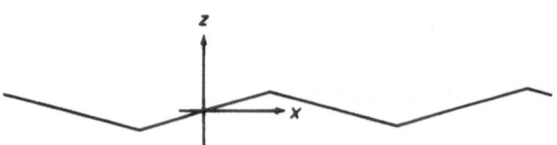

Figure 17.4   Cross section of piecewise flat initial surface.

of such a system, with the acceleration replaced by an equivalent gravitational field, is water in the upper part of an enclosure supported by air pressure in the lower part. If the interface is exactly plane and level, the system is in an equilibrium configuration, but an unstable one, in which waves on the surface have a time-dependent factor of the form $e^{\pm at}$, rather than the sine or cosine of at, as for normal water waves. Again the coefficient a increases without limit (if surface tension is absent) as the wavelength decreases.

17.16 A Conjecture on Piecewise Analytic Initial-Value Problems of Fluid Dynamics

In the absence of existence and uniqueness proofs, it is convenient to adopt a working hypothesis or conjecture, to provide a background for future investigations. A reasonable conjecture for the initial-value problems of fluid dynamics would seem to be that if the initial data are piecewise analytic, there exists a unique piecewise analytic solution, at least for some finite time interval. It is assumed that the equation of state is piecewise analytic. The remarks above on Helmholtz and Taylor instabilities indicate that it would not suffice in all cases to replace "piecewise analytic" by "piecewise smooth." To say that the initial data are piecewise analytic means that space can be divided into cells, in each of which the flow functions are analytic at $t = 0$. The cells are separated by analytic surfaces, across which the functions have simple jumps, and the surfaces are bound by analytic curves, which are the edges of the cells. It is not clear what kind of singularities of the functions or of the surfaces can be permitted at the edges and corners of the cells (a few examples of such singularities will be described in the next section); hence, even the formulation of the conjecture must be left somewhat vague. To say that the solution is piecewise analytic means that space-time can be similarly divided into cells, and so on; it doesn't mean of course that the subdivision of space remains for all time the same as for $t = 0$, because shocks and other singularities can form during the flow. In the Riemann problem, the space (one dimensional) is divided into two regions of analyticity ($x > 0$ and $x < 0$) at $t = 0$, but into five regions, according to Figure 17-2, for $t > 0$.

17.17 Singularities of Flows

Several examples of singularities are found in the phenomenon of Mach reflection. A plane steady shock moves down a shock tube from left to right and encounters a wedge or ramp on the floor of the shock tube, as shown in Figure 17-5. A disturbance starts when the shock front touches the tip of the wedge and then spreads out from that point; the disturbance has a curved secondary shock as its frontier. Under certain conditions of wedge angle and primary shock strength (see Bleakney and Taub 1949 and Duff 1962), the primary and secondary shocks meet at a so-called *triple point P* above the

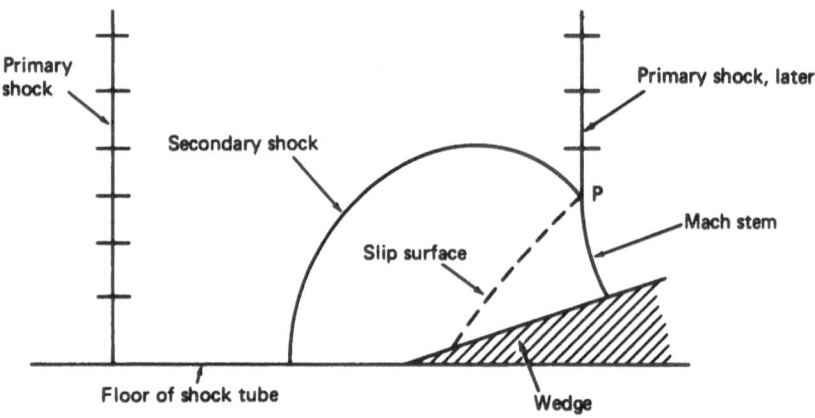

Figure 17.5 Mach reflection.

wedge; from that point down to the wedge, the primary and secondary shocks are coalesced into a single shock (also curved), called the *Mach stem*. Running downstream from the triple point, there is also a slip surface, shown dashed. A fluid element immediately below the slip surface has been compressed by a single shock, and a fluid element immediately above has been compressed to the same final pressure (the pressure is continuous across the slip surface) by two shocks in succession. The compression by a single shock produces a larger entropy increase, because the entropy increase produced by a given shock varies roughly as the cube of the shock strength. Hence, the temperature and the specific volume are higher just below the slip surface than just above, hence, the speed of flow away from the shocks is higher, just below, to achieve mass conservation.

The flow obeys a similarity law, because of the absence of any constant length in the formulation of the problem. (The length and transverse dimensions of the shock tube are clearly irrelevant, and may be thought of as infinite, for purposes of discussion.) Namely, the entire flow pattern at a time t_2 (measured from the instant when the primary shock touched the tip of the wedge) is the same as that at a time t_1, after rescaling in the ratio t_1/t_2. In other words, u, v, p, and ρ depend only on x/t and y/t, where x and y are cartesian coordinates with their origin at the tip of the wedge.

It seems evident that the shocks and the slip surface must be analytic curves (except perhaps at the triple point P itself), and that the flow variables in each region must be analytic functions of $\xi = x/t$ and $\eta = y/t$. This suggests that, on the curves, the coordinates $\xi - \xi_0$ and $\eta - \eta_0$ should be power series in arc length σ (arclength in the similarity variables: $d\sigma^2 = d\xi^2 + d\eta^2$), where ξ_0 and η_0 are the coordinates of P, and that in each of the two regions bounded by the curves, u, v, p, and ρ should be power series in the two variables $\xi - \xi_0$ and $\eta - \eta_0$. These functions are equal to known constants in each of the regions before and behind the primary shock. It is easy to see, however, that no solution of this power-series form can satisfy the jump conditions, the differential equations, and the similarity law. The belief has been expressed that one or more of the curves may have infinite curvature at P, like the graph of the function $y = |x|^{3/2}$ at the origin. This suggests trying power series expansions in fractional powers. This attempt also fails. Expansions have been tried that include logarithms, and they also fail. The nature of the singularity at the point P is unknown, at present.

At the tip of the wedge there is an instance of flow in a corner. In the irrotational incompressible case, this is known to involve fractional powers of ζ and η (see Landau and Lifshitz 1959, Section 10, Problem 6), and presumably the same is true here, even though the flow is rotational and the fluid is compressible.

These two examples and many others show that the formulation of the piecewise analytic initial-value problem must allow for singularities in addition to simple jumps, but the appropriate class of singularities is not clear at this time.

Appendix to Chapter 17 (Parts A–E)—The Detached Shock Problem

The power series solution of the Cauchy problem was a device introduced by Cauchy and later by Kovalevski solely for proving the existence of a solution. Until the development of fast automatic computers, it was out of the question to use the power series method for actual numerical calculations, except in rather trivial problems. Even when a modern computer is available, several technical difficulties must be overcome to make the method practical. Successful methods were developed by Richtmyer (1957) and Lewis (1959), who introduced techniques on three different levels (arithmetic, algebraic, and analytic), which made the power series method usable and highly accurate in the problem (detached shock calculation) on which it was tested. These techniques—significance arithmetic, automatic algebraic manipulation of the power series, and analytic continuation of the solution—are described in this appendix, after a brief description of the detached shock problem.

17.A The Problem

The aim is to calculate the steady, two-dimensional flow of a compressible fluid when analytic Cauchy data are given on an analytic curve, which is not characteristic. The flow quantities (pressure, density, velocity components, etc.) are represented by power series in two space variables or curvilinear coordinates, one of which is constant on the given curve. The point about which the expansion is made is a point on the curve. The determination of the expansion coefficients from the given data (values on the curve) and the partial differential equations of fluid dynamics is then a formal calculation, which is made feasible by computer programs for manipulation of power series. When sufficiently many coefficients (say, a few hundred for each flow quantity) have been computed, the series are summed, to calculate the flow in a neighborhood of the point about which the expansion was made. The validity of the procedure can be tested empirically by (1) comparing the results obtained at a given point in the flow from expansions about two or more neighboring points on the curve, (2) testing Bernoulli's law, and (3) checking the constancy of entropy along a streamline.

The particular problem of this class that was chosen for study is that of the inviscid flow between a detached shock and a rotationally symmetric blunt body like the nose of a projectile traveling at high Mach number through air. The general characteristics of this flow are well known; the relationships between the shock, the body, and the streamlines, are indicated in Figure 17-6; a frame of reference has been chosen in which the body is at rest. The stand-off distance is of the order of one tenth the radius of curvature of the body, the actual distance depending on the Mach number and the equation of state. Near the axis, the flow behind the shock is subsonic; farther downstream it is supersonic. Under usual conditions, the boundary layer that builds up adjacent to the body is of negligible thickness, hence the assumption of inviscid flow throughout the region is justified.

As in several other methods of attack on this problem, we start by assuming a shape for the shock, with the idea of determining from the calculation what

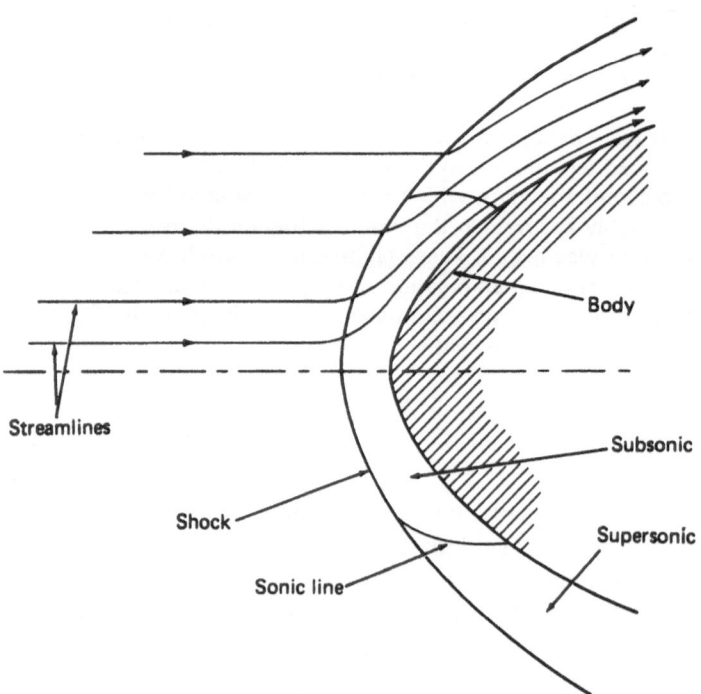

Figure 17.6 A detached shock.

shape of body will produce such a shock. If the pressure, density, and velocity of the air ahead of the shock are known, then their values are determined immediately behind the shock by the Rankine–Hugoniot jump conditions. If the assumed shock surface and the equation of state are analytic, then the flow quantities immediately behind the shock (the Cauchy data) are analytic functions of a suitable coordinate along the shock and can be expanded in powers of that coordinate. A preliminary computer program calculates the coefficients of those expansions; the main program calculates the remaining coefficients of the full expansion in the two variables from the partial differential equations; a final program then calculates various quantities, by summing the power series, at any desired points in the flow field.

The mathematical formulation is as follows: let z and r be cylindrical coordinates. It is assumed that the flow is symmetric about the z axis and that the shape of the shock is given by the equation

$$z = G(r) = G_0 + G_2 r^2 + G_4 r^4 + \cdots. \qquad (17.A\text{-}1)$$

In the region ahead of the shock, that is, for $z < G(r)$ (to the left in Figure 17-6), the flow quantities are constant, and the velocity is parallel to the z axis. The air is assumed to be a perfect gas with the equation of state

$$p = (\gamma - 1)\mathscr{E}\rho \qquad (17.A\text{-}2)$$

in the notation of Section 17.2. In suitable units, the conditions ahead are

$$
\begin{aligned}
p &= 1, \\
\rho &= 1, \\
u &= U, \\
v &= 0,
\end{aligned}
\qquad (17.A\text{-}3)
$$

Here, u and v are the axial and radial components of the fluid velocity, and U is a constant. The sound speed ahead is $\sqrt{\gamma p/\rho} = \sqrt{\gamma}$, hence the Mach number is $M = U/\sqrt{\gamma}$.

Behind the shock, the flow variables depend on z and r. It is convenient to introduce curvilinear coordinates x and y defined by

$$
\begin{aligned}
x &= z - G(r), & z &= x + g(y), \\
y &= r - r_0, & r &= r_0 + y,
\end{aligned}
\qquad (17.A\text{-}4)
$$

where r_0 is a positive constant, and $g(y) = G(r_0 + y)$; y is a radial coordinate measured from the radius r_0 of the point on the shock about which the power series expansions is to be made, and x is a coordinate measured parallel to the axis, but from the shock, rather than from a fixed plane. The shock is given by $x = 0$. If f stands for any one of the flow variables u, v, p, ρ then $f = f(x, y)$; $f(0, y)$ denotes the value immediately behind the shock, i.e., the limit of $f(x, y)$ as $x \downarrow 0$.

To describe the inclination of the shock at a point $(0, y)$, let α be the angle between the shock normal and the direction of the incident flow. Then,

$$\sin \alpha = \frac{g'(y)}{\sqrt{1 + g'(y)^2}}, \qquad \cos \alpha = \frac{1}{\sqrt{1 + g'(y)^2}}$$

Immediately behind the shock, the flow quantities are determined, as functions of y, by the Rankine–Hugoniot jump conditions of Section 17.5, which take the following form

$$\frac{U}{\sqrt{1 + g'^2}} = \sqrt{\frac{p - 1}{1 - 1/\rho}},$$

$$p = \frac{\theta p - 1}{\theta - \rho}, \quad \text{where } \theta = \frac{\gamma + 1}{\gamma - 1}, \tag{17.A-5}$$

$$\frac{(U - u) + g'v}{\sqrt{1 + g'^2}} = \sqrt{(p - 1)(1 - 1/\rho)}, \qquad \frac{(U - u)g' - v}{\sqrt{1 + g'^2}} = 0,$$

where g' stands for $g'(y)$ and u, v, p, ρ stand for $u(0, y), v(0, y), p(0, y), \rho(0, y)$. Since $g'(y)$ is known, the above equations can be solved to give u, v, p, ρ as functions of y, for $x = 0$.

For $x \geq 0$, the quantities $u(x, y), v(x, y), p(x, y),$ and $\rho(x, y)$, satisfy the partial differential equations

$$\rho\left[(u - vg')\frac{\partial u}{\partial x} + v\frac{\partial u}{\partial y}\right] + \frac{\partial p}{\partial x} = 0$$

$$\rho\left[(u - vg')\frac{\partial v}{\partial x} + v\frac{\partial v}{\partial y}\right] + \frac{\partial p}{\partial y} - g'\frac{\partial p}{\partial x} = 0,$$

$$\frac{\partial}{\partial x}\rho u + \left(\frac{\partial}{\partial y} - g'\frac{\partial}{\partial x}\right)\rho v + \frac{\rho v}{r_0 + y} = 0 \tag{17.A-6}$$

$$\gamma p\left[(u - vg')\frac{\partial \rho}{\partial x} + v\frac{\partial \rho}{\partial y}\right] - \rho\left[(u - vg')\frac{\partial p}{\partial x} + v\frac{\partial p}{\partial y}\right] = 0.$$

The first two of these are the equations of motion, the third is the equation of continuity, and the fourth is the energy equation, from which $\mathscr{E}$ has been eliminated by use of the equation of state (17.A-2). The quantities in the square brackets are derivatives along the streamlines; the terms in vg' come from the nonorthogonality of the coordinate system.

Equations (17.A-5) and (17.A-6) determine a Cauchy problem. In terms of its solution, the stream function ψ is given, except for a factor 2π, by the equation

$$\psi(x, y) = \tfrac{1}{2}(r_0 + y)^2 U - (r_0 + y)\int_0^x \rho(x', y)v(x', y)dx'. \tag{17.A-7}$$

$2\pi\psi$ is the mass of fluid flowing per unit time through a circle in a plane perpendicular to the z axis with its center on the z axis and passing through the point x, y in the plane of Figure 17-6. The streamline $\psi = 0$ consists of two parts: the z axis and a curve intersecting the z axis at a positive value of x; this curve determines the surface of the body. The problem, then, is to find this streamline and the flow between it and the shock.

17.B Ill-Posedness of the Problem

The problem is ill posed, in that the solution does not depend continuously on the Cauchy data. If the functions u, v, p, and ρ, at $x = 0$, are subjected to a small perturbation, the resulting perturbation of the solution generally increases, as x increases away from zero. In fact, a perturbation can be found which increases, as x increases, at any given exponential rate. That is connected with the elliptic character of the partial differential equations in the subsonic region (see Courant and Friedrichs 1948, Section 22). The simplest example of such an instability of an elliptic equation is provided by Laplace's equation, which has solutions of the form $e^{kx} \sin ky$, for arbitrarily large k. As is well known, the appropriate (well-posed) boundary value problems of Laplace's equation (the Dirichlet and Neumann problems) have a single function prescribed on a closed curve, rather than two functions on an initial curve.

To investigate the growth of perturbations in a similar but simpler problem, the equations (17.A-6) are first written in terms of two plane Cartesian coordinates, x, y, whereupon the terms containing g' and r_0 are omitted:

$$u \frac{\partial u}{\partial x} + v \frac{\partial u}{\partial y} + \frac{1}{\rho} \frac{\partial p}{\partial x} = 0,$$

$$u \frac{\partial v}{\partial x} + v \frac{\partial v}{\partial y} + \frac{1}{\rho} \frac{\partial p}{\partial y} = 0,$$

$$\frac{\partial(\rho u)}{\partial x} + \frac{\partial(\rho v)}{\partial y} = 0,$$

$$\gamma p \left(u \frac{\partial \rho}{\partial x} + v \frac{\partial \rho}{\partial y} \right) - \rho \left(u \frac{\partial p}{\partial x} + v \frac{\partial p}{\partial y} \right) = 0.$$

(17.B-1)

Second, the equations are linearized by writing $u_0 + u_1$ for u (and similarly for the other flow variables), where u_1 is a small and rapidly varying perturbation (like $\varepsilon e^{kx} \sin ky$ for Laplace's equation). It is assumed that u_1 can be neglected in comparison with u_0, and $(\nabla u_0)/u_0$ in comparison with $(\nabla u_1)/u_1$, etc. If a perturbation of the form

$$u_1 = \hat{u} \exp[i(k_1 x + k_2 y)],$$

$$v_1 = \hat{v} \exp[i(k_1 x + k_2 y)],$$

$$p_1 = \hat{p} \exp[i(k_1 x + k_2 y)],$$

$$\rho_1 = \hat{\rho} \exp[i(k_1 x + k_2 y)],$$

(17.B-2)

is substituted into (17.B-1), it is found that the constants $\hat{u}$, $\hat{v}$, $\hat{p}$, $\hat{\rho}$, k_1, and k_2, must satisfy the equations

$$\rho_0(\mathbf{v}_0 \cdot \mathbf{k})\hat{\mathbf{v}} + \hat{p}\mathbf{k} = 0,$$

$$(\mathbf{v}_0 \cdot \mathbf{k})\hat{\rho} + \rho_0(\mathbf{k} \cdot \hat{\mathbf{v}}) = 0, \qquad (17.B\text{-}3)$$

$$(\mathbf{v}_0 \cdot \mathbf{k})(\gamma p_0 \hat{\rho} - \rho_0 \hat{p}) = 0.$$

Here $\mathbf{v}_0$, $\hat{\mathbf{v}}$, and $\mathbf{k}$ are the vectors

$$\begin{pmatrix} u_0 \\ v_0 \end{pmatrix}, \quad \begin{pmatrix} \hat{u} \\ \hat{v} \end{pmatrix}, \quad \begin{pmatrix} k_1 \\ k_2 \end{pmatrix}.$$

These equations have solutions of several kinds. In particular, if $\mathbf{v}_0 \cdot \mathbf{k} \neq 0$, then

$$\hat{p} = \frac{\gamma p_0}{\rho_0} \hat{\rho} = c^2 \hat{\rho},$$

where c is the sound speed. Then $\hat{\mathbf{v}}$ and $\hat{p}$ can be eliminated from (17.B-3) to give

$$c^2 \hat{\rho} \mathbf{k} \cdot \mathbf{k} = (\mathbf{v}_0 \cdot \mathbf{k})^2 \hat{\rho},$$

hence

$$c^2(k_1^2 + k_2^2) - (u_0 k_1 + v_0 k_2)^2 = 0.$$

If this last is regarded as a quadratic equation for the ratio k_1/k_2, the discriminant is

$$c^2(u_0^2 + v_0^2 - c^2) = c^2(\|\mathbf{u}_0\|^2 - c^2).$$

Hence, if $\|u_0\| > c$, i.e., if the flow is supersonic, then k_1/k_2 is real, and the solution is stable, whereas, if $\|u_0\| < c$, i.e., if the flow is subsonic, then k_1 is complex, for k_2 real, and there are solutions of the form

$$e^{ik_1 x} \sin k_2 y, \qquad e^{i\bar{k}_1 x} \sin k_2 y,$$

one of which grows exponentially with increasing x.

The physical problem, in which the body rather than the shock is given in advance, is well posed, just as the Dirichlet problem of Laplace's equation is well posed. In the corresponding mathematical formulation, the shock front and the flow behind it have to be solved for simultaneously. There are at present no computational methods for such implicit problems, except rough time-dependent methods in which one hopes that the flow will eventually settle down to a steady state.

The ill-posedness is the price that one pays for inverting the problem and treating the shock as given. Although that causes difficulties for finite-difference methods, it causes none for the power-series method. Furthermore, the fear was at one time expressed that the sonic line, which separates the subsonic and supersonic regions, would be associated with some sort of singularity or instability of the flow. However, the flow turns out to be smooth across the sonic line, and the power series method is completely insensitive to the subsonic or supersonic character of the flow.

17.C The Power Series Method

If f stands for any one of the flow variables u, v, p, or ρ, we write

$$f = f(x, y) = \sum_{(jk)} f_{jk} x^j y^k,$$

summed over all pairs (j, k) of nonnegative integers; the f_{jk} are the coefficients of the expansion, and the expansion applies in the region $x \geq 0$.

Equation (17.A-5) can be solved for the quantities u, v, p, ρ, as functions of y, for $x = 0$, in terms of the function $g'(y)$. In the process, the irrationalities represented by the radical signs disappear; hence one obtains power series for $u(0, y)$, etc., in terms of the power series for $g(y)$. This part of the calculation follows standard lines and need not be described in detail. As a result of it, a dozen or so of the coefficients are computed for each function and are available in the computer for the further calculation; they are the coefficients u_{0k}, v_{0k}, p_{0k}, ρ_{0k}, for $k = 0, 1, \ldots, K$. Starting with these, the main calculation determines all the coefficients u_{jk}, v_{jk}, p_{jk}, and ρ_{jk} for which $j + k \leq K$. In the numerical tests, K was taken, in different problems, as 10, 14, and 19; the total number of terms for each power series was then $\frac{1}{2}(K + 1)(K + 2)$, or 66, 120, and 210, respectively, for each function.

Equations (17.A-6) are first solved for the x derivatives:

$$\frac{\partial p}{\partial x} = \frac{(u - vg')\rho v \left[\dfrac{\partial p}{\partial y} + \dfrac{\gamma p}{y + r_0} \right] + \gamma p \left[\rho u \dfrac{\partial v}{\partial y} + g' \dfrac{\partial p}{\partial y} - \rho v \dfrac{\partial u}{\partial y} \right]}{\gamma p (1 + g'^2) - \rho (u - vg')^2}$$

$$\frac{\partial u}{\partial x} = -\frac{\dfrac{\partial p}{\partial x} + \rho v \dfrac{\partial u}{\partial y}}{\rho(u - vg')}$$

(17.C-1)

$$\frac{\partial v}{\partial x} = -\frac{\dfrac{\partial p}{\partial y} - g' \dfrac{\partial p}{\partial x} + \rho v \dfrac{\partial v}{\partial y}}{\rho(u - vg')}$$

$$\frac{\partial \rho}{\rho x} = \frac{\rho \left[(u - vg') \dfrac{\partial p}{\partial x} + v \dfrac{\partial p}{\partial y} \right] - \gamma p v \dfrac{\partial \rho}{\partial y}}{\gamma p (u - vg')}.$$

The general idea of the method can be explained with the aid of Figure 17-7. Each dot represents a pair of index values (j, k) for which we wish eventually

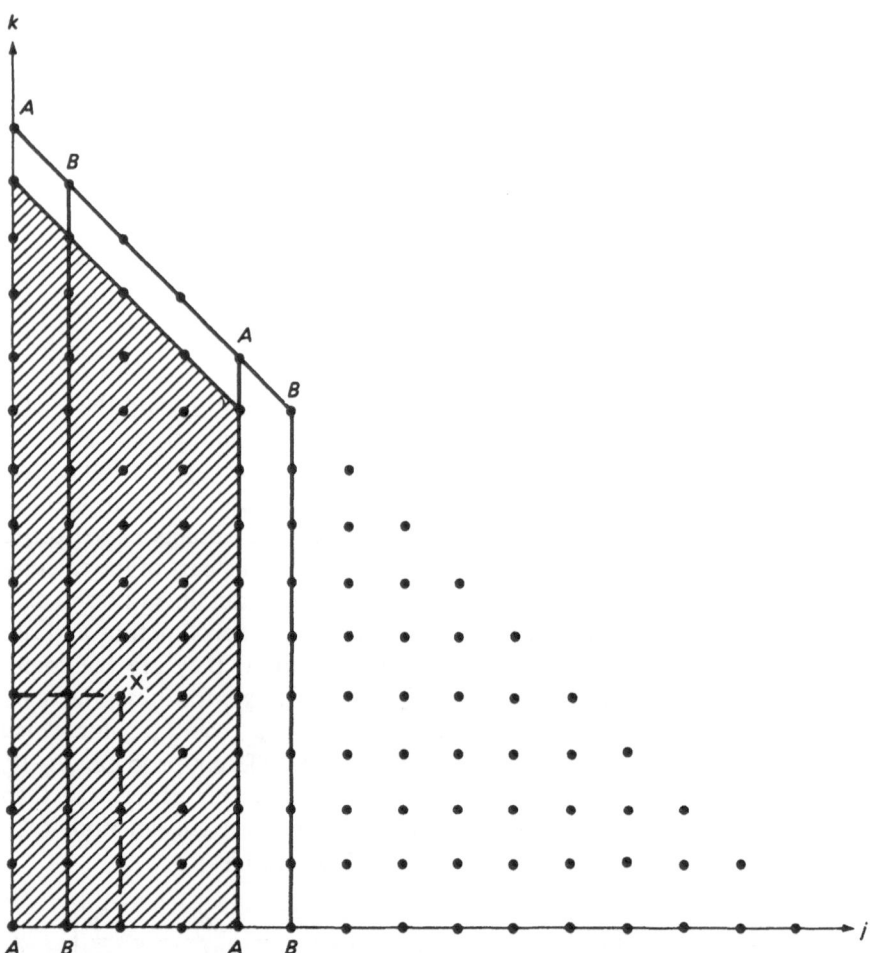

Figure 17.7 The schema of the calculation.

to know the coefficients u_{jk}, etc. Suppose that at some stage of the calculation all coefficients of u, v, p, and ρ are known for which $j + k \leq K$ and $j \leq j'$ for some j' in the interval $[0, K]$. The corresponding dots in Figure 17.C lie in the region $AAAA$(j' is $= 4$ for the case shown). For brevity, it will be said that the *functions* u, v, p, ρ, are known in that region. Such a region has the property (i.e., shape) that if two functions are known in it their product or sum or difference or quotient can be found in the same region by formal multiplication, addition, subtraction, or division, of power series. Addition and subtraction are trivial. In multiplication, the coefficient at point X in the figure, for example, of the product of two functions f and g involves the coefficients of f and of g at points lying in the dotted rectangle, of which the point X and the origin are opposite corners. A similar remark holds for division, which is performed by the rules of long division for polynomials (in two unknowns), as taught in elementary algebra. Subroutines were programmed for performing these operations, and also subroutines for formal (i.e., term-by-term) differentiation with respect to y and integration with respect to x.

If u, v, p, and ρ are known in the region $AAAA$, then their derivatives with respect to y can be found (by the differentiation subroutine) in the shaded region of the drawing, which is the same as the region $AAAA$, but shifted downward unit distance because differentiation reduces the exponent of y in each term of the series. By use of the other subroutines, the complete expression on the right of the first equation of (17.C-1) can then be found in the shaded region, and then similarly the expressions on the right of the other equations. At this stage, the four partial derivatives $\partial p/\partial x$, $\partial u/\partial x$, $\partial v/\partial x$, and $\partial p/\partial x$ are known in the shaded region. Term-by-term integration with respect to x then gives u, v, p, ρ in the region $BBBB$, which is the same as the shaded region but shifted unit distance to the right. If the original coefficients for $x = 0$ are again included, the functions are now known in a region of the same kind as the region $AAAA$ mentioned at the beginning of the paragraph, but with the inequality $j \leq j'$ replaced by $j \leq j' + 1$. This completes what will be called one cycle of the calculation.

The information available initially from the Cauchy data gives a special case in which the region $AAAA$ reduces to part of the k axis, corresponding to $j' = 0$. Therefore, after K cycles of the calculation, as described above, the functions are known in the entire triangular region $j + k \leq K$.

The subroutines referred to above were incorporated in a special computer programming language, each instruction of which corresponds to one of the formal operations on power series. The partial differential equations (17.C-1) were then written out directly in that language, the steps being taken in the order described in the preceding paragraph.

The language also contains a provision for starting a new cycle when the previous cycle is finished, if $j' < K$, and for stopping, when $j' = K$. Slightly less than a hundred instructions in this language were required to program the equations (17.C-1).

The use of the subroutines and the special programming language must be regarded as more than merely a labor-saving device. The program in effect constructs and uses the recurrence relations that determine the coefficients of the higher terms of the series from those of the earlier terms. Because we are dealing with power series in two variables and with a rather complicated fourth order system of nonlinear partial differential equations, it is clear that the recurrence relations must be very complicated, and this idea is confirmed by the fact that many million arithmetic steps are required, using these recurrence relations, to compute the coefficients of the four flow functions up to order 10. It is probably safe to say that it is beyond human capability to write down these recurrence relations (at least without mistakes); hence, the use of formal algebraic manipulation by computer is an essential part of the practical utilization of the Cauchy–Kovalevski power-series method.

17.D Significance Arithmetic

It was found necessary to use the so-called significance arithmetic in the calculations described above, not merely in order to have some estimate of the accuracy of the results, but in order to get results at all. The significance index actually controls the course of the calculation at one point in the program.

It is recalled that, in the days of hand computing, where the result of each arithmetic step had to be recorded on paper, it was customary to underline, in each number, that digit that was considered to be the last significant one. After each arithmetic operation, the number of significant digits of the result was determined from those of the operands by a simple set of rules. In this way, a warning was provided in the case of serious loss of significance by cancellations in a long calculation. This procedure was regarded as an essential part of good computing practice, and it is perhaps astonishing that it has not been provided for in automatic computers, either in the hardware or in the software, with a very few exceptions.

Two types of significance arithmetic have been used in this work. They have been described by Richtmyer (1957, 1960), by Gray and Harrison (1959), and by Ashenhurst and Metropolis (1959). In the first type, each floating point number is represented by three quantities: a fraction or mantissa f, an exponent e, and a significance index s, all combined in one computer word. The fraction f is such that $\frac{1}{2} \le |f| < 1$, and it is represented as a signed binary number having l binary places (where l is of the order of 40), e is a signed integer (say, $-200 < e < 200$), and s is an integer in the interval $0 \le s \le l$. [Some of the work was done on a decimal machine, but the binary terminology will be used.] The triple (f, e, s) represents the number

$$x = (f \pm 2^{-s}) \cdot 2^e. \qquad (17.D\text{-}1)$$

The number of presumably correct digits in f is $\approx s$. The rules for determining s are the same as those used earlier in hand computing. Before two numbers are added or subtracted, the one with the smaller exponent e is denormalized; that is, if Δe is the difference of the two exponents, the smaller number (f, e, s) is replaced by (f', e', s'), where $f' = 2^{-\Delta e}f$, $e' = e + \Delta e$, $s' = \min(s + \Delta e, l)$. Then, the two numbers have the same exponent, and the fractions can be added or subtracted. The resulting new fraction, say f_1, lies in the interval $-2 < f_1 < 2$, and the new significance index s_1 is set equal to the smaller of the two indices before the addition or subtraction. Lastly, the result is normalized: if $1 \le |f_1| < 2$, then f_1, e_1, and s_1 are replaced by $\frac{1}{2}f_1$, $e_1 + 1$, $\min(s_1 + 1, l)$, whereas if $|f_1| < \frac{1}{2}$, then f_1 is shifted to the left, relative to the binary point, and simultaneously e_1 and s_1 are reduced by 1 for each place of shift, until $|f_1|$ is in the interval $[\frac{1}{2}, 1)$, or until s_1 becomes zero in the process, in which case e_1 is reduced only by the same amount as s_1. [A result of the form $x = (f, e, 0)$ can occur; it represents a number ≈ 0 with an uncertainty $\approx \pm 2^e$, and the uncertainty must be taken into account if x is used later in the calculation. Hence, the value of e is meaningful. When

$s = 0$, the fraction f, which consists of totally insignificant debris, is also set $= 0$, to prevent an apparent (but false) regain of significance when two or more such numbers are added.] Multiplication and division are performed in the usual normalized floating point manner, and the significance index of the result is set equal to the smaller of the two indices before the operation. Division by a number x with $s = 0$ is forbidden, because, according to (17.D-1), zero is one of the possible values of x.

The other type of significance arithmetic that was used in the detached shock problem is the unnormalized arithmetic devised by Metropolis. Instead of keeping a significance index s, one supplies each floating-point number with just enough leading zeros so that all the remaining digits are significant. (Rather, the digits after the leading zeros include a fixed number of *guard digits*, at the extreme right of the number, to guard against contamination of truly significant digits by accumulation of rounding errors. However, that is merely a point of view—the machine has no way of knowing which digits are considered to be guard digits rather than significant ones.) There is no normalization after addition or subtraction. After multiplication or division, the result is so shifted (in a double length shifting register) as to acquire exactly the same number of leading zeros as the less accurate of the two operands, and the exponent is adjusted accordingly.

Both types were found to be effective in the power series work.

Elementary examples show that, when a quantity x loses all significance in a calculation (whether done in significance arithmetic or not), subsequent quantities that depend on x can be not merely inaccurate but of the wrong order of magnitude. Suppose one calculates e^{-x}, for $x = 20$, by the power series for e^{-x}. [That would be foolish, since there are much better ways of calculating e^{-x}, but one doesn't generally know the best ways in very complicated problems.] Suppose the calculation is done using ten decimal places. The partial sum S_n of the series exceeds 10^7 in magnitude, for $n \approx 20$, the rounding error is $\approx 10^{-3}$, and this error cannot be reduced by adding further terms of the series, even if the further terms are highly accurate. Consequently, as n increases, S_n does not approach e^{-20}, which is $< 10^{-8}$, but remains too large by a factor $\approx 10^5$.

Now suppose that a function $f(z)$ is to be obtained from a series $\sum a_n z^n$ whose coefficients a_n are themselves obtained from a lengthy calculation that is increasingly subject to loss of significance with increasing n. Then, for n larger than some n_0, the a_n may be of zero significance and too large by increasing factors, as n increases. Then, the series $\sum a_n z^n$ behaves like a semiconvergent series: the error of the partial sum, as an approximation to $f(z)$, decreases for a while, as more terms are added, but then gets larger again. It is important to stop the summation at $n \approx n_0$.

For the power series in the two variables x and y, in the detached shock problem, it is found that the coefficients $u_{j\,k}$ etc. are significant in a region $\mathfrak{R}$ of the j, k plane bounded by the j and k axes and a rather vaguely defined curve that cuts out a roughly triangular region. Hence, the series $\sum u_{j\,k} x^j y^k$ ought to be summed only over $\mathfrak{R}$. In a calculation of this complexity, there is no way of determining $\mathfrak{R}$ in advance, but when significance arithmetic is used, the remedy is very simple. The subroutine that sums the series examines the significance index s of each coefficient u_{jk} and omits the corresponding term from the series when $s = 0$. When this provision was put into the program, the accuracy obtained jumped suddenly from around one decimal place to around six decimal places, for reasonably small x and y (less than about half the distance from the shock to the body).

17.E Analytic Continuation

The domain of convergence of a power series expansion $\sum_{(j\,k)} x^j y^k$ of a function $f = f(x, y)$ is determined by the singularities of f, regarded as an analytic function of two complex variables. Very little is known about the singularities of the functions u, v, p, and ρ, in the detached shock problem, but there is considerable evidence to indicate that the domain of convergence is too small to include real values of x and y with x sufficiently large positive to reach the body (the streamline $\psi = 0$). Therefore, G. E. Lewis (1959) devised methods for analytic continuation of the solution. He found three methods, one of which will be described in this section.

Suppose that x_0, y_0 is a point in the domain of convergence of the series for u, v, p, and ρ, where $x_0 > 0$, and call $x - x_0 = \xi$, $y - y_0 = \eta$. Then, the equation $\xi = 0$ $(x = x_0)$ determines a curve $\mathscr{C}'$ parallel to the shock in the flow region. In some neighborhood of the point x_0, y_0, the functions are given on $\mathscr{C}'$ by the calculation described above. Hence, one may consider a new Cauchy problem, in which

$$u(x_0, y_0 + \eta)$$

$$v(x_0, y_0 + \eta)$$

$$p(x_0, y_0 + \eta)$$

$$\rho(x_0, y_0 + \eta)$$

are assumed to be known functions of η on $\mathscr{C}'$, and in which the partial differential equations are used once more, just as above, to expand $u(x_0 + \xi, y_0 + \eta)$, etc., in powers of ξ and η, say as

$$u(x_0 + \xi, y_0 + \eta) = \sum_{(j\,k)} \hat{u}_{j\,k} \xi^j \eta^k.$$

The initial data for the new expansion, namely the coefficients $\hat{u}_{0k}$, etc., are given in terms of the coefficients u_{jk}, etc., of the previous expansion, as follows:

$$u(x_0, y_0 + \eta) = \sum_{(k)} \hat{u}_{0\,k} \eta^k = \sum_{(j\,k)} u_{j\,k} x_0^j (y_0 + \eta)^k.$$

For simplicity, y_0 can be taken as zero (then, the new expansion point is at the same distance from the axis of the body as the old one); then

$$\hat{u}_{0\,k} = \sum_{(j)} u_{j\,k} x_0^j,$$

and similarly for the other functions v, p, and ρ.

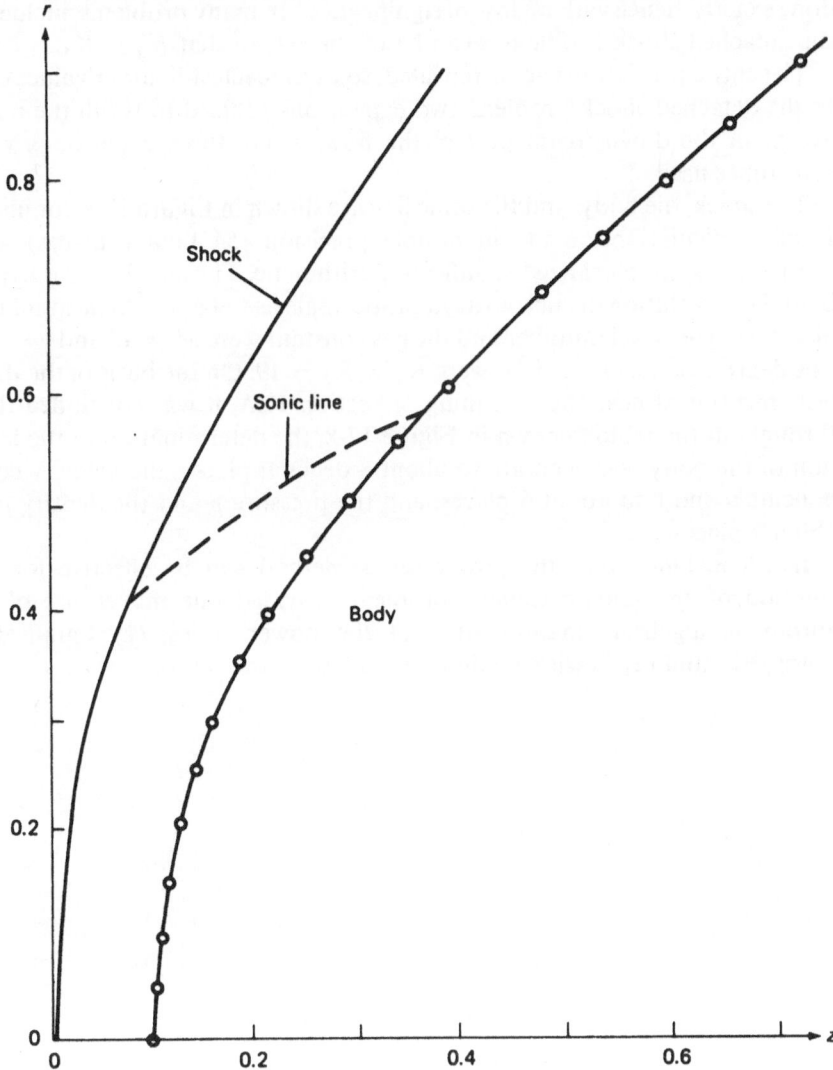

Figure 17.8 Rough sketch of the body determined by Lewis.

Let K_1 be the degree of the first expansion, so that all terms with $j + k \le K_1$ are included (K_1 was called K earlier), and let K_2 be the degree of the second expansion. Lewis proved that, if a certain restriction is put on the choice of the point x_0, y_0, there is a constant s ($0 < s \le 1$) such that if we let K_1 and $K_2 \to \infty$, with K_2 always taken as the integer part of sK_1, then the second expansion converges to the solution of the Cauchy problem within the domain of convergence of the power series expansion of that solution about x_0, y_0; hence, true analytic continuation is obtained. This is essentially because the Cauchy data used for the second expansion are of increasing accuracy, as $K_1 \to \infty$. (The theorem assumes that all arithmetic steps are done exactly, hence with no loss of significance.) In many problems, including the detached shock problem, s can be taken $= 1$, so that $K_2 = K_1$.

The entire procedure can be repeated, so as to reach still larger values of x. In the detached shock problem, two expansions sufficed to reach the body, except in the downstream part of the flow, where three expansions were sometimes used.

The shock, the body, and the sonic line are shown in Figure 17-8, for one of Lewis's calculations, made in double precision (54 binary places) with Metropolis's unnormalized significance arithmetic. The shock was a hyperboloid of revolution of such an asymptotic angle as to be just sonic at infinite distances. The Mach number and the gas constant were $M = 12$ and $\gamma = 1.4$. The degrees of the expansions were $K_1 = K_2 = 19$. On the basis of the three tests mentioned near the beginning of Section 17.A, it was concluded that, throughout the region shown in Figure 17-8, the determination of the location of the body was accurate to about 8 decimal places, the velocity components u and v to about 6 places, and the pressure p and the density ρ to about 7 places.

It is concluded that the power series method can be effective for the solution of an analytic Cauchy problem provided one makes use of (1) automatic algebraic manipulation of the power series, (2) significance arithmetic, and (3) Lewis's method of analytic continuation.

References

Abramowitz, M., and Stegun, I. A. (1964): *Handbook of Mathematical Functions.* National Bureau of Standards, Applied Mathematics Series, Washington, D.C.

Akhiezer, N. I., and Glazman, I. M. (1961, 1963): *Theory of Linear Operators in Hilbert Space,* Vol. I (1961), Vol. II (1963). Frederick Ungar Publishing Co., New York.

Arnol'd, V. I. (1973): *Ordinary Differential Equations.* M.I.T. Press, Cambridge, Mass.

Ashenhurst, R. L., and Metropolis, N. C. (1959): Unnormalized floating point arithmetic. *J. Assoc. Comput. Mach.,* vol. **6**, pp. 415–428.

Baernstein, A. (1971): Representation of holomorphic functions by boundary integrals. *Trans. Amer. Math. Soc.,* vol. **160**, pp. 27–37.

Banach, S. (1955): *Théorie des Opérations Linéaires.* Chelsea Publishing Co., New York.

Barut, A. O. (1967): *The Theory of the Scattering Matrix.* Macmillan, London.

Bers, L. (1958): *Mathematical Aspects of Subsonic and Transonic Gas Dynamics.* John Wiley and Sons, New York.

Bethe, H. A., and Salpeter, E. E. (1957): *Quantum Mechanics of One- and Two-Electron Atoms.* Springer-Verlag.

Birkhoff, G. (1962): Helmholtz and Taylor instability. Proceedings of Symposia in Appl. Math., Amer. Math. Soc., vol. **92**, p. 13 ff.

Birkhoff, G., and Rota, G. (1962): *Ordinary Differential Equations.* Ginn and Co., Waltham, Mass.

Bleakney, W., and Taub, A. H. (1949): Interaction of shock waves. *Rev. Mod. Physics,* vol. **21**, pp. 584–605.

Carleson, L. (1966): On the convergence and growth of partial sums of Fourier series. *Acta Math.* vol. **116**, pp. 135–157.

Coddington, E. A., and Levinson, N. (1955): *Theory of Ordinary Differential Equations.* McGraw-Hill, New York.

Courant, R., and Friedrichs, K. O. (1948): *Supersonic Flow and Shock Waves.* Interscience, New York.

Courant, R., and Hilbert, D. (1953, 1962): *Methods of Mathematical Physics, Volumes I and II.* Interscience, New York.

DiPrima, R. C., and Habetler, G. J. (1969): A completeness theorem for non-selfadjoint eigenvalue problems in hydrodynamic stability. *Arch. Rat. Mech and Anal.* vol. **34**, pp. 218–227.

Dirac, P. A. M. (1930, 1935, 1947, 1958): *The Principles of Quantum Mechanics,* Editions 1, 2, 3, and 4. Clarendon Press, Oxford.

Donoghue, Wm. F. (1969): *Distributions and Fourier Transforms.* Acad. Press, New York.

Duff, R. E. (1962): Slip line instability. Proceedings of Symposia in Appl. Math., Amer. Math. Soc. vol. **13,** p. 77ff.

Dunford, N., and Schwartz, J. T. (1958): *Linear Operators Part I: General Theory.* Interscience, New York.

Dym, H., and McKean, H. P. (1972): *Fourier Series and Integrals.* Academic Press, New York.

Fefferman, C. (1971): On the divergence of multiple Fourier series. *Bull. Amer. Math. Soc.* vol. **77,** pp. 191–195.

Fefferman, C. (1971): On the convergence of multiple Fourier series. *Bull. Amer. Math. Soc.* vol. **77,** pp. 744–745.

Feller, W. (1950, 1966): *An Introduction to Probability Theory and its Applications Volumes I and II.* John Wiley and Sons, New York.

Friedman, A. (1969): *Partial Differential Equations.* Holt, Rinehart, and Winston, New York.

Gantmacher, F. R. (1959): *The Theory of Matrices, Volumes I and II.* Chelsea Pub. Co., New York.

Garabedian, P. R. (1964): *Partial Differential Equations.* John Wiley and Sons, New York.

Gel'fand, I. M., and Shilov, G. E. (1964, 1968, 1967): *Generalized Functions Volumes 1, 2, and 3.* Academic Press, New York.

Gel'fand, I. M., and Vilenkin, N. Ya. (1964): *Generalized Functions Volume 4.* Academic Press, New York.

Gel'fand, I. M., Graev, M. I., and Vilenkin, N. Ya. (1966): *Generalized Functions Volume 5.* Academic Press, New York.

Gray, H. L., and Harrison, C. (1959): *Normalized floating-point arithmetic with an index of significance.* Proc. Eastern Joint computer Conference.

Gross, L. (1966): The Cauchy problem for the coupled Maxwell and Dirac equations. *Comm. Pure Appl. Math.,* vol. **19,** pp. 1–15.

Gustafson, K., and Johnson, G. (1974): On the absolutely continuous subspace of a self-adjoint operator. *Helv. Physica Acta,* vol. **47,** pp. 163–166.

Gustafson, K., and Rejto, P. A. (1973): Some essentially self-adjoint Dirac operators with spherically symmetric potentials. *Israel Jour. Math.,* vol. **14,** pp. 63–75.

Halmos, P. R. (1951): *Introduction to Hilbert Space and the Theory of Spectral Multiplicity.* Chelsea Pub. Co., New York.

Hille, E. (1962): *Analytic Function Theory Volume II.* Ginn and Co., Boston.

Hille, E., and Phillips, R. S. (1957): Functional Analysis and Semi-groups. *Amer. Math. Soc.*

Hörmander, L. (1969): *Linear Partial Differential Operators.* Springer-Verlag.

John, F. (1971): *Partial Differential Equations.* Springer-Verlag.

Johnson, G. (1968): Harmonic functions on the unit disc I and II. *Ill. Jour. Math.*, vol. 12, pp. 366–396.

Jordan, P., and von Neumann, J. (1935): On inner products in linear metric spaces. *Ann. of Math.*, vol. 36, pp. 719–723.

Jörgens, K. (1967): Zur Spektraltheorie der Schrödingeroperatoren. *Math. Zeitschr.*, vol. 96, pp. 355–372.

Jörgens, K. (1970): *Lineare Integraloperatoren.* B. G. Teubner, Stuttgart.

Jörgens, K., and Rellich, F. (1976): *Eigenwerttheorie gewöhnlicher Differentialoperatoren* (bearbeitet von J. Weidmann). Springer-Verlag.

Jörgens, K., and Weidmann, J. (1973): *Spectral Properties of Hamiltonian Operators.* Springer-Verlag.

Jost, R. (1965): *The General Theory of Quantized Fields.* American Mathematical Society, Providence, RI.

Kantorovich, L. V., and Akilov, G. P. (1964): *Functional Analysis in Normed Spaces.* MacMillan, New York.

Kato, T. (1966): *Perturbation Theory for Linear Operators.* Springer-Verlag.

Kelley, J. L. (1955): *General Topology.* D. Van Nostrand, Princeton, NJ.

Lanczos, C. (1956): *Applied Analysis.* Prentice Hall, Englewood Cliffs, NJ.

Landau, L. D., and Lifshitz, E. M. (1959): *Fluid Mechanics.* Pergamon Press, London.

Lax, P. D. (1954): Weak solutions of nonlinear hyperbolic equations and their numerical computation. *Comm. Pure Appl. Math.*, vol. 7, pp. 159–193.

Lax, P. D. (1957): Hyperbolic systems of conservation laws II. *Comm. Pure Appl. Math.*, vol. 10, pp. 537–566.

Lax, P. D., and Phillips, R. S. (1967): *Scattering Theory.* Academic Press, New York.

Lewis, G. E. (1959): *Analytic Continuation Using Numerical Methods.* Ph.D. Thesis, New York University; also in *Methods in Computational Physics*, Academic Press, New York, vol. 4, pp. 45–81 (1965).

MacDuffee, C. C. (1946): *The Theory of Matrices.* Chelsea Pub. Co., New York.

Magnus, W., and Oberhettinger, F. (1943): *Formulas and Theorems for the Functions of Mathematical Physics.* Chelsea Pub. Co., New York.

Meisters, G. H. (1971): Translation-invariant linear forms and a formula for the Dirac measure. *J. Functional Anal.*, vol. 8, pp. 173–188.

Morse, P. M., and Feshbach, H. (1953): *Methods of Theoretical Physics, Volumes I and II.* McGraw-Hill, New York.

Natanson, I. P. (1955): *Theory of Functions of a Real Variable.* Frederick Ungar, New York.

Neumann, J. von (1929): Allgemeine Eigenwerttheorie Hermitescher Funktionaloperatoren. *Math. Annalen*, vol. 102, pp. 49–131.

Neumann, J. von (1931): Die Eindeutigkeit der Schrödingerschen Operatoren. *Math. Annalen*, vol. 104, pp. 570–578.

Rejto, P. A. (1971): On reducing subspaces for one-electron Dirac operators. *Israel J. Math.*, vol. 9, pp. 111–143.

Richtmyer, R. D. (1957): *Detached shock calculations by power series.* Report NYO-7973, Courant Institute, New York University; also in *Annals of the New York Academy of Sciences*, vol. 86, pp. 828–842.

Richtmyer, R. D. (1960): *Flow diagrams and the estimation of significance*. Report TID 6199, New York University.

Richtmyer, R. D., and Morton, K. W. (1967): *Difference Methods for Initial-Value Problems*. Wiley-Interscience, New York.

Riesz, F., and Sz. Nagy, B. (1953): *Leçons d'Analyse Fonctionelle*. Akadémiai Kiadó, Budapest.

Roos, B. W., and Sangren, W. C. (1962): Spectral theory of Dirac's radial relativistic wave equation. *J. Math. Phys.*, vol. **3**, pp. 882–890.

Sattinger, D. H. (1970): The mathematical problem of hydrodynamic stability. *Jour. Math. and Mech.*, vol. **19**, pp. 797–817.

Schwartz, L. (1950, 1951): *Théorie des Distributions, Tomes I et II*. Hermann Cie, Paris.

Sobolev, S. L. (1963): *Applications of Functional Analysis in Mathematical Physics*. Amer. Math. Soc. Translations.

Solovay, R. M. (1970): A model of set theory in which every set of reals is Lebesgue measurable. *Ann. of Math.* (2), vol. **92**, pp. 1–56.

Spanier, J., and Gelbard, E. M. (1969): *Monte Carlo Principles and Neutron Transport Problems*. Addison Wesley, Reading, Mass.

Stone, M. H. (1932): *Linear Transformations in Hilbert Space and Their Applications to Analysis*. American Mathematical Society, Providence, RI.

Taylor, A. E. (1958): *Introduction to Functional Analysis*. John Wiley and Sons, New York.

Taylor, J. R. (1972): *Scattering Theory*. John Wiley and Sons, New York.

Thron, W. (1966): *Topological Structures*. Holt, Rinehart, and Winston, New York.

Titchmarsh, E. C. (1946): *Eigenfunction Expansions Associated with Second Order Differential Equations*. Clarendon Press, Oxford.

Wasow, W. (1976): *Asymptotic Expansions for Ordinary Differential Equations*. Krieger Pub. Co., Huntington, N.Y.

Weidmann, J. (1971): Oszillationsmethoden für Systeme gewöhnlicher Differential-gleichungen. *Math. Z.*, vol. **119**, pp. 349–373.

Weinberger, H. F. (1965): *A First Course in Partial Differential Equations*. Blaisdell Pub. Co., New York.

Werner, P. (1969): Bemerkungen zur Theorie der L_p Räume. *Journal für die reine und angewandte Mathematik*, vol. **239/240**, pp. 401–434.

Whittaker, E. T., and Watson, G. N. (1927): *A Course of Modern Analysis*, Fourth Edition. Cambridge Univ. Press.

Wiener, N. (1930): Generalized harmonic analysis. *Acta Math.*, vol. **55**, pp. 117–258.

Wright, J. D. M. (1973): All operators on a Hilbert space are bounded. *Bull. Amer. Math. Soc.*, vol. **79**, pp. 1247–1250.

Zygmund, A. (1952): *Trigonometrical Series*. Chelsea Pub. Co., New York.

Index

Texts and Monographs in Physics

Edited by **W. Beiglböck, M. Goldhaber, E. Lieb,** and **W. Thirring**

Texts and Monographs in Physics includes books from any field of physics that might be used as basic texts for advanced training and higher education in physics, especially for lectures and seminars at the graduate level.

A Springer-Verlag Journal

Communications in Mathematical Physics

Editor-in-Chief J. Glimm, New York, New York

Editorial Board H. Araki, Kyoto, Japan
R. Geroch, Chicago, Illinois
R. Haag, Hamburg, Federal Republic of Germany
W. Hunziker, Zurich, Switzerland
A. Jaffe, Cambridge, Massachusetts
J.L. Lebowitz, New York, New York
E. Lieb, Princeton, New Jersey
J. Moser, New York, New York
R. Stora, Marseille, France

Communications in Mathematical Physics is devoted to such topics as general relativity, equilibrium and nonequilibrium statistical mechanics, foundations of quantum mechanics, Lagrangian quantum field theory, and constructive quantum field theory. Mathematical papers are featured if they are relevant to physics.

Contact Springer-Verlag for subscription information.

Lecture Notes in Physics

Managing Editor **W. Beiglböck**

This series reports on new developments in physical research and teaching—quickly, informally, and at a high level. The type of material considered for publication includes preliminary drafts of original papers and monographs, lectures on a new field, or presenting a new angle on a classical field, collections of seminar papers, and reports of meetings.

Vol. 10 J.M. Stewart, **Non-Equilibrium-Relativistic Kinetic Theory.** 1971. iii, 113p.

Vol. 13 M. Ryan, **Hamiltonian Cosmology.** 1972, vii, 169p.

Vol. 14 **Methods of Local and Global Differential Geometry in General Relativity.** Edited by D. Farnsworth, J. Fink, J. Porter, and A. Thompson. 1972. v, 188p.

Vol. 44 R.A. Breuer, **Gravitational Perturbation Theory and Synchrotron Radiation.** 1975. vi, 196p.

Vol. 46 E.J. Flaherty, **Hermitian and Kählerian Geometry in Relativity.** 1976. viii, 365p.

 **Springer-Verlag New York Inc.**
175 Fifth Avenue
New York, NY 10010